EARTH CONSTRUCTION SERIES

EARTH CONSTRUCTION

Modern social housing scheme at the Domaine de la Terre in the new town of Isle d'Abeau, France, using compressed earth blocks, rammed earth and straw-clay (Thierry Joffroy, CRATerre-EAG)

EARTH CONSTRUCTION SERIES

EARTH CONSTRUCTION

A COMPREHENSIVE GUIDE

Hugo Houben and Hubert Guillaud

CRATerre-EAG

Practical Action Publishing Ltd
25 Albert Street, Rugby, CV21 2SD, Warwickshire, UK
www.practicalactionpublishing.com

CRATerre-EAG
Maison Levrat, Rue du Lac, BP 53, F-38092, Villefontaine cedex, France

Cover picture: The earth constructed city of Habban Yemen, (K.H. Bochow)

This book was originally published by Editions Parenthèses (Marseille) as
Traite de construction en terre de CRATerre by Hugo Houben and Hubert Guillard

Practical Action Publishing edition 1994

ISBN 13 Paperback: 9781853391934
ISBN Library Ebook: 9781780444826
Book DOI: http://dx.doi.org/10.3362/9781780444826

A catalogue record for this book is available from the British Library.

CRATerre-EAG, The Internation Centre for Earth Construction and School of Agriculture at Genoble, has accumulated a vast knowledge of earth construction through experience over twenty years in more than fifty countries. Its activities include research, technical assistance, dissemination and training in economic construction, industrialization and preservation. As a member of the international Building Advisory Service and Information Network (BASIN), CRATerre advises on every aspect of the various earth construction technologies at all levels.

Since 1974, Practical Action Publishing has published and disseminated books and information in support of international development work throughout the world. Practical Action Publishing is a trading name of Practical Action Publishing Ltd (Company Reg. No. 1159018), the wholly owned publishing company of Practical Action. Practical Action Publishing trades only in support of its parent charity objectives and any profits are covenanted back to Practical Action (Charity Reg. No. 247257, Group VAT Registration No. 880 9924 76).

Typeset by J&L Composition Ltd, Filey, North Yorkshire

Authors

Hugo Houben, CRATerre-EAG
Hubert Guillaud, CRATerre-EAG

Scientific Direction

Patrice Doat, CRATerre-EAG

Co-Authors for the chapter on earthquakes

Michel Dayre, University of Grenoble
Pierre-Yves Bard, University of Grenoble
Guy Perrier, University of Grenoble
Julio Vargas, Catholic University of Lima

Co-Author for the chapter on conservation

Alejandro Alva, ICCROM

Assistants

Titane Galer, CRATerre-EAG
Marie-France Ruault, CRATerre-EAG

Drawings
Fabienne Dath

Translation

Albert Gompers
Janice Schilderman

Editing
Theo Schilderman, ITDG

This work has been published thanks to the financial support of:

The School of Architecture of Grenoble/The French Ministry of Infrastructure, Transport and Tourism Directorate of Architecture and Urbanism

This book is based on the study realized in the context of the research programme 'Earth construction technologies appropriate to developing countries', implemented by the Post Graduate Centre Human Settlements of the K.U. Leuven, Heverlee, under the co-ordination of Han Verschure; in collaboration with the Centre for Architectural Research, UC Louvain-la-Neuve, under the co-ordination of François Mabardi; and CRATerre in France, Peru and Belgium, under the co-ordination of Hugo Houben, and financed by GACD, the General Administration for Development Co-operation, Brussels, with support from UNCHS, the United Nations Centre for Human Settlements (Habitat), Nairobi.

ACKNOWLEDGEMENTS

We would like to thank everybody who has contributed to the creation and the realization of this book, which we want to consider above all as a 'technology survey'. Therefore, our warmest gratitude goes in the first place to all researchers and builders who, over the past years, have worked patiently and often in the margins of technical and architectural fashion in their furthering of earth construction. We are no more than the heirs of those pioneers; without their work, we would have been unable to accomplish as much as we have. This patient and watchful survey work is one of the major vocations of CRATerre, which will be continued and widened in order to guarantee a broader dissemination of knowledge to a larger public of potential designers and builders. We also want to express our sincere gratitude to all directors and staff of the large international organizations, non-governmental organizations, national institutions, and numerous private companies who, over the past fifteen years, have expressed their confidence and support, both morally and materially. We are especially grateful to our supervising institutions for research and education, who have most directly contributed to the new dynamics of earth construction via the reinforcement of our team, the renewal of aid to research and experiments, and credits to university and professional education. And we particularly thank those organizations and institutions who, morally and financially, have participated in the creation of this work, which could not have been published without their help.

Hugo Houben, Hubert Guillaud

Preface

The international United Nations conference on Human Settlements – Habitat 1 (Vancouver 1976) – called for the development and promotion of building materials and techniques that would be suitable for local conditions. The emphasis was placed on the production and use of audiovisual educational tools, documentation and the promotion of debates provoking the creativity and self-development of local communities. Internationally it was accepted that the dissemination of information and the decentralization of knowledge are the major success factors of appropriate use of a technology. The conference also stressed that, among all the materials used by man over the years, earth remained the one most frequently used by low-income populations in developing countries. Recent research has shown the great potential of earth to respond to the enormous housing need of millions of people. This acknowledgement argues for the development of a more profound knowledge of earth as a material and its technology, as yet too often poorly used, to allow a progressive and substantial improvement in its performances. All the research efforts developed during the last few years are now beginning to show tangible results. The large number of experienced scientists, technologists, architects and builders who have worked on this are finally able to utilize their knowledge, that is, to shift from small-scale experiments to large-scale project activity, including production management. This new phase in earth construction will undoubtedly contribute to an increase in its use and to the improvement of the living conditions of the most disadvantaged populations.

However, to ensure that earth construction contributes positively to the solution of the low-cost housing crisis, it is essential to become aware of its advantages and requirements. If the technical aspects are essential in terms of construction, the vast number of practices and skills still have to be taken into account and disseminated in order to achieve the maximum effect. For that reason, we welcome the publication of *Earth Construction*. This is an ambitious project, aiming to grasp the totality of a field that is constantly changing, rather than to set out concrete facts, which will rapidly become fixed and obsolete. This project is more than a simple 'state-of-the-art' manual, it is in itself part of a constantly-changing knowledge. The manual, setting out a common language and the scientific and technological culture of earth construction, will not only help the scientific community to take stock of this area of study, but also to respond to the expectation of non-scientists desiring to understand the issues.

This publication does not aim to 'define' knowledge but rather it hopes to stimulate continuing intensive research, i.e. test methods, practical rules and building regulations. It also deals with the problem of questioning the entire production process, both upstream and downstream, and in this sense induces a veritable 'philosophy' of earth construction. This work is therefore a first in the study of local building materials – a technology suitable for sustainable production is now the object of a professional approach. We do hope that this long-term task, first initiated with the publication of the first volume of the *Encyclopaedia of Earth Construction*, may be a source of inspiration and will instigate similar undertakings in other fields of technology, to benefit a larger community which may desire to know, understand and use the tools of its own progress and development.

MAURICE FICKELSON
Secretary General of RILEM (International Union of Test and Research Laboratories for Materials and Construction).

PROF. DR GY SEBESTYEN
Secretary General of CIB (International Council of Building for research, study and documentation).

Foreword

For fifteen years, the CRATerre team of the Grenoble School of Architecture has been undertaking the considerable task of updating the scientific and technical knowledge of unbaked earth. This patient and continuous investment has been supported by a will to modernize this age-old and modest building material in order to propose an alternative to the prolific architecture that is costly in scarce energy and foreign exchange, particularly in the poor regions of the Third World. This aim was also linked to a strong wish to re-establish the necessary dialogue between the architect and the users who have been gradually deprived of any legitimate participation in the conception and realization of their living environment by the imposition of modern building techniques, which are too sophisticated and in many cases not accessible.

World Habitat Day, commemorated on October 3rd, 1988, in the Palace of the United Nations at Geneva, confirmed our obligation to house a quarter of the world's population decently. The housing needs of this one billion homeless people amounts to hundreds of millions of houses. The national decision-makers of those countries, as well as the whole international community, are really powerless in the face of this overwhelming reality; they will have to mobilize the widest range of useful human and material resources, since it is not only industrialized materials which can guarantee a massive and rapid access to decent housing by all people. We will also, therefore, have to count on unbaked earth and on the vast technical and architectural potential that builders have always been aware of.

As a result of development in the field of architectural research, university and professional education, the implementation of demonstration projects, international expertise and, within the latter, a large dissemination of knowledge and skills, CRATerre now presents us with a global and coherent action plan which can give a real future to earth architecture. *Earth Construction* reveals an ethics strengthened by an aim to communicate skills meant to stimulate control of their own environment by people themselves. This book offers a unique and complete collection of the current knowledge in this new technological area backed up by the well-proven field experience of CRATerre. It offers the tools for action, which can be helped by theoretical as well as practical advice – elements which are all essential for a useful extension of new knowledge.

JEAN DETHIER
Architect consultant to the
Georges Pompidou Centre, Paris.

To Jacques Chaudoir (1948–81) who, in Algeria, brought me to the insight that the practice of earth construction could be professional.

Hugo Houben

To Professor Dr Jean Pieri, my grandfather, looking for a positive and peaceful relationship between humanism, science and technology.

Hubert Guillaud

Introduction

The approach used and the form given to this book have been determined by a desire to put forward the widest possible range of solutions for earth architecture, with a view to facilitating informed decision-making. A knowledge of the real potentialities of the material means that they can be used to greatest effect, and, more important still, their misuse avoided. The latter is of key importance, if the capabilities of earth are not to be denied because of failed demonstrations due to ignorance, or even negligence.

Building in earth can be approached from the same high levels of technology and science, as other construction technologies. The current research effort in this field is the proof of this. There are virtually no limitations on the use of the material, if users are aware of how to profit from the wide range of its qualities and ameliorate its defects.

Both traditional applications, which represent the accumulated knowledge and experience of centuries of use, and modern earth construction, which has been able to introduce considerable sophistication and highly advanced technical research, are rich in possibilities and are easily adapted to numerous different contexts. It is difficult, if not impossible, to evaluate the relative importance of the technological approach compared to that of traditional architecture and vice versa. In fact many traditional codes of good practice conform to modern standards: 'scientific' know-how combines with traditional savoir-faire and it appears that it can only confirm the correctness of traditional solutions to technical and architectural problems. These two approaches have their inherent qualities and deficiencies and are similar in their effectiveness as well as in their inconsistencies.

The object of this book is not to make any fundamentally new contribution to earth architecture but rather to encourage reflection about building in earth taking the entire production process into consideration. This is a question of providing the information upon which logical and sensible decision-making can be based, with a permanent exchange between the global approach to the various problems, which takes cultural, social, and economic parameters into account as well merely technical factors, and an approach which concentrates on the details.

At first sight building in earth may seem like an enormous jumbled puzzle. However, long and slow work shows that the puzzle can be solved by assembling and ordering the pieces, and patiently working out where they fit.

Our first wish has been to assemble the many bits of information scattered throughout the abundant general literature, classify, and order them, and finally simplify them so that they are accessible to a wide range of interests and abilities. To this end texts and illustrations have been deliberately kept simple and deal only with the essential. All problems which could arise are dealt with at various levels as the process proceeds: in the decision-making phase, passing through the logical and ordered planning stage, to the design and, finally, the realization of the project. This desire to order the accumulated knowledge should, if these ambitious goals are attained, make this knowledge easily accessible, and most of all give it didactic form. As this manual is not only intended for people involved in actual work on site, but also for giving backers and other persons involved the necessary training, applications are also discussed.

This book is furthermore intended to serve as a practical manual and teaching handbook. It is aimed at all persons involved in earth construction projects: decision-makers and planners, building inspectors, architects and engineers, technicians of all levels, building promoters, bricklayers, and sub-contractors. Apart from these it is our hope that this work will also be consulted by students and members of the public in search of information.

Inevitably the desire to classify and simplify information, so that the essential problems are discussed at every level, will make some readers wish for more detailed information. It is hoped that such readers will make use of the numerous bibliographical references, which will guide them to more specialized information, or to get in touch with the authors. Even so this primer is the first of a collection of specialized works, currently in preparation, which will go into more detail on the various subjects and matters mentioned here.

It is our sincerest hope that this book will be a source of inspiration for training programmes, and encourage effective and logical construction practice.

Hugo Houben, Hubert Guillaud
Grenoble, January 1989.

Contents

Traditional earthen housing in Yemen (Thierry Joffroy, CRATerre-EAG)

1. EARTH CONSTRUCTION

Ever since man learnt to build homes and cities around 10,000 years ago, earth has undoubtedly been one of the most widely-used construction materials in the world.

There is hardly an inhabited continent, and perhaps not even a country, which does not have a heritage of buildings in unbaked earth, and even nowadays more than a third of all humanity lives in a home built of earth.

To build in earth is to build with a material which we trample with our feet every day of the week. However, earth can only be used for construction purposes if it has inherently good cohesion, provided by the presence of clay which acts as a natural binder.

Earth architecture is a surviving witness to the history and culture of humanity, particularly in regions where familiar landscapes have been richly endowed with earth structures.

In the tradition of earth construction we can recognize numerous building methods with an infinity of varieties reflecting the identity of the locations and cultures. In fact 12 main methods of using earth as a construction material are recognized. Among these, seven are very commonly used and represent the main classes of technique.

Adobe Sun-baked earth brick is more commonly known as adobe or adobe brick, and is made using a thick malleable mud to which straw is often added. Traditionally adobes were shaped by hand, in wood or metal moulds, but nowadays the use of machines is widespread.

Rammed earth The earth is compacted in formwork. In many countries, wooden forms and rammers are used. The technique makes it possible to build monolithic walls in compacted earth.

Straw-clay (or clay-straw) In this technique the soil used is very clayey and is dispersed in water to form a greasy slip which is then added to the straw. The role of the earth is to bind the straw together. Straw-clay can be easily adapted to the pre-fabrication of various building components, such as bricks, insulating panels and flooring blocks.

Wattle and daub A bearing structure, usually wooden, is filled with a daubed lattice or netting woven from vegetable matter. An extremely clayey earth is used which is mixed with a straw or other sort of vegetable fibre to prevent shrinkage upon drying.

Direct shaping This very ancient technique is widely used in many countries. It makes use of a plastic earth and makes it possible to model forms directly without using any kind of mould or form-work. Only the hands of the builder are required.

Compressed earth blocks The manual production of earth blocks by compressing them in small wooden or steel moulds has been practised for centuries. Nowadays the process has been mechanized and a variety of presses are used, including manual and hydraulic presses and completely integrated plants. Products range from accurately shaped solid, cellular and hollow bricks to flooring and paving elements.

Cob Basically the cob procedure consists of stacking earth balls on top of one another and lightly tamping them with the hands or feet, to form monolithic walls. The earth is reinforced by the addition of fibres, usually straw from various types of cereal or other types of vegetable fibre, such as grass and twigs.

Nowadays it is the techniques of adobe, rammed earth, and the compressed block which are the most widespread and have reached extremely high scientific and technological levels. It is perhaps regrettable that these three techniques now dominate the field to the detriment of the others, which are still of a certain interest.

A wide range of applications

The earth construction techniques mentioned above are highly flexible and permit the construction of a wide variety of components and construction systems.

- Foundations
- Base course
- Walls and pillars
- Openings
- Floors and pavements
- Flat and pitched roofs
- Arches and domes
- Tiles
- Insulating elements
- Stairs
- Flues and chimneys
- Built-in furniture
- Ventilation elements
- and so on.

These are of course not the only elements which can be built in earth. There are numerous other applications which lie outside the exclusive field of home construction:

- Gutters
- Footways
- Canals and reservoirs
- Bridges and aqueducts
- Car parks and landing strips
- Roads
- Dams
- Embankments
- and many, many more.

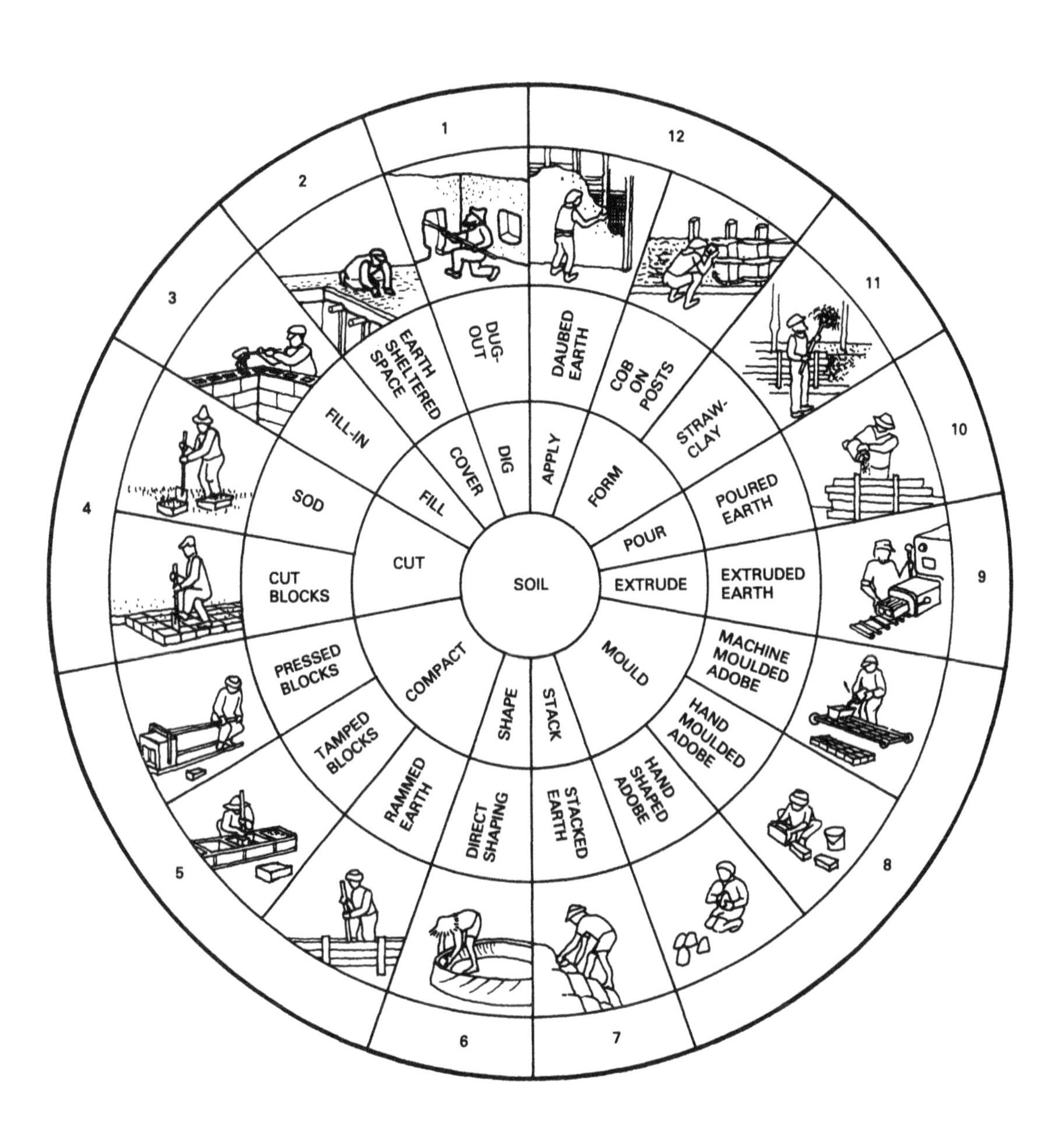
SOIL
DIG
COVER
FILL
CUT
COMPACT
SHAPE
STACK
MOULD
EXTRUDE
POUR
FORM
APPLY
DUG-OUT
EARTH SHELTERED SPACE
FILL-IN
SOD
CUT BLOCKS
PRESSED BLOCKS
TAMPED BLOCKS
RAMMED EARTH
DIRECT SHAPING
STACKED EARTH
HAND SHAPED ADOBE
HAND MOULDED ADOBE
MACHINE MOULDED ADOBE
EXTRUDED EARTH
POURED EARTH
STRAW-CLAY
COB ON POSTS
DAUBED EARTH
1
2
3
4
5
6
7
8
9
10
11
12

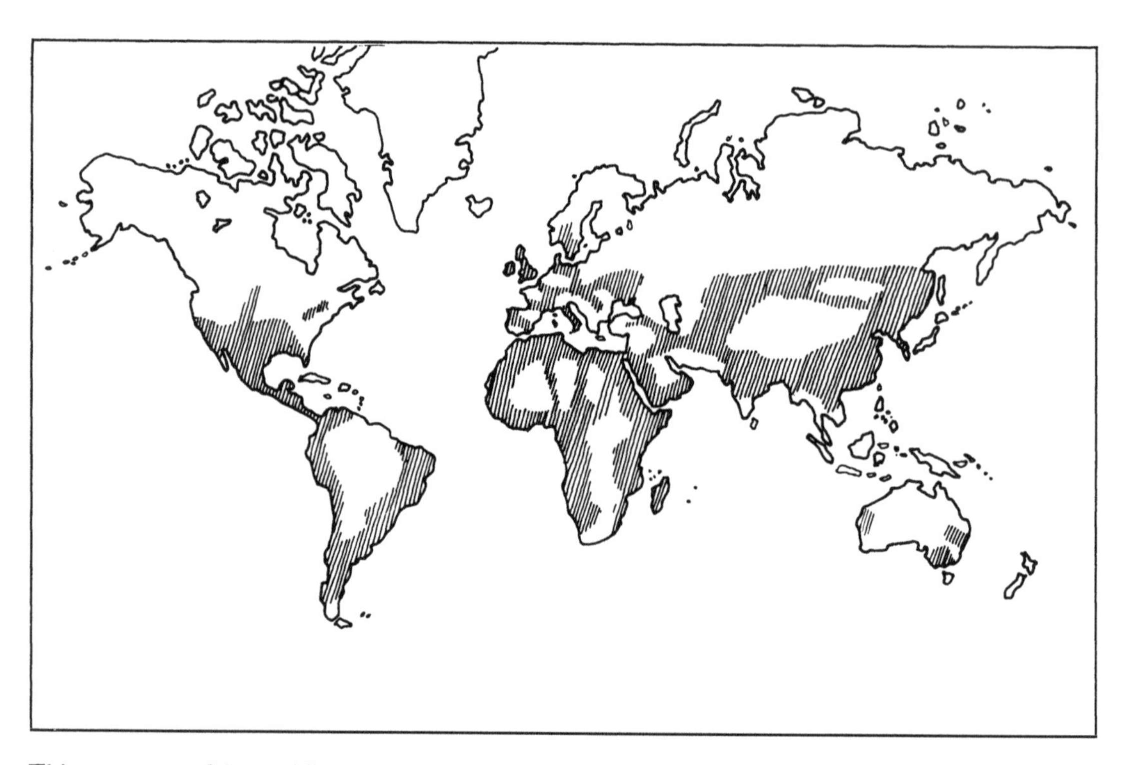

Thirty per cent of the world's population, or nearly 1,500,000,000 human beings, live in a home in unbaked earth. Roughly 50% of the population of developing countries, the majority of rural populations, and at least 20% of urban and suburban populations live in earth homes. It may even be that these figures understate the true situation. The writings of several authors tend to confirm this point of view. For example it has been observed that in Peru 60% of homes are built in adobe or rammed earth. In Kigali, the capital of Rwanda, 38% of housing is built in unbaked earth, while in India the 1971 census shows that 72.2% of all buildings are in earth, that is to say 67 million houses inhabited by nearly 375 million people. To return to the Africa continent, we see that the great majority of rural, and even urban, structures are in *banco* (West Africa), *thobe* (Egypt and North Africa), *dagga* (South-east Africa) or in *leuh* (Morocco). This understandable diversity of these names also reflects the great variety of construction techniques and the very refined knowledge of the engineering qualities offered by earth, which have been known to man since earliest times.

From the most humble rented hut to granaries in a riot of shapes, from palaces on the banks of the Niger to the ksour and kasbah of Southern Morocco, from the fortified homes of the Somba tribe in Benin, to the shell huts of the Mousgoum tribe in Cameroon, from town houses to the mosques of Mali (at Djenné and Mopti) – the earth architecture of the African continent reflects the spirit of the site, the material and the builder. Earth architecture has also deep roots in the Middle East: in Iran, the heart of Ancient Persia; Iraq, cradle of the Sumerian civilization; Afghanistan, North and South Yemen. The techniques of the barrel vault and dome were perfected in Iran, as the ancient centres of Bam, Yazd, Seojan and Tabriz bear witness. At Shibam in South Yemen, there are cob buildings which are more than ten storeys high. In China, in the provinces of Henan, Shanxi, and Gansu, more than 10 million people live in homes dug out of the loess layer.

Passing on to Inner Mongolia, in Hebei and Jilin, at Sichuan and Hunan, rural dwellings are for the most part built in daub, adobe, or rammed earth. Rounded yurts have taken permanent form in earth in Mongolia, rectangular houses with anything up to seven bays built in the north-eastern provinces, large residences with inner courtyards in Zhejiang Province as well as the multistorey rammed earth houses, the concentric-ring dwellings of the Hakka of the Central Plateau (Fujian Province) the construction of which was continued until 1954–5. In Europe earth dwellings continue to be a feature of the countryside in several countries, including Sweden, Denmark, Germany, the United Kingdom (in Devon in particular), Spain and Portugal. In France, 15% of the population, for the most part rural, live in homes built in rammed earth, adobe, or daub.

Here we can also point to the immense growth in popularity of adobe as a building material in the American South-West. In 1980 close to 176,000 houses in unbaked earth were built in the United States of America, and 97% of these were in the South-West. In California construction in adobe increases by about 30% per year. In New Mexico, 48 adobe producers turn out more than four million blocks every year. It is estimated that do-it-yourselfers produce a similar number of adobes each year.

Confronted with a twin crisis of economy and energy the industrialized countries have started a dialogue on the revival of earth as a building material and are backing research programmes and developing applications. The United States have officially recognized the use of adobe and rammmed earth by integrating these construction techniques into national and regional standards. The investment in research programmes on the thermophysical characteristics of earth runs into millions of dollars. France has outlined the areas in which research into earth as a building material should be concentrated in the coming years. The budget devoted to research into these priority areas is close to 24 million francs, 83% of which will go into operational research and training. The construction of 72 experimental homes (Domaine de la Terre) is evidence of the commitment of the country to the use of real projects for research purposes. Programmes of research have also been started in the Federal Republic of Germany, Switzerland, and Belgium. National and international colloquia have multiplied not to forget the appearance of a specialist training course for architects and building technicians (e.g. France).

In the developing countries, construction in earth appears to be an effective means of building homes in the short term so that the greatest number of people can be housed, while at the same time encouraging the use of local resources for building materials, the training of building technicians and craftsmen, and the creation of jobs. 'A material is not attractive in itself but rather in terms of what it can do for society as a whole', as John Turner so succinctly puts it. The demand for homes in the developing countries at the present time is immense. Recent statistics (Cache) show that no less than 36,000,000 homes must be built by the year 2000 for the urban population of Africa alone. Feasibility studies indicate that the population affected by this immense programme have no other choice than to use local materials, which in the majority of cases means using earth. We can expect that 20% of urban dwellings in the developing countries, that is to say some 7,000,000 units, will be built in earth in the coming twenty years. This means a construction rate of 350,000 units per year.

That this is an underestimate can be seen when the numbers of required in rural areas are added. Thus, regardless of whether one is for or against, earth will continue to be one of the foremost building materials in the years to come. What at one time was purely and simply a necessity for the poorer peoples of the world may now very well become a matter of preference. Here it is worth noting the exemplary attitude expressed by the Peruvian newspaper *Correo* in its issue of January 31, 1978, 'Adobe will become the star of the Peruvian architecture'. Similarly the Mali newspaper, *L'Essor*, wrote in 1980 of the urgency 'of reversing the current trend in building . . . and setting up realistic programmes for home construction and town planning making use of earth, water, sun, and above all Mali manpower'. The late Mrs Indira Gandhi also spoke in this vein, witness this statement made in 1980, 'All the new houses are built for energy consumption . . . whereas our old houses are not. You can't keep all of it (the old technology) . . . but I think a lot of it can be adapted and made more efficient'.

So it seems that the universal use of earth for construction will continue to figure in the future of all mankind.

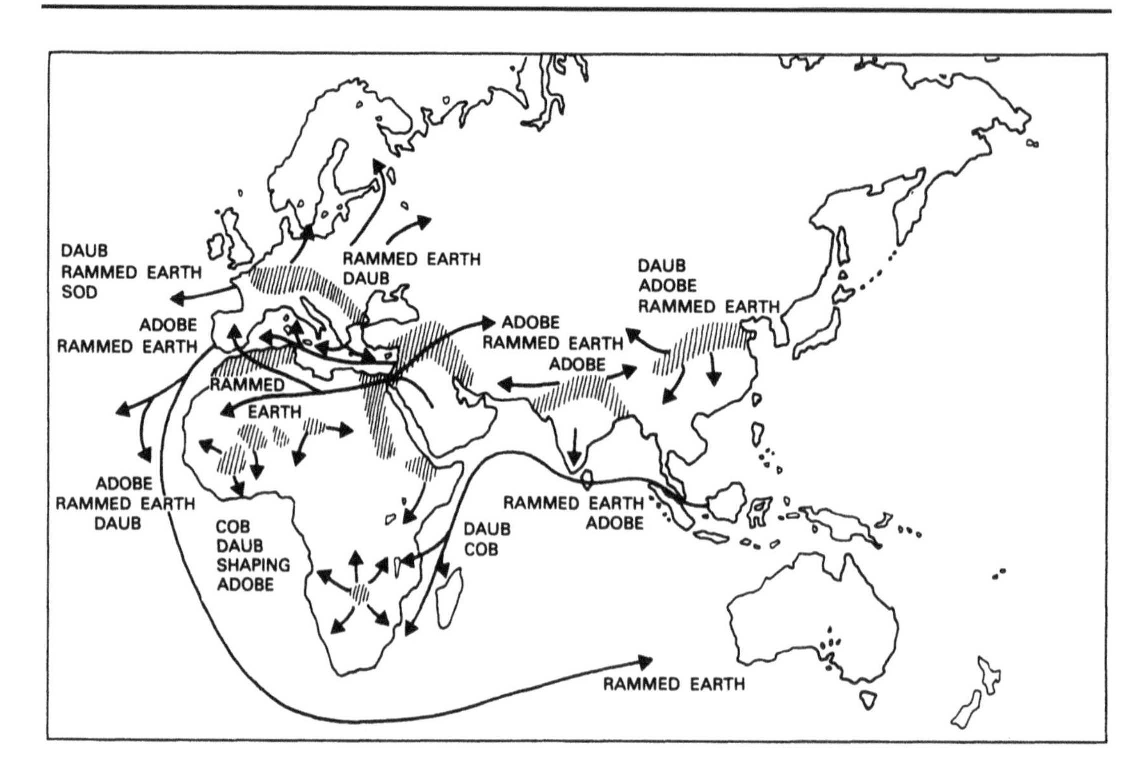

The history of building in earth is not well-documented. Interest in this material, often regarded as being inferior and archaic, has been eclipsed by that devoted to stone and wood, the 'noble' materials. Nevertheless it is earth which is associated with the decisive eras of the urban revolution and which served the daily requirements as much as the prestige of the greatest civilizations of ancient times. Archaeological findings at many a site have borne witness to this. The debris of time has not been able to hide the accumulated evidence. Our age is rescuing the remains. Ruins are laid bare and marked out, categorized, protected and restored. The further we go back in history the more earth seems to be the material of choice of building man, from the most distant past down to the present day. Construction in earth was developed independently in all the main cradles of civilization: in the lower valleys of the Tigris and Euphrates, on the banks of the Nile, the Indus and the Huangho. These fertile regions favoured the settlement by communities of hunter-gatherers, which was the essential precondition to the transformation into a farming culture.

These clayey and sandy alluvial soils, mixed with straw from cultivated cereals, gave man his first solid and durable construction material, which the change to sedentary habits made necessary. This was the mud brick. But even before the Neolithic era, the semi-permanent shelters of hunter-gatherers were constructed partly in earth. Huts made of branches covered with clayey soil undoubtedly formed a feature of the Mesolithic landscape, the age in which mankind domesticated small livestock. Prior to this, hunting peoples built huts made from branches covered with animal skins. Evidence for this comes from the Olduvai site in Tanzania and the Molodova site in the Ukraine. Whatever contacts there may have been between different civilizations using earth as a building material, we may be fairly sure that the use of the material arose in each region independently. The findings from archaeology justify both a regional and global approach, while other disciplines (history, anthropology, ethnology) help us determine the direction of cultural influences from the study of migrations, conflicts, and cultural contacts. These links, both assumed and confirmed, make it possible, albeit cautiously, to evaluate the likelihood of the passing on of construction techniques. Here we will briefly examine the major centres of civilization, reconstructing a minimal portion of the available material.

Africa

The role played by the African continent in human evolution has been immense. Mankind probably made his debut on the world stage in Africa (Rift Valley and the Olduvai Gorge). Africa has also been the home of Egyptian civilization, which flourished for close to three millennia. Homes constructed from woven reeds and branches covered with clay or filled with clods of earth were used in the first human settlements found at the Merimd and Fayum sites in the Nile Delta, which date from the fifth millennium before Christ. Advanced civilization makes its appearance in Africa with the establishment of Egyptian dynasties (2900 BC). The Nile valley provided the prime building material: clayey silt mixed with the sand of the bordering desert to which the straw from cultivated grains was added. At first this material was shaped by hand, but subsequently moulds were used to form unbaked bricks which were allowed to dry in the sun. The first mastabas (an oblong tomb) for royalty and high-ranking officials were built using mud bricks. Their outside walls were stepped, perhaps in imitation of Mesopotamian structures.

Excavations at Saqqarah and Abydos indicate the development towards sloping bricks walls, which were subsequently covered with stone. Earth was not eclipsed by stone, the eternal material, which was first used in the limestone temple at Saqqarah, constructed by Imohotep. Rather it was reserved for civil architecture, the mansions and palaces of nobles and kings as well as for rural dwellings. J.P. Lauer has clearly shown that the stone shell at Saqqarah imitates the techniques and forms of traditional sun-baked brick and clay-daubed wattle. Earth was immortalized by stone. The written or painted decoration of funerary monuments confirms the use of sun-baked brick up till the most recent eras of Egyptian civilization. Archaeological traces are, however, rare as the material has not stood up to well to time. The main remains left to us are at the Tel el-Amarna (Echetaton) site in Middle Egypt, and at the necropolis at Thebes. Furthermore, there are the little pottery 'houses of the soul' found in tombs, which give us an idea of the sort of dwelling ordinary people lived in. In the Tel el-Amarna centre, built in the New Kingdom (1552–1070 BC), we find the homes of artisans and nobles, palaces and temples, all built in sun-baked brick. The ordinary dwelling often consisted of one or more rooms with earthen walls dressed with chalk. The richer homes of the nobility were spacious, enormous reception rooms connecting with various chambers and a drawing room with a central pillar, as well as lavatories.

At the bottom of the garden there were ancillary buildings (storehouses, stables, and kitchens) and the servant quarters, also in sun-baked brick. At Deir el-Medineh, a village for artisans who worked on the necropolis at Thebes, all the houses are terraced, and were built in mud brick on stone foundations. Each house offers in succession a reception room, a dayroom, a sleeping room, and a kitchen. A stair provides access to a flat roof. This village in sun-baked brick was inhabited and occupied for 400 years by generation after generation of royal artisans. The Egyptian builders also developed the art of brick vaulting, as can be seen in Lower Nubia, between Luxor and Asswan (the Ramesseum granary, about 1200 BC).

Although brilliant, Egyptian civilization remained extremely conservative and isolated. In the long run it had only a limited influence on those African cultures which developed semi-permanent homes consisting of huts made from branches covered with clay, or made entirely from earth. Numerous shapes and techniques were used. The northern regions of the African continent were influenced by successive Mediterranean civilizations, which had the effect of spreading the use of sun-baked brick and rammed earth. East Africa was affected by people coming from the Indian Ocean (Melanesians) who used daub and direct shaping. The migration of the Kushite and the influence of the Axum Kingdom (3rd to 8th C. AD) from Nubia as far as Kenya may have spread the use of sun-baked brick. Islam, however, had a far greater influence. Beginning in the 11th century, it brought about profound changes in the appearance of the ancient centres of Africa and introduced the architecture of the mosque. These were for the most part built in earth, using the techniques available locally, which could vary from direct shaping, daub, or sun-baked brick. Among the most beautiful examples are the mosques of San, Djenné, and Mopti in Mali, which have served as models for neighbouring countries. Despite all these real and theoretical influences, the enormous African continent has nourished specifically African cultures which have mastered and perfected the art of building in sun-baked earth. These techniques were spread by the kingdoms of Ghana (8th–11th C.), the Malinké kings of Mali (13th C.), the Songhay kingdom (14–16th C.) and the Hausa States (10th–19th C.), right up to modern times.

The oldest settlements in Europe date from the 6th millennium before Christ. The primitive dwellings of the Aegean coast, Thessaly (Argissa, Nea-Nicodemia, Sesklo) were of woven wood and clay which subsequently evolved into groups of square structures built from sun-dried brick. On the Sesklo site, the homes found in the upper levels were built with daub and dried brick, and had an oblong floor plan and an upper floor (4600 BC). The form evolved into the megaron which came to occupy a leading position in Greek architecture. This type of home was carried into inland Europe, to be replaced by structures in wood and earth in the North. These were the homes typical of the Danube cultures which were widespread in Central Europe throughout the Bronze Age (1800–570 BC). Excavations at Köln-Lindenthal (Germany) have unearthed traces of wood and earth cabins with four naves, which sometimes were as much as 25m long and 8m wide. In the Aegean the Mycenae built numerous fortresses in the late Bronze age under pressure of attacks by the Dorians. 'Cyclopean' walls in stone replace unbaked earth, which was reserved for the protected homes in the acropolis (Tyrinthe). In the same era the relative isolation of the island of Crete favoured the development of the Minoan civilization.

The superb palaces of Knossos, Phaistos, and Mallia used unbaked brick jointly with tuff, gypsum, schist, marble and wood, and these materials were painted dark red, deep blue and various ochre tints.

Archaeological diggings at Acrotiri on Thera, an island close to Crete, and the famous ceramic maquettes in the Herakleion Museum, dating from the Middle Minoan period (1900–1600 BC) confirm the importance of half-timbered construction in domestic architecture. Houses had one or two floors. They were timber-framed and infilled with daub or sun-baked brick. On the Greek mainland, the dark age which ensued after the Dorian invasions (1100–700 BC) were marked by a return to wattle covered with mud. In the 8th century the country reorganized into regional states. The work of R.V. Nicolls shows that at Smyrna homes had apses and were protected by thick outer walls in earth brick. These oval houses measuring 3×5m with earth brick walls, coated on the outside, were not provided with foundations or a base. The apsidal megaron evolved into the oblong megaron. By the end of the 3rd century BC in Athens, a city had grown up at the foot of the splendid Acropolis of Phidias, consisting of dense quarters built in earth brick, with tiled or thatched roofs, giving the impression of a village, according to Dicearch. The use of 'Pentadoron' and 'Tetradoron' bricks continued until the 1st century BC, as Vitruvius observed, 'The walls of the Temple of Jupiter and the chapels of that of Hercules . . ., the house of Croesus at Sardes as well as the palace of the mighty king Mausolus at Halicarnassus were all built in earth brick.' The first settlements on the banks of ancient Phoenicia date from the seventh millennium before Christ. Digs carried out about 20 years ago in the south of Syria's western regions appear to indicate the influence of the surrounding cultures (Jericho-Munhata) which developed the use of earth brick at a very early date (the Tell de Ras Shamra site). As the use of Tyre stone reached a peak (Tyre, Sidon, Ugarit) earth was reserved for use on flat roofs, rammed and rolled in successive layers over laths, or for rural structures.

It seems that builders in these regions developed the earth brick dome at a very early date, by resorting to corbelling, for the construction of conical grain silos. It is not known if the rammed earth tradition which can be observed in Lebanon and Syria are of such distant origins. This technique, however, was used for the construction of the Punic villages when the Phoenicians transferred their civilization to the shores of the Central Mediterranean (founding of Carthage in 820 BC). Pliny describes the method in his Natural History, 'What must we say about the rammed earth walls which we can see in Barbary (Carthage) and in Spain, where they are called moulded walls, as the earth is moulded between two boards . . .; there is no cement or mortar which is harder than this earth; . . . the watchtowers and lookouts built by Hannibal in Spain . . . are in rammed earth.' Excavations at Carthage on the hill of Byrsa have confirmed that rammed earth was used for the construction of homes. According to Strabon six storey buildings in rammed earth brick, sometimes limewashed or faced with marble, were common in this vast city of 700,000 souls in the 2nd century BC.

Rome was no more than a large farming village when it was caught up by Hellenic influences reaching it indirectly through the Etruscans in the early 7th century. The dwellings consisting of wooden huts rendered with earth and covered with thatch, on the seven hills, slowly gave way to oblong houses with earth brick walls. This material was used for the first sacred and public buildings of the Republic (4th and 3rd C.) but was soon displaced by tuff and marble, and later by travertine, which was the most prestigious material of Imperial Rome. Sun-baked brick (lidio) continued to be the material used in more modest houses and for the dwellings of the poor, until the Augustan era. Vitruvius refers to it in his *De architectura*. Indeed he had a considerable regard for the material, saying '(it is) extremely useful as (it) does not load the walls.' He recommends its use, '. . . on condition that the necessary care is taken to use it properly'. Unbaked brick was, however, limited to use outside the towns by laws which forbade thick walls. In Gallic Celtic territory, the Iron Age (750 BC–50 AD) settlements took the form of the 'oppidum', with small houses in wood, daub, and cob behind fortifications. In Mediterranean Gaul, Hellenic and Carthaginian influences introduced unbaked brick and rammed earth, an influence which Vitruvius reports observing at Massalia (Marseilles) and which has been confirmed by excavations at the La Lagaste and Entremont sites. Before the Romans spread the use of baked brick and blockwork throughout the Empire, unbaked earth was widely used in Cisalpine Gaul in rural and even in urban structures. Digs at Lugdunum (Lyons) have recovered timber frame structures which were infilled with dried earth brick and daub. The Latin verb *pinsare* for the action of ramming earth (the French *pisé*) suggests that the technique was known to the Romans, although excavations have revealed very few traces of it. The periods of decline of the Middle Ages (the Dark Ages) witnessed a return to the rudimentary structures in wood split lengthways and in daub.

These techniques predominated in rural areas until the Late Middle Ages until growing skill in carpentry introduced the timber frame building with infillings in daub or brick. Even so it was not until the 18th century that building in raw earth, cob, rammed earth, and brick was able to re-establish itself in most of the rural areas of the countries of Europe, and once established persisted during the entire 19th century. The ideas of the Enlightenment as transmitted by the Physiocrats, who wished to improve the miserable conditions of country life, were of key importance to the advance of rammed earth techniques. One of the most enthusiastic promoters of rammed earth was François Cointereaux, who saw it as a way of providing cheap, healthy, and durable housing. His writings (72 pamphlets) were translated into various languages and were distributed in Germany, Denmark, the United States, and even in Australia, and undoubtedly played a role in the spread of the techniques in these countries. In Europe, building in earth continued until the 1950s, having experienced an astonishing revival in the years following the Second World War, a period in which there was a dearth of industrial materials and an urgent need to rehouse thousands of homeless people. The techniques were developed systematically in Germany and training centres were set up, with the result that several thousand homes were built in earth. Since then, the desire for modernity has eclipsed the use of earth. Nowadays, however, the cost of energy has risen to such heights that research into and the use of earth as a building material has once again become a matter of lively debate.

The Middle East

Archaeological excavations in the Middle East have brought quantities of information to light regarding the development of earth-built housing in those regions, from Neolithic times onward. The city of Jericho covered four hectares. The oldest dwellings (8000 BC) were round, with stone base structures surmounted by walls in earth bricks in the form of hand-shaped loaves. At Mureybet, in Syria, the upper levels of the excavations revealed four-cornered buildings, laid out next to one another in a chequer-board pattern. The Tell Hassuna site, in southern Iraq, appears to confirm the supposition that the first bricks were moulded to a parallelepiped shape. The Obeidian era (5000 to 3200 BC) saw the appearance of a monumental architecture, that of the future temple cities of the Uruk era (3200 to 2800 BC). The first religious temples were constructed during the 3rd millennium (the temple of Eanna at Uruk, the temple of Enki at Eridu) in pasty brick, without the use of mortar. Houses at Ur were built in earth. The rooms were on two levels and gave out onto a courtyard open to the sky. At Assur and Mari, the architecture of the palace of the Isin-Larsa period (2015–1516 BC) had earth-brick walls which added to the fortifications.

Assyria dominated the Near East until the 6th century BC. Nineveh with its immense walls pierced by 15 gates was built entirely in earth. In the palace of Sargon II at Khorsabad earth-brick was used together with the finest of materials, such as ivory, sandalwood, ebony, tamarisk, marble, basalt, gold and silver. The galleries of the palace were vaulted, with keyed barrel vaults in earth bricks with a pronounced slanting shape. Babylon continued this tradition of construction in earth. The famous Ishtar Gate decorated with blue-glazed reliefs is the first stage on the processional route leading to the temple of Marduk, which is overshadowed by the 90m high Ziggurat of Etemenanki. The Babylonians were the first on the path towards the development of reinforcement techniques for earth structures. Twisted ropes of reeds, as thick as an arm, interlaced and traversed the cores of the sun-baked bricks of the ziggurats. The Elam civilization, contemporaneous with that of the Sumerians, was established in south-west Iran and grew rapidly. Earth was the primary building material of the Elamites, whose lands included the plains of Khuzistan, and the clayey plateaux rising in terraces from them. The unbaked brick architecture of Susa easily bears comparison with that of the cities of Mesopotamia, with the proud ziggurat of Choga Zanbil reaching a height of no less than 53m. The period of the Indo-European invasions in the 2nd millennium before Christ has left only a few traces for the archaeological record, which remains scanty until the establishment of the Kingdom of the Medes (9th to the 6th century BC). The Medean centres of religion and administration were girt with thick, sharply indented unbaked brick. At the Medean capital of Ekbatana, now entombed by the modern town of Hamadan, it has been found that unbaked brick was used for the construction of load-bearing walls and columns. Dried bricks were laid using a mortar of clay. The conquests of Cyrus the Great and his successors pushed the frontiers of the Persian Achaemenian Empire to the banks of the Indus in the East, Anatolia in the North, and to Egypt and Libya in the West. At Parsagadae (546 BC) columns of stone together with walls of unbaked bricks were used and the principle of the hypostyle type of hall was introduced. Similarly at Persepolis, all buildings had high rooms with the roof supported by columns which were flanked by galleries and entrance halls at the doors. The ruins are impressive despite the collapse of the shafts of many of the columns, the cedar roofs, and the disappearance of the walls in unbaked brick faced with polished stone panels, which were erected between the huge doors with lintels decorated with an Egyptian style gorge. The fortifications of the Achaemenid cities such as Susa were built of earth. The Persian style of architecture reached its peak with the mastery of the techniques for building vaults and domes. Vaults were built in sections of inclined barrelling and domes were built on pendentives or squinches. The vault and the dome were not reserved for use in palace architecture and were employed widely in domestic architecture, being passed down the ages to Iran of the present day. Countless fortified sites, ghost towns with their abandoned homes (fortress of Bam in south-east Iran) and modern towns (Tabriz, Seojan, Ispahan) bear witness to Persian earth architecture.

The Far East

We must now turn our attention to the Far East and the New World. We will first complete our survey of the Old World. In India, at the same time as the flowering of Ur and Babylon, numerous cities grew up on the banks of the Indus. Neolithic sites in Baluchistan (Mehrarch) have been dated as stemming from the 7th millennium before Christ. From these sites it has been possible to determine that the inhabitants lived in dwellings built in unbaked brick. We know in pre-Harappan times the ability to design encircling walls for villages which were built entirely in mud brick was developed (Kalibangan I). Of the 250 known settlements left to tell us of this Indic civilization (2500–1800 BC) two were of particular importance. These were Mohenjo Daro and Harappa, metropolises covering nearly 850,000m^2 each. The town of Mohenjo Daro is unique in that it was built on two mounds. The Western mound was occupied by earth brick buildings built on a platform of beaten earth surrounded by a fortified wall. These were probably the public buildings, baths and granaries. The Eastern mound was covered with groups of houses which had an inner courtyards and which fronted on long streets. These buildings were constructed in earth brick and faced with oven-baked brick. This technique is typical of the Harappan civilization of Vedic times, and can also be seen at Harappa, particularly in the remains of the city granaries' foundations.

In China neolithic farming communities appeared in the 5th millennium BC and settled in the north and north-western regions on the loess plateaux through which the rivers have cut. The first dwellings were dug out in the loess. They were 3 metres deep and 2m wide at the bottom closing as they approached the surface. These excavations were of the 'pocket' type and were circular or oval in shape. Later on the association of wood and earth gradually freed mankind from his trogolodytic existence in the ground. Banpo is one of the best known sites of the Yangshao culture. This settlement was established on a loess bank overlooking the tributaries of the Huangho and was clearly surrounded by a ditch and an embankment in earth. The remains which have been uncovered have revealed dozens of houses with round, oval, and square floor plans, varying in cross-section from 3 to 5m, and which were all built to face the centre of the village. These dwellings, built partially above ground level and partially underground, give the impression of huts buried in earth. A large half-subterranean house measuring 11 by 10 metres seems to have been built at the centre of the Banpo village. Archaeologists think that it was probably a clan house. Its groundplan is marked by an earth wall about one metre thick and 50cm high, the outer surface of which has been burned and hardened. This round-topped wall served to support a heavy wooden structure which covered the house. The definitive raising of the wood and earth dwelling above the level of the soil and the transition to a rectangular plan surrounded by fortifications in rammed earth took place during the Shang era, while the Yin palaces of this era were built on platforms of rammed earth (excavations at Zhegzhan and Anyang). These were long houses with a wooden framework divided by fire-hardened compressed daub partitions. These principles were conserved during the Chou dynasty (12th to 5th centuries BC). In the period of the Warring States (5th to 3rd centuries) work was begun on the Great Wall, many sections of which were in rammed earth. This 6000km long structure was eventually completed by the Ming Emperors (15th to 17th centuries).

The appearance of earth brick as an infill for timber frame building or in load-bearing structures seems to date from the Han dynasty (3rd century BC to the 3rd century AD), and the technique was used in urban architecture under the Eastern Han (1st to 3rd centuries). The Chinese town adopted a square plan divided into square sections occupied by palaces and residential quarters. The walls were in rammed earth pierced by the town's gates. Beijing retained its double walls until 1950. From the Ming Dynasty until modern times Chinese architecture has continued to build in wood infilled with daub or cob, and in earth brick. It seems, moreover, that the Chinese developed the use of rammed earth for construction purposes after the Three Kingdoms period (221 to 581 AD). Earth rammed in formwork or in long split wood casings held in place by poles driven into the ground allowed the construction of multi-storey fortified farms. These farms were usually oblong or circular, and examples of this tradition can be seen to this today among the Hakka people of the Central Plateau.

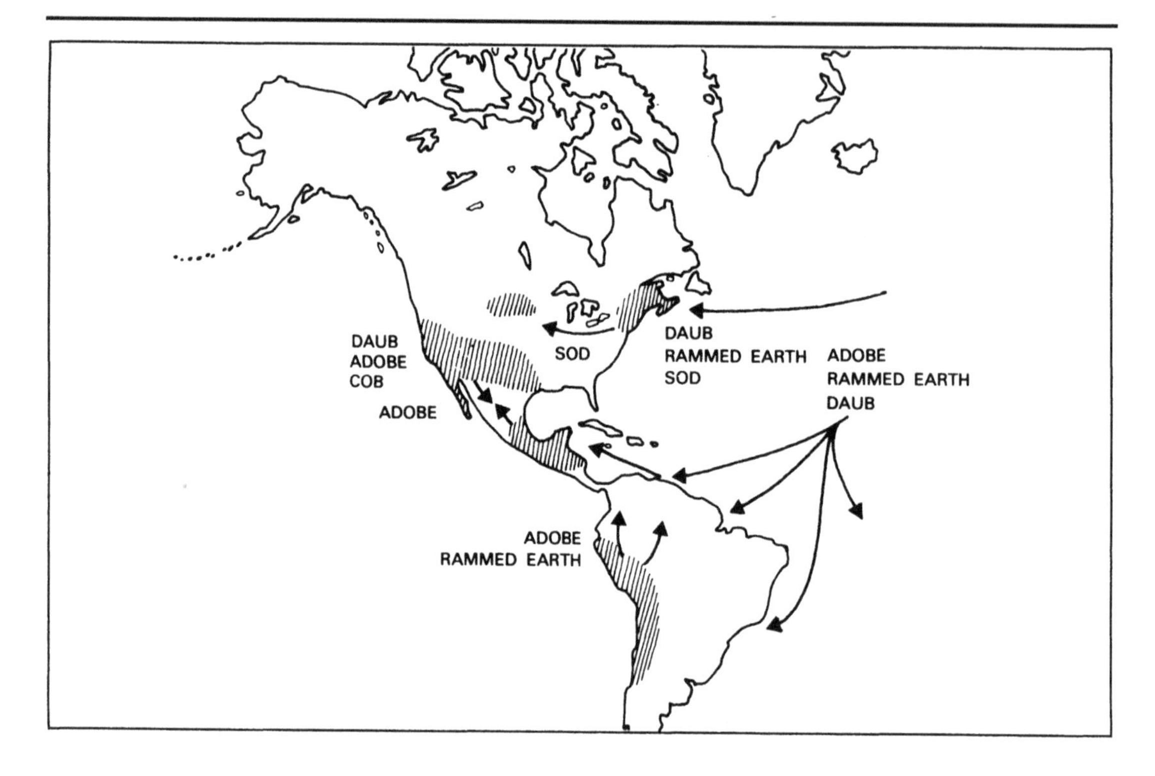

On the American continent the nomadic pattern of life of groups of hunter-gatherers lasted several thousand years before agriculture was to make its appearance. It was in Central America that the domestication of maize permitted the establishment of the first permanent pre-Columbian villages. In Mesoamerica numerous centres of civilization sprang up during the Formative period (1200 BC to 300 AD) typified by a complex society with towns based on religious centres. The Olmecs (Tres Zapôtes-La Venta) and the Zapotecs (Monte Alban) reached this stage in about 800 BC onwards. The La Venta site was dominated by a pyramid built in earth, no less than 65 metres wide and 35 metres high. Housing appear to have been an open system of small rectangular buildings constructed in light materials, such as wood, daub, or balls of earth, covered with palms. The use of sun-baked brick appeared between 500 BC and 600 AD, depending on the degree of complexity and hierarchy of the society. Stone was used for facing *teocali*, the name for the earth mounds and temples built upon them. In the Classical period (300–900 AD) the Pyramid of the Sun was built at Teotihuacan on a square base with a side of 225m and rising to a height of 63m. This structure was faced with lava and was built around a core of two million tons of rammed earth. The temples bordering the four kilometre long 'Way of the Dead' were designed along the same lines. Excavations on the outskirts of the ceremonial centres have uncovered various dense residential areas. Remains of sub-foundations in masonry, levelled to the ground, suggest that the upper walls were in earth brick rendered with lime. In the 13th century, the Aztecs captured the marshy islands in Lake Texcoco (site of Mexico City) and slowly built their city of Tenochtitlan, the splendid city described by the contemporary chroniclers of the conquering Spanish. It was organized into four administrative regions centred on the cultural hub. The residential districts covered close to 1000 hectares. The palaces had an administrative floor as well as a floor for receiving visitors and living quarters. The coloured houses had a single floor and terrace roofs. They were for the most part built entirely in earth bricks, which were whitened with lime. Stone was reserved for the palaces, religious buildings and defensive structures. The superb city of Tenochtitlan was razed by the troops of Hernan Cortés in 1521.

In South America, the use of earth predominated in the alluvial plains and on the coast which, unlike the mountainous areas, lacked copious supplies of stone. The most ancient *huaca* (funeral totem) of the Andean region consist of heaps of stones, which subsequently became pyramids designed as a shell of earth brick which was filled with pebbles and then levelled off with rammed earth (Rio Seco, 1600 BC). In Peru, at the Cero Sechin site, there is a Chavin (1000–200 BC) conical temple in earth brick encircled by a ring of engraved stela. The use of earth was used widely by the Mochica culture (2nd–8th C. AD) of the Pacific Coast. The Mochica irrigation canals were true embanked aqueducts in rammed earth and unbaked brick, and on the river Moche stand the biggest pyramids ever built in unbaked brick, the Huaca del Sol and the Huaca de la Luna. Their internal structure consists of close-set pillars in parallelepiped-shaped earth brick. Chan Chan, the capital of the Chimu Empire, was built in the 11th century entirely in dried brick. Girt with an enormous earthen wall, the whole city covers an area of some 20km^2 and includes a dozen or so walled palaces.

In the Von Tschudi quarter, the adobe walls are decorated with lattice work and zoomorphic reliefs in moulded clay. The majority of the mountain cities (Cuzco, Pisac, Machu Picchu) were built at the height of the Inca Civilization (1493–1525) using enormous stone blocks. On the Andean coast, however, earth continued to be used. In the Rio Pisco valley, the city of Tambo Colorado was built entirely from cubic blocks of unbaked earth.

The walls were rendered with clay brightly coloured in yellow or red. The great majority of rural housing, including the dwellings of the curaca (village chiefs) and the tucricuc (administrator), was undoubtedly built in earth. In the Rimac valley, rich residences in adobe and tapia (rammed earth) have recently been restored. Nowadays, adobe and tapia continue to be the predominant building materials in Central and South America.

In North America, the Indian cultures of the South-West developed the use of earth for construction purposes at a very early date. The dugout homes of the Hohokam (Snaketown, Arizona) are built in wood and covered in earth (Colonial period, 500–110 AD). Under the impulse of the Anasazi (1100–1450), the Hohokam built homes at ground level and subsequently with upper floors entirely in earth. The Casa Grande site reveals walls constructed from banks of cob which are 1.5m long, 1.2m thick, and 60cm high. The Mogollon culture developed a dwelling of a characteristic shape rather like a short pipe, which was built in wood and covered in earth. Even so the Anasazi culture, which represents the common heritage of numerous Indian tribes of the American South-West (Hopi, Zuni, Acoma, and Pueblo), has left the most remains.

The Basket Maker I period (0–500 AD) is characteristic for its circular dugouts known as shallow pit houses, which were made from wooden poles and branches covered with earth. With the Basket Maker II culture (500–700 AD), these dwellings became rectangular in shape and assumed the shape of a truncated pyramid, but continued to be built in wood and earth (Mesa Verde). With the Pueblo I and II phases (700–1100 AD) the structures of the above-ground dwellers became more solid, and walls were coated with daub (Wattle houses) or with balls of earth filling the gaps in the framework (Jacal houses). Spreading out from the canyon sites of New Mexico (Mesa Verde, Chaco Canyon, 1100–1300), the Anasazi migrated to other mesa or valley sites. The alluvial banks of the Rio Grande and the Rio Puerco offered loams and sands suitable for use as building material. The architecture of the Pueblo Indians shows how perfect was their mastery of earth-brick technique. At Taos, the tiered houses form a gently tapering pyramid. The adobe walls are rendered with earth mixed with finely chopped straw, or balls of earth pressed and smoothed by hand. The roofs with their projecting rafters are covered with brushwood and then finished with rammed earth. This highly developed home served as the cultural model for the Hispano-Mexican adobe architecture built since then in American South-West. Nowadays adobe and rammed earth are key elements in the tremendous development which 'solar' architecture is experiencing in all these countries.

References
Danzen Liu, *La Maison Chinoise*. Paris, Berger Levraut, 1980.
Encyclopedie d'archèologie de Cambridge. Paris, éditions du Fanal, 1981.
Galdieri, E. 'The use of raw clay in historic buildings: economical limitation of technological choice? In *IIIrd International symposium on mud brick (adobe) preservation*, Ankara, Icom-Icomos, 1980.
Le Grand Atlas de l'Architecture Mondiale. Paris, Encyclopédia Universalis, 1981.
Le Grand Atlas de l'Histoire Mondiale, Paris, Encyclopedia Universalis, 1981.
Smith, E. *Adobe Bricks in New Mexico*. Socorro, New Mexico Bureau of Mines and Mineral Resources, 1982.
Uppal, I.S. 'Des abris durables et bon marchés.' In *Bâtiment Build International* Paris, CSTB, 1972.

'Passive solar' house in France, built of prefabricated cob blocks: the thick walls store solar energy (Architect: Maryvonne Rigourd) (Thierry Joffroy, CRATerre-EAG)

2. SOIL

Soil is a stage in a long process of deterioration of the parent rock and its physico-chemical evolution. Depending on the parent rock and climatic conditions soil appears in an infinity of forms possessing an endless variety of characteristics.

It is essential to be aware of the properties of a soil before using it for construction. These properties fall into four main categories: grain (particle) size distribution, plasticity, compressibility, and cohesion.

Soils must be classified in order to rationalize and optimize the exploitation of knowledge of their properties. Nowadays, the most useful classifications are those due to engineering geology and soil science.

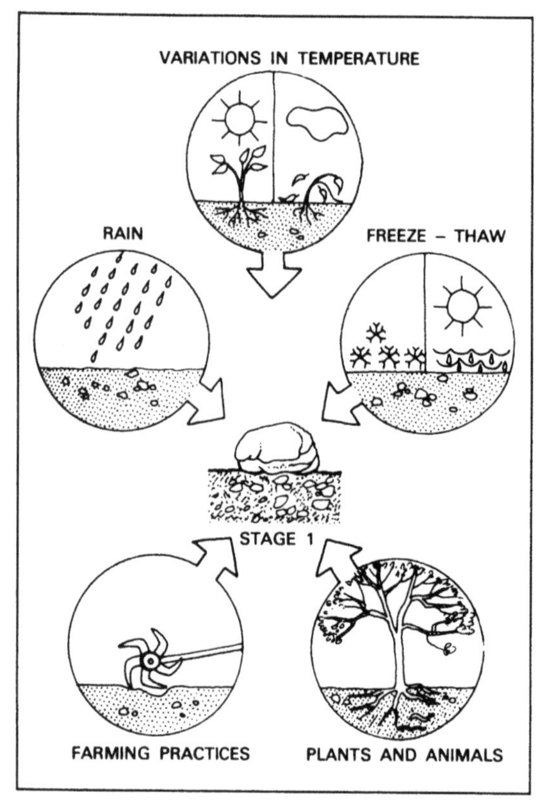

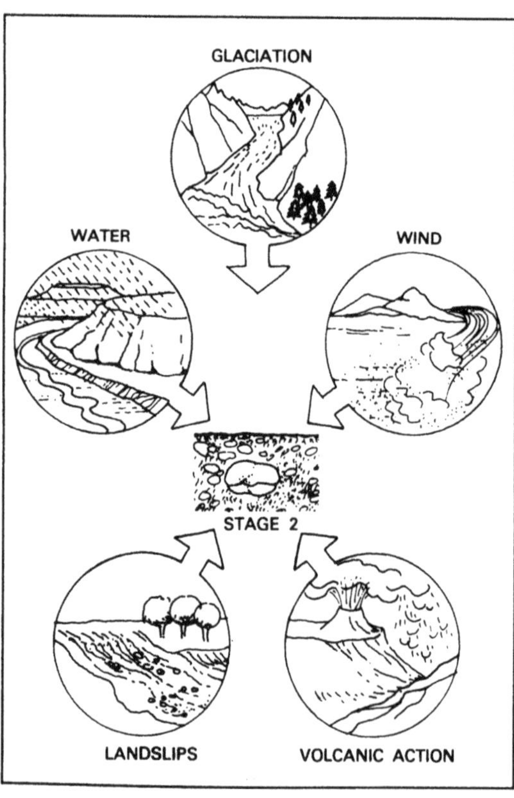

Definition

The ground is the solid part of our planet. At its surface it becomes soil – a loose material of varying thickness, which supports vegetation, and bears humanity and its structures. Soil is the result of the transformation of the underlying parent rock under the influence of a range of physical, chemical, and biological processes related to biological and climatic conditions and to animal and plant life.

The formation and development of a soil is the result of the more or less simultaneous interaction of three different processes.

1. Process: Transformation of the parent rock

When the parent rock is laid bare by erosion, a number of climatic factors immediately start to act on it: sun, rain, frost and wind. The parent rock – which may be hard (e.g. granite, schist, sandstone), soft (e.g. chalk, marl, clay), or be loose (e.g. sands, scree, loess) – is cracked, broken up into smaller components, and dissociated. Finally, climatic factors bring about chemical changes. The result of this process is a mixture of elements, minerals which have dissociated together with those which have not undergone any transformation: pieces of stone, gravels, sands, and powdery silts; an amalgam or 'alteration complex' resulting from the chemical alteration of minerals: a clayey paste, coloured ferrous oxides, more or less soluble salts of Ca, Mg, K, Na, and others.

2. Process: Further alteration by organic material

The dissociated and altered soil, made up of minerals and elements, is then colonized by flora and fauna which enrich it with chemical and organic substances known collectively as humus. The properties of the humus differ with the climate, the parent rock, and the vegetation. Under its influence and that of the climate the minerals in the soil continue to be altered. The new undeveloped soil is homogeneous and its physical, chemical and biological characteristics continue to be constant.

3. Process: Vertical leaching of soluble minerals

In a rainy climate soluble minerals migrate downwards. This process is known as leaching. In dry climates with high evaporation rates, soluble minerals tend to migrate to the surface enriching the soil. This migration of elements, which may be hastened or slowed by the climate, the permeability of the soil, and by the kind of humus formed, creates more or less distinct layers in the soil and determines the horizons, which constitute the profile, which is the object of study by pedologists, or soil scientists. (Pedology studies the physical, chemical and biological characteristics of soils.)

Soils can be divided into main groups. There are the young or 'undeveloped' soils, which are shallow soils, not much different from the underlying rock, and are often made up of a single horizon. The others are the 'developed' soils, which are deep, and typified by a succession of leached and enriched horizons.

Essentially the origin of a soil is largely determined by the nature of the parent rock, the climate, the vegetation, and the topography.

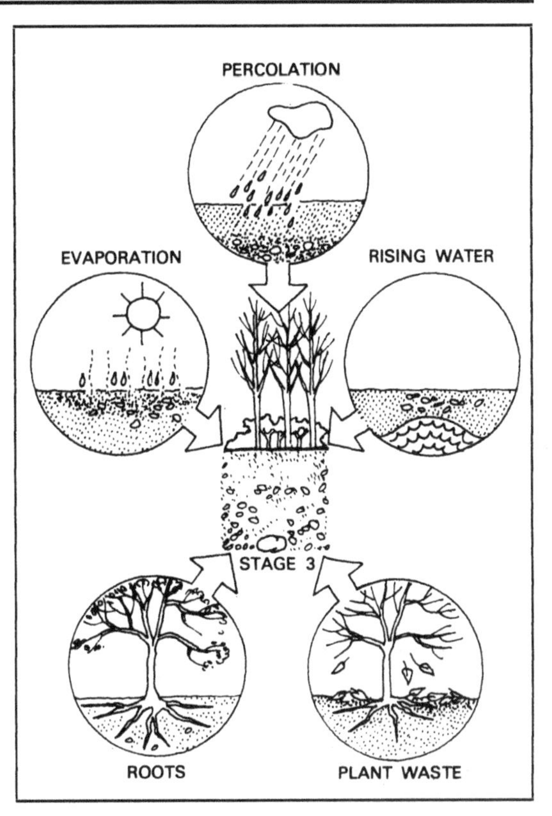

Main horizons

A cross-section of the ground makes it possible to observe the various soil layers.

A0 An organic layer of only partially decomposed organic material covering the mineral soil (more than 30% of organic material).
A1 Mixed horizon, a mixture of organic materials (- 30%) and minerals.
A2 The 'eluvial' horizon, poor in organic materials, clays and ferrous oxides often leached out, discoloured.
A3 Transition between the 'eluvial' and 'illuvial' regions, region where colloids start to accumulate.
B1 Ferrous layer containing organic material and oxides of Fe and Al (sesquioxides)
B2 The 'illuvial' horizon, enriched by the accumulation of clays and ferrous oxides.
B3 Transition layer between B and C.
C Broken original material.
R Unaltered parent rock.

Numerous sub-classifications have been adopted to describe specific conditions. The determination of horizons allows the scientific classification of soils taking the entire profile of the soil into account.

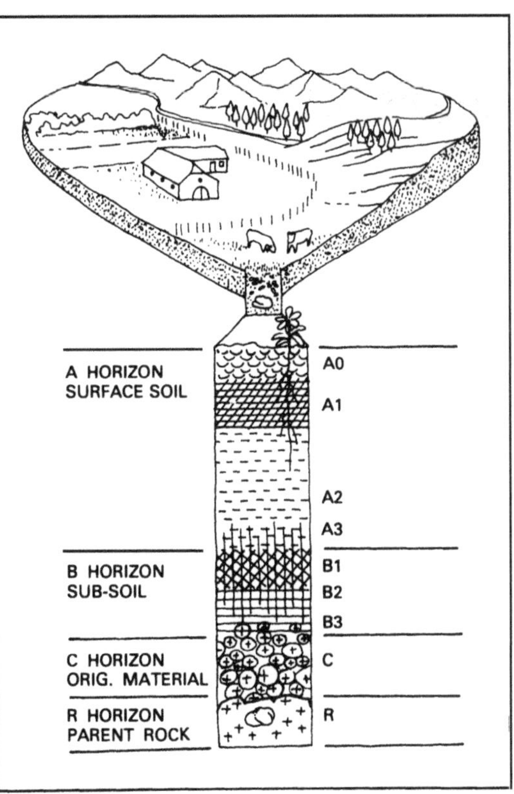

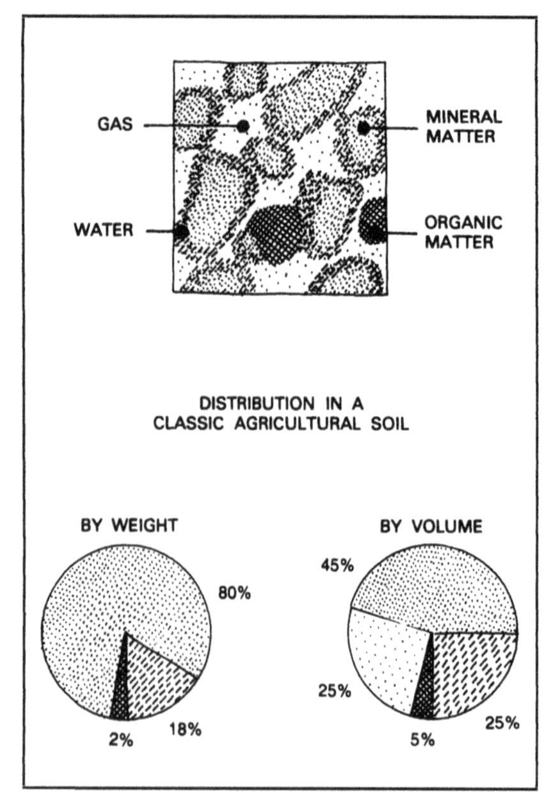

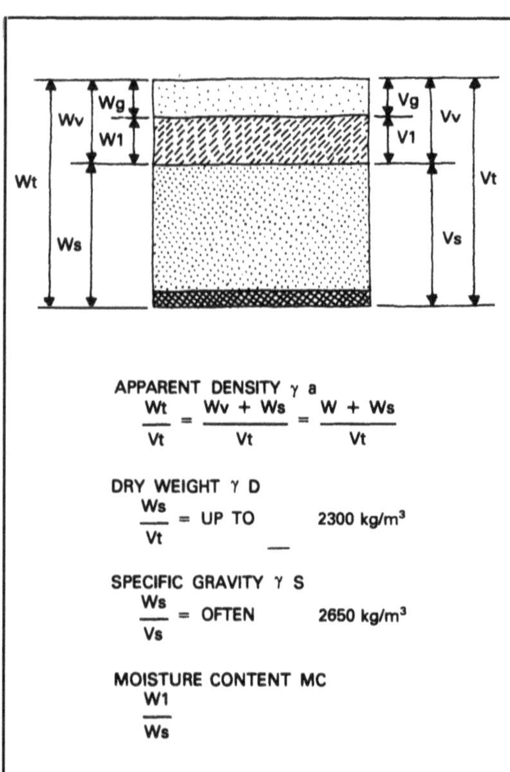

Earth is made up of a number of substances:

- gases: primarily air;
- liquids: primarily water;
- solids: mineral and organic materials.

1. Gaseous constituents

These form the internal atmosphere of the soil. They fill the voids in the soil and come from the outside atmosphere, organic life in the soil and the decomposition of organic material. Air's components include nitrogen, oxygen, and carbon dioxide. Gases resulting from organic decay, and respiration of living creatures include carbon dioxide, hydrogen, and methane.

2. Liquid constituents

These are the solution of a soil. These constituents are soluble in water and come from the rain and atmospheric conditions (mist, relative humidity), mankind, the weathering of rock and the decay of organic material. The liquid constituents are: water, soluble substances dissolved in this water such as organic compounds (sugars, alcohols, organic acids) and mineral compounds (acids, bases and salts partly dissociated into ions Ca^{++}, Mg^{++}; K^+, Na^+, PO_4^{---}, SO_4^{--}, CO_3^{--}, NO_3^{--}, etc.).

3. Solid constituents

These are insoluble in water.

Organic constituents or organic substances from plant and animal life in or carried into the soil. Four groups can be specified:

- Living plants and animals: bacteria, fungi, algae, higher plants, protozoans, worms, insects, etc.
- Animal wastes, undecomposed dead plants and animals.
- Decomposing organic matter under attack by microbes present in the soil or the 'transition products'.
- Humus, the stable colloidal fraction of organic matter, which decomposes only very slowly.

Mineral constituents or physical constituents, resulting from the dissociation of parent rock or from the acts of mankind. The 'sandy elements', the result of the dissociation of the parent rock, are either fragments of rock (stones and gravel) or the minerals making up these rocks (sands and silts). They have they same composition as these minerals and may be silicas, silicates, or limestones.

The respective proportions and distribution of these constituents determine the structure and texture of the soil which in turn determine its properties.

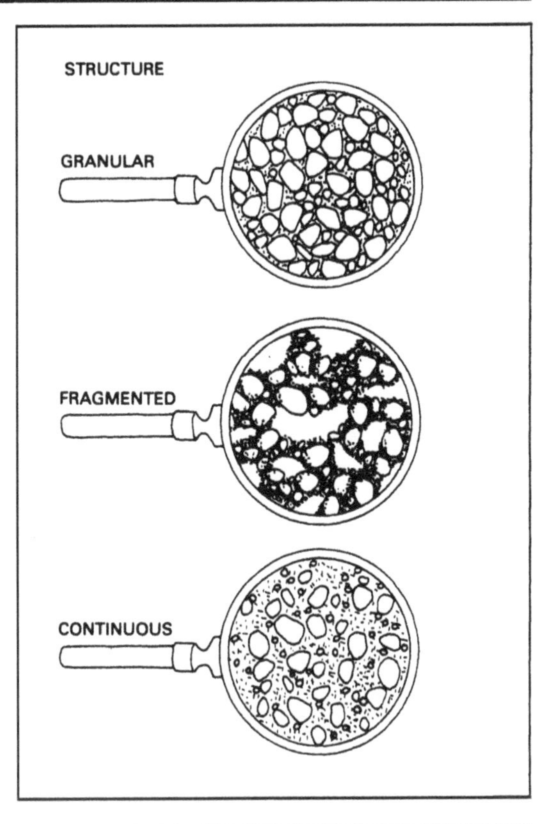

Structure

The constituents of the soil are more or less evenly arranged, disturbed, or bonded. The way in which the solid components are assembled at any given moment determines the structure of the soil and thus affects the circulation of water and air and other physical properties. Three types of structures are generally recognized:

1. **Granular structure** Like gravel; very little bonding by clay between the inert elements.

2. **Fragmented structure** Crumbly structure; bonding by clay between accretions of gravel which are then bonded to one another.

3. **Continuous structure** Pudding-like structure; the inert elements are held in a mass of clay (and silt).

Texture

This reflects the particle sizes contained in the soil. The texture influences the properties of the soil as each particle fraction with a specific set of characteristics can define those of the soil if the fraction is present in adequate quantities. Ten per cent clay for example is enough to give a soil cohesive and plastic properties. Soils with 40 to 50% clay fines give a soil the properties of a clay. Five major texture types are recognized:

1. **Organic soil** e.g. peat;

2. **Gravel soil** Gravel and pebbles predominate;

3. **Sandy soil** Sand predominates; has the appearance of mortar;

4. **Silty soil** Silt predominates; fine soil with low cohesion and with a silky appearance;

5. **Clayey soil** Clay predominates; extremely cohesive soil, sticky and malleable when wet.

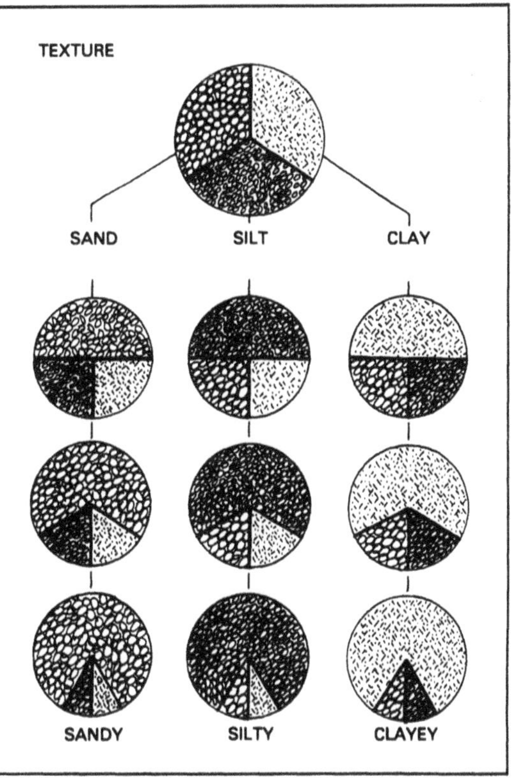

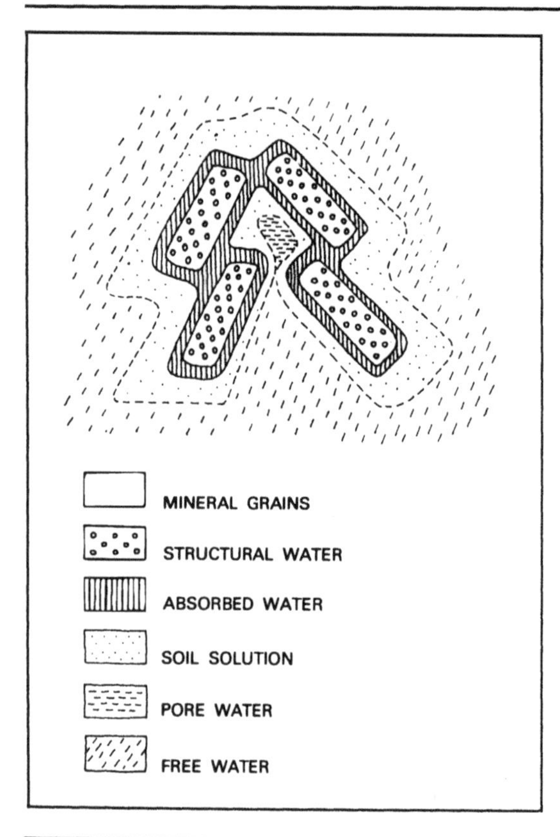

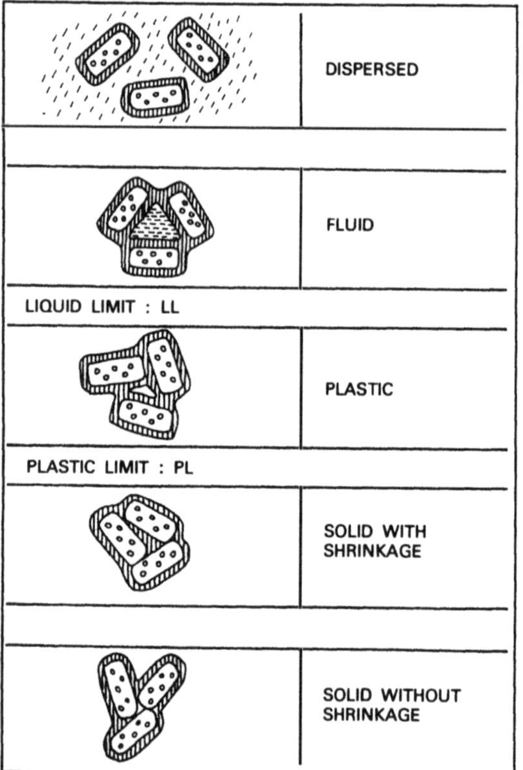

The properties of a soil vary according to its air and moisture content.

Air

Air makes no contribution to the strength of the soil and the air content should, if possible, be reduced. Air in the soil also entraps various micro-organisms such as bacteria and moulds, and these can lead to the destruction of the organic components of construction materials. Pockets of air form channels which allow the penetration of water in the form of water vapour. The moisture content may vary from one part of the material to another. The relative humidity of the water vapour contained in the soil varies according to type, differences in moisture content and temperature. Soils in tropical and arid regions undergo variations in moisture content and degrees of temperature change which amplify the movements of water vapour.

Water

The water penetrating the soil and retained by it falls into different categories. They play a major role in the determination of its properties.

1. Free water Moves under the effect of gravity or capillary action at the dictates of movements of the groundwater and the daily variations of atmospheric pressure and temperature. Water which accumulates in the fine pores on the surface of grains is not absorbed. It can be eliminated at normal room temperatures.

2. Pore water Retained in extremely fine pores where the capillarity is greater than hydrodynamic forces. This water can be eliminated at normal temperatures but only after long drying or kiln drying between 50 and 120°C.

3. Soil solution Takes the form of a film around the solid grains and is held on their surfaces by polar and electrostatic forces, and ionic hydration. It can be eliminated at normal room temperatures.

4. Absorbed water Takes the form a very thin film covering both external and internal surfaces. The forces retaining this water are so powerful that it cannot be moved. Heating to temperatures of between 100 and 200°C will eliminate this water.

5. Structural water Not really water at all as it represents hydroxyl groups which form solid crystalline networks. Heating to temperatures in the region of 600°C will tend to eliminate this water.

Effect of water

Differences in free water, pore water, and soil solution can change the physical properties of the soil. In rough sand pore water predominates, while in fine clayey soil it is soil solution which predominates. The structural and hydraulic properties of this latter type of soil are influenced by the thickness of the film of soil water, which in turn is affected by the elements dissolved in the water.

1. Effects due to liquidity

Cohesion Fine grains (silts and clays) owe their cohesiveness partly to the films of water linking them. These cohesive forces are of two types, those due to water tension at the air/water interface (large grains) and those due to the interaction of particles of clay and molecules of polarized water.

Suction The forces arising from the surface hydration of particles combine with surface tension to create water suction which increases with the reduction of the water content.

Swell At the surface of the clays, the absorbing forces acting on the water molecules are high. The absorbed layers swell as the clay becomes damper. The soil increases in volume.

Shrinkage The shrinking of clays is usually the result of the evaporation of the water.

Plasticity After reaching its elastic limit a well hydrated cohesive soil can deform without breaking. The plasticity resulting from the lubricating effect of the films of water between the grain depends on the size, the grain shape and the chemical make-up of their surfaces.

2. Effects due to solvent action

Soluble salts In solution these dissociate into metallic cations of Na^+, Mg^+, Ca^+ and Al^+, which are absorbed into the grain surfaces. Sulphates of Na, Mg and Ca affect the earth by crystallizing and making it brittle.

Organic matter Can influence the redistribution of the mineral elements in the soil; this is particularly the case for iron.

The 12 states of hydration of soil

Rocky concretion Monolithic agglomerations of coarse material; compact and heavy soil which is difficult to cut.

Friable concretion Agglomerations of crumbly or decomposed material, including peat and sod, which is easy to cut.

Solid concretion Completely dry earth, in large pieces or solid lumps.

Friable aggregation Absolutely dry soil in powder form.

Dry soil: Soil characterized by a naturally low humidity (4 to 10%); it is dry rather than moist to the touch.

Moist soil Soil unmistakably moist to the touch (8 to 18%), but cannot be shaped because of its lack of plasticity.

Solid paste An earth ball (moisture content 15 to 30%) which flattens only slightly when dropped from a height of one metre is formed by powerful kneading with the fingers.

Semi-solid paste Only slight finger pressure is sufficient to form an earth ball (moisture content 15 to 30%) which flattens slightly but does not disintegrate when dropped from a height of 1m.

Semi-soft paste With this very homogenous material it is easy to shape an earth ball which is neither markedly sticky nor soiling (moisture content 15 to 30%) that flattens markedly without disintegrating when dropped from a height of 1m.

Soft paste This kind of soil is so adhesive and soiling (moisture content 20 to 35%) that it is extremely difficult, if not impossible, to make balls from it.

Mud This kind of soil is saturated with water and forms a viscous, more or less liquid mass.

Slurry This consists of a suspension of clayey earth in water and constitutes a highly liquid, fluid binder.

The limit of the plastic state, or **plastic limit** (PL) is situated approximately between the states of solid paste and semi-solid paste. The limit of the liquid state, or **liquid limit** (LL) is situated approximately between the states of soft paste and mud.

*Values for moisture content are only indicative and vary greatly according to soil type.

The solid fraction of soil is made up of minerals resulting from the physical dissociation and chemical alteration of the underlying parent rock, and of organic matter, which are basically the more or less decomposed remains of plant and animal organisms.

Organic matter

Usually, organic matter is concentrated in the surface horizon of the soil, over a depth of between 5 and 35cm. Sometimes organic matter may contain visible plant components. Elsewhere the decomposition of the plant structure is so advanced that a black material is encountered. This is called humus.

Recently decomposed organic matter has different properties than humus. It consists of macrograins or fibres which are relatively inert from the physical or chemical point of view.

Humus is colloidal and acid, with a very high cation exchange capacity, and the ability to absorb water, which increases its volume.

Organic matter has an open and spongy structure and has only low mechanical strength. The moisture content may be very high (from 100 to 500%) eliminating all mechanical stability. The acidity of the organic components tends to trigger acid reactions with water in the soil, which may lead to the corrosive attack of materials with which it comes into contact. The concentration and type of organic matter has marked effects on the characteristics of a natural soil once it exceeds 2 to 4%.

Mineral matter

The mineral or inorganic components of a soil, usually represents by far the greatest part of the soil.

Two groups of minerals may be distinguished:

Unweathered minerals or incompletely weathered minerals. These minerals are identical in composition to the parent rock from which they derive. They include pebbles and gravels, sands and clays. These are the sandy elements.

Weathered minerals The result of the chemical weathering of the minerals of the parent rock and typical for their extreme smallness (less than 2μ). Because of their smallness these grains of weathered mineral have the appearance of a sticky paste if they are wetted. They are called colloids, which derives from the French *colle* (glue) and means 'gluelike'. They were given this name because they form the binder in the soil. The main binders are the clays and for this reason engineering geology refers to them as the clayey fraction rather than the colloidal fraction.

Sandy elements

These may be either silicas, silicates, or limestones.

Silicas Resist chemical weathering. They are grains of quartz resulting from the disintegration of sandstone and crystalline rocks. They are found in both the larger and the finer fractions.

Silicates The chemical weathering of these proceeds continuously but is very slow. They are made of grains of mica, feldspar and other free minerals resulting from the disintegration of crystalline rocks such as granite and volcanic rocks. The weathering of these elements is increasingly effective as grain size is reduced.

Limestones This is that part of the sandy soil which is made from calcium carbonate. A distinction must be between soils created on a limestone parent rock and the soils formed on a non-limestone parent rock. Limestone is not always present in the soil but all soils contain calcium fixed on clay in the form of calcium ions or in soil solution in the form of soluble calcium salts.

In order to facilitate their identification, the mineral components have been divided into grain fractions.

1. Pebbles

Pebbles range in size from 20 to 200mm. They form a rough material which is the result of the disintegration of the parent rock from which they draw their basic characteristics. They may also have been carried from elsewhere. Young pebbles still have sharp corners. Severely weathered pebbles are rounded as well as those which have been carried by water courses or glaciers.

2. Gravel

Gravel ranges in size from 2 to 20mm. It is made up of small grains of rough material, which are the result of the disintegration of the parent rock and pebbles. They may also have been carried by water courses and thus be rounded, though angular gravel also exists. Gravel constitutes the skeleton of the soil and imposes a limit to its capillarity and shrinkage.

3. Sands

Sand ranges in size from 0.06 to 2mm. It is often made up of particles of silica or quartz. Some beach sand contains calcium carbonate (shell fragments). The sandy component of a soil is marked by its high internal friction. Sand grains lack cohesion because of the weak effect of films of water close to their surfaces. The low adsorption of these surfaces limits swell and shrinkage. The open structure and permeability are typical of sands.

4. Silt

The grain size of silt ranges from 0.002 (2μ) to 0.06mm. From the physical and chemical point of view the silt component is virtually identical to the sand component, the only difference being one of size. Silt gives soil stability by increasing its internal friction. The films of water between the particles grant a certain degree of cohesion to silty soil. Because of their high degree of permeability silty soils are very sensitive to frost. They are subject to small-scale swell and shrinkage.

5. Clays

Clay grains are smaller than 2μ. They differ from other grains in their chemical composition and physical properties. In chemical terms they are hydrated alumino-silicates formed by the leaching process acting on the primary minerals in rock. Physically speaking clays very often assume a platy elongated shape. Their specific surface is infinitely greater than that of rougher round or angular particles. Clays are very susceptible to swell and shrinkage.

6. Colloids

Sandy material is often coated with a sort of gluey paste which sticks it into aggregate. This gluey paste is made up of 'colloids', the dimensions of which are less than 2μ. Some of these are the result of the weathering of the parent rock. These are mineral colloids, the chief of which is clay. Clay is not the only mineral colloid. It is often mixed with very fine quartz grains (1 to 2μ), with hydrated silica, extremely fine crystals of limestone and magnesium colloids, as well as colloidal iron and aluminium oxides which go under the name of sesquioxides (sesqui = half more; indeed, in Fe_2O_3 each Fe has half an O more than FeO). Other colloids result from the decomposition of organic matter. These are the organic colloids: humus and bacterial glues.

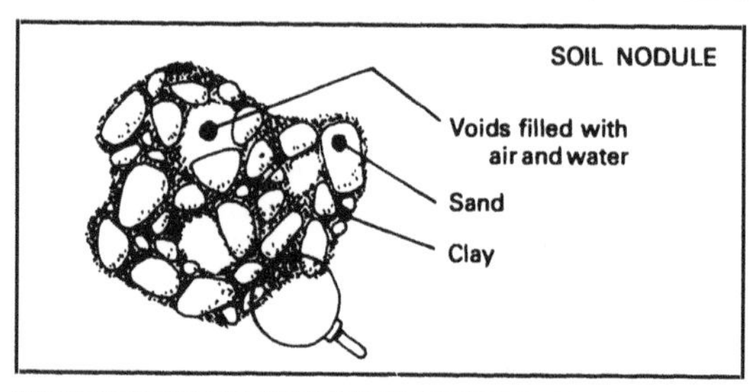

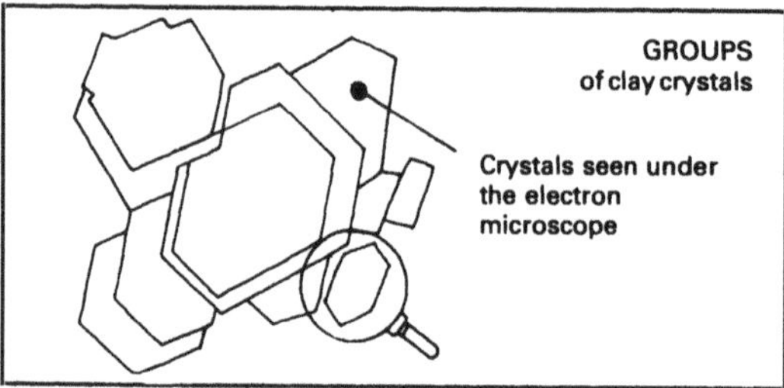

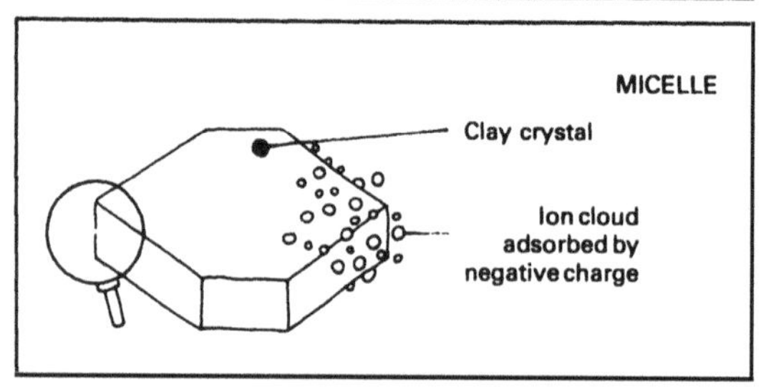

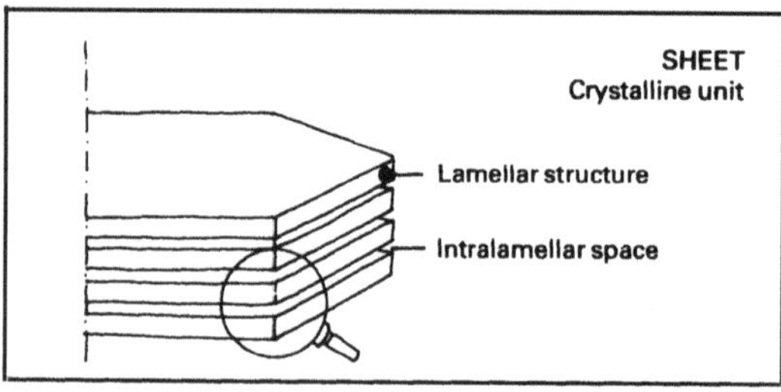

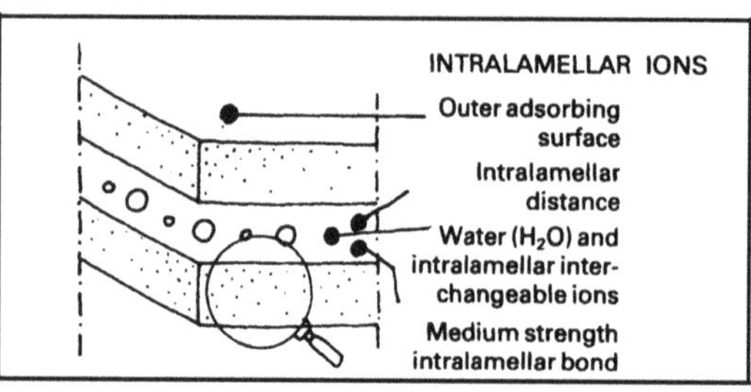

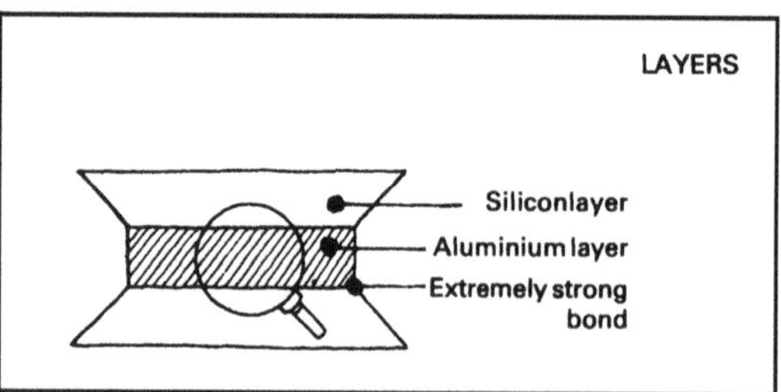

Origin

Clay is the result of the chemical weathering of rock, and of silicates in particular (feldspar, micas, amphiboles, and pyroxenes, etc.).

Structure

Large clay molecules (or micelles) are fine crystals of an irregular or hexagonal shape. The latter is the most common, but there are others. These included pseudo-hexagonal wafers, cylindrical or hollow tubular fibres, thick tablets or discs. Clay micelles are made up of thin sheets or leaves, which is why argillaceous minerals are referred to as phyllite ('phyllon' = leaf in Greek). Like mica they form part of a group known as phyllosilicates.

Each micelle is made up of several tens or even hundreds of sheets, the structure of which determines that of the minerals as well the immediate properties of the crystal, including adsorbent properties analogous to those of colloids. These sheets have a chemical make-up which varies according to the type of clay and the degree of hydration as well as their thickness and spacing, namely from 7 to 20 Angström (1Å = 10 millionth of a mm). Their size is of the order of 0.01 to 1 micron. Some sheets are made of silica (atoms of silicon surrounded by oxygen atoms), and others are made of alumina (aluminium atoms surrounded by oxygen atoms and OH groups).

Other clays exist, however, whose base is not Si and Al but Si and Mg or Si and Fe. Nevertheless the alumino-silicates represent 74% of the earth's crust (1) and we will be considering these exclusively.

Like most silicates the structure of argillaceous minerals is determined by the way in the oxygen and hydroxyl groups are arranged. These may be located in tetrahedral or octahedral cavities.

The physics and chemistry of clays are highly complex because of the innumerable electrical phenomena which argillaceous minerals are subject to.

Main groups

There are several families of clay minerals, and a few dozen groups. Three main types, however, make up the most frequently encountered clays: kaolinites, illites, and montmorillonites.

1. Kaolinites

The sheets are made up of a layer of oxygen tetrahedrons with a silicon centre and a layer of oxygen (or hydroxide) octahedrons with an aluminium centre. Kaolinite is negatively charged only at the edge of the sheets and its ion fixing capacity is low. The distance between sheets is constant, being 7Å. The thickness of the crystals is between 0.005 and 2μ. The outer area OA is between 10 and 30m^2 per gram. The internal area IA = 0. Kaolinite is generally stable in contact with water.

2. Illites

This group has a three layer structure: one mainly aluminous octahedral layer, between two mainly silicaceous tetrahedral layers. Mg or Fe ions may partly replace the Al ions in the aluminous layer, and Al ions may substitute for Si in the silica layer. As the sheet is non-saturated, the negative charges are balanced by K ions, which bond the sheets. The distance between sheets is 10Å, and the thickness of the crystals is between 0.005 and 0.05μ. OA is 80m^2/g and IA is 800m^2/g. Illite is not particularly stable when in contact with water and suffers swell.

3. Montmorillonites

The structure of this group is comparable to that of illite, but substitution takes place in the octahedral alumina layer: Al ions may be replaced by Mg, Fe, Mn, Ni etc. The sheets are not electrically neutral and are weakly linked. The ions between the sheets are not K ions but exchangeable cations (Na, Ca) and water molecules. The distance between sheets ranges from 14 to 20Å. The thickness of the crystals lies between 0.001 and 0.02μ. OA is 80m^2/g and IA is 800m^2/g. Montmorillonite is not stable when in contact with water and suffers severely from swell.

4. Others

There are wide range of other clays, such as chlorite, muscovite, halloysite, vermiculite, sepiolite, attapulgite, etc. as well as the interstratified materials, which are complex combinations of various types of clay.

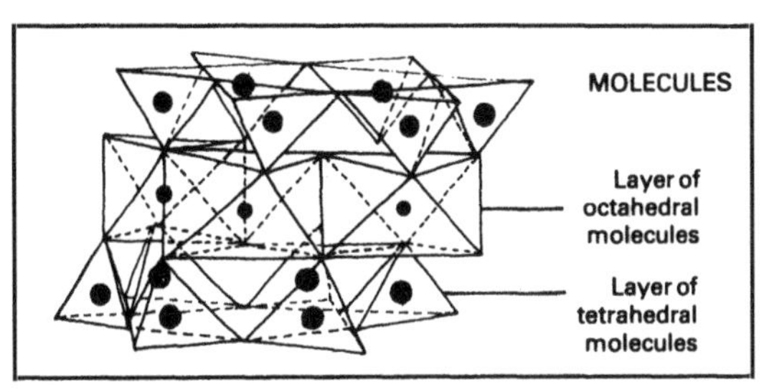

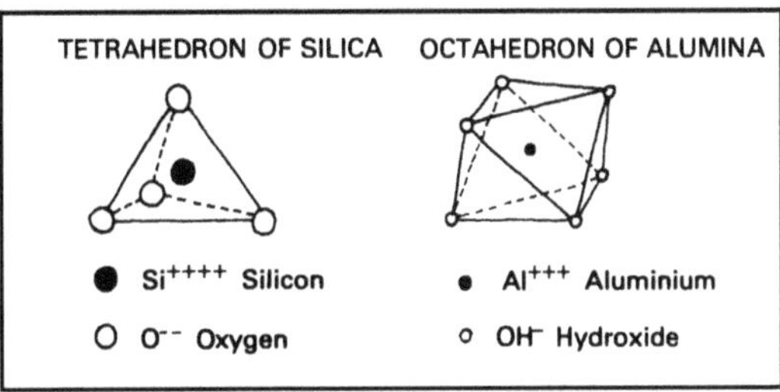

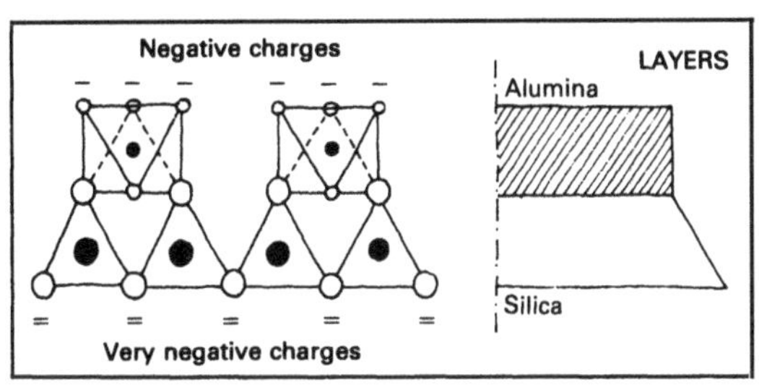

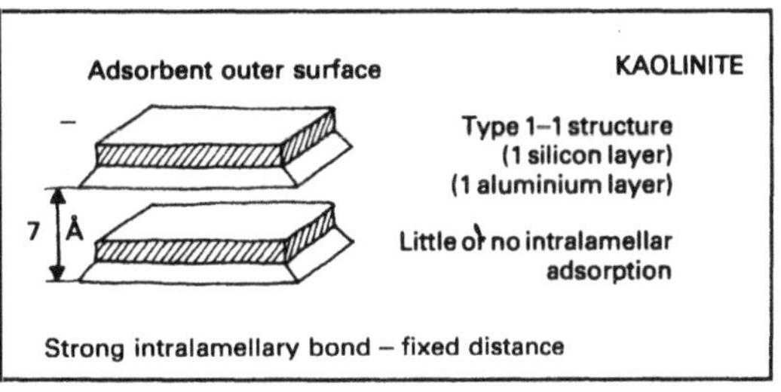

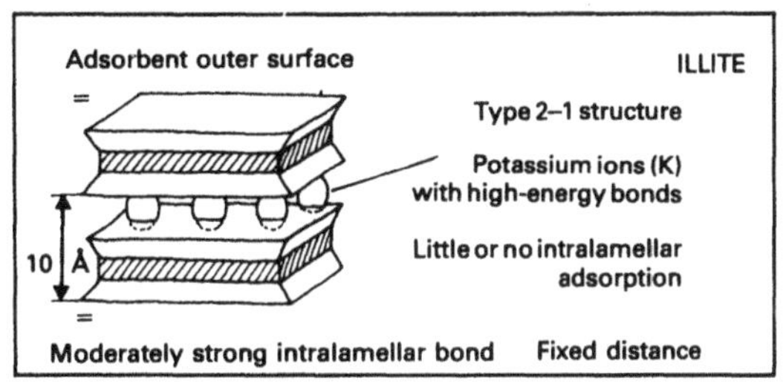

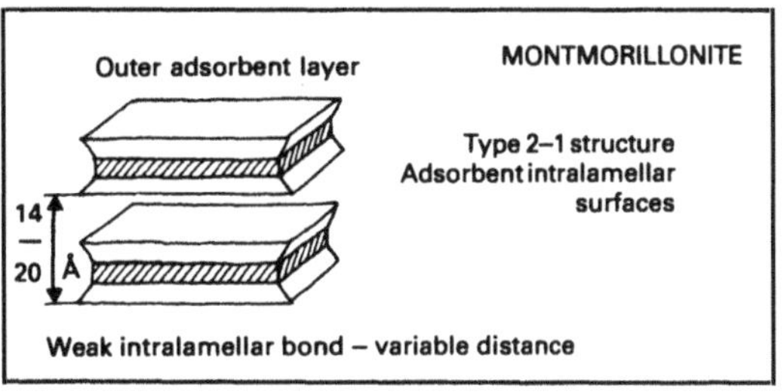

Internal

There are a number of different binding forces which act to give coherence to soil grains, causing them to aggregate. These forces vary immensely in force and depend on how each separate fraction of components behaves.

This makes it impossible to quote a single value indicative of the cohesive force. For the concretions of the sandy fraction, breaking the weakest links tends to produce smaller aggregations but with a higher degree of internal cohesion. Driven to its extremes, the breakdown of aggregates will result in elementary mineral particles whose cohesive force is analogous to that of crystalline minerals. Mineral particles with a diameter $> 2\ \mu$ are thus highly cohesive: their chemical bond is of the interatomic type in three dimensions. These quartz or feldspar minerals constitute the skeleton of the soil, which is highly resistant to deformation when they are present in sufficient quantity to touch one another.

For the colloidal fraction, on the other hand, particles of a diameter $< 2\ \mu$ have an essentially two-dimensional bond (structure in sheets). This bond is of the intermolecular type (i.e. physical and having moderate or weak force). These are mainly the phyllosilicates such as the kaolinites, illites and the montmorillonites, micas, hydrous aluminium oxides, etc.

External

Clay plays the role of cement. It holds the inert grains together and to a great extent provides the cohesion of the soil. The cohesive forces of the clay micelles thus merit particular study. Although it is not possible to speak of a single force, it can be said that the binding forces are by and large electrostatic. Numerous forces of attraction and repulsion come into play but it is nevertheless the electrostatic forces which are most significant.

1. Electrostatic forces

The micelles of clay are not electrically neutral, they may be negatively or positively charged.

Negative charges can arise in two different ways:

- unfilled valencies, either at the end of cracked sheets (oxygen atom) or on the outer flat surfaces as the result of the dissociation of an H ion in an OH group (kaolinite sheet).
- substitution: Si and Al atoms can be replaced by atoms of a lower valency: e.g. in a sheet of silica, atoms of Si^{++++} are replaced by atoms of Al^{+++}, the valency of which is lower. A negative charge uncompensated by an oxygen atom will appear. In the same way an Al^{+++} atom in an aluminium sheet may be replaced by an Mg^{++} atom.

Positive charges are less common than negative charges. They may appear either at the point of breakage of a sheet, if a break exposes a silica molecule or aluminium atom where the positive charge is no longer balanced by an oxygen atom or a group of OH, or as a result of the frequent association between the clay and Fe or Al hydroxides releasing OH ions by dissociation.

Mechanisms As a result of the attractive forces a surface to side bond may be established between the clay micelles. However, a surface-surface bond or a side-side bond with negative charge may also be established. The flocculation theory (the opposite of dispersion) helps to explain these phenomena. The water contained in the soil is a bonding agent. It is loaded with positive ions, or cations (Na^{+}, Ca^{++}, Al^{+++}), in large enough quantities to balance the negative charge of the grains: overall the system is electrically neutral. Depending on their degree of hydration, cations give rise to oriented chains of water molecules. When a hydrated cation is close to a particle, the two sets of water molecules link the ion and the surface of the particle. Similarly, an ion can act as a bridge between two adjacent particles of clay.

2. Other forces

The bonds between clay micelles, between clay and inert grains, and between inert grains themselves imply the existence of other forces. Here we can point to the importance of cementation, capillarity, and electromagnetic forces.

Cementation is the result of precipitation and dissolution cycles and is manifested in the form of 'bridges' of binder between particles of similar or differing composition. The main agents of cementation are calcite, silica, ferric oxide, bacterial adhesives, etc.

Capillarity is similar to cementation, although the capillary bond is not as rigid and remains reversible. When a fluid impregnates a grain, it will tend to follow the capillary paths or channels. Extreme tension may be required to separate particles bound by capillarity. Apart from the forces of attraction resulting from interface tensions, water may act as an agent transmitting force between grains or as a di-electric agent. The powerful contraction forces due to capillary action may reinforce the action of less powerful forces and reduce the interparticle distance. Cohesive capillary forces act mainly on the inert grains and not the micelles of clay.

Electromagnetic forces are van der Waals' forces. Although these are weak, they play a significant role in the binding mechanism of clay micelles and inert grains; they create a film of oriented micelles, thus increasing friction. Van der Waals' forces chiefly affect micelles of a normal size, but also appear to play a role in the cohesion of smaller micelles.

Friction is the result of surface roughness. This roughness may be on the atomic or molecular scale depending on microscopic disorientations and macroscopic striation. Observation under the electron microscrope shows that clayey surfaces can be bound by friction.

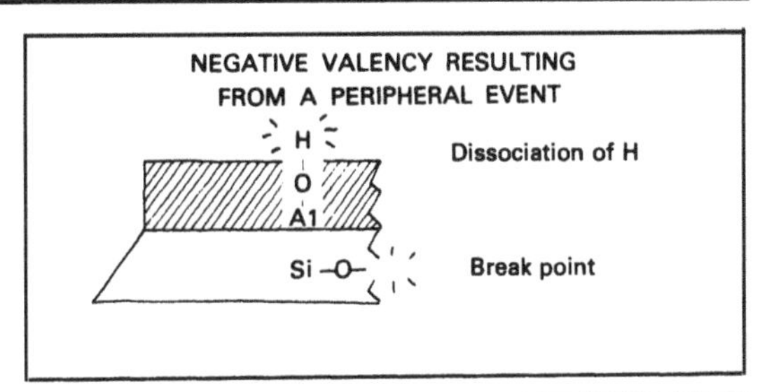

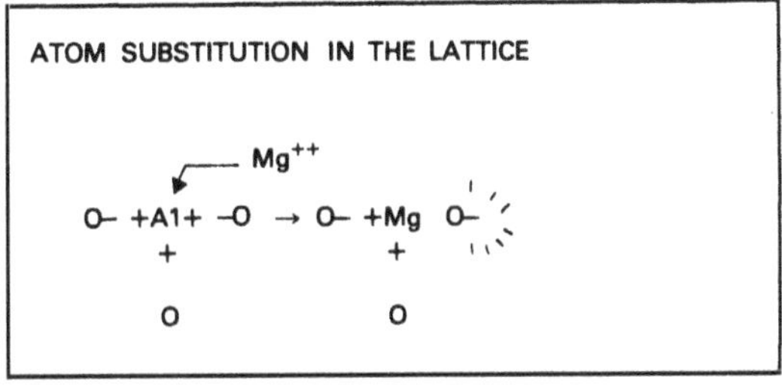

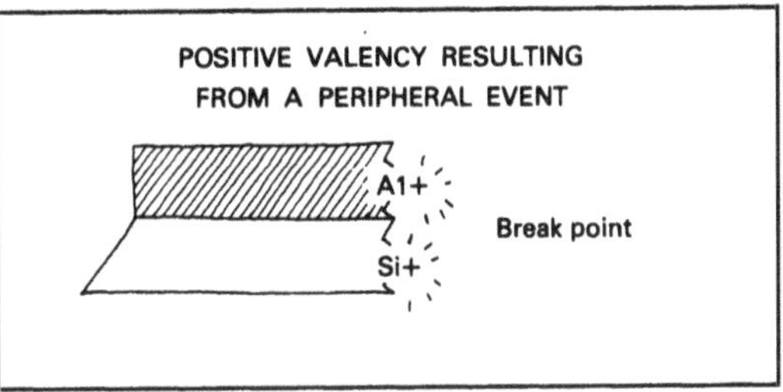

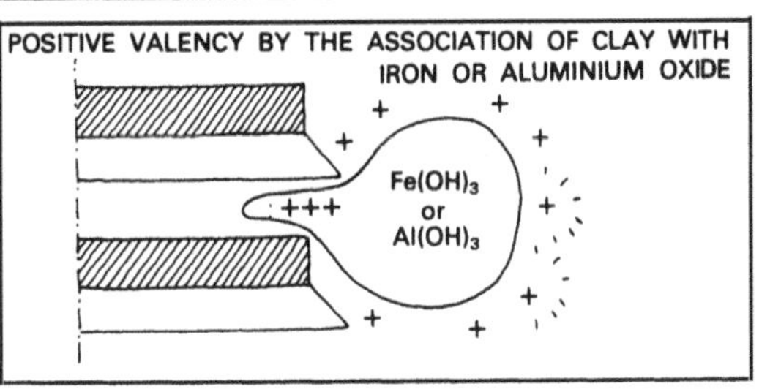

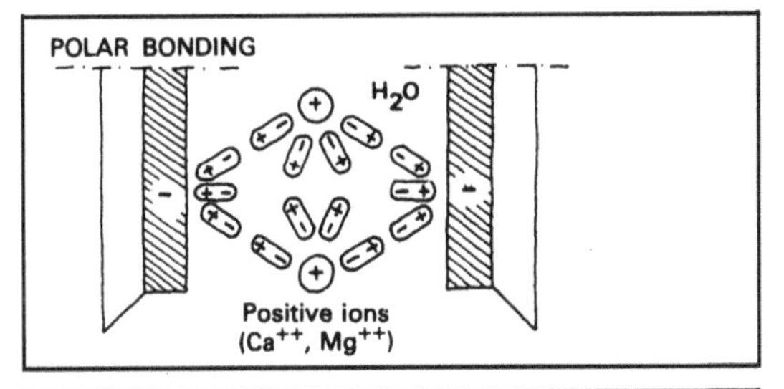

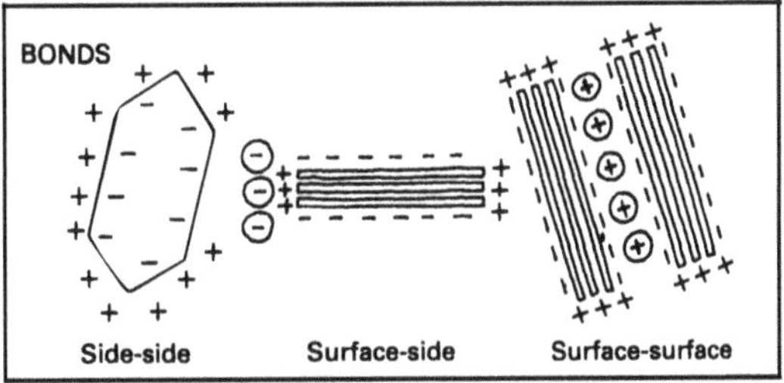

Properties may vary considerably from one soil to another. They depend on the complex nature of the mix of the various granular fractions. The characteristics of the soil depend on the proportions of pebbles, gravel, sand, silt, clays, colloids, organic matter, water and gas. It is often the predominant fraction in a soil which governs the fundamental properties of a material.

Properties of a chemical nature

These depend on the chemical constitution of the components of the soil. Among these components the most influential from the chemical point of view are the salts, which may be either soluble or insoluble. The high salinity of a soil may induce extremely marked chemical properties.

These properties are also influenced by the mineralogical characteristics of the minerals involved and their chemical composition, and the nature and quantity of the organic matter. These unstable components, themselves, in the course of chemical and biochemical development can cause the development of the structure of the soil itself, producing precipitates of differing natures, colloid and various types of adhesive, humic, and bacterial pastes. Similarly, the quantity of oxides of iron, magnesium, or calcium, carbonates and sulphates can characterize the soil from a chemical point of view. Sulphate of calcium in particular when subject to hydration can swell disastrously; its solubility in water (selenite water) may increase the sensitivity of clays. Metallic oxides can be very influential. For example in lateritic soils, iron oxide can speed certain solidification processes. Similarly abundant aluminium oxide can reduce the ability to withstand ageing. The pH of a soil is also important, as it gives an indication of H^+ and OH^- ions and consequently of its acid or basic nature.

Physical properties

Soils have numerous physical properties. They constitute a guide to the suitability of a soil for construction purposes.

Colour The colour range of soils is extremely wide and may range from white to black passing through beige, yellow or red ochre, orange, red brown, grey, and even blue and green.

Break-up This is the ability of a soil to be easily broken up. Soils with a dominant sand fraction are easily loosened while highly clayey soils can only be loosened with considerable difficulty.

Structural stability This refers to the solidity of the soil structure, indicative of its resistance to deterioration.

Adhesion This is the ability of a soil at a given level of humidity to stick to other objects, tools in particular. It increases with humidity until it reaches a maximum and subsequently diminishes.

Apparent bulk density This refers to the soil as a whole and is expressed in kg/m^3.

Specific bulk density Refers to the density of the constituents of the soil themselves. Expressed in kg/m^3. To give some examples: micas and feldspars have a specific density of between 2600 and 2700kg/m^3, sands between 2600 and 3000kg/m^3, and clays 2500kg/m^3.

Moisture content The amount of water contained in the soil, either in the natural state, or after manipulation and drying. It is expressed as a percentage and defines the different hydrous states of the soil.

Porosity or voids ratio The volume of voids in the earth expressed as a percentage of the total volume. There is a relation between porosity and specific density; for example, for a silt with a density of between 1600 and 1800kg/m^3, porosity is lower than 40%.

Capillarity or pF measures the force exerted by the suction of the water and is expressed in g/cm^2 or in atmospheres. pF is the decimal logarithm of this pressure. The moister the soil the greater the suction will be and the less water will be retained by the soil. The drier the soil the more the suction will increase.

Power of absorption A property possessed by clay, humus, and the clay-humus complex which permits them to retain on their surfaces electropositive, as well electronegative, ions drawn from the soil solution. The fixing of ions is explained by the negative and positive charges around the clay sheets and the humus micelles. The number of positive charges adsorbed by 100g of materials ranges from 20 to 90×10^{20} for kaolinite, from 120 to 240×10^{20} for illite, and from 360 to 500×10^{20} for the montmorillonites.

Capillary diffusion The displacement of water held in the soil.

Permeability Refers to the speed of passage of fluids through the soil, and depends on its texture but more especially on its structure. It is expressed in cm/hour; e.g. a clayey silt with low permeability may have a value of 0.6cm/h and a highly permeable sandy soil a value of 50 to 60cm/hour (unworked soils).

Specific heat The amount of heat required to raise the temperature of one unit of mass of a soil through 1°C. It is expressed in Joule/(kgK). Water has a value of 4186.8J/(kgK), sand a value of 800J/(kgK), and clay a value of 963J/(kgK).

Specific area A measure which is applied primarily to clays and which makes it possible to estimate the chemical activity of ion exchange. Expressed in cm^2/g, coarse sand has a specific area of $23cm^2/g$, silts have a specific area of $454cm^2/g$, while clays rise to $800m^2/g$.

Total exchange capacity (T) Refers to the maximum quantity of cations of all kinds that a soil is capable of retaining. This measure represents the totality of negative charges available in the soil for fixing metallic cations or H^+ ions. It is expressed in milliequivalents or m.e.q. for 100g of soil. The equivalent of a body is the ratio of atomic mass to the valency of the body. The m.e.q. is one thousandth of this. To give an example: a soil with a T of 30m.e.q. can retain in Ca: 30m.e.q. $\times$ 40/2 = 600m.e.q. of Ca/100g of soil. The value of T for a soil is stable as it depends on the rate and the nature of colloids which are unable to vary greatly. T is high for clay and humus soils and low for sandy soils.

Saturation rate (V) The ratio of the sum of the exchangeable bases to total exchange capacity. It is expressed as a percentage: $V = S/T \times 100$. This rate varies immensely from soil to soil. It is dependent on the availability of cations in the parent rock, the frequency and volume of the supply of cations (Ca in particular) and the extent of leaching. Soils rich in active limestone have a rate V close to 80 or 90%, while soils formed on sandy parent rock have a rate V often lower than 20%.

Linear contraction Provides a measure of the reduction of size of a mass of worked clay soil after drying and is often expressed as a percentage of the initial size. Kaolinites have a linear shrinkage upon drying of the order of 3 to 10%, illites of the order 4 to 11%, while values for the montmorillonites range from 12 to 23%.

Dry strength Shear strength in the dry condition reaches very different values depending on the clays and depends on the distribution and size of the grains, the degree of their perfection and crystallinity but also of the nature of the exchangeable ions. Kaolinite have a strength of the order of 0.07 to 5MPa, values for the illites range from 1.5 to 7MPa and montmorillonites from 2 to 6MPa.

Fundamental properties

An exhaustive study of the properties of the soil is not always necessary. Above all it is important to be aware of certain basic properties, which are:

1. Texture or grain size distribution of the soil in percentages of pebble, gravels, sands, silts, clays, and colloids.

2. Plasticity or ease of shaping the soil.

3. Compactibility or the soil's potential to reduce its porosity to a minimum.

4. Cohesion or the ability of soil grains to remain in association.

1. Texture

Also referred to as the grain or particle size distribution of a soil, it represents the percentage content of the different grain sizes. The texture of a soil is determined by the sieving of the rougher grains: pebbles, gravel, sands and silts, and by sedimentation for the fine clayey materials. The classification of grain sizes adopted by a large number of labora- tories based on the ASTM-AFNOR standards are as follows:

>V:PEBBLES:	200 mm –	20 mm
V:GRAVEL:	20 mm –	2 mm
IV:COARSE SAND:	2 mm –	0.2 mm
III:FINE SAND:	0.2 mm –	0.06 mm
II:SILTS:	0.06 mm –	0.02 mm
IIA:FINE SILTS:	0.02 mm –	0.002 mm
I:CLAYS:	0.002 mm –	0 mm

The texture of the soil is plotted on a grain size distribution chart similar to that shown in diagram 'G'.

It is often more useful to adopt a similar classification, i.e., the decimal classification

PEBBLES:	200 mm –	20 mm
GRAVEL:	20 mm –	2 mm
COARSE SAND:	2 mm –	0.2 mm
FINE SAND:	0.2 mm –	0.02 mm
SILT:	0.02 mm –	0.002 mm
CLAY:	0.002 mm –	0 mm

2. Plasticity

Plasticity refers to the ability of a soil to submit to deformation without elastic failure characterized by cracking or disintegration.

The plasticity of a soil as well as the limits between differing states of consistency are determined by making measurements of the Atterberg limits.

These measurements are carried out on the 'fine mortar' size of the soil (ø of grains < 0.4mm). The quantity of water, expressed as a percentage, which corresponds to the limit of the transition between the state of fluid consistency and the state of plastic consistency is called the Liquid Limit (LL). The transition between the plastic and the solid state is called the Plastic Limit (PL). At LL the soil starts to manifest a certain resistance to shearing. At PL the soil stops being plastic and becomes brittle.

The Plasticity Index (PI) equal to LL – PL determines the range of plastic behaviour of the soil. The combination of the LL and the PL specifies the sensitivity of the soil to variations in humidity. The plastic properties of a soil are represented in the plasticity diagram 'P'.

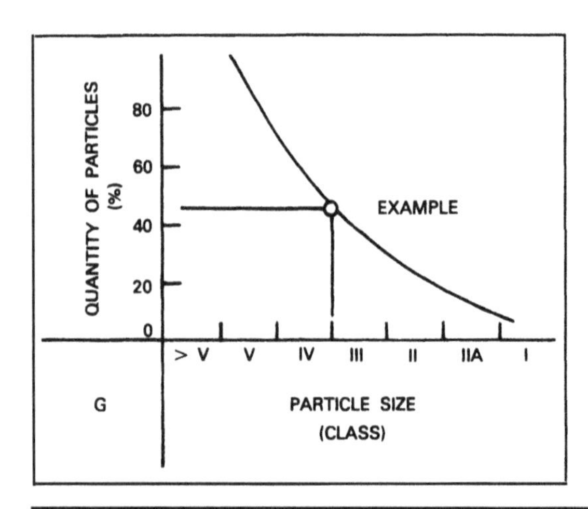

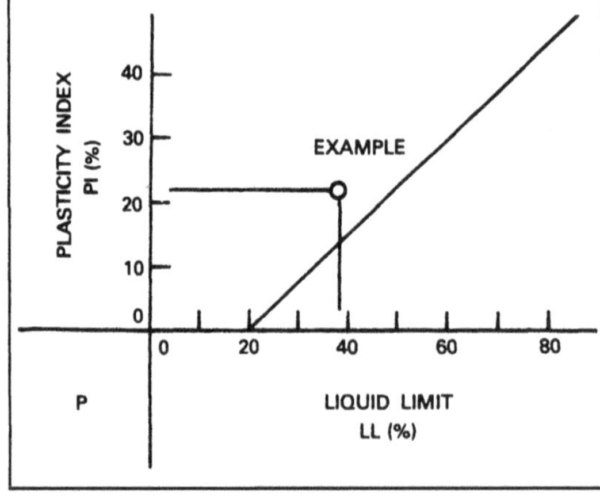

3. Compactibility

The compactibility of a soil defines its ability to be compacted to a maximum for a given compacting energy and degree of humidity (optimal moisture content or OMC). When a volume of soil is subjected to the action of a force the material is compressed and the voids ratio decreases. As the density of a soil is increased its porosity is reduced, and less water can penetrate it. This property is the result of the interpenetration of the grains, which in turn results in a reduction of the disturbance of the structure as the result of water action.

The compactibility of a soil is measured by the Proctor compaction test. The compactibility of a soil can be represented on a compactibility diagram such as diagram 'C' below, where Optimal Moisture Content is plotted against Optimal Dry Density for any given compression energy.

4. Cohesion

The cohesion of a soil is an expression of the capacity of its grains to remain together when a tensile stress is imposed on the material. The cohesion of a soil depends on the adhesive or cementation properties of its coarse mortar (grain size of ø < 2mm) which binds the inert grains together.

This property thus contributes to the quantity and adhesive quality of the clays.

The coarse mortars may be classed as follows:

A Sandy mortar
B Lean mortar
C Average mortar
D Fat mortar
E Clays

Cohesion is measured by a Tensile Test in the moist condition sometimes also referred to the '8' Test. Cohesion is plotted in a tensile strength chart 'T'.

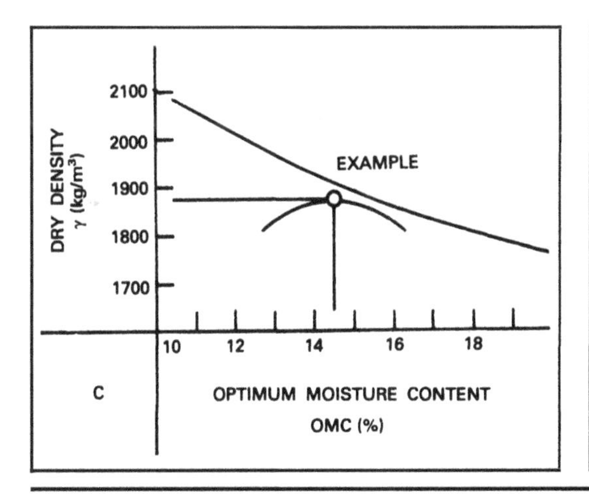

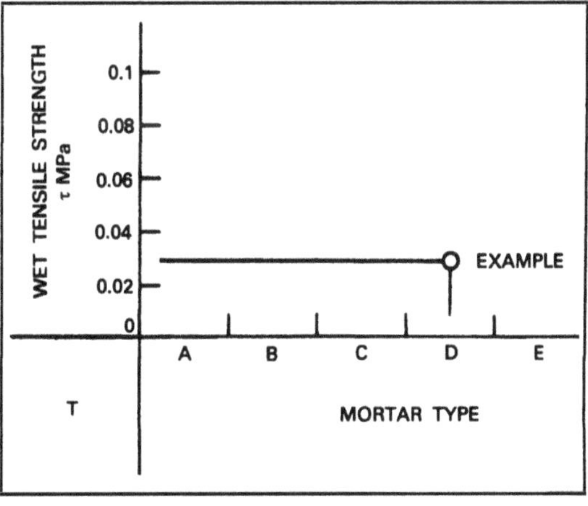

The engineering geology classification is the most suited to earth construction. It classes soils by:

- Grain size distribution (direct measurement)
- Plasticity (direct measurement)
- Compactibility (indirect measurement)
- Cohesion (indirect measurement)
- Quantity of organic matter.

It does not consider:

- The hydrous state of the soil;
- The in-situ density of the soil
- Gaseous and liquid components.

There are numerous different soil engineering classifications. These differences can give rise to confusion. In general it is better to refer to regional classifications suited to local conditions. No system has been specifially adapted to building in earth. The systems shown here are among the most attractive for our purposes. They have been slightly simplified and adapted to tests specifically recommended for construction in earth.

AASHO M 145 ENGINEERING SOILS CLASSIFICATION (USA)

AASHO CLASSIFICATION	**GRANULAR MATERIAL** (35% or less of the grains pass through a 0.08m sieve)							**SILT-CLAY MATERIAL** (More than 35% of grains pass through a 0.08mm sieve)			
CLASSIFICATION	A-1		A-3	A-2				A-4	A-5	A-6	A-7
BY GROUP	A-1-a	A-1-b		A-2-4	A-2-5	A-2-6	A-2-7				A-7-5 A-7-6
% OF GRAINS PASSING THROUGH A SIEVE OF											
2mm	15 max	25 max	10 max	35 max	35 max	35 max	35 max	36 min	36 min	36 min	36 min
0.4mm	30max	50 max	51 max	–	–	–	–	–	–	–	–
0.08mm	50 max	–	–	–	–	–	–	–	–	–	–
LL	–		–	40 max	41 max	40 max	41 min	40 max	41 min	40 max	41 min
PI	6 max		N.P.	10 max	10 max	11 min	11 min	10 max	10 max	11 min	11*min
MAIN MATERIALS	stone fragments, gravels and coarse sands		fine sands	gravels and silty or clayey sands				silty soils		clayey soils	

* For A-7-5 PI $\leq$ LL-30
For A-7-6 PI $>$ LL-30

N.P. = Non-plastic

CORRESPONDENCE OF SYMBOLS	
AASHO	USCS
A-1 A-2-4, A-2-5 A-3	GW, GP, GM, SW, SP, SM GM, GC, SM, SC SP
A-2-4, A-2-5 A-4 A-5	GM, GC, SM, SC CL, ML ML, MH, OH
A-6 A-7	CL, CH OH, MH, CH
	OL Pt

GEOTECHNICAL CLASSIFICATION USCS SYSTEM				SYMBOLS	DESCRIPTION
More than half the elements have a diameter greater than 0.08mm GRANULAR SOILS	More than half the elements > 0.08mm have a diameter > 2mm GRAVELLY SOILS	Without fines	All diameters are represented, none predominates	GW	Clean gravel Well graded
			One grain size of one grain fraction predominates	GP	Clean gravel Poorly graded
		With fines	Fine elements have no cohesion	GM	Silty gravel
			Fine elements have cohesion	GC	Clayey gravel
	More than half the elements > 0.08mm have a diameter < 2mm SANDY SOILS	Without fines	All diameters are represented, non predominates	SW	Clean sand Well graded
			One grain size or one grain fraction predominates	SP	Clean sand Poorly graded
		With fines	Fine elements have no cohesion	SM	Silty sand
			Fine elements have cohesion	SC	Clayey sand
More than half the elements have a diameter less than 0.08mm FINE SOILS – CLAY AND SILT	CLAY AND SILT SOILS — Liquid Limit LL < 50%	Above A		CL	Low-plasticity clay
		Under A	Inorganic	ML	Low-plasticity silt
			Organic material	OL	Organic silt and clays with low plasticity
	CLAY AND SILT SOILS — Liquid Limit LL > 50%	Above A		CH	Highly plastic clay
		Under A	Inorganic	MH	Highly plastic silt
			Organic material	OH	Highly plastic organic silt and clay
Organic matter predominates. Can be recognized by smell, dark colour, fibrous texture, low moist density				Pt	Peat and other highly organic soils

NOTES

- This classification is a simplified and modified version.
- Elements with a diameter greater than 60mm are not taken into account.
- The weight of the grain fraction can be estimated.
- The grain sizes correspond to grain size distribution chart.

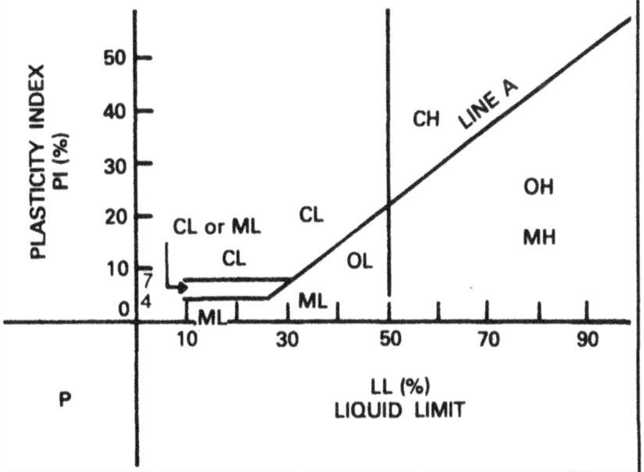

PLASTICITY INDEX PI (%)
50
40
30
20
10
7
4
0
CH
LINE A
OH
MH
CL
CL or ML
CL
OL
ML
ML
10
30
50
70
90
P
LL (%)
LIQUID LIMIT

The classification used by modern soil science takes the whole profile of the soil into account and stresses formation and development processes by considering:
- the degree of development and differentiation of the soil;
- the method of formation and weathering of clays;
- the basic physical and chemical processes which have resulted in the creation of the soil. These are often related to organic matter.

The classification due to P. Duchaufour (Centre de Pédologie du CNRS, France) reflects the modern trend. Tables showing how this classification relates to others, in particular the FAO classification, can be found in the specialist literature.

The Duchaufour Classification (simplified version)

Division 1: Soils whose origin is very closely related to the evolution of organic matter

Class 1 – Immature soils

- climatic: desert soils, frozen soils or cryosols (AC or AR profile – very little organic matter);
- erosion: regosols, lithosols (AC profile – clear humic horizon);
- deposited: alluvial and colluvial soils (deposited by volcanoes or water courses) (AC profile – clear humic horizon).

Class 2 – Soils with little horizon differentiation, desaturated humic soils (AC profile)
These have a profile uniformly coloured by the humus. The humus is rich in organic metal complexes which rapidly become insoluble.
- poor in aluminium: rankers (on crystalline rock)
- rich in aluminium: andosols (on volcanic rock)

Class 3 – Calcimagnesian soils

Typified by the arrest of humification in an early stage by active limestome. Large quantities of immature humus in the profile.
- humic: rendzines (dark humic horizon, thick – well represented in the Mediterranean);
- low humic: brunified calcimagnesian soils (brown calcareous, brown calcic soils)
- very humic: humo-calcareous soils and humic lithocalcic soils (little or no active carbonate – mountains)

Class 4 – Isohumic soils

Typical for the thorough integration, by biological means, of organic matter which has been stabilized by prolonged climatic maturation.
- saturated complex: chernozems, chestnut soils, grey forest soils (dark colour – steppes);
- desaturated complex: brunizems (prairie soil in moist continental climates);
- in increasingly arid climates: reddish chestnut soils, sierozems (grey).

Class 5 – Vertisols

Soils containing expansive clay: thorough integration, by 'Vertic movements' (through the cracks) of extremely stable organic mineral complexes and the dark colour (expansive clay – stable humus). Marked dry season.
- Vertisols (dark) (30 to 40% expansive clay – black cotton soils);
- Vertic soils (coloured) (montmorillonites, semi-expansive interstratified soils).

Class 6 – Brunified soils with indistinct A (B) C or ABC profiles

Characterized by a thinnish humus layer, resulting primarily from its becoming insoluble as a result of sufficient quantities of free iron, which forms a 'ferric bridge' with the clay.

- brown soils with a B-weathering horizon;
- eluviated soils with an accumulation of clay in the B horizon;
- eluviated continental or boreal soils.

Class 7 – Podsolized soils

Immature organic matter forming mobile organic-mineral complexes. Very distinct sandy or gravelly horizons: A2 light and B dark.
- podzolic soils and podsols which are not, or only slightly, hydromorphic
- podzolic soils and podsols hydromorphic (with water table)

Division 2: Soils whose origin:
- is fairly independent of the evolution of organic matter;
- is on the other hand closely related to fairly moist warm climates, and to the special behaviour of iron and aluminium oxides (sesquioxides)

Class 9 – Fersiallitic soils

Development of iron oxides towards 'rubefaction'. Clay formation of the 2/1 type dominates (transformation and new formation). Mediterranean and dry tropical climates.
- incomplete rubefaction: brunified fersiallitic red soils (terra rossa)
- complete rubefaction, saturated or virtually saturated complex fersiallitic red soil;
- partial depletion and desaturation of the complex: fersiallitic acid soils

Class 10 – Ferruginous soils

Abundant crystallized iron oxides (goethite and haematite), as yet incomplete weathering of primary minerals: 1/1 newly formed clay dominates (kaolinite) occurring, however, together with 2/1 clays.
- incomplete weathering: ferruginous soils;
- weathering approaching completion: ferrisols.

Class 11 – Ferrallitic soils

Total weathering of primary minerals (except quartz). 1/1 clays only. High sesquioxide content. Crystallized iron and aluminium oxides. Occurrence is in the wettest tropical regions.
- ferrallitic soils in the strict sense (predominantly kaolinite);
- ferrallitic soils hydromorphic

Division 3: Soils whose origin is related to local conditions

Class 8 – Hydromorphic soils (Appearance affected by water)

These are soils whose appearance is temporarily or permanently affected by an excess of water. Oxidation reduction of iron related to the permanent or temporary presence of a water table.
- marked oxidation reduction (presence of a water table): pseudogley, stagnogley, gley (grey colour, often accompanied by yellow, red, or rust-coloured marks).
- mild oxidation reduction: hydromorphic as the result of capillary rise in a clayey material and surface depletion of clay: pelosols, planosols.

Class 12 – Sodic soils

These are common in the Mediterranean region and in dry subtropical regions. Sodic soils are soils whose evolution is marked by the presence of soluble salts, chloride, sulphate, etc. or by the presence of Na^+ in two forms:
- saline: saline soils
- exchangeable sodium: alkaline soils.

TABLE OF EQUIVALENTS BETWEEN THE FAO AND DUCHAUFOUR CLASSIFICATION SYSTEMS

FAO	DUCHAUFOUR	FAO	DUCHAUFOUR
Fluvisols	Alluvial and colluvial soils	Chernozems	Chernozems
Gleysols	Gley	Cambisols	Undeveloped soils
Regosols	Regosols		Brown soils (tropical)
Lithosols	Lithosols		Iron soils (non-sandy)
Arenosols	Ferrallitic soils (sandy)	Luvisols	Leached soils
Rendzinas	Rendzines		Ferrisols
Rankers	Rankers	Planosols	Planosols
Andosols	Andosols	Aerisols	Ferallitic soils (extremely dry)
Vertisols	Vertisols		
Solontchak	Saline soils	Nitrosols	Ferallitic soils (moderately dry)
Solonetz	Alkaline soils		
Yermosols	Sierozems	Ferralsols	Ferallitic soils (subarid not dry)
Xerosols	Brown soils		
Castanozems	Reddish chestnut soils	Ferrisols	Ferallitic soils
	Chestnut soils		Iron soils
		Histosols	Peat soils

There are a number of very specific soil types whose names are specific to each discipline: agriculture, soil engineering, and soil science. The following names are those most commonly encountered in the literature.

1. Laterites

In wet climates in the tropics and subtropics, the disintegration of the parent rock and the chemical weathering associated with leaching and evaporation leads to an accumulation of sesquioxides in the B horizon (particularly of iron). Lateritic soils are characterized by highly advanced disintegration and by a concentration of metallic hydroxides. Some laterites are richer in aluminium components: these are the bauxites. Lateritic soils have only a thin layer of organic material. Depending on their location, laterites may be soft and composed of clay or sand, or on the other hand they may be hard and contain quantities of pebbles. They are typical for their hardening on exposure to the air. Apart from these general characteristics, no precise and exclusive definition of laterites has been worked out as yet. The chemical ratio $SiO_2/Al_2O_3 < 1.33$ has long been used, although often disputed. Recent work suggests that this ratio may be closer to 2. It is usually lower but may be also be higher. Soil science replaces the general term laterite with compound terms which reflect the specific nature of the soil: fersiallitic soils, ferrallitic soils, and ferrisols. Even nowadays the general definition still used dates from 1807 and is due to Buchanan, who also suggested the name 'Laterite', from the Latin word 'later' meaning brick.

Nevertheless we will use the definition due to Mukerji, 'Laterites are highly weathered soils, which contain large, though extremely variable, proportions of iron oxide and aluminium, as well as quartz and other minerals. They are found in large quantities in the tropics and subtropics, usually just below the surfaces of vast open plains and clearings in heavy rainfall regions. The natural state varies from a compact concretion to crumbly soil. It may be of many colours: ochre, red, brown, violet, and black. The material is easy to cut. It hardens quickly in air, and becomes quite resistant to meteorological agents'.

These thus are the essential properties of induration (rapid and significant hardening). For plinthite, which is a type of laterite, induration takes place rapidly, powerfully and irreversibly. It is fairly rare (e.g. India, Burkina Faso). Influenced by how much iron there is in the parent rock, the humidity of the soil and the lie of the land, the degree of rubefaction (slow dehydration of ferrous oxides and crystallization into Fe_2O_3 (haematite)), the colour of laterites can vary from virtually black, to rust, dark red, and red in cases of extreme drying (haematite Fe_2O_3). If the drying-out process is less extreme the result is red ochre (goethite $Fe_2O_3H_2O$). Stilpnosiderite ($Fe_2O_3H_2O$) gives a yellow ochre colour in damp surroundings. If aluminium preponderates colours such as light red, pale pink, and ochre result. Gibbsite ($Al(OH)^3$), boehmite ($AlOOH$), and diaspore ($H\text{-}AlO_2$) have little colour, or are greyish and transparent.

The physical properties of the laterites vary tremendously. Bulk density ranges from 2500 to 3600kg/m^3. Their hardness rises with the concentration of iron oxides and is associated with an increasingly dark colour, and advanced induration can lead to the formation of a hard outer shell. These ferrous casings may be anything from a few centimetres to more than a metre thick.

2. Terra rossa and terra fusca

These are very slowly decarbonating clays formed over thick layers of hard limestone which were poor in clay, prior to the last ice age, when a mediterranean or even tropical climate dominated. Terra rossa is red, and is distinguished from terra fusca by its degree of rubefaction. Terra rossa can be found in most mediterranean regions. Further to the north, they are fossil soils which become slowly browner at their surface (as the result of the brown colour of a clay-iron mix). These brown soils are still red deep down.

3. Black cotton soils

These soils are found in wet tropical regions and occur over volcanic rock, such as basalt. They are most often referred to as 'black cotton soils', because of their dark colour (black, or deep grey or brown) and from the fact that cotton is often grown on these soils (e.g. in India). They are rich in calcium carbonate and are extremely clayey. The dominant clays are the montmorillonites, which have a very high ion exchange capacity. Up to 90% of the clays may have a diameter of less than 0.15μ. They are noted for their remarkable swell in moist conditions and equally severe shrinkage upon drying. When dry these soils are extremely hard. Close to 500,000km^2 of the Indian sub-continent are covered by them. They are also encountered in Northern Argentina, and in various African countries. In Morocco they are called 'tirs' (the plains of Gharb and Loukkos). Other names include Regur soils (India), Margalitic soils (Indonesia), Black Turfs (English-speaking Africa). The liquid limit LL of black cotton soils is of the order of 35 to 120% and the plasticity index PI can range from 10 to 80% or more. Their linear shrinkage lies classically between 8 and 18%. To give some examples: the Moroccan 'tir' has an LL of between 50 and 70%, a PI of 30 to 35%, and their shrinkage is between 10 and 12%. In Sudan the 'Badobe' soils have an LL of between 47 and 93%, a PI of between 13 and 58%, and a shrinkage of 8 to 18%. Montmorillonite clays often predominate in the black cotton soils. Frequently these are A-7-6 and A-7-5 soils.

4. Loess

Loess is a wind deposit. It is fine and homogenous with a silty texture, low in sand content, with between 10 to 20% of calcium carbonate. The material originates in desert areas (e.g. in China the loess comes from the Gobi desert) or, in areas close to large glaciers, in moraine deposits (as is the case in Europe). The loess layers range in thickness from several tens of centimetres to 10 or 20 metres. Loess is extremely friable. It can be easily excavated (e.g. dugout dwellings in Northern China or in Tunisia).

5. Clayey rocks

They constitute the majority (80%) of sedimentary rocks. Of these, the plastic clays of the phyllosilicate group are the best example. But they include also the shales, silico-aluminous rocks better known as schists or slates, which are plastic in a humid state, and also marl, clayholding minerals mixed with carbonated particles. When these rocks are held together by clay, they constitute a soil. When they are held together by a natural cement (such as calcareous tufa), they are rocks. This distinction is important, because the same regional term can indicate very different materials. The term 'Tepetate' in Mexico indicates both 'Calcrete' (gravel cemented by calcareous tufa) as very calcareous soils or rocks, or even chalk. The same applies to the USA, with the terms 'Caliche' or 'Chalk' or to Tunisia with the 'Torba'. Other terms are equally confusing: marl, tufa, chalk, mergel, Bhata, Dhandla . . . One can often notice high contents of calcium carbonates ($CaCO_3$), of 50 to 75%, and colours varying from ochre to white. The plasticity index decreases with the increase of carbonates. In Tunisia, for instance, the PI is 20% at 75% $CaCO_3$, and 13% at 90% $CaCO_3$. The granularity of these soils is hard to determine, because they harden and dissolve badly in water. However, they remain quite crumbly when dry.

6. Saline soils

These soils are rich in sodium chloride (NaCl) or in sodium sulphate (Na_2SO_4). They are encountered mainly in dry areas with semi-desert, steppe or tropical dry climates where high evaporation rates prevent natural drainage processes. They may also be found where the underlying soil has a high salt content or the groundwater contains salt. In arid climates, these soils are often close to the large saline subdesert depressions (the Sebkhas or Chotts of Northern Africa, the Playas of North America, the Takyr of Central Asia) and in the major irrigated valleys of Egypt, Libya, Israel, Syria, Iraq, and Turkey. In moist climates saline soils are only encountered near the sea (the polders of temperate climates and mangrove swamps in equatorial regions).

7. Alluvial soils

These soils border rivers and streams in the wider valleys. They are rich in minerals and are subject to a continuing weathering process. Their texture varies, though they are usually filtering, with the finest material on their surface (fine sand, silts, clays), and become coarser with depth. Their colour varies from brown ochre on higher ground, grey on the flood plain, and black in marshy areas.

8. Peat

This material is the result of the decomposition of vegetable matter exposed to the air, often in shallow lakes or in marshland. Peat is usually dark brown in colour and contains plant matter, which can be clearly identified as such, and very few minerals.

MAP OF THE MAIN SOILS RECOGNIZED BY SOIL SCIENCE

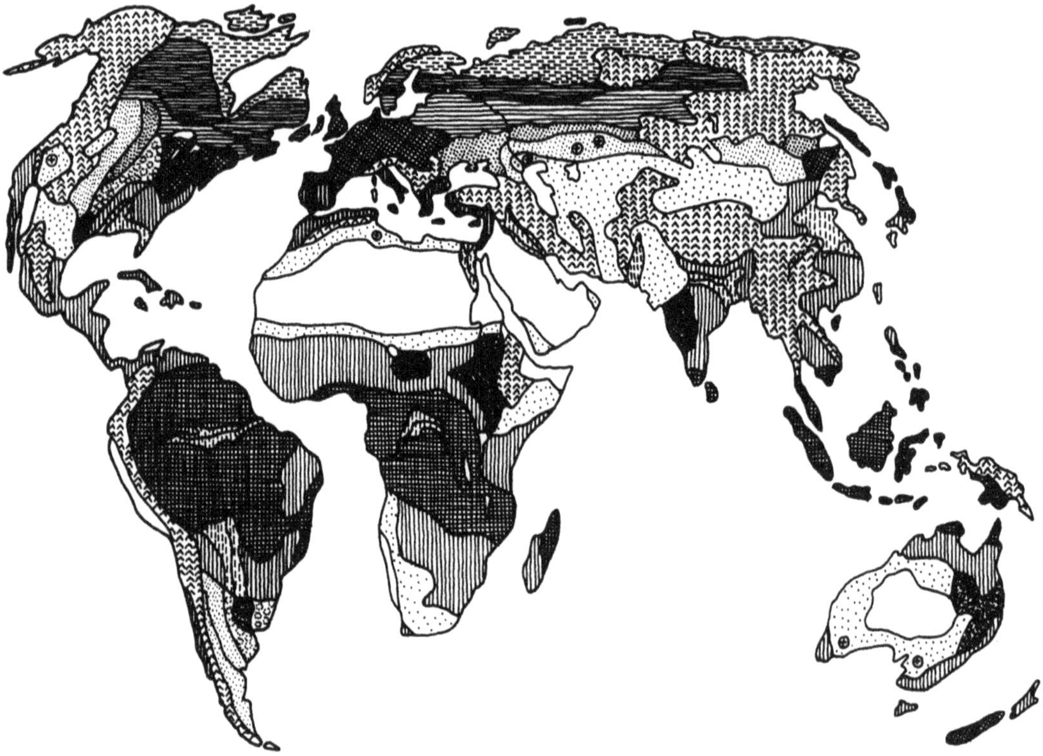

MOUNTAINS: various types of mountain soils

DESERTS

SEMIDESERT AREAS: cold (sierozem), warm (grey and brown sub-arid soils, chestnut soils, iso-humic tropical soils, etc . . .)

LIGHT STEPPE: chestnut soils, burozems

DENSE STEPPE (chernozems)

PRAIRIE (brunizems)

TEMPERATE ZONES: deciduous forest (brown soils)

TRANSITION BOREAL ZONE: mixed deciduous and coniferous forest (podsols and leached boreal soils, grey forest soils)

BOREAL ZONE: coniferous forest (podsols)

TUNDRA and boreal peat soils

SUBTROPICAL AND MEDITERRANEAN WITH A DRY SEASON (predominantly fersiallitic soil)

SUBTROPICAL HUMID AND TROPICAL WITH A DRY SEASON (ferruginous soils, ferrisols)

HUMID EQUATORIAL ZONE: dense forest (ferallitic soils and ferrisols predominate)

HYDROMORPHIC INTRAZONAL SOILS (alluvia, gleys, planosols)

DEVELOPING INTRAZONAL SOILS (mainly greenish soils)

INTRAZONAL SODIUM SOILS (soils containing sodium salts or sodium compounds)

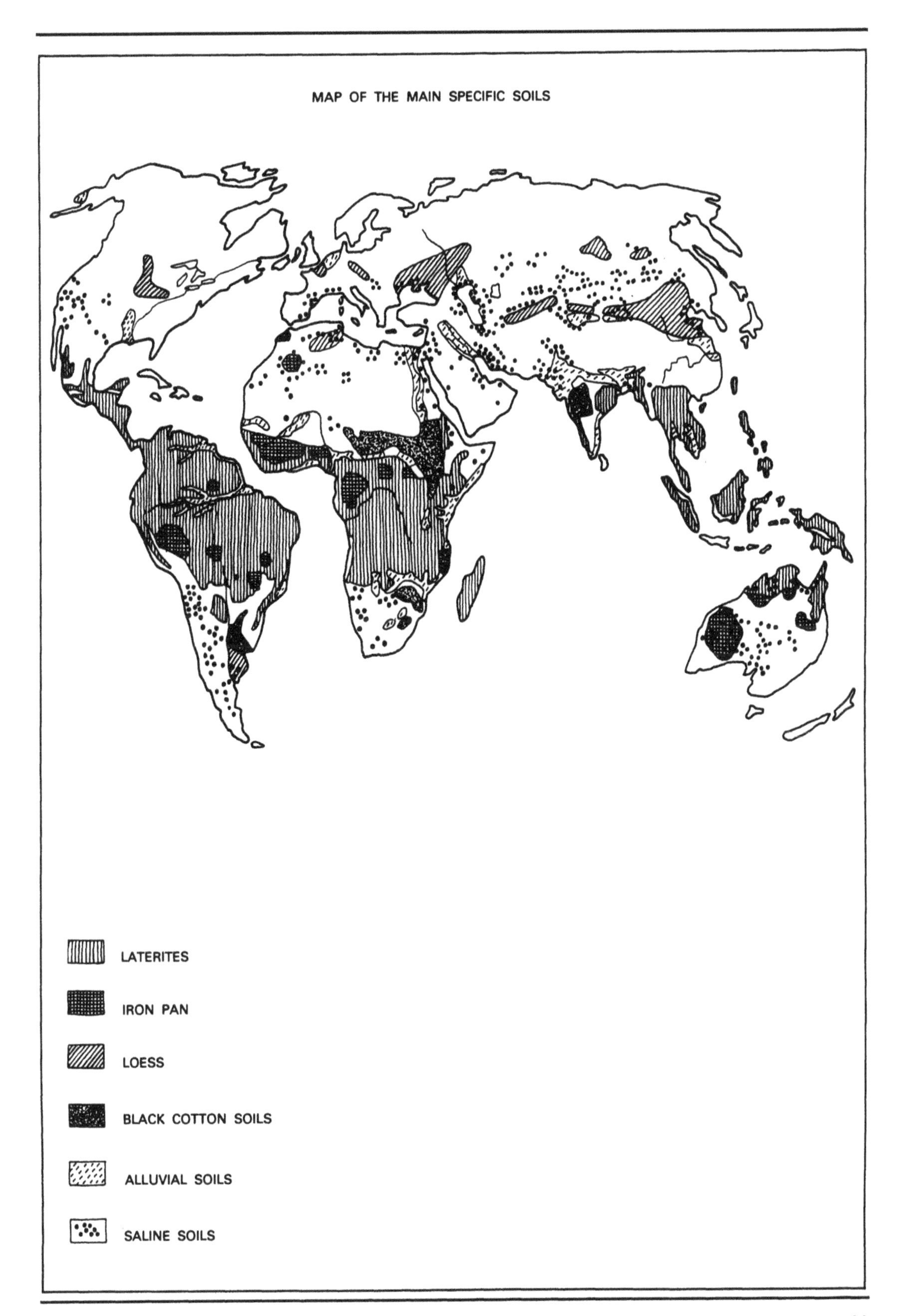
MAP OF THE MAIN SPECIFIC SOILS
LATERITES
IRON PAN
LOESS
BLACK COTTON SOILS
ALLUVIAL SOILS
SALINE SOILS

Bunnett, R.B. *Physical Geography in Diagrams*. London, Longman, 1977.
Cytryn, S. *Soil Construction*. Jerusalem, the Weizman Science Press of Israel, 1957.
Duchaufour, P. *Atlas Écologique des Sols du Monde*. Paris, Masson, 1976.
Dunlap, W.A. *US Air Force Soil Stabilization Index System*. New Mexico, Air Force Weapons Laboratory, 1975.
El Fadil, A.A. *Stabilised Soil Blocks for Low-cost Housing in Sudan*. Hatfield Polytechnic, 1982.
Foth, H.D.; Turk, L.M. *Fundamentos de la Ciencia del Suelo*. México, CESSA, 1981.
GATE. *Lehmarchitektur*. Rückblick-Ausblick. Frankfurt am Main, GATE, 1981.
Graetz, H.A. *et al. Suelos y Fertilización*. México, editorial Trillas, 1982.
Grim, R.E. *Applied Clay Mineralogy*. New York, McGraw-Hill, 1962.
HRB. 'Soil Stabilization: multiple aspects'. In *Highway Research Record*, Washington, 1970.
Ingles, O.G. 'Bonding Forces in Soils'. In *The first conference of the Australian Road Research Board*, 1962.
Ingles, O.G.; Metcalf, J.B. *Soil Stabilization*. Sydney, Butterworths, 1972.
Legrand, C. CSTC. Priv. com. Limelette, 1984.
Maignien, R. *Compte Rendu de Recherches sur les Latérites*. Paris, UNESCO, 1966.
Mukerji, K.; Bahlmann, H. *Laterit zum Bauen*. Starnberg, IFT, 1978.
PCA. *Soil-cement Laboratory Handbook*. Stokie, PCA, 1971.
Pour la science. *Les Phénomènes Naturels*. Paris, Berlin, 1981.
Puplampu, A.B. *Highways and Foundations in Black Cotton Soils*. Addis Abada, UN; ECA, 1973.
Ringsholt, T.; Hansen, T.C. 'Lateritic Soil as a Raw Material for Building Blocks'. In *American Ceramic Society Bulletin*, 1978.
Road Research Laboratory DSIR. *Soil Mechanics for Road Engineers*. London, HMSO, 1958.
Smith, E. *Adobe Bricks in New Mexico*. Socorro, New Mexico Bureau of Mines and Mineral Resources, 1982.
Soltner, D. *Les Bases de la Production Végétale*. Angers, Collection sciences et techniques agricoles, 1982.
Whitten, D.G.A.; Brooks, J.R.V. *The Penguin Dictionary of Geology*. Middlesex, Penguin, 1981.

Compressed earth block building with a flat roof covered with stabilized earth in Auroville, India (architect: Suhasini Ayer-Guigan, Serge Maini Craterre-Eag)

3. SOIL IDENTIFICATION

An essential preliminary to any decision involving a reliable choice of technology for processing soil into a building material is the correct identification of the soil.

There are numerous identification tests which can be carried out on soil; relatively few of these, however, give results which can be directly interpreted in terms of suitability for use in construction.

The tests which are directly useful can be divided into indicator and laboratory tests. These two types of test provide information which is indispensable to good decisions about the use of a building soil. The indicator tests must, however, be carried out first, as they can provide valuable information about the usefulness of carrying out laboratory tests, which are more sophisticated, take longer to carry out, and above all are more exacting to carry out.

Identification procedures

Because it is naturally so diverse, soil can be difficult to identify. If economies in the production of unburnt earth materials and their use in construction are to be achieved, the soil must be identified with some degree of precision. Where simple work is involved identification based on experience may be adequate but even so care must be taken that all the indicators used in the identification process are in agreement. If there are any apparent contradictions, the identification should be confirmed by laboratory tests. Where major works are involved the identification should be precise enough to allow the selection of suitable quality controls and to eliminate those which serve no purpose, full identification of the soil being a tiresome task. A good basic identification of the soil can thus ensure considerable gains in time and money. However, the utility of specific quality controls will be judged with reference to a knowledge of the properties of the material and its main components, their fundamental physical and mechanical characteristics, and to trustworthy reference materials (e.g. tables, nomograms). It should always been borne in mind that soil is a complex material and that identification alone is not enough to provide an absolute guarantee of its correct use in construction work. Various tests which evaluate the mechanical performance of the construction material will also be necessary.

The general procedure detailed below is not exhaustive and may be completed by other procedures. Local knowledge should be sought out and traditional know-how heeded, while procedures borrowed from other disciplines such as geology, agronomy, and soil science can serve as a guide to interpretation.

There are three basic steps in identifying and classifying a soil:

1st stage: identification of the basic characteristics and properties of the components of the soil which are likely to affect the mechanical behaviour of the material; these are the preliminary site analyses, and are visual and manual.

2nd stage: a description of the soil must be drawn up recording the basic characteristics and properties which have been identified by the preliminary analyses. This descriptive information is necessary in order to differentiate the analysed soil from any broader descriptive group.

3rd stage: if field analysis has not permitted accurate enough classification, laboratory analysis will be carried out. This step is only required if very exact analysis is required, e.g. soils of a very specific nature, mineralogical details and so on. The soil may then be assigned to a group or even to a subgroup and given a classification symbol.

Sources of information

Before going out to the site, it is advisable to consult the information which has been preserved or recorded in the form of maps and descriptive accounts with respect to geology, soil science, geography, surveys, hydrology, rainfall, vegetation, agriculture, road networks and so on. A comparison of these data provides preliminary information which can form a valuable guide to the field work. If necessary, local specialists in the disciplines referred to may be consulted in order to obtain an insight in the available information. Further information can often also be obtained from regional farm research stations, research centres, universities, civil engineering, mine and natural resource advisory services, as well as from civil engineering contractor, and others.

The identification file card, filing

Each sample taken in the field is given an 'identity card'. This is a file card containing as much information as possible about the sample such as the place and date of collection, the site involved, the person requesting the sampling, sample number and core number, sampling depth, name of the sampler or corer, weight, special remarks, and so forth. This identity card is completed as more information becomes available and forms a file for the sample, and contains the typical name of the soil, its group symbol, texture, structure, the grain shape, the maximum diameter, its plasticity, mineralogy, smell, colour, state of hydration, compactness, compressibility, cohesion, etc.

Soil identification apparatus

The apparatus required for identifying soils may be rudimentary – a few tools and instruments of the commonest sort such as knives, and various flasks and containers; or relatively sophisticated – a fully equipped laboratory, which could cost several hundred thousand US dollars to set up. However, there is also a compromise solution, using moderately priced apparatus: a makeshift laboratory, or even a mobile laboratory installed in a small truck. There is also the portable site laboratory packed in a suitcase. This type of compact laboratory is very practical and allows the most essential tests to be carried out. The apparatus contained in such a field kit should make it possible to test for the following: brilliance, adhesion, decantation, sedimentation, grain size, plasticity, compactibility (not absolutely necessary), cohesion, mineralogy, and chemistry.

The soil identification apparatus referred to here should by implication be suitable for carrying out the less sophisticated series of tests and trials. It must be realized that these will above all be field tests, involving apparatus and tools such as a small pick, knives and spatulas, several recipients for materials which must be kept, a graduated flask, moulds for the linear and volumetric shrinkage tests, a pocket rule, and so on.

1. Collecting samples

A manual auger or mechanical version mounted on a lorry can be used. Augers allow rapid sampling to considerable depths. With extensions depths of between 5 and 6 metres can be reached, and without extensions depths of between 0.6 and 0.7m. Normal sizes for augers range from 6 to 25cm. They weigh about 5kg increasing by 3kg with every 1m extension. The main drawback of the tool is the risk of mixing surface layers with those at a greater depth.

An alternative is to dig a hole with a side of 1m to a depth of 2m. The hole should be properly oriented with respect to the sun in order to facilitate observation. Precautions must also taken to ensure the safety of the labourers as there is a cave-in risk when working in poorly cohesive material. The excavated earth is removed in its entirety and no samples are taken from it. The soil used in the analysis is taken from one of the sides of the holes by digging sideways into the wall. Samples can also be taken from natural slopes where the dip of the soil layers is clearly visible. Care should be taken to remove all the vegetation and organic matter from the surface.

2. Sample weight

In principle, 1.5kg of soil is enough for all basic identification tests, except for compactibility tests, which require 6 to 10kg. If at least one brick measuring 29.5×14×9cm is to be tested, 10kg of soil is required. The quantity of soil to be taken for the sample will depend on the number and type of tests to be carried out, on the degree of precision required as this may make a double test series necessary, on the expenses and difficulties involved, as the cost of a test is often related to the quality of the soil under test, and finally on grain size, as a large grain soil requires a larger sample than a fine soil.

3. Sample quality

The sample must be representative of the quality of the soil under test. In order to ensure that the sample is representative, an effort must be made to ensure that certain general principles are observed.

- Care must be taken to avoid soil contamination due the mixing of different sampling horizons.
- Take nothing and add nothing to the sample. Do not try to improve its natural state.
- Take samples from only a very restricted area.
- If the soil is heterogenous, do not try to take an 'average' but take more samples from each different spot.
- In order to divide a sample up, place it in the shape of a cone on a clean tissue; flatten it and divide it into four. Discard two of the opposing sectors, then shape the material into a cone again and repeat the operation until the desired quantity is obtained.

4. Sample packing

The samples are packed into recipients or waterproof bags which cannot be broken or ruptured during transport. If the original state of hydration is to be maintained, packing in paraffin wax can be recommended. The containers should be carefully labelled, the identifying label being packed inside the container in order to prevent it being lost, blurred, or altered during handling.

Field work requires a number of rapid identification tests to help in determining what soils are likely to be suitable for construction purposes. These simple field tests make it possible to evaluate some of the properties of the material and to determine the suitability of the soil for construction purposes. These tests are somewhat empirical; and they should be repeated to ensure that more than a superficial impression is gained. From these tests it can be seen whether further laboratory testing is justified.

1. Visual examination

The dry soil is examined with the naked eye to estimate the relative proportions of the sandy and fine fractions. Large stones, gravel, and coarse sand are removed in order to facilitate evaluation (this operation must also be carried out for all the following tests). The fines fraction is made up of grain sizes with a diameter of less than 0.08mm. This diameter lies at the limit of the resolving power of the human eye.

2. Smell test

The soil should be smelt immediately after removal. If it smells musty it contains organic matter. This smell will become stronger if the soil is heated or wetted.

3. Nibble test

The tester nibbles a pinch of soil, crushing it lightly between the teeth. The soil is sandy if it grinds between the teeth with a disagreeable sensation. Silty soil can be ground between the teeth but without giving a disagreeable sensation. Clayey soil gives a smooth or floury sensation, and a small piece of it is sticky when applied to the tongue. Of course care should be taken that it is safe to place any such samples in the mouth.

4. Touch test

After removing the largest grains, crumble the soil by rubbing the sample between the fingers and the palm of the hand. The soil is sandy if a rough sensation is felt and has no cohesion when moist. The soil is silty if it gives a slightly rough sensation and is moderately cohesive when moistened. The soil is clayey if when dry it contains lumps or concretions which resist crushing and if it becomes plastic and sticky when it is moistened.

5. Washing test

Wash the hands with the slightly moistened soil. The soil is sandy if the hands easily rinse clean. The soil is silty if it appears to be powdery and the hands can be rinsed clean without any great difficulty. The soil is clayey if it gives a soapy sensation and the hands can be rinsed clean only with difficulty.

6. Lustre test

A slightly moist ball of earth is cut in two with a knife. If the freshly revealed surface is dull, the soil will be predominantly silty. A shiny surface on the other hand indicates the presence of a plastic clayey soil.

7. Adhesion test

Take a mass of moist soil but which does not stick to the fingers and stick a spatula or knife into it. The soil is extremely clayey if the spatula penetrates it only with difficulty and soil sticks to it upon withdrawal. The soil is moderately clayey if the spatula can be pushed into it without great difficulty and soil sticks to it upon withdrawal. The soil contains only a little clay if the spatula can be pushed into it without encountering any resistance at all, even if the spatula is dirty upon withdrawal.

8. Sedimentation

The foregoing tests make it possible to form an idea of the texture of the soil and the relative size of the different fractions, as well as of the quality of the fine fraction. To obtain a more exact idea of the soil fractions a simplified sedimention test can be carried out in the field. The apparatus required need only be simple: a transparent cylindrical glass bottle, with a flat bottom and a capacity of at least one litre and with a neck wide enough to get a hand in, but small enough to be closed off with the palm.

The test procedure is as follows:

- fill the bottle a quarter full with soil;
- fill the remaining three-quarters with water;
- leave the bottle to stand so that the soil is soaked (the soaking can be facilitated by disturbing the soil manually);
- shake the bottle vigorously;
- decant the murky water;
- shake again after an hour and decant again;

– after a further 45 minutes, it will be seen that the sand has been deposited on the bottom of the bottle. Above it is a layer of silt and above the silt a layer of clay. On the surface of the water floats organic debris, while any very fine colloids will remain in suspension in the water. Normally eight hours are allowed to go by before measuring the different layers precipitated. First of all the overall depth of the sediment (100%) is measured, without including the depth of clear water covering them, and then measure each separate layer.

This measurement of the respective depths of the sediment, which makes it possible to estimate the percentage of each grain fraction, is slightly distorted by the fact that the silt and clay fractions will have expanded and thus appear slightly larger than they really are.

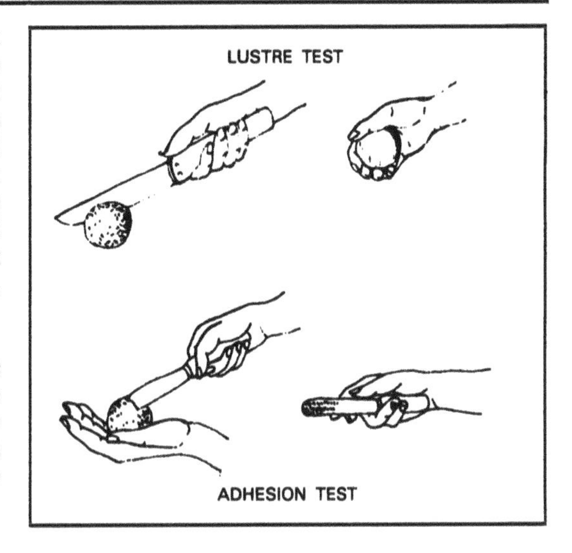

9. Shrinkage

The linear shrinkage test, or Alcock's test, is performed with the help of a wooden box, 60cm long, 4cm wide and 4cm deep. The inside surfaces of the box are greased before being filled with moist soil with optimum moisture content (OMC). The soil is pressed into the corners of the box with a small wooden spatula which is also used to smooth the surface. The filled box is exposed to the sun for a period of three days, or left in the shade for seven days. After this period the hardened and dried mass of soil is pushed to one end of the box and the total shrinkage of the soil is measured from the soil to the other end of the box.

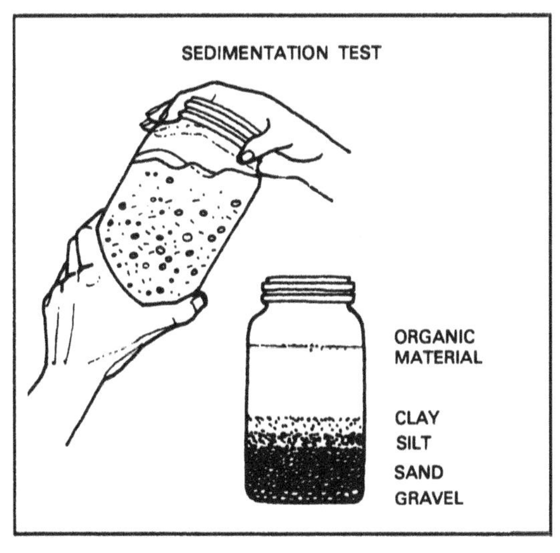

It is now known whether the soil contains large or small amounts of coarse material, or large or small amounts of fines. It is also possible to determine the relative proportions of clay and silts in the fines, and to determine the presence of organic material. These thus are tests which can be carried out with the means immediately available. They may lack precision but are very useful when working in difficult conditions far away from all laboratory equipment.

Nevertheless, when these tests are systematically and stringently carried out it is possible to make fairly accurate estimates of the quality of the soil being contemplated for use in construction.

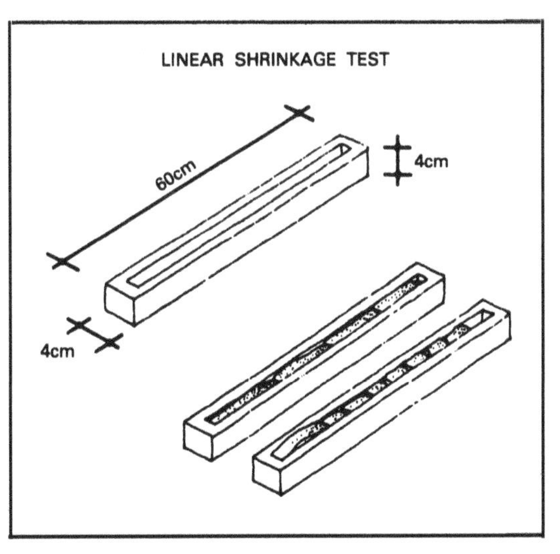

When preliminary field analyses do not give sufficiently satisfactory results, it is advisable to carry out another series of tests based primarily on the observation of the soil texture, its plasticity and cohesion. These tests require no sophisticated equipment, other than a few recipients, a length of flexible tubing, and a spoon. This second series of tests should make it possible to classify the soil from the engineering geologist's point of view.

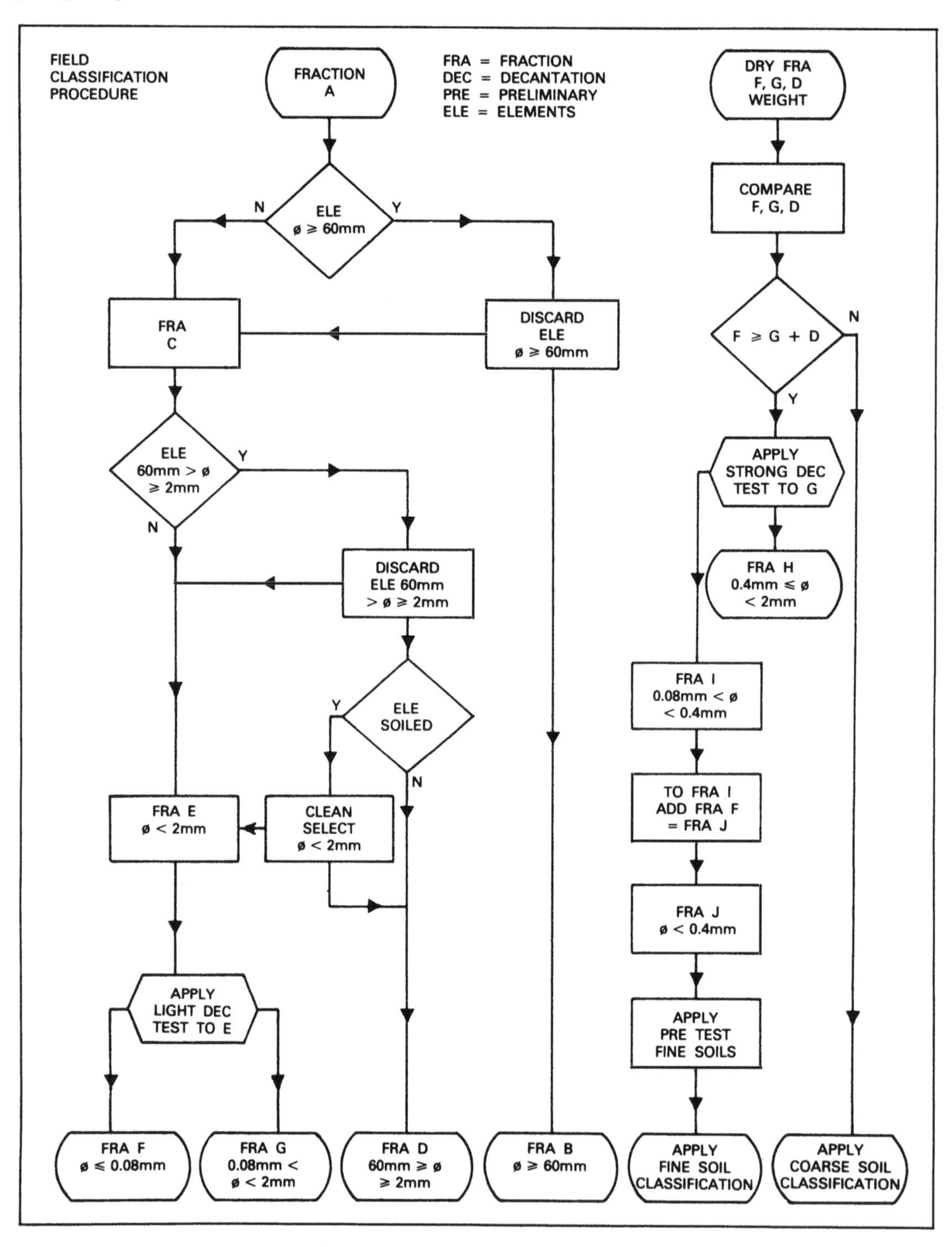

Decantation

The test procedure starts in the same way as that for the sedimentation test. The bottle containing the soil sample and the water is shaken vigorously, and is then decanted. The water and the suspended material are siphoned off into a recipient using a flexible tube with an inside diameter of at least 0.5cm. This operation may be repeated several times. The excess water, which may still contain different grain fractions, is evaporated off. Decantation is not a particularly accurate separation method but is more effective than the simple visual evaluation of two fractions.

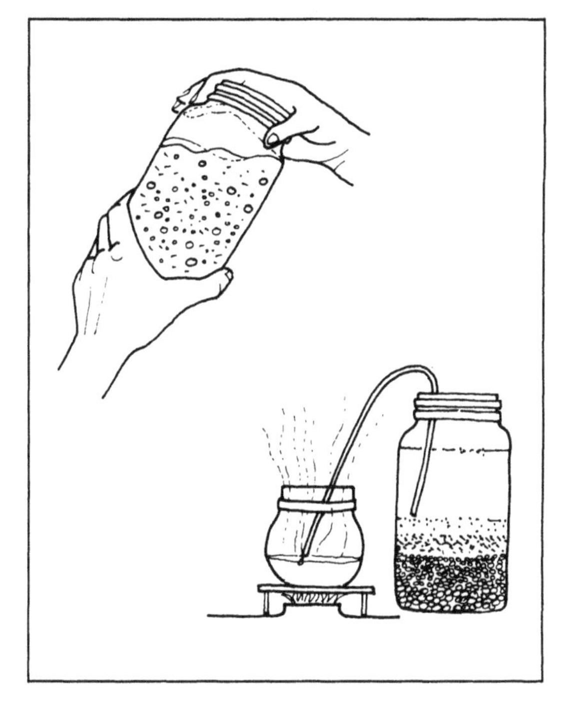

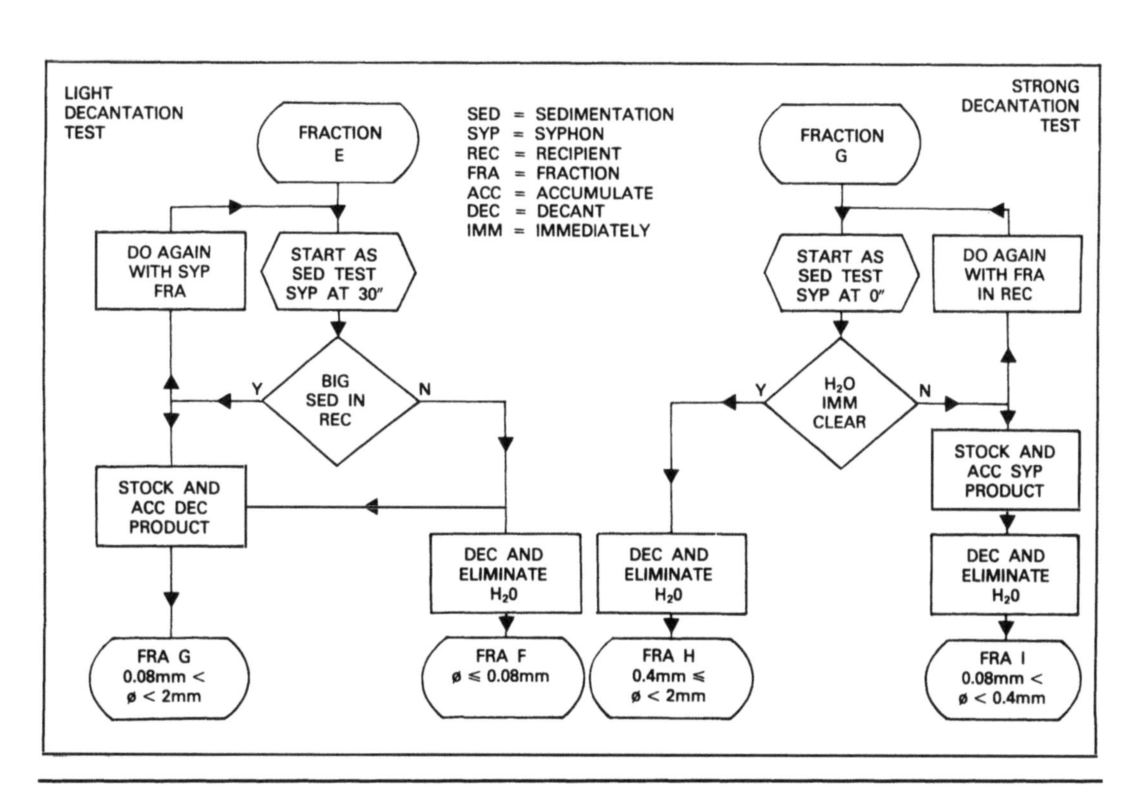

The analyses appearing below were carried out on the 'fine mortar' (diam. < 0.4mm) graded by sieving, or by the decantation test on the grain fraction with a diameter of 2mm.

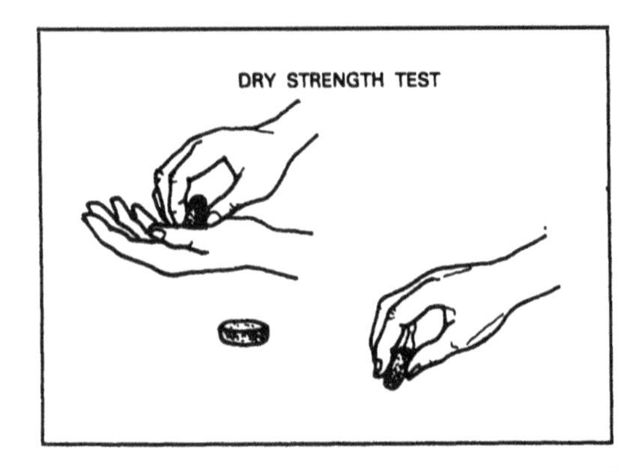

1. Dry strength test

- Prepare two or three pats of soft soil.
- Dry the pats in the sun or in an oven until they have completely dried.
- Break the soil pat and attempt to pulverize it between thumb and index finger.
- Estimate the strength of the pat, interpret.

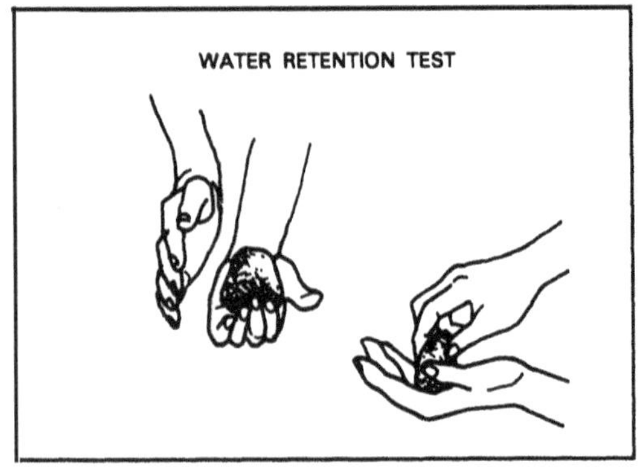

2. Water retention test

- Prepare a ball of 'fine mortar' of 2 or 3cm in diameter.
- Moisten the ball so that it sticks together but does not stick to the fingers.
- Slighty flatten the ball and hold it in the palm of the extended hand. Hit the palm of the hand holding the ball vigorously so that the water runs out. The appearance of the ball may be smooth, shiny or greasy.
- Next press the ball flat between index finger and thumb and observe the reactions, interpret.

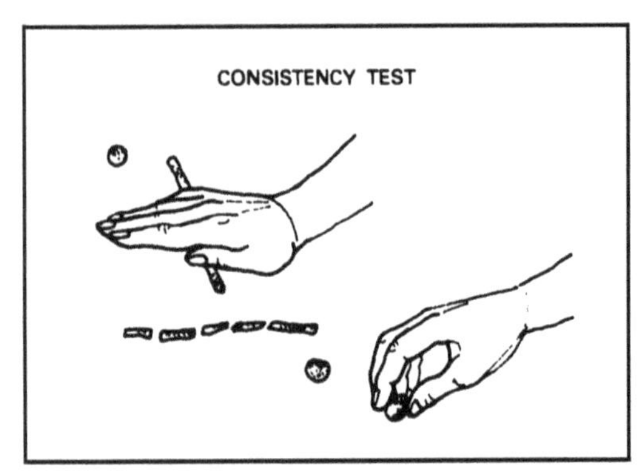

3. Consistency test

- Prepare a ball of 'fine mortar' of 2 or 3cm in diameter.
- Moisten the ball so that it can be modelled without being sticky.
- Roll the ball on a flat clean surface until a thread is slowly formed.
- If the thread breaks before its diameter is reduced to 3mm, the soil is too dry; add water.
- The thread should break when its diameter is 3mm.
- When the thread breaks, make it into a small ball again and crush it between thumb and index finger, interpret.

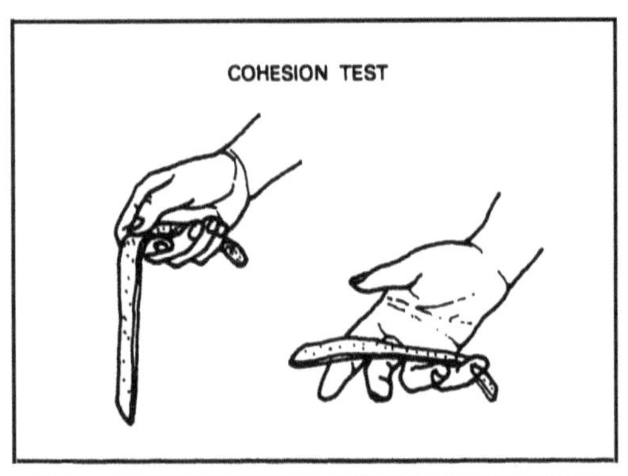

4. Cohesion test

- Make a roll of soil about the size of a sausage with a diameter of 12mm.
- The soil should not be sticky and should be capable of being shaped so that it makes a continuous thread 3mm in diameter.
- Place the thread in the palm of the hand.
- Starting at one end flatten it between index finger and thumb to form a ribbon of between 3 and 6mm in width (handle with care) and which is as long as possible.
- Measure the length obtained before the ribbon breaks. Interpret the result.

Observations	Interpretations
High dry strength	• A dry pat is very difficult to break. When it does, it breaks with a snap, like a dry biscuit. The soil cannot be crushed between thumb and forefinger, it can merely be crumbled, though without reducing it to dust: **almost pure clay.**
Moderate dry strength	• A pat is not too difficult to break. It can be crushed to powder between thumb and forefinger after a little effort: **silty or sandy clay.**
Low dry strength	• A pat can be easily broken and can be reduced to powder between thumb and forefinger without any difficulty at all: **silt or fine sand, low clay content.**
Rapid reaction	• 5 or 6 blows are enough to bring the water to the surface. When pressed the water disappears and the ball crumbles: **very fine sand or coarse silt.**
Slow reaction	• 20 to 30 blows are needed to bring the water to the surface. When pressed the ball does not show any cracking nor does it crumble; it flattens: **slightly plastic silt or silty clay.**
Very slow reaction or no reaction at all.	• No water appears on the surface. When pressed the ball retains its shiny appearance: **clayey soil.**
Hard thread	• The small reconstituted ball is difficult to crush, does not crack nor crumble: **high clay content.**
Medium hard thread	• The small reconstituted ball tends to crack and crumble: **low clay content.**
Fragile thread	• It is impossible to make a little ball from the thread without it breaking or crumbling: **high sand or silt content, very little clay.**
Soft or spongy thread	• The thread and the reconstituted ball has a soft or spongy feel: **organic soil.**
Long ribbon (25 to 30cm)	• **High clay content.**
Short ribbon (5 to 10cm obtained with difficulty)	• **Low clay content.**
No ribbon at all	• **Very low clay content.**

Sieving

This consists of passing the soil through a series of standardized sieves set on top of one another with the finest sieve at the bottom and observing the grain fractions retained by each sieve.

1. Method

Sieve analysis is carried out on the grain fraction with a diameter greater than 0.08mm. The amount of soil required is about 800g. Less is required for fine soils, while more must be used for coarse soils (2 to 3kg). The equipment required for the test is a set of standardized sieves (square holes) or screens (round holes), two holders containing the sieves, a 500ml squeeze wash bottle, a 2 to 5kg balance with an accuracy of least 0.1g, a gas boiling ring, artist's brushes, a pan and trays, a spatula, asbestos gloves, and an oven (optional). The sieve sample is dried until it reaches its constant weight, which is then recorded. Sieving is carried out sieve by sieve, in water. The rinsing makes it possible to wash and to separate the fine material from the sand and gravel. The rinsed fines are recovered on each occasion in a receptacle and transferred to another receptacle containing the following sieve and so on until all the sieves have been tried. The material rejected by each sieve is dried and weighed and its dry weight recorded. When sifting fines (smaller than 0.4mm) the material is agitated with an artist's brush. The flexible wash bottle allows the careful rinsing off of the material retained by each sieve.

2. Evaluation of the method

Although there are a large number of variations on the method, sieve analysis is a fairly reliable procedure as the results remain more or the less the same regardless of the exact method used. Differences between sieves and screens can also be fairly large although there is a tendency towards the abandonment of screens in favour of sieves. Even so grain-size distribution curves vary, which makes it difficult to compare results. Nowadays all types of conventional diagrams can be used without risking incorrect interpretation. One drawback lies in the fact that grain fractions are expressed by weight while soil texture is more often expressed as a volume. Indeed the specific gravity of fractions of fine sand and of clay are very different, but special criteria for interpretation have consequently been introduced. Drying on the boiling ring can cause changes in mineral structure if the temperature is allowed to rise too high, while drying in a drying cabinet is very slow. To avoid these drawbacks a method has been developed which weighs the retained material when wet making use of the displacement method, thus eliminating the need for drying.

3. Simplified method

There is also a simplified sieve analysis method, which makes use of siphoning. The fraction of the material which has passed through a 2mm or 5mm sieve is poured out into a 20cm graduated flask and allowed to settle for 20 minutes. The material still in suspension at the end of this period is siphoned off, allowed to dry and weighed. The sedimented material is passed through a series of sieves. The material retained by each sieve is dried and weighed.

4. Coefficients

Any given grain diameter is referred to by the letter 'D' accompanied by a figure which indicates the percentage passing through the sieve. For example D50 = 2mm means that 50% by weight of the soil is composed of grains with a diameter of less than 2mm. In order to determine if the grain size of the soil is suitable, coefficients have been laid down which give an indication of the shape and the slope of the grain-size distribution curve:

coeff. of uniformity: $CU = D60/D10$,

coeff. of curvature: $CC = (D30)^2/(D10 \times D60)$.

Sedimentation

Sieving gives only an incomplete picture of grain sizes. While it may be enough for the majority of road works, it is not entirely suitable for building in soil as the latter technique demands an analysis of the fines with a diameter of < 0.08mm. An analysis of these elements can be made by means of sedimentation. The principle of this technique is the fact that particles in suspension in water tend to fall at different rates. The largest grains settle first, while the finest settle last. Changes in density are measured at regular intervals at a given height (reduction of density as the liquid clears). When the speed of descent of the various grain sizes is known the proportions of the different grain sizes can be calculated.

1. Method

For sedimentation analysis the following apparatus is required: two graduated 1000ml flasks, with a diameter of 5.5cm, 1 hydrometer graduated from 995 to 1050, a thermometer and a chronometer. The soil fraction with a diameter of less than 0.08mm requiring analysis is first prepared. 20g of material are taken from the fines which have been dried after sieving and mixed with 20cc of dispersant. After dispersion it is essential to check that the solution is not acid (pH > 9.5) as there is a danger that the clays will flocculate. The solution is stirred with an agitator for 3 minutes and left to stand for a period of 18 hours. Measurements can start after this waiting period. The solution in the first flask is again stirred for 3 minutes and after waiting 45 seconds the hydrometer is placed in the solution. Without removing the hydrometer make measurements after 1 minute and after 2 minutes, using the chronometer to check the elapsed time. The subsequent measurements should be taken after 5 minutes, 10 mins, 30 mins, 1 hour, 2h., 5h., and 24h. The hydrometer should be placed in the solution about 15 seconds before taking a reading, and it should be taken out as quickly as possible in order to make the control reading on the control sample. It is extremely important to ensure that the temperature of the two samples is the same when making the measurements as any difference may seriously affect the quality of the test. The test is not used for grain sizes below 0.001mm as turbulence phenomena and dispersion start to affect the sedimentation process adversely.

2. Evaluation of the method

The measurements made in sedimentation tests take only a few moments but from beginning to end the test lasts close to 48 hours. Various problems may arise with the first measurements, and when the results are plotted on the grain size distribution chart it will be seen that the sedimentation curve does not correspond well to the sieving curve. In that case the sedimentation test will have to be repeated.

3. Dispersant

The addition of a dispersant to the fines in suspension prior to analysis is absolutely essential. Various compounds may be used but one of the most widely employed is sodium hexametaphosphate in the proportion of 20g per litre of water. The solution ought to be well mixed and used immediately. Other compounds can be used as well but then it is imperative to recalibrate the test against the method using sodium hexametaphosphate. Alternative dispersing agents, used in the proportion of 1g per cc of distilled water include:

- gum arabic;
- bicarbonate of soda;
- sodium silicate (window cleaning product);
- bases or basic salts such as: washing soda, ammonia, sodium silicates in solution, and soda ash in solution.

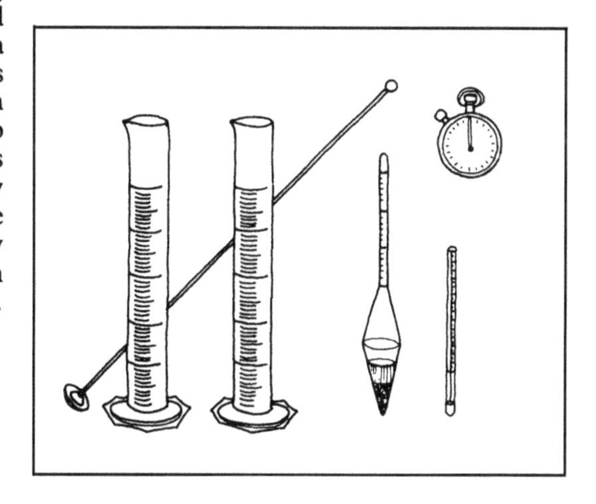

CORRESPONDENCE OF SIEVE IDENTIFICATION

EUROPEAN SIEVES (SQUARE HOLES) A.F.N.O.R. XII–501 (mm)	EUROPEAN SIEVES (ROUND HOLES) A.F.N.O.R. XII–501 (mm)		PRACTICAL MODUL A.F.N.O.R. XII–501 (N°)	UNITED KINGDOM SIEVES B.S. 410 (N°)		UNITED STATES SIEVES A.S.T.M.E. 11/26 (N°)	
20	25	20.0, 16.0	44	3/4", 1/2"		3/4", 1/2"	
10	12.5	10.0, 8.0	41	3/8"		3/8"	4
5	6.3	5.0, 4.0, 3.15	38	5	6, 7, 8	5	6, 7, 8
2	2.5	2.0, 1.6	34	10	12, 14, 16, 18	10	12, 14, 16, 18
1	1.25	1.0, 0.8	31	20	22, 25, 30, 36, 44	20	30, 40
0.5	0.63	0.5, 0.4, 0.315	28	50	52, 60, 70, 85	50	60, 70, 80
0.2	0.25	0.2, 0.16	24	100	120, 150, 170	100	120, 140, 170
0.1	0.125	0.1, 0.08	21	200	240	200	250, 270, 325
0.05	0.063		18	300			
0.02			14				
0.01			11				
0.005			8				
0.002			4, 0				

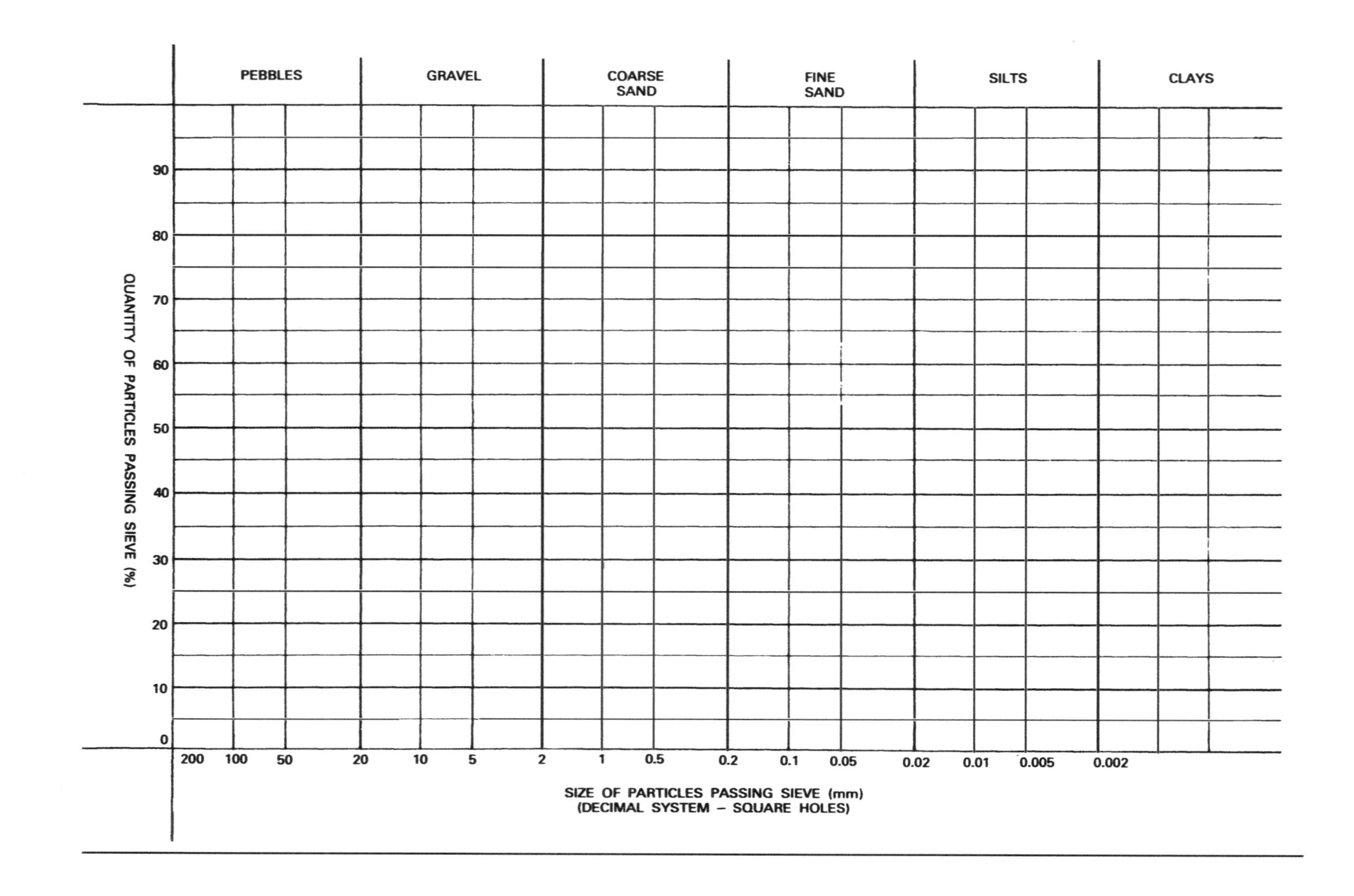
PEBBLES
GRAVEL
COARSE SAND
FINE SAND
SILTS
CLAYS
QUANTITY OF PARTICLES PASSING SIEVE (%)
90
80
70
60
50
40
30
20
10
0
200
100
50
20
10
5
2
1
0.5
0.2
0.1
0.05
0.02
0.01
0.005
0.002
SIZE OF PARTICLES PASSING SIEVE (mm)
(DECIMAL SYSTEM – SQUARE HOLES)

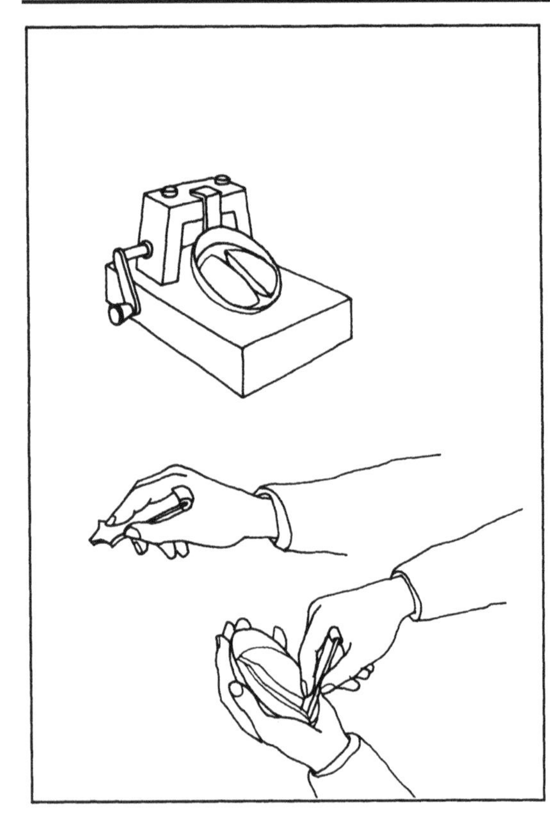

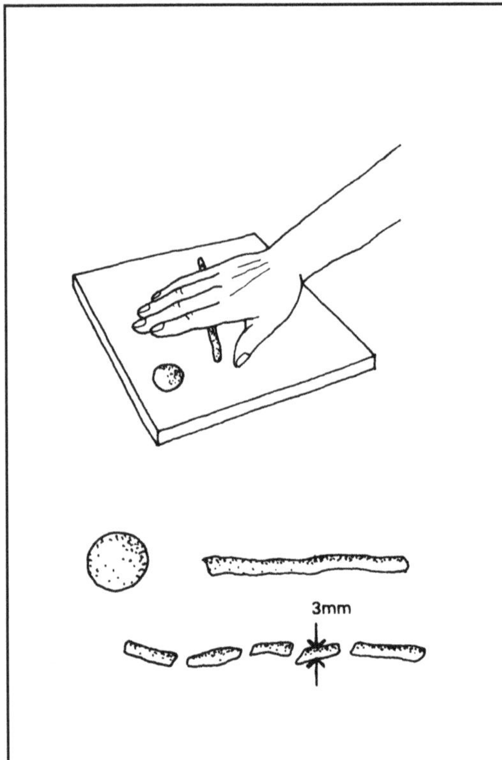

A soil may have different states of consistency. It may, for example, be liquid, plastic or solid. The Swedish research worker, Atterberg, defined these states, which correspond to different moisture contents, and boundaries dividing them by limits and indexes expressed in the percentage moisture content by weight. Five limits can be measured:

- the liquid limit;
- the plastic limit;
- the shrinkage limit;
- the adsorption limit;
- and the limit of adhesion.

The most important of these limits are the first two, while the other three, interesting as they are, are only rarely used. The determination of the Atterberg limits is usually carried out on the 'fine mortar' fraction of the soil which can pass through a 0.4mm sieve, water having little effect on the consistency of larger grain sizes.

1. Method

Liquid Limit (LL) This is the transition from the plastic state to the liquid state. LL is measured with the Casagrande apparatus. Between 50 and 70g of previously prepared fine mortar is spread in the cup (maximum thickness = 1cm) and divided into two parts by a standard axial groove (min. length = 4cm). LL is the water content expressed as a percentage by weight of the material after drying in an oven at 105°C at which the groove closes over a length of 1cm under the influence of 25 blows produced by allowing the cup to drop from a height of 1cm onto a hard surface.

Plastic Limit (PL) This is the transition from the plastic state to the solid state with shrinkage. PL is the moisture content expressed as a percentage by weight of the material after drying in a cabinet at 105°C at which a thread of fine mortar breaks into sections of between 1 and 2cm long when the thread is reduced to a diameter of 3mm. The thread should be between 5 and 6cm long.

Plasticity Index (PI) An indicator of the plasticity of the soil: PI = LL – PL. The greater PL becomes, the greater the swell when the soil is moistened and its shrinkage when it dries. PI is thus a measure of the likelihood of the material becoming deformed.

2. Evaluation of the method

The results of Atterberg limit tests depend on how carefully the procedures are followed, and the 'touch' of the tester. These tests are indeed empirical in nature but have proved to be enormously useful.

3. Interaction of parameters

The two most important parameters in the analysis of the plasticity of a soil are its texture and the mineralogical nature of its clays. In fact it is just as much the quantity as the quality of the clay fraction which influences the plasticity of the soil and consequently the Atterberg limits. Diagrams of Atterberg limits delineate areas which qualify the soils under consideration. Because there is a mathematical relationship between LL, PL, and PI it is enough to draw up a diagram for two of these parameters with PI on the ordinate for LL to be given on the abscissa.

1. **Cohesion** Depends on the moisture content, and this can be very high with small grain sizes. Cohesion is high when the moisture content is less than PL: a virtually solid state where the water becomes a binder the viscosity of which affects cohesion. At PL, the quantity of water is such that it possesses the properties of free water. When the moisture content increases, the apparent cohesion (pore water pressure, surface friction) falls to nothing. At LL, the true cohesion itself disappears. Thus PI = LL – PL describes the cohesion of the soil; but comparative trials have demonstrated that PI > $\frac{1}{4}$ of LL represents the quantity of water necessary to overcome existing true cohesion, thus specifying the degree of true cohesion of the surface of the particles, which also depends on the mineralogical nature of the clays.

2. **Coefficient of Activity (CA)**

$$CA = \frac{PI}{\text{\% clays with a diameter of } 2\mu}$$

This coefficient indicates the activity of clays.

If the activity of the clay is considered in relation to its quantity, a value can be put on the expansivity of the mortar. Apparently contradictory results may be obtained: extremely active clays associated with low expansivity of the mortar, if the latter contains only a small proportion of the clay.

- CA < 0.75 : inactive (I);
- 0.75 < CA < 1.25: average activity (AA);
- 1.25 < CA < 2 : active (A);
- CA < 2 : very active (VA).

3. **Other relationships** The Atterberg limits are also to some extent related to other parameters such as:

- maximum dry density;
- optimum moisture content;
- linear shrinkage;
- compressive strength.

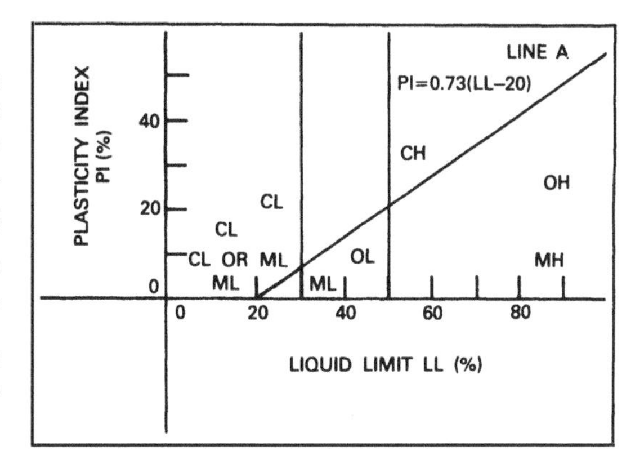

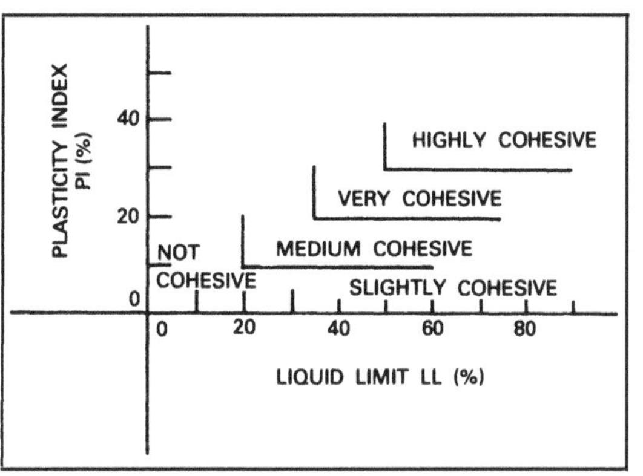

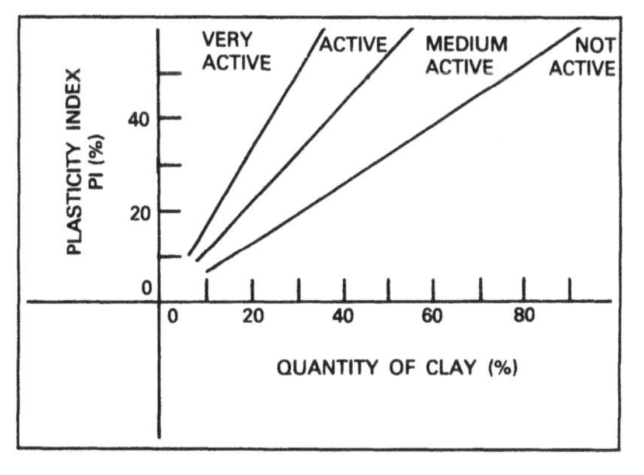

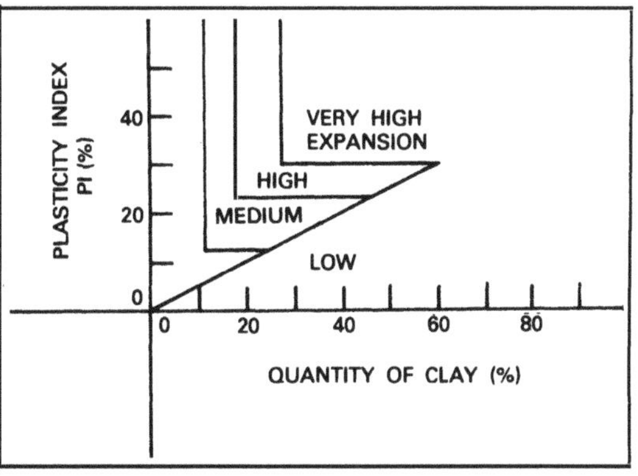

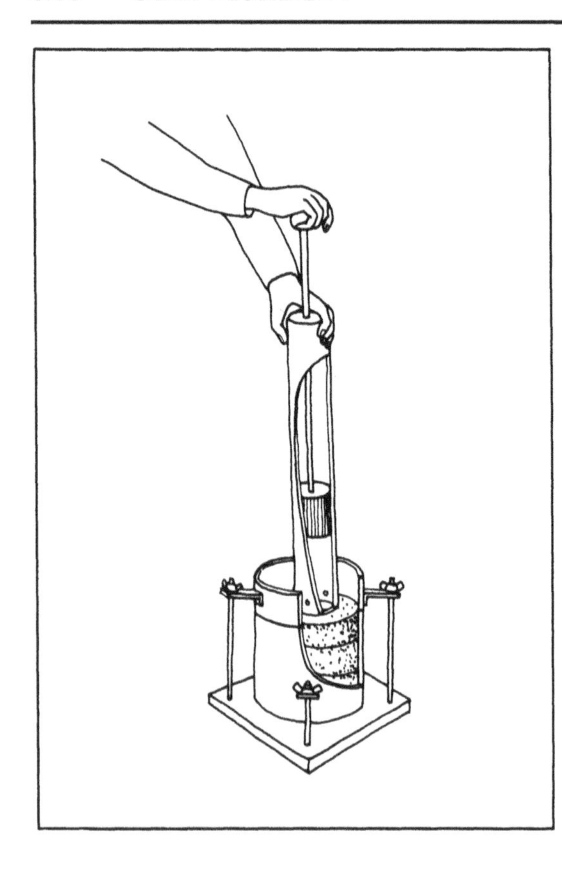

STANDARD PROCTOR TEST AASHO	
TAMPER WEIGHT	2.496kg
TAMPER DIAMETER	5.08cm
DROP HEIGHT	30.5cm
MOULD VOLUME	949cm^3
MOULD DIAMETER	10.16cm
MOULD HEIGHT	11.70cm
SOIL WEIGHT	1.5kg
ENERGY	6J/cm^3
BLOWS PER LAYER	25
LAYER HEIGHT	4cm
LAYERS PER SAMPLE	3

If soil compaction is to be effective, it must be carried out on material whose moisture content is such that it ensures good lubrication of the grains so that they can be rearranged to occupy the least space possible. In fact, when the moisture content is too high, the soil may swell, and the pressure of the compacting equipment will be absorbed by the water, which cannot be driven from between the grains. On the other hand, when the moisture content is too low, the grains will be insufficiently lubricated, and the soil cannot be compacted to its minimum volume. The optimum moisture content (OMC) at which the maximum dry density can be obtained is determined by the Proctor compaction test (after the American contractor who perfected the test). The results are recorded on a graph where the Y axis represents dry density, γd, expressed in kg per m^3 and with the moisture content L expressed in percentage by weight on the X axis. The three main variables involved in obtaining the maximum dry density are: texture, moisture content, and compaction energy.

1. Method

The Proctor compaction test is in principle carried out on that fraction of the soil which passes through a 5mm sieve but can tolerate grains of up to 25mm in size. A soil sample of which the moisture content is known (weighed and compared with the dry weight obtained after drying in an oven at 105°C) is placed in a standard cylindrical mould (which itself has been weighed in advance). Compaction is carried out on three layers of equal thickness and in such a way that the compacting force is equally distributed over the surface of the layers with a standard weight falling 25 times from a predetermined height. Each time the operation has been completed the mould and the sample is weighed and the results are recorded on a Proctor diagram as a curve passing through the points drawn from the test results. From this curve it is possible to read off the maximum value for γd and the optimum moisture content. The most usual Proctor tests are the AASHO standard Proctor test and the AASHO modified Proctor standard.

2. Evaluation

This test is only applicable when the soil construction technique used involves dynamic compaction such as rammed earth or blocks compacted by means of a tamper.

In theory at least, it is not applicable to blocks compacted statically or dynamically by vibration or hammering. Nevertheless it has been found that test values of the standard test correspond very closely to those obtained in manual presses. The test is not valid for techniques such as adobe, cob, etc.

3. Influences

Compaction energy As compaction energy increases so does the dry density, while the OMC is reduced. In general the Proctor curve is more pronounced with increasing compaction and is flatter when the compaction energy is lower. Above the OMC and when the volume of air is negligible increasing compaction has little effect on dry density (γs), while below the OMC and for significant volumes of air the effect of increasing compaction is very evident. The elimination of air can never be totally achieved and thus nor can the maximum theoretical dry density be achieved (often fixed at 2650kg/m^3).

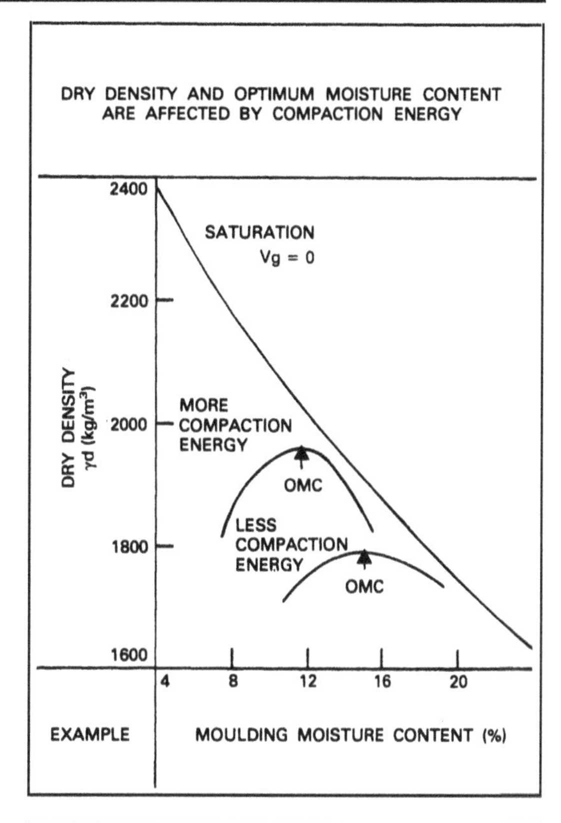

Moisture content Soil is difficult to compact at low moisture contents. Low values for γd are obtained and the volumes of air (Va) are high. An increase in moisture content lubricates the soil and makes it more workable: values for γd are higher and Va is reduced. γd max is obtained at the OMC. If moisture content is increased, Va will be decreased but the combination of water and air will prevent any appreciable reduction of the volume of air. The total volume of air and water increases and γd is reduced.

Texture γd max. depends on the type of the soil and its main characteristics where:

- mean grain size exceeds 50% (D50) on the grain size distribution curve.

- Clayey soils: γd max. = 2000kg/m^3.
- Sandy soils: γd max. = 2200kg/m^3.
- Gravelly soils: γd max. = 2500kg/m^3.

On average γd max. of a compacted soil ranges from 1700 to 2300 kg/m^3.

- grain size distribution: when grain size is uniform, porosity is higher and sensitivity to moisture content is reduced. The Proctor curve is then flattened. When there is a wide distribution of grain sizes the Proctor curve is more pronounced.

- the % of grains with a diameter Ø < 0.08mm makes variations in γd with water content more pronounced, and results in sharper curves.

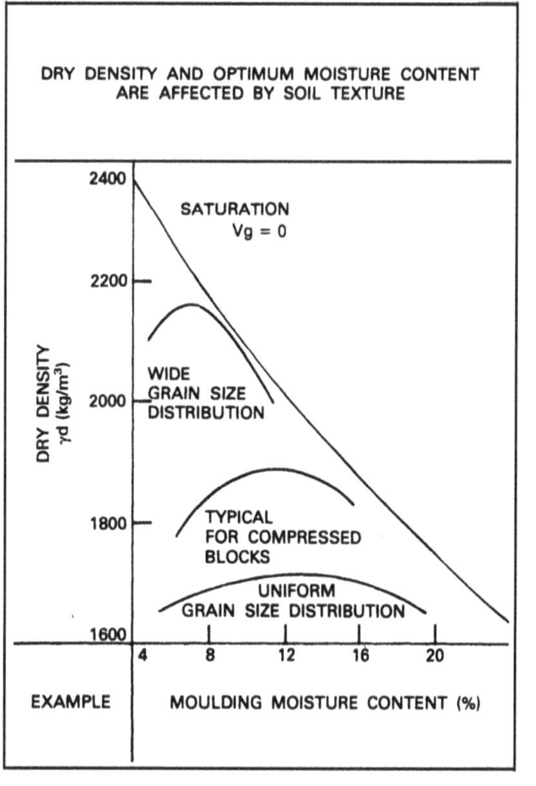

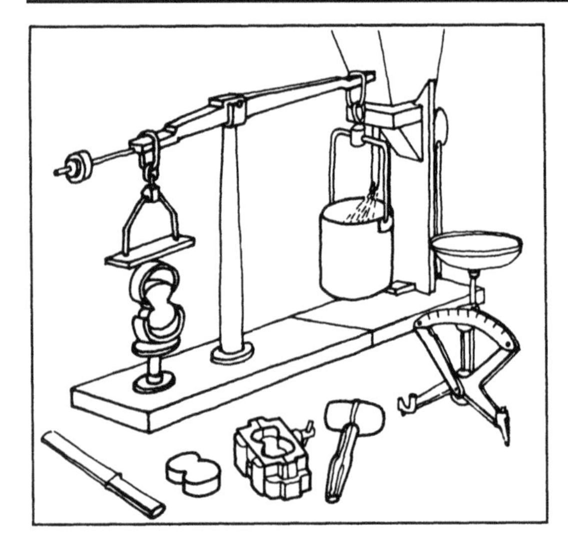

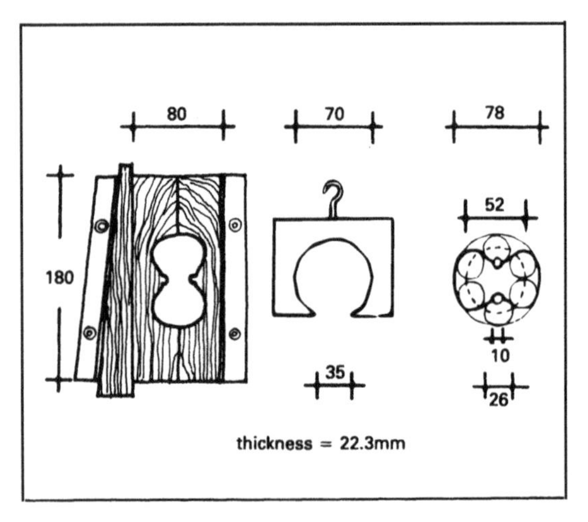

The cohesion test, which is also referred to as the wet tensile test or the 8 test (from the shape of the test sample), was developed in Germany by Niemeyer. It was first published in 1944 and was subsequently adopted by Wagner, and finally taken up in the DIN system of standards in 1956.

The soil is made up of two main grain fractions: inert material (Ø > 2mm) and coarse mortar (Ø < 2mm). If the coarse mortar is examined in order to obtain some idea of the mutual cohesion of grains, a tensile strength test must be carried out. In order to shorten the duration of this test it is carried out on a wet sample, which saves drying time. The test can be performed either in the laboratory or in the field. In the laboratory a balance with automatic loading can be used or a standing bracket with an automatic loading and cut-off device, or an entirely automatic apparatus with an automatic recorder. In the field a few planks and nails, bits of wire and suitable recipients are enough, and the sample is suspended from the branch of a tree, a table, a door frame, or similar. The sand used to load the sample can be replaced by a liquid such as water, or sump oil. The distance that the load falls should be kept to a minimum.

1. Method

The grain fraction with a diameter greater than 2mm is eliminated by sieving. The test is thus carried out on the coarse mortar after drying and crushing. The soil is crushed on a metallic plate (60 × 60cm) using a hammer (2.5 × 2.5cm), adding water at regular intervals, until a compact pancake with a plastic consistency is obtained. This pancake is then lifted with a knife and sliced into broad strips which are placed vertically on edge next to one another, and then again crushed with the hammer. The operation is repeated with a new pancake being obtained, which in turn is cut into slices and again crushed until the structure of the pancake is quite uniform on its lower surface and of the same humidity throughout. The soil, which must be perfectly dry prior to preparation, should then be allowed to rest for 12 to 24 hours in order to obtain the best possible dispersion of the moisture and consequently the best possible cohesion of the grains.

A sample of 200g of the prepared soil is taken. The density of this soil is further increased by kneading and pressing it several times agains a metal plate. The material is then rolled into a ball with a diameter of 50mm.

Prolonged manipulation should be avoided as this may result in local variations in the consistency of the ball. For lean soils the diameter of the ball should be reduced by between 0.5 and 1mm.

For fat soil the diameter should be increased by between 0.5 and 1mm. The diameter is to be checked with a ring. The next step is to allow the ball to fall from a height of 2m onto a smooth hard surface. The right consistency has been obtained when the flattened part of the ball is exactly 50mm thick. If it is larger or smaller the preparation must be corrected by allowing it to dry where Ø < 50mm (soil too plastic) or by adding a small quantity of water with a spray where Ø > 50mm (soil too dry).

The sample is then given its test form by shaping it in a lightly oiled figure-of-eight mould (section 22.5mm × 22.3mm = $5cm^2$); the prepared soil is vigorously tamped with a metal rod 20mm in diameter, in three fractions, until no further increase in density is observed.

The two sides of the sample are then equalized with a knife, without moistening them. The sample is removed forthwith by allowing the mould to drop on a hard surface from a height of 10cm. At least three samples should be prepared.

The sample is suspended from the measuring apparatus by a steel or hardwood hook with an inside diameter of 70mm, and an opening of 35mm (for steel hooks the section should be 22mm). Hanging from the sample is another hook which carries the recipient for the breaking load. The loading sand (0 = 0.2mm to 1mm) is supplied from a hopper at a rate of 3000g in 4 minutes; i.e. 12.5g per second or a maximum of 750g per minute, until the sample breaks. If the broken section contains a foreign body, the prepared soil is not uniform and the measurement must be rejected. The result for cohesion is the mean of three measurements with values which do not diverge by more than 10%. It is expressed in Pa (1mbar = 100Pa).

QUALIFICATION OF THE COHESION				
COHESION	MORTAR	QUALIFICATION	MASS ($cN/5cm^2$) ($g/5cm^2$)	TENSILE STRENGTH Pa
VERY LOW COHESION	SANDY	VERY LEAN	200 – 300	4000 – 6000
LOW COHESION	LEAN	VERY LEAN LEAN	300 – 400 400 – 550	6000 – 8000 8000 – 11000
MEDIUM COHESION	MEDIUM	NEARLY LEAN NEARLY FAT	550 – 750 750 – 1000	11000 – 15000 15000 – 20000
HIGH COHESION	FAT	FAT VERY FAT	1000 – 1350 1350 – 1800	20000 – 27000 27000 – 36000
VERY HIGH COHESION	CLAYEY	LEAN FAT VERY FAT	1800 – 2400 2400 – 3200 3200 – 4500	36000 – 48000 48000 – 64000 64000 – 90000

Mineralogical analysis

After determining the grain distribution of the soil, it is useful to study the minerology of the fine fraction in order to be able to determine the volumetric stability of the soil and its cohesion. This knowledge is essential when it is intended to stabilize the soil (physico-chemical reactions of minerals). Among the hundreds of minerals included in the fine fraction of the material, fewer than ten are of any great importance and directly concerned with construction in soil. Simple soil tests, based on visual observations followed by an Emerson test, allow preliminary evaluations to be made with a reasonable degree of accuracy. These tests, which cost little and are very effective, make it possible to select the samples which should be subjected to further laboratory analysis and can permit considerable savings in time and money in carrying out mineralogical analyses.

X-rays and DTA analysis

These extremely accurate analyses suffer the disadvantage of being very expensive (≃ $100 per sample). The samples analysed are very small (sieved to 40μ) and must be representative of the soil used. Only fully equipped laboratories can perform and interpret these analyses, which are chiefly qualitative and are rarely quantitative. The analysis of X-ray diffraction of heated samples makes it possible to measure the characteristic height and position of the diffraction fringes according to the scanning of the corresponding diffraction angles. In differential thermal analysis (DTA), the sample is heated from 20 to 900°C. Water losses and the effects they bring about, which are typical of various minerals, are recorded at these different temperatures. These expensive and sophisticated analyses must be limited to exceptional cases.

Observation of the environment

- Turbid water with a brownish yellow or brownish red colour: montmorillonites, illites, salts in the soil.
- Clear water: calcium, magnesium, very acid soil with a high iron content, sands.
- Clear water with mottled or bluish deposits: non-saline kaolins.
- Eroded gulleys or natural tunnels: saline clays, often montmorillonites.
- As above but less marked: kaolinites.
- Earth slips: kaolinites, chlorites.
- Surface micro-relief: montmorillonites.
- Granite pebbles: kaolinites, micas.
- Basalt pebbles and poorly drained landscape: montmorillonites.
- As above but well drained: kaolinites.
- Sandstone pebbles: kaolinites.
- Shale pebbles: montmorillonites or illites, perhaps saline soil.
- Calcareous pebbles: alkaline montmorillonites and chlorites with diverse properties.

Profile analysis

- Mottled clays, with patches of red, orange, white: kaolinites.
- Mottled clays, with patches of yellow, orange, and grey: montmorillonites.
- Clays in muted colours, dark greys to blacks: montmorillonites.
- Brown and reddish-brown clays: illite in appreciable quantities and a few montmorillonites.
- White and light-grey clays: kaolinites and bauxites.
- Discrete highly reflective micrograins: mica soils.
- Soft pieces easily dissolved in acid, widely distributed: carbonates.
- Hard pieces, reddish brown: iron ore, laterites.
- Numerous wide, long, deep cracks with little space between them (5 or 6cm at the most): illites rich in calcium and montmorillonites.
- Same type of cracks but 30cm or more apart: illites.
- Clayey soils, crumbly and of open texture, with a lot of clay: presence of carbonates, allophanes or kaolins, but never montmorillonites and rarely illite.
- The same type of soils as the previous ones, but of a black colour: organic soils, peat.
- The same type of soils, but with very little clay: carbonates, loams and sands.

Environmental and profile analysis

Observations made on natural and artificial slopes, trenches, holes with a predominantly clay mineral content; the colours of the soil and the clarity of surface water in the immediate vicinity is examined.

Emerson test

This test is performed on a small lump of unworked soil, about the size of a bean, taken directly from the soil it is planned to use. The sample is immersed in a translucent recipient containing extremely pure water – distilled water, or rain-water is suitable. No dispersant is added to the water. The behaviour of the immersed lump is observed after a few minutes in accordance with the procedure indicated in the diagram. This test makes a rough evaluation of the clay type possible.

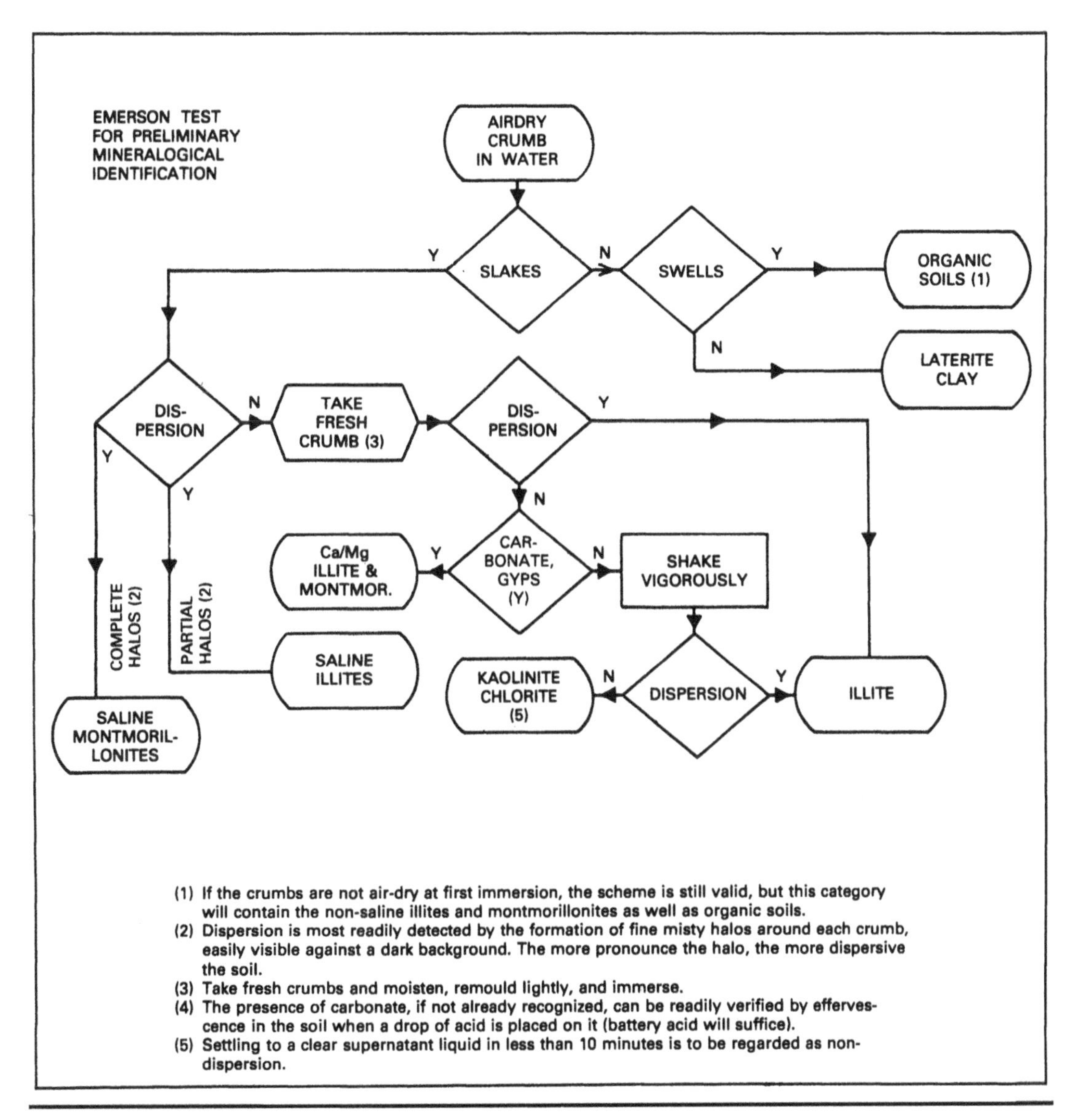

(1) If the crumbs are not air-dry at first immersion, the scheme is still valid, but this category will contain the non-saline illites and montmorillonites as well as organic soils.
(2) Dispersion is most readily detected by the formation of fine misty halos around each crumb, easily visible against a dark background. The more pronounce the halo, the more dispersive the soil.
(3) Take fresh crumbs and moisten, remould lightly, and immerse.
(4) The presence of carbonate, if not already recognized, can be readily verified by efferves-cence in the soil when a drop of acid is placed on it (battery acid will suffice).
(5) Settling to a clear supernatant liquid in less than 10 minutes is to be regarded as non-dispersion.

Chemical analysis

Laboratories which carry out chemical analyses of soil provide their answer in the form of a list of the various chemicals present and their quantity expressed as a percentage. A typical list includes the following:

- Iron oxides;
- Magnesium oxides;
- Aluminium oxides;
- Calcium oxides (*);
- Carbonates;
- Sulphates;
- Soluble and insoluble salts;
- Loss on ignition (*);
- Chemically bound water(*).

(*): These items of the chemical balance are not interpreted.

It may be important to learn about other items such as:

- the nature of organic matter and humus;
- pH (acidic or basic);
- ion exchange capacity.

The chemical analysis methods used in the laboratory are widely used but are still fairly complex. Not all of them are suitable for use in the field. Simple field applications are necessary a priori. They give reasonably accurate results and indicate whether it is worth proceeding to a further chemical analysis in the laboratory. These field tests make it possible to determine the presence of soluble salts and consequently the pH of a soil. If it is acid, it is because organic matter is present, such as iron salts. If it is basic, it is because it contains carbonates, sulphates, chlorides, and similar.

1. Soluble salts

The amount of these items present in a soil sample can be estimated fairly accurately by lessivage.

The sample is first of all dried in an oven and weighed and then lessivated with hot water. Afterwards it is dried again and reweighed. The amount of soluble salts present is given by the difference between the two weights.

2. pH

Methyl red is poured into a recipient containing a mixture of soil and water which has been prepared by stirring and then allowing to stand. The liquid above the soil turns red if the soil is acid and yellow if basic. Indicator paper can also be used.

3. Acid salts

A sample of the soil is stirred with distilled water in a recipient, and then allowed to stand. Indicator paper sensitive to acid salts is then used to detect their presence.

4. Organic matter and humus

The concentration of these in the soil can be determined from their musty odour or their colour which may be blackish, dark brown, blue or dark grey, or even dark green. Tests using standard colour cards or simplified colouring tests can also be used.

The standard test used for the detection of organic matter consists of stirring a mixture of soil and sodium hydroxide solution and then comparing the colour with a standard solution of tannic acid. Although this test is somewhat limited, it has nevertheless been accepted by the ASTM and the BSI. A humus detection test can also be carried out as follows:

- Prepare a solution of 300 to 400ml of sodium hydroxide (NaOH) or potassium hydroxide (KOH) diluted to 3% (9 to 12g). If no sodium hydroxide or potassium hydroxide is available, milk of lime can be prepared by sprinkling 1kg of fat lime into 3 litres of water. The milk of lime is collected after standing overnight by siphoning off.
- Between 50 and 100g of dried crushed soil is introduced into the prepared solution and mixed with care by shaking vigorously.
- The mix is allowed to stand for at least 24 hours, and 48 hours where milk of lime is used.
- After renewed vigorous shaking and a short rest period, the colour of the surrounding water is then observed: brown or black means a high humus content; while a pale pink or neutral colour means low humus content.

5. Basic salts

The mixture of soil and pure water turns violet when a drop (1%) of phenolphtalein is added to it – half a glass of soil + two-thirds of water. Basic salts can be also be tested for with indicator paper.

6. Carbonates

The presence of carbonates in a soil can be tested for by allowing a drop of 5% nitric acid or 1:3 hydrochloric acid (HCl) (1 vol HCl to 3 vol of distilled water) to fall on the soil sample. Any effervescence betrays the presence of carbonate.

Great care should be taken when handling these solutions, which can cause unpleasant burns to the skin.

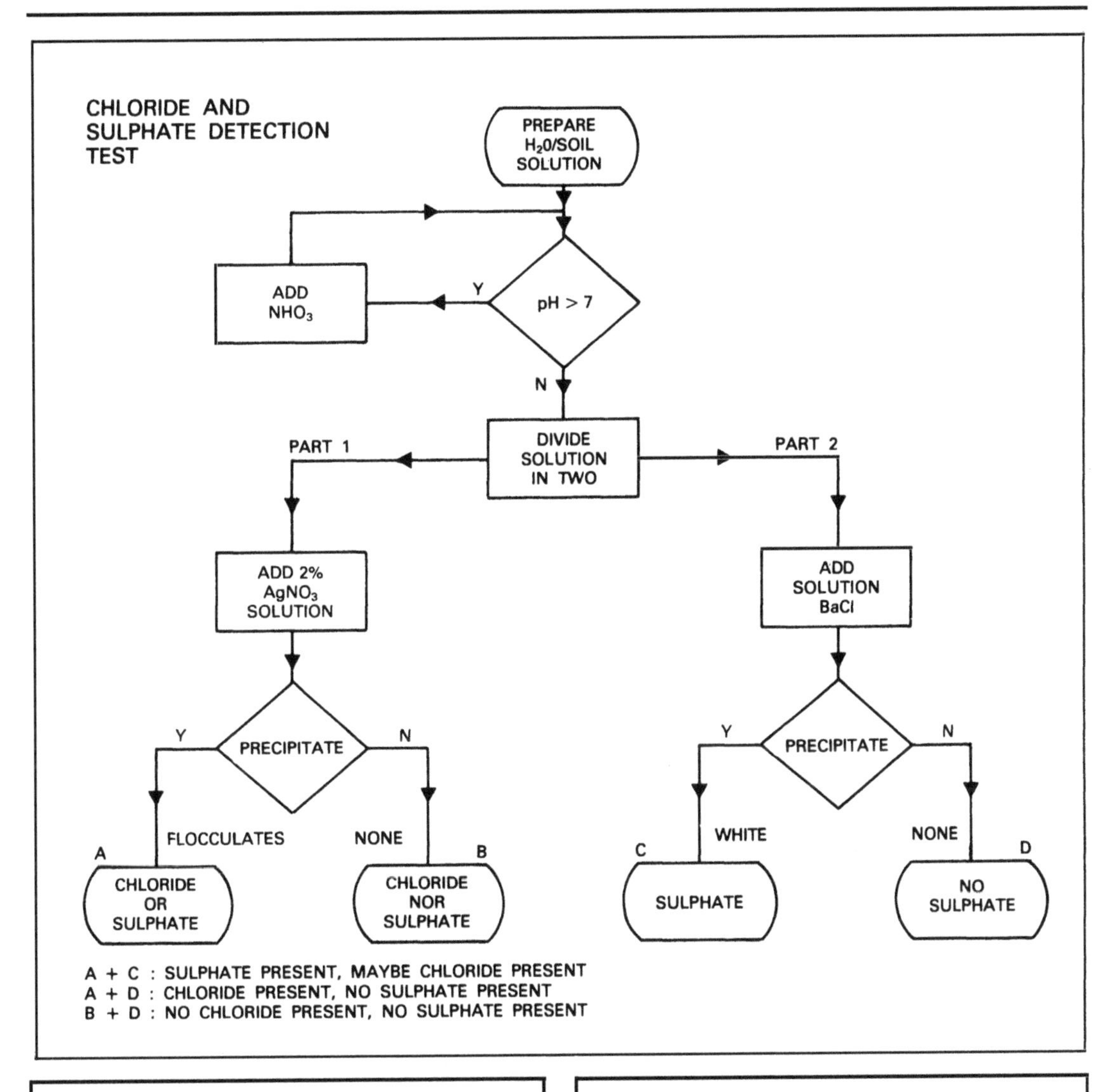

Chloride content

Fill a graduated burette with silver nitrate to the zero line. Pour 100ml of a solution of soil and distilled water into a graduated flask. Pour 3 drops of phenolphtalein/TA into it. If water turns pink, add oxalic acid drop by drop until the colour is gone. Using the burette add silver nitrate until a brick red colour is obtained. Let A be the number of degrees of silver nitrate used. Refill the burette to the zero line and repeat the operation with deionized water (control test). Let B be the number of degrees of $AgNO_3$ used. The difference between A and B is the concentration of chloride in the tested solution, expressed in degrees. This number multiplied by 7.1 gives the ion concentration of chloride expressed in mg/l.

Sulphate content

Prepare 10 standard solutions of distilled water and barium chloride (BaCl) in test tubes, prepared in the following percentages: 0.05, 0.1, 0.15, 0.2, 0.25, 0.3, 0.35, 0.4, and 0.5%. Dissolve 10g of soil in the distilled water, filter and then add a drop of barium chloride. The presence of the sulphates is indicated by a white precipitate. Compare the test with reference test tubes containing known quantities of sulphates and an equal quantity of barium chloride.

GEOTECHNICAL CLASSIFICATION USCS SYSTEM LABORATORY PROCEDURE				SYMBOLS	DESCRIPTION
More than half the elements have a diameter greater than 0.08mm COARSE SOILS	More than half the elements > 0.08mm have a diameter > 2mm GRAVELLY SOILS	Without fines	All diameters are represented, none predominates	GW	Clean gravel Well graded
			One grain size or one grain fraction predominates	GP	Clean gravel Poorly graded
		With fines	Fine elements have no cohesion	GM	Silty gravel
			Fine elements have cohesion	GC	Clayey gravel
	More than half the elements > 0.08mm have a diameter < 2mm SANDY SOILS	Without fines	All diameters are represented, none predominates	SW	Clean sand Well graded
			One grain size or one grain fraction predominates	SP	Clean sand Poorly graded
		With fines	Fine elements have no cohesion	SM	Silty sand
			Fine elements have cohesion	SC	Clayey sand
More than half the elements have a diameter less than 0.08mm FINE SOILS - CLAY AND SILT	CLAY AND SILT SOILS: Liquid Limit LL < 50%	Above A		CL	Low-plasticity clay
		Under A	Inorganic	ML	Low-plasticity silt
			Organic material	OL	Organic silt and clays with low plasticity
	CLAY AND SILT SOILS: Liquid Limit LL > 50%	Above A		CH	Highly plastic clay
		Under A	Inorganic	MH	Highly plastic silt
			Organic material	OH	Highly plastic organic silt and clay
Organic matter predominates. Can be recognized by smell, dark colour, fibrous texture, low moist density				Pt	Peat and other highly organic soils

NOTES

- This classification is a simplified and modified version
- Elements with a diameter greater than 60mm are not taken into account
- The weight of the grain fraction can be estimated
- The grain sizes correspond to grain size distribution chart.

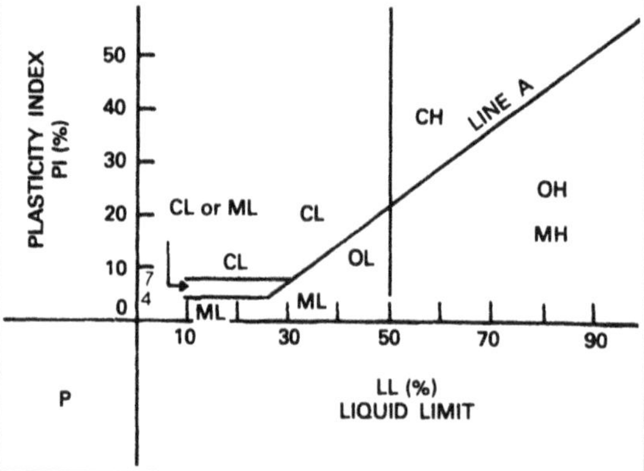

PLASTICITY INDEX PI (%)
50
40
30
20
10
7
4
0
CL or ML
CL
ML
CL
OL
ML
CH
LINE A
OH
MH
10
30
50
70
90
P
LL (%) LIQUID LIMIT

<table>
<tr><td colspan="5">GEOTECHNICAL CLASSIFICATION USCS SYSTEM
FIELD TEST PROCEDURE</td><td>SYMBOLS</td><td>DESCRIPTION</td></tr>
<tr><td rowspan="8">More than half the elements have a diameter greater than 0.08mm
$F < G + D$</td><td rowspan="4">More than half the elements $> 0.08mm$ have a diameter $\geq 2mm$
$G < D$</td><td rowspan="2">$F <<< G + D$</td><td colspan="2">All diameters are represented, non predominates</td><td>GW</td><td>Clean gravel
Well graded</td></tr>
<tr><td colspan="2">One grain size or one grain fraction predominates</td><td>GP</td><td>Clean gravel
Poorly graded</td></tr>
<tr><td rowspan="2">$F << G + D$</td><td colspan="2">Fine elements have no cohesion</td><td>GM</td><td>Silty gravel</td></tr>
<tr><td colspan="2">Fine elements have cohesion</td><td>GC</td><td>Clayey gravel</td></tr>
<tr><td rowspan="4">More than half the elements $> 0.08mm$ have a diameter $< 2mm$
$G > D$</td><td rowspan="2">$F <<< G + D$</td><td colspan="2">All diameters are represented, non predominates</td><td>SW</td><td>Clean sand
Well graded</td></tr>
<tr><td colspan="2">One grain size or one grain fraction predominates</td><td>SP</td><td>Clean sand
Poorly graded</td></tr>
<tr><td rowspan="2">$F << G + D$</td><td colspan="2">Fine elements have no cohesion</td><td>SM</td><td>Silty sand</td></tr>
<tr><td colspan="2">Fine elements have cohesion</td><td>SC</td><td>Clayey sand</td></tr>
<tr><td rowspan="6">More than half the elements have a diameter less than 0.08mm
$F \geq G + D$</td><td>MODERATE TO HIGH</td><td>NONE TO VERY SLOW</td><td>MEDIUM HARD</td><td>SHORT TO LONG</td><td>CL</td><td>Low-plasticity clay</td></tr>
<tr><td>NONE TO LOW</td><td>RAPID TO SLOW</td><td>FRAGILE TO NONE</td><td>SHORT TO NONE</td><td>ML</td><td>Low-plasticity silt</td></tr>
<tr><td>LOW TO MODERATE</td><td>SLOW</td><td>SOFT OR SPONGY</td><td>SHORT TO NONE</td><td>OL</td><td>Organic silt and clays with low plasticity</td></tr>
<tr><td>HIGH TO VERY HIGH</td><td>NONE</td><td>HARD</td><td>LONG</td><td>CH</td><td>Highly plastic clay</td></tr>
<tr><td>LOW TO MODERATE</td><td>SLOW TO NONE</td><td>MEDIUM HARD</td><td>SHORT</td><td>MH</td><td>Highly plastic silt</td></tr>
<tr><td>MODERATE TO HIGH</td><td>NONE TO VERY SLOW</td><td>FRAGILE TO SOFT</td><td>SHORT</td><td>OH</td><td>Highly plastic organic silt and clay</td></tr>
<tr><td>TESTS</td><td>DRY STRENGTH</td><td>WATER RETENTION</td><td>CONSIS-TENCY</td><td>COHESION</td><td></td><td></td></tr>
</table>

NOTES

- The preliminary geotechnical classification of a soil according to P&CH and USCS is based on visual observations and proceeds by elimination, starting on the left of the classification table and going to the right, where the symbol of the group is obtained. A description has to complete the general symbolic designation; for instance CL: Low-plasticity clay, sandy with predominantly fine sands, presence of a few gravels.
- Many soils have characteristics which do not classify them in a precise group but in two of more groups or at the limit of two groups, due to the size of their grains or to their plasticity. Some distinct classifications use double symbols from two large groups, linked by a hyphen, for those soils; e.g. GW-GC, SC-SL, ML-CL, etc.

3.13 REFERENCES

Agib, A.R.A.; El Jack, S.A. 'Foundations on Expansive Soils'. in *Digest*, Khartoum, NBRS, 1976.
Beidatsch, A. *Wohnhaüser aus Lehm*. Berlin, Hermann Hübener, 1946.
Bertram, G.E.; La Baugh WM.C. *Soil Tests*. Washington, American Road Builders' Association, 1964.
Bureau of reclamation. *Earth Manual*. Washington, US Department of the Interior, 1974.
CINVA. *Le Béton de Terre Stabilisé; Son Emploi dans la Construction*. New York, UN, 1964.
Department of Housing and Urban Development. 'Earth for Homes'. In *Ideas and Methods Exchange*, Washington, Office of International Affairs, 1955.
Doat, P. *et al. Construire En Terre*. Paris, éditions Alternatives et Parallèles, 1979.
Hays, A. *De la Terre pour Bâtir. Manuel Pratique*. Grenoble, UPAG, 1979.
Hernández Ruiz, L.E.; Márquez Luna, J.A. *Cartilla de Pruebes de Campo Para Selección de Tierras en la Fabricación de Adobes*. Mexico, CONESCAL, 1983.
Houben, H. *Technologie du Béton de Terre Stabilisé pour l'Habitat*. Sidi Bel Abbes, CPR, 1974.
Ingles, O.G.; Metcalf, J.B. *Soil Stabilization*. Sydney, Butterworths, 1972.
Knaupe, W. *Erdbau*. Düsseldorf, Bertelsmann, 1952.
Markus, T.A. *et al. Stabilised Soil*. Glasgow, University of Strathclyde, 1979.
Peltier, R. *Manuel du Laboratoire Routier*. Paris, Dunod, 1969.
Pollack, E.; Richter, E. *Technik des Lehmbaues*. Berlin, Verlag Technik, 1952.
Post, G.; Londe, P. *Les Barrages en Terre Compactée*. Paris, Geuthier-Villars, 1953.
Sharma, S.K.;Vasudeva, P.N. *Making Soil Stabilized Bricks*. Chandigarh, Punjab Eng. College.
Van Olphen, H. *An Introduction to Clay Colloid Chemistry*. New York, John Wiley and Sons, 1977.
Vatan, A. *Manuel de Sédimentologie*. Paris, Technip, 1967.
Volhard, F. *Leichtlehmbau*. Karlsruhe, CF Müller GmbH, 1983.
Wagner, W. *Anleitung zur Untersuchung und Beurteilung von Baulehmen*. Dotzheim, Hessischen Lehmbaudienst Wiesbaden, 1947.
Wolfskill, L.A. *et al. Bâtir en Terre*. Paris, CRET.

Janadriyah exhibition centre of the Royal Commission for Jubail and Yanbu, Saudi Arabia, built of compressed earth blocks (architect: Ibrahim Aba Al-Khail, CRATerre-EAG) (Thierry Joffroy, CRATerre-EAG)

4. SOIL STABILIZATION

It is possible to considerably improve the characteristics of many types of soil by adding stabilizers to them. Even so, every variety of soil requires an appropriate stabilizer. Currently more than a hundred products are used for stabilizing building soils and earth. These stabilizers can be used in the bulk of earth walls or in their 'skin' as a surface protection. Stabilization is very ancient, but it was only in 1920 that a scientific approach could be developed. Major research was carried out in the three decades immediately after the Second World War, and work still goes on nowadays. Despite the research effort, soil stabilization is not an exact science and there is no 'miracle' stabilizer which can applied indiscriminately.

The best known and most practical stabilization methods are increasing the density of the soil by compaction, reinforcing the soil with fibres, or adding cement, lime, and bitumen. Many other products exist, which have either been suggested or even tried. Most of these products do not, however, have the benefit of an adequate research effort. The way in which they stabilize and their efficiency are not adequately known. When dealing with a stabilization problem, it is essentially a matter of choosing the best product from a wide variety of possibilities, many of which cannot even be considered either because of their ineffectiveness or because of their prohibitive cost.

At present, there is a trend to systematic stabilization; often, sad to say, using uncontrolled methods. This is regrettable as it must be borne in mind that stabilization only helps to improve the characteristics of a soil or to provide it with characteristics which it may lack. It is particularly unfortunate that many practitioners of systematic stabilization do not know, or do not appreciate, the original characteristics of a soil, and start about stabilizing soil with undue haste when it is not particularly useful.

Fundamental problem

Building in soil implies a choice between three main approaches:
- Using the soil available on the site and adapting the project as far as possible to the quality of the soil;
- Using another soil more suited to the requirements of the project but which has to be brought to the site;
- Modifying the local soil so that it is better suited to the requirements of the project.

The third possibility is generally referred to as soil stabilization and comprises the entirety of techniques permitting the improvement of the properties of the soil.

Definition

Stabilizing a soil implies the modification of the properties of a soil-water-air system in order to obtain lasting properties which are compatible with a particular application. Stabilization is nevertheless a complex problem, as an extremely large number of parameters are involved. In fact a knowledge of the following is necessary:
- the properties of the soil requiring stabilization;
- the planned improvements;
- project economy: costs and delays involved in soil stabilization;
- the soil construction techniques chosen for the project and the system of construction;
- maintenance of the completed project; maintenance cost.

The improvement of the properties of the soil by stabilization will be successful if the procedure used is compatible with the imperatives of the programme: in particular the cost of and delays in construction, and the cost of maintenance.

Objectives

Action can be be directed at only two characteristics of the soil itself, i.e. its texture and structure. There are three courses of action which can be aimed at texture and structure:
- Reducing the volume of interstitial voids: acts on porosity;
- Filling the voids which cannot be eliminated: acts on permeability;
- Improvement of the bonding between grains: acts on mechanical strength.

The main objects aimed at are:
- Achieving better mechanical characteristics: improving dry and wet compressive strength; tensile strength and shearing strength;
- Achieving better cohesion;
- Reducing porosity and changes in volume: shrink and swell due to water;
- Improvement of resistance to wind and rain erosion: reduction of surface abrasion and waterproofing.

Procedures

There are three basic stabilization procedures:

1. Mechanical stabilization The compaction of the soil resulting in changes in its density, mechanical strength, compressibility, permeability and porosity.

2. Physical stabilization The properties of the soil can be modified by acting on its texture: e.g. the controlled mixing of different grain fractions. Other techniques can involve heat treatment, drying or freezing, electrical treatment, electro-osmosis to improve the draining qualities of the soil, and giving new structural qualities.

3. Chemical stabilization Other materials or chemicals are added to the soil thus modifying its properties, either by a physico-chemical reaction between the grains and the materials or the added product, or by creating a matrix which binds or coats the grains. A physico-chemical reaction can lead to the formation of a new material: for example a pozzolana resulting from a reaction between clay and lime.

When is stabilization required?

Stabilization is not compulsory. It can be ignored quite satisfactorily and the soil can be used for construction without stabilizing.

Nevertheless, there is clearly a tendency at present to the oversystematic use of stabilization, which is regarded as a universal panacea for all problems. This attitude is unfortunate as stabilization can involve considerable extra costs, ranging from 30 to 50% of the final cost of the materials. Furthermore stabilization complicates the production of the material: such as longer preliminary studies of the behaviour of the material.

It is thus advisable to insist that stabilization is only used when absolutely essential and that it should be avoided where economic resources are limited.

If the dangers of exposure to water are taken into consideration, we can say:

Do not stabilize when the material is not exposed to water: sheltered walls, dressed walls, inside walls, architecture designed to accommodate the exigencies of soil as a building material.

Stabilize when the material is very exposed: poorly designed architecture which ignores the rules of building with earth; or requirements imposed by the site: such as a damp site, walls exposed to driving rain, and so on.

However, there may be other reasons for stabilizing the material such as:

- the improvement of the soil's compressive strength;
- raising the material's bulk density, or even reducing it.

Methods of stabilization

- Where the soil is not disturbed stabilization is usually carried out by grouting or impregnation. This technique is, however, rarely used for earth structures, and is by and large limited to public works, engineering structures, improvement of foundations, and for the preservation of historic monuments.
- When the soil is reworked any of numerous stabilization procedures can be used and the literature describes various classification systems. These may class the method by the sort of stabilizer, animal or vegetable, mineral or synthetic, etc. or by the appearance of the stabilizer: powder, fibres, platy, paste, liquid, etc.

Classification can be simplified by enumerating the six basic methods of stabilization:
- raising density
- reinforcement
- linking
- binding
- waterproofing
- water repellent treatment.

The stabilizer does not necessarily act by one of these methods alone but may combine several of them.

<table>
<tr><th colspan="7">STABILIZATION MODES FOR DISTURBED SOILS</th></tr>
<tr><th colspan="2">STABILIZER</th><th>NATURE</th><th>METHOD</th><th>MODE</th><th>PRINCIPLE</th><th>SYMBOL</th></tr>
<tr><td colspan="3">WITHOUT STABILIZER</td><td>MECHANICAL</td><td rowspan="2">DENSIFICATION</td><td rowspan="2">CREATE A DENSE MEDIUM, BLOCKING PORES AND CAPILLARITY</td><td rowspan="2"></td></tr>
<tr><td rowspan="6">WITH STABILIZER</td><td rowspan="2">INERT STABILIZER</td><td>MINERALS</td><td rowspan="2">PHYSICAL</td></tr>
<tr><td>FIBRES</td><td>REINFORCEMENT</td><td>CREATE AN ANISOTROPIC NETWORK, LIMITING MOVEMENT</td><td></td></tr>
<tr><td rowspan="4">PHYSICO-CHEMICAL STABILIZER</td><td rowspan="2">BINDERS</td><td rowspan="4">CHEMICAL</td><td>CEMENTATION</td><td>CREATE AN INERT MATRIX OPPOSING MOVEMENT</td><td></td></tr>
<tr><td>LINKAGE</td><td>CREATE STABLE CHEMICAL BONDS BETWEEN CLAY CRISTALS</td><td></td></tr>
<tr><td rowspan="2">WATER-PROOFERS</td><td>IMPERVIOUSNESS</td><td>SURROUND EARTH PARTICLES WITH A WATERPROOF FILM</td><td></td></tr>
<tr><td>WATERPROOFING</td><td>ELIMINATE ABSORPTION AND ADSORPTION</td><td></td></tr>
</table>

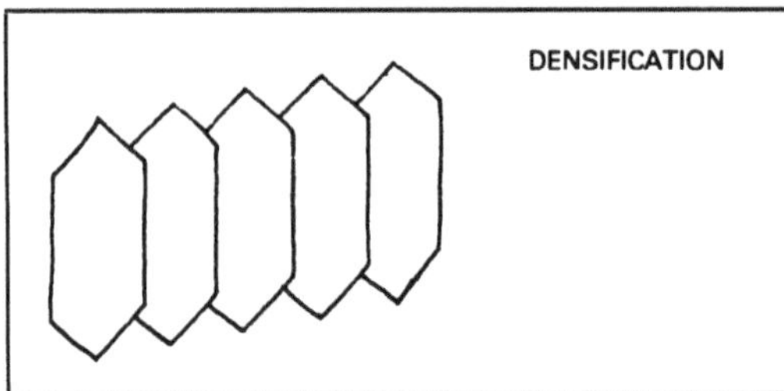

There are two different ways in which density can be increased:
1. By mechanically manipulating the soil, so that a maximum of air can be eliminated; by kneading and compressing the soil. Grain size distribution is not affected but its structure is changed because the grains are redistributed. The

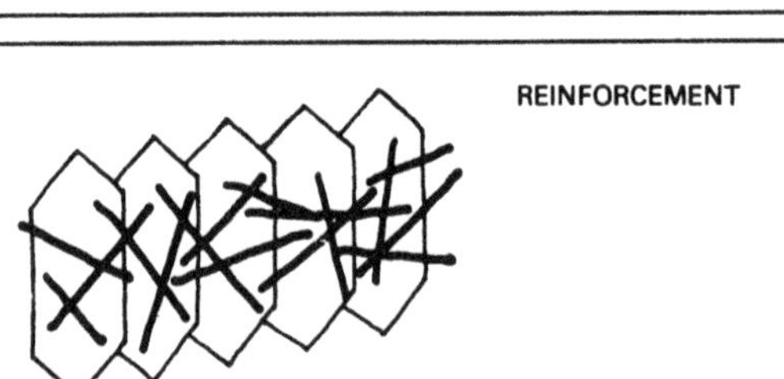

If there are reasons for not acting on the grain size distribution of the soil and if the material remains over sensitive to movements induced by various causes such as compression, tension, water action, thermal expansion, and so on, these movements can be counteracted with an armature. This armature can made be of a wide variety of fibres: animal,

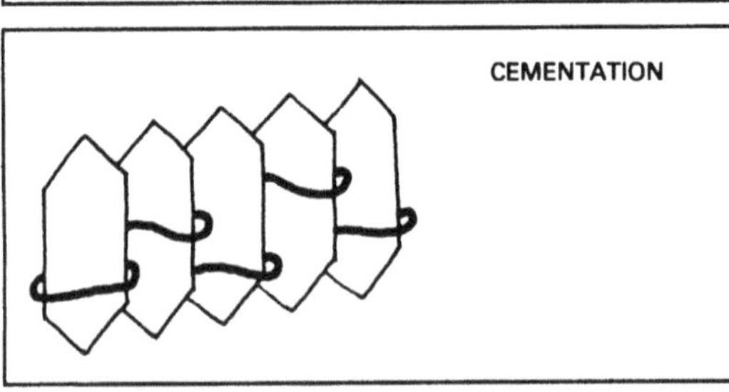

A three-dimensional matrix can be introduced in the soil. Strong and inert it resists all movements of the soil. This is in essence the consolidating action of cementing. It results in the filling of the voids with an insoluble binder which coats the grains and holds them in an inert matrix. The main stabilizer which acts by this mechanism is Portland

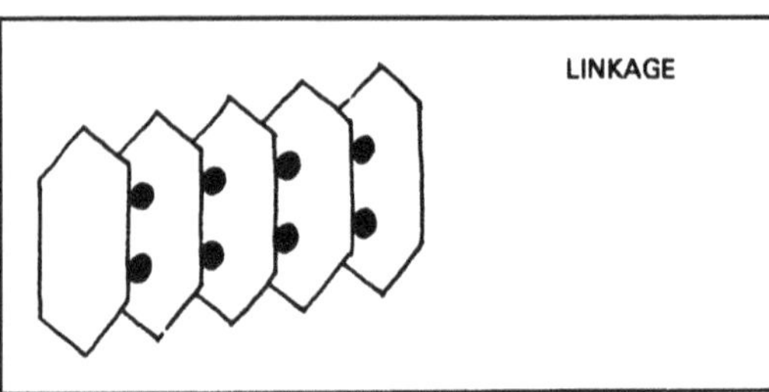

In this case the inert matrix introduced in the soil includes the clays. Two mechanisms are known and they give the same result.
1. An inert matrix is created by the clays: use is made of the negative and positive charges of the clay plates or their chemical composition to bind them together by means of a stabilizer, which plays

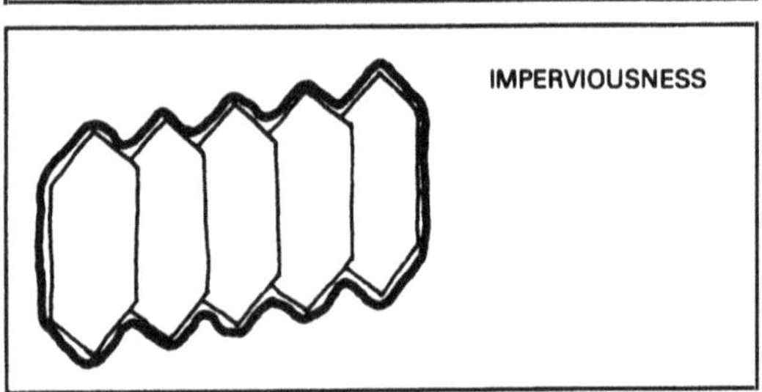

This stabilization method helps to reduce water erosion, swell, and shrink when the material is subjected to successive wetting and drying cycles. There are two possible waterproofing methods:
1. All the voids, pores, cracks and microcracks are filled with a material which is unaffected by water. Bitumen is one of the best examples of a product

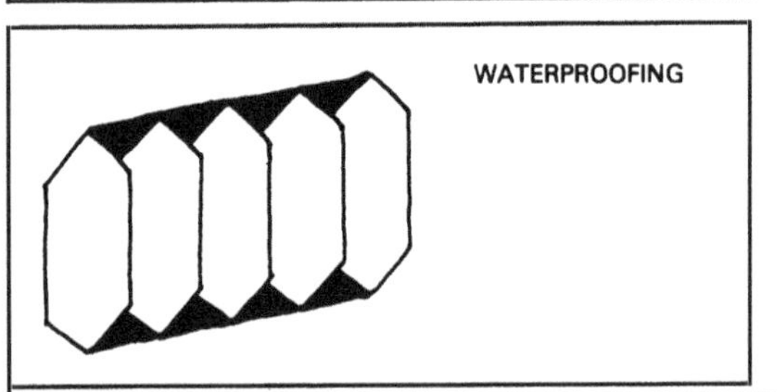

Here the object of the procedure is the movement of water and water vapour in the soil. This can be achieved by either changing the nature of this water, or reducing the sensitivity of the clay plates to water.
Three systems are used:
1. The modification of the state of the pore water:

soil is not simply compressed in its original state, it is first ground to make it more uniform, and then compressed. After the grinding phase, it is possible – but not essential – to make use of dispersants or waxes which may facilitate compaction.
2. By filling the voids as far possible with other grains. If this second method is to be used, grain size distribution must be perfect: the void left between each group of grains is filled by another group of grains. This method acts directly on grain size distribution.

vegetable, mineral or synthetic. The fibre armature has its effect at the macroscopic level, that is to say at the level of grain aggregations rather than at the level of the individual grains.

cement. Similar results can be achieved with electrolytic solutions of sodium silicate salts or with certain resins and adhesives. Viewed as a chemical reaction, the essential characteristic of this stabilization mechanism is that the formation of the inert matrix is relatively independent of the clay. In fact, the main consolidating reactions take place in the stabilizer itself and between the stabilizer and the sandy fraction of the soil. Even so a secondary reaction between the stabilizer and the clayey fraction may be observed. The quantity and quality of the clay has an effect on the efficiency of the stabilization procedure and may alter the mechanical behaviour of the material.

the role of binder or catalyst for this bond. Certain chemical stabilizers act in this way, including certain acids, polymers, flocculants, etc.
2. An inert matrix is formed by the clay. A stabilizer reacts with the clay and precipitates a new inert and insoluble material, which is a sort of cement. This is a pozzolanic reaction and is obtained primarily with lime. The reaction proceeds slowly and depends essentially on the quantity and quality of the clay.

which acts in this way. This stabilization method is particularly suitable for sandy soils with a good stability for their volume and which are not much affected by the movement of water. It is equally suitable for silty and clayey soils which need larger amounts of stabilizer because of their greater specific surface.
2. A material is dispersed in the soil which expands upon the slightest contact with water and prevents the infiltration of pores. A typical material of this type is bentonite.

by drying the soil by introducing calcium chloride into it. This raises surface tension, reduces the vapour pressure of the water and the evaporation rate and also reduces variation in moisture content.
2. Ion exchange: ions are replaced by others until the ions are very well fixed to the clay plates and the water can no longer dilute them. Certain acids can give rise to this phenomenon.
3. Molecules are fixed on one of the extremities of the clay plates on the outside of compact aggregates. The other ends of these molecules are water repellent. Certain quaternary amines and resins work in this way.

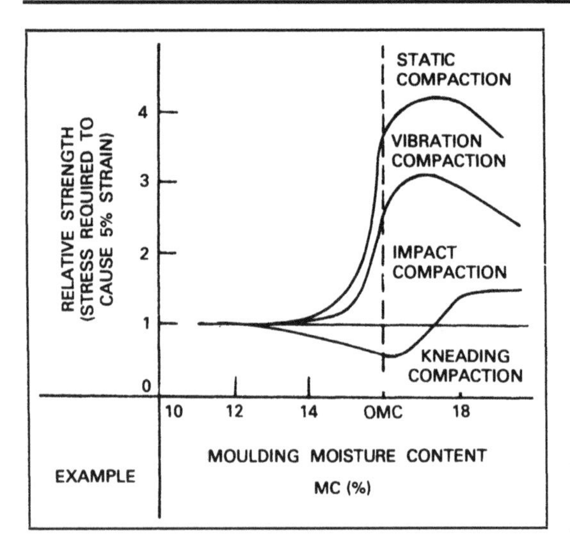

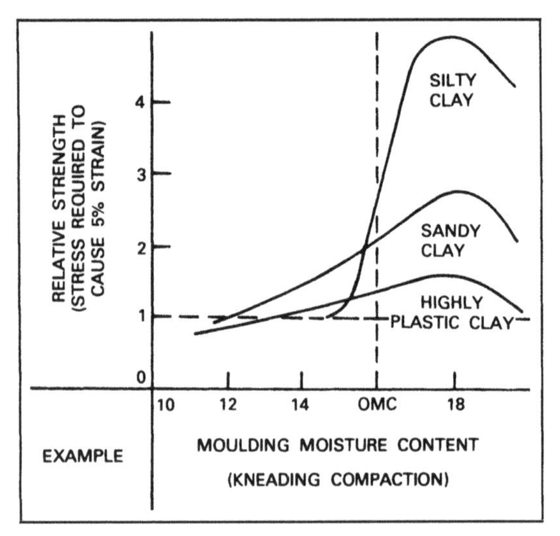

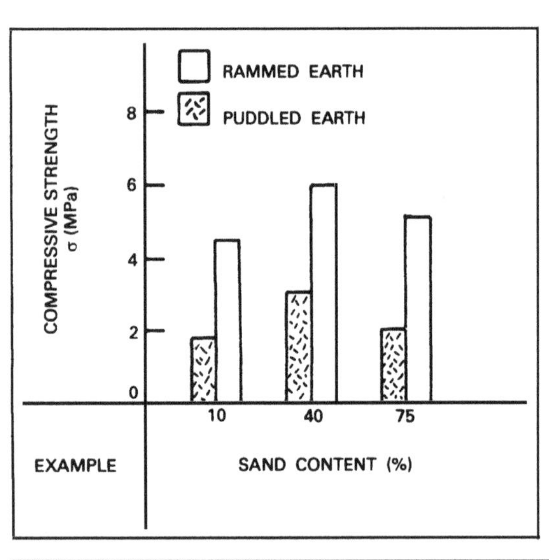

There is a direct relationship between dry density and compressive strength for all earth materials. The more compact the material is the higher the value becomes. In the breaking process, the strain always takes the path of least energy, which is usually marked by the largest pores. It is thus best for the pores in the material to be very small and extremely well distributed. Naturally compact soils which can be cut as blocks are rare. More usually the available soil must be worked. The care taken in the working process, which influences pore distribution and their size, has a direct effect on the behaviour of the material.

Methods of compaction

There are four fundamental methods:

- Static compaction;
- Dynamic compaction by vibration;
- Dynamic compaction by impact;
- Compaction by kneading.

There is an Optimum Moisture Content (OMC) for each of these methods and there is a corresponding optimum dry density.

Each method of compaction has its advantages and drawbacks. Kneading for example favours good pore distribution and uniform size in the final product, which helps to reduce permeability. However shear strength is lower when the compacted material is on the moist side, than when other means of compaction are used. The grain size distribution of the soil also plays a great part in determining this strength.

Compaction parameters

1. Compaction energy Whatever the type of soil and method of compaction, greater compaction energy reduces the OMC and increases dry density. But higher compaction energies can have harmful consequences such as the layering of blocks.

2. Grain size distribution Close grading does not yield high compaction rates. In contrast, wide grading gives compaction curves with marked maxima.

Effects of compaction

Two main types of clay structure are observed depending on the magnitude of the forces of repulsion or attraction:

1. Dispersed structure The plates of clay are distant from one another because the repelling forces predominate. They have the tendency to be parallel. The dispersed state corresponds to high moisture contents, to the right of the optimum value on the compaction curve.

2. Flocculate structure The plates of clay approach one another and form sharp angles to another because the attractive forces predominate. The flocculate state corresponds to low moisture contents, to the left of the optimum value on the compaction curve.

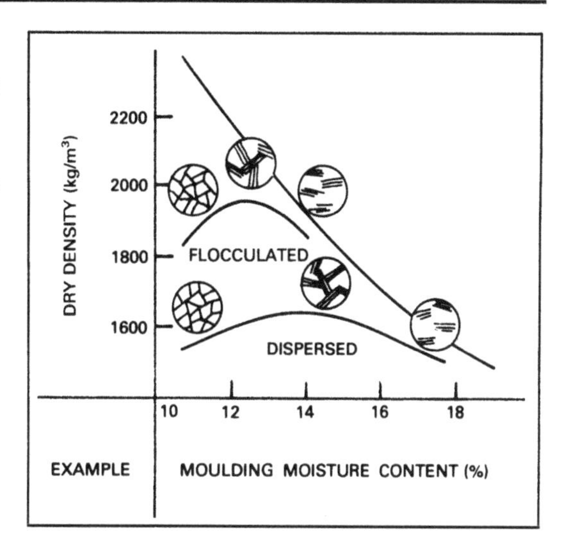

Thus, optimum compaction is the state at which the attracting forces are adequate to allow proper compaction while the repelling forces facilitate the orderly arrangement of the grains. Compaction carried out in good conditions allows reductions in permeability and compressibility to be achieved as well as limiting water absorption and swell in moist conditions. It also improves the initial and long-term mechanical strength of the material.

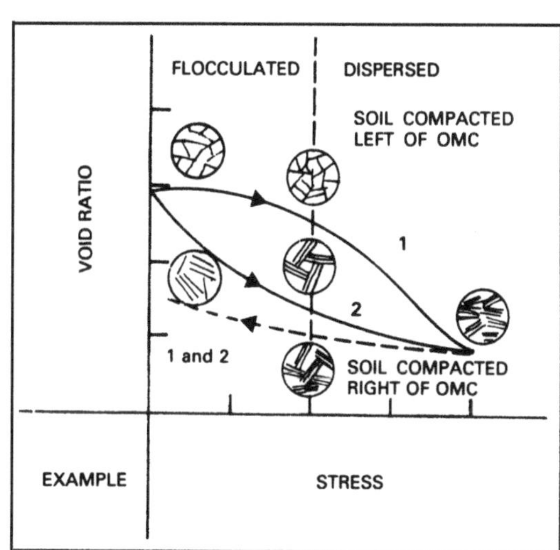

1. Compressibility

As long as the stress exerted does not exceed a certain value, a material with a flocculate structure is less compressible than the same material with a dispersed structure. The two materials tend to move towards the same state and the same compactibility, if the compaction energy becomes enough to bring about the rearrangement of the grains.

2. Absorption of water and swell in moist conditions

These are more significant for material compacted in the flocculate state and less for the dispersed state.

3. Excessive compaction

When a soil is close to saturation the incompressibility of water renders any effort to increase compaction illusory, as this has no effect on the arrangement of the grains.

Note Compaction has a great effect on the success or failure of all stabilization methods. However, it should not be forgotten that improvements obtained by compaction are virtually suppressed in moist environments.

INFLUENCE OF MOULDING MOISTURE (MC) ON:	COMPACTED AT	
	MC < OMC	MC > OMC
STRUCTURE	FLOCCULATED	DISPERSED
SWELLING	MORE	LESS
ABSORPTION	MORE	LESS
PERMEABILITY	MORE	LESS
GREEN STRENGTH	MORE	LESS
FINAL STRENGTH	MORE	LESS
DENSITY	MORE	LESS

Desired grain size distribution

In order to obtain the greatest mechanical strength and resistance to water action the following must be done:

- reduce the voids ratio;
- increase contacts between grains.

For spherical grains it is possible to calculate the relative proportion of each grain fraction of a different diameter, when arranged in the densest possible environment. The Fuller formula is used:

$$p = 100\,(d/D)^n.$$

where
p = the proportion of grains of a given diameter
d = the diameter of grains for a given value of p
D = the largest grain diameter
n = the grading coefficient
When the grains are entirely spherical n = 0.5.

In soil, however, while sand and gravel may perhaps have a shape which approximates the spherical, clays tend to elongate. Furthermore in road engineering, where sandy soil is often used, the lack of sphericity is corrected by using an n of 0.33. In earth construction an n of 0.20 to 0.25 is assumed.

Correction of grain size distribution

It is possible to correct an excessively large or small content of gravel, sand, or fines either by adding the missing fractions or by reducing the excessive fractions.

1. Excessive gravel content

To correct an excessively high gravel content, it is enough to remove the coarsest elements by sieving. The removal by hand of the largest stones may perhaps be necessary.

2. Excessive fines content

Soils of this sort can be improved by washing fine material out. Nevertheless this technique is very difficult to control as there is a danger of removing all fines. A better approach is to wash a certain quantity of the soil first, allow it to dry, and then mix it with the original soil. This is, however, a delicate procedure. Furthermore, it is usually better to mix the original soil with a soil containing more coarse grains, and which contains neither fines nor grains with a size greater than that permitted.

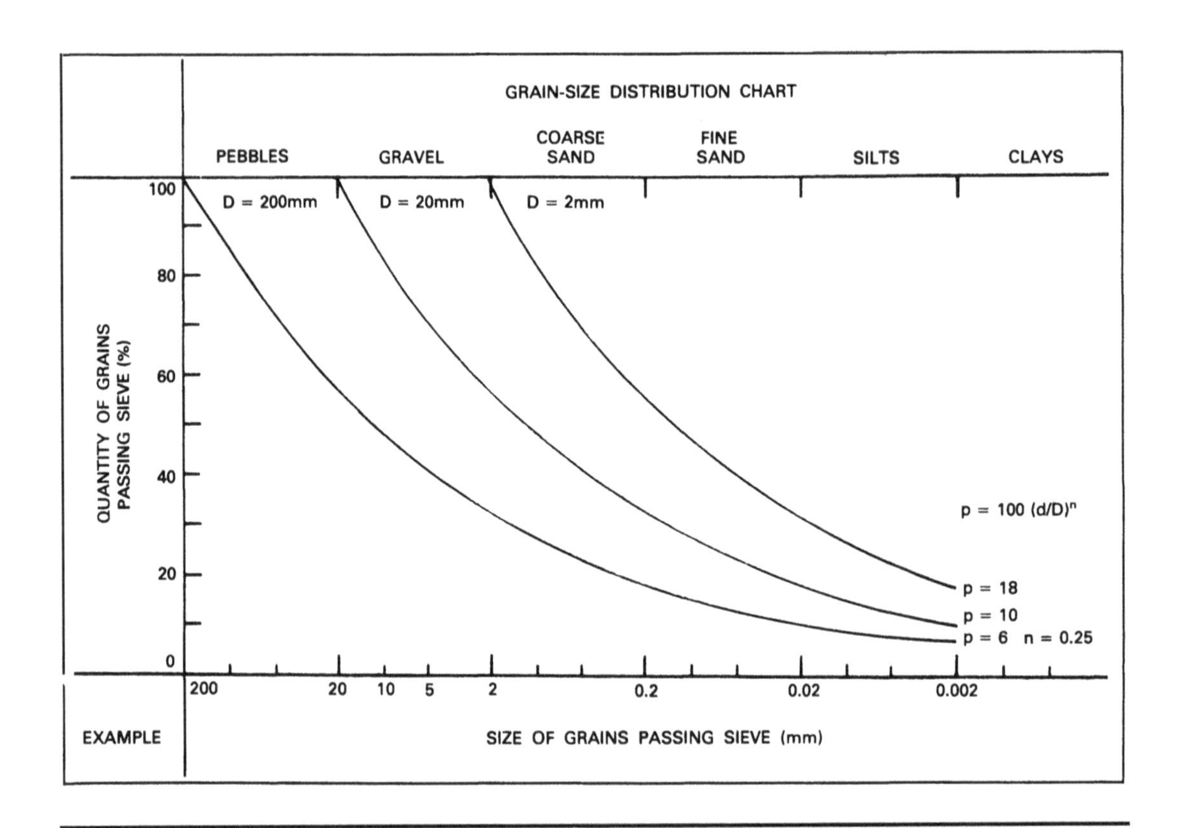

3. Discontinuously graded soil

Discontinuously graded soil can be quickly recognized from its typical grain size curve. There are in fact two characteristic curves.

– The curve may be flat for a particular grain fraction, which means that this fraction is missing in the analysed soil. It is then advisable to add the elements of this missing fraction in the correct proportions.

– The curve rises very abruptly for a particular grain fraction. In this case this fraction is too abundant in the soil analysed. It is then advisable to remove a part of this fraction by screening or by balancing the excess by adding to the other fractions. This procedure should never be carried out unless there is no other way of obtaining the desired quality of material.

4. Very sandy or very clayey soil

If the available soils are very different and markedly sandy or clayey, it will be necessary to mix them. The procedure is to draw the curves for the sandy and clayey soils and the desired optimal curve all on the grain size distribution curve. The lowest point on the sandy soil curve and the highest point on the clayey soil curve are joined by a straight line. The ordinate of the point of intersection of this line and the optimal curve gives an indication of what proportion of fine soil is to be mixed with the coarse soil in order to obtain a texture which approaches that of the optimal curve.

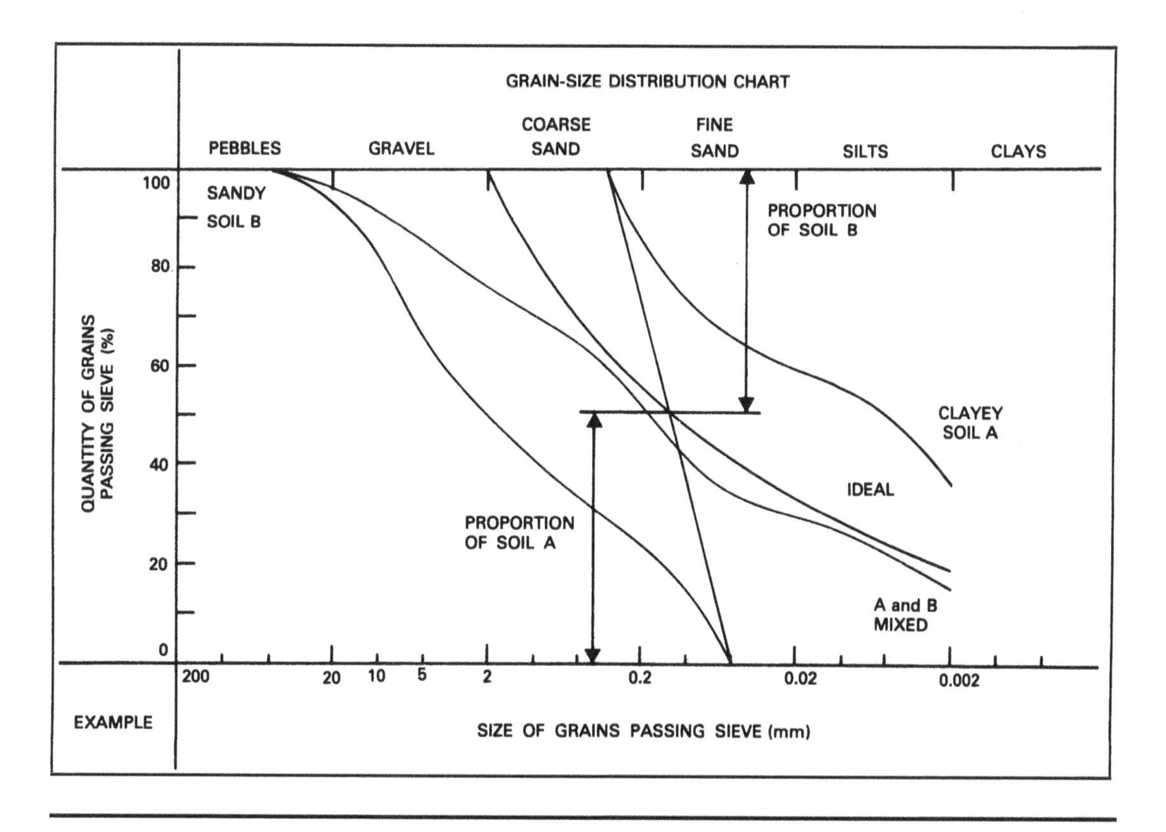

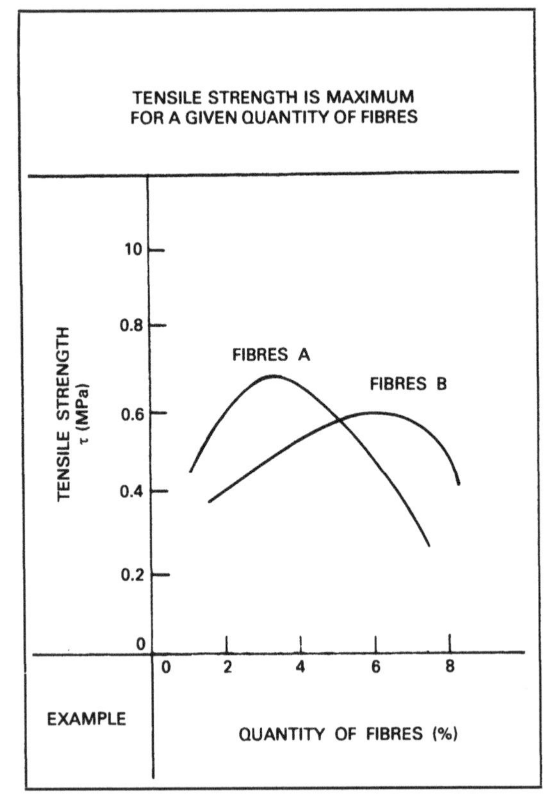

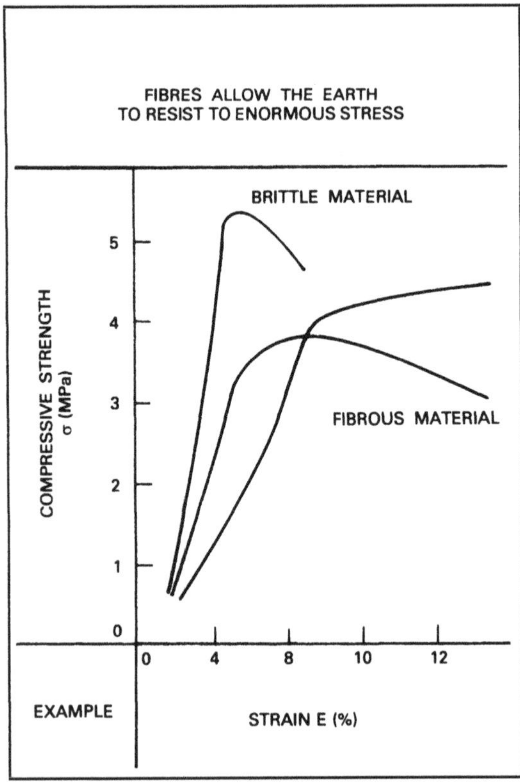

Stabilization by means of fibre reinforcement, very often straw, is widely used throughout the world. Straw should in fact be regarded as a structural reinforcement agent, similar to gravel. Nowadays, even in the most modern and industrial settings, such as in adobe production in the US, straw combined with bitumen is often added to the soil. This method of stabilization is interesting because it can be adapted to various methods of execution, with soil used in the liquid or plastic state, or even with compaction techniques. Fibres are mainly employed for making kneaded blocks with largely clayey soils and which usually suffer a considerable measure of shrink. Craft production of straw-stabilized adobe brick varies greatly, and quite apart from this straw is also used for daub, clay-straw, cob, compressed blocks and rammed earth.

Role of fibre

- Fibres hinder cracking upon drying by distributing the tension arising from the shrinkage of the clay throughout the bulk of the material.
- Fibres accelerate drying because they improve the drainage of moisture towards the outer surface through the channels afforded by the fibres. On the other hand fibres increase absorption in the presence of water.
- Fibres lighten the material. The volume of straw is often very large, reducing the bulk density and improving its insulating properties.
- Fibres increase tensile strength. This is undoubtedly the most interesting property of fibres.

Mechanisms

Earth materials reinforced with fibres can stand up very well to cracking and resist crack propagation. Considered at the level of a potential crack, the fibre opposes the formation of crack in step with the increase in the stress. The degree of shear strength depends largely on the tensile strength of the fibres. Apart from this, good compressive strength can also be achieved with fibre reinforcement, which depends both on the quantity of fibre used and on the initial compressive strength of the soil, the initial tensile strength of the fibres and the internal friction between the fibres and the soil. Some research seems to suggest that preliminary rotting of the straw in the soil for a period of several weeks produces lactic acid which has a secondary effect on the efficiency of the stabilizing action.

Results

With respect to dry compressive strength, the addition of fibres such as straw permits an increase in strength of at least 15% in relation to the material without fibres. In very exceptional cases, the fibres make hardly any difference at all to the compressive strength, as for example is the case with very sandy material. Fibre-stabilized blocks can withstand high stresses as they absorb a fairly high amount of applied energy. This makes their use particularly attractive in earthquake regions. The addition of fibres makes a fundamental difference to the behaviour of the bricks beyond the failure point. Where non-reinforced materials crack into bits, reinforced blocks stay in one piece and will continue to increase in compressive strength beyond the failure point of non-reinforced blocks.

Practical aspects

The strength of reinforced blocks depends on the quantity of the fibres added but there is an optimal quantity which should not be exceeded. This is because an over-large quantity reduces density too much, while the number of contact points between the fibre and soil, which are responsible for transmitting stress, becomes too low and the strength of the block is reduced. The minimum proportion for satisfactory results starts at 4% by volume. Quantities of between 20 and 30kg per m^3 are very common. The straw should by preference be chopped into stalks of between 4 and 6cm long.

The best results are obtained when the stalks are scattered in all directions in the soil. Excessively long stalks, in parallel, do not yield good results. Nor are good results obtained when fibres are concentrated in specific spots, or in nests which may happen when too much fibre is used.

Fibres can also be used in together with other stabilizers such as cement, lime, and bitumen. If straw is used with bitumen, the bitumen must be first added to the soil and thoroughly mixed before then adding the straw. If the operations are carried out in a difference sequence there is a danger that the straw and the bitumen will conglomerate independently of the soil.

The fibres contained in the soil will be preserved without deterioration on condition that the material is kept dry. If the material remains in a moist environment for too long, there is a danger that the fibres will rot. On the other hand an alternating wetting and drying cycle will not encourage rotting as long as proper drying is certain. Analysis of very ancient material (e.g. adobe from the Egypt of the Pharaohs) has clearly demonstrated this.

Fibres may also be attacked by rodents and harmful insects, such as termites, particularly when wet.

Research

Very few organizations have made serious studies of fibre-stabilized soil. Some research has been carried out by the University of Tehran (Iran), the CSTB (France) and the IFE University in Nigeria.

Sorts of fibres

1. **Plant fibres** Straw of all kinds: barley, rye, hard and soft wheat, winter barley, lavender. Chaff of cereal crops: wheat, rice, barley, etc. Light filler such as sawdust and shavings. Other suitable vegetable fibres include hay, hemp, millet, cane bagasse, coir fibres, sisal, manilla, elephant grass, and fibres of the bamboo palm and hibiscus, as well as the left-overs after scutching flax.

2. **Animal fibres** Fur and hair from livestock.

3. **Synthetic fibres** Cellophane, steel, and glass wool/fibres.

Historical background

The first attempts at using cement as a stabilizer were made for road-building purposes in the USA in 1915. J.H. Amies applied for two patents for techniques using cement in 1917 and 1920. Cement stabilization was developed independently in Germany in the 1920s. In the USA cement stabilization has been used increasingly both for road and runway construction after about 1935. Applications of cement-stabilized material have multiplied enormously since then and it is used worldwide as much for public works as for construction. Nowadays knowledge of the technique and material is very complete.

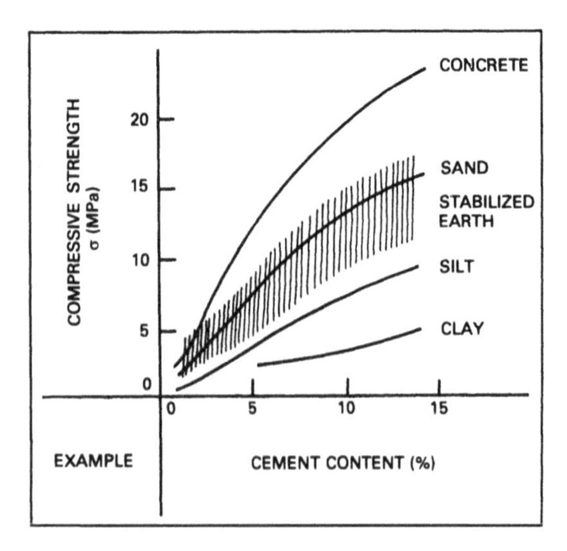

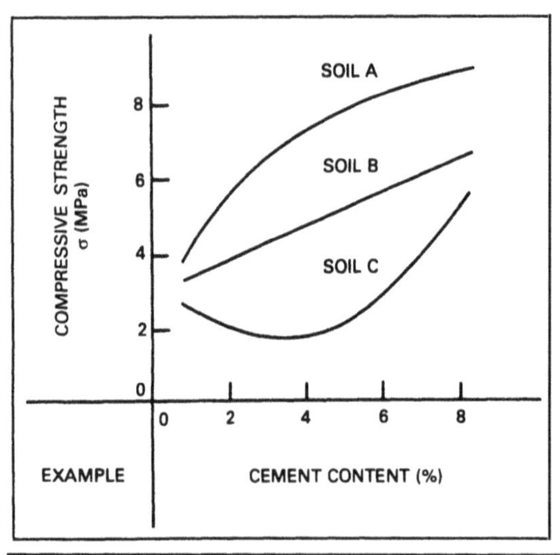

Stabilization mechanisms

Hydrated cement reacts in two different ways in soil:

– It may react with itself or with the sandy skeleton. This results respectively in the formation of a pure hydrated cement mortar, or the formation of a conventional mortar.
– It may undergo three-phase reaction with the clay:

1. Hydration sets off the formation of cement gels on the surface of the clay aggregations. The lime which comes free during the hydration of the cement tends to react with the clay. The lime is quickly used up and the clay starts to degenerate;
2. Hydration proceeds and encourages the disaggregation of the clay aggregates. The latter are deeply penetrated by the cement gels;
3. The cement gels and the clay aggregates becomely intimately entwined. Hydration goes on but more slowly.

In fact three mixed structures are obtained:

– An inert sandy matrix bound with cement:
– A matrix of stabilized clay;
– A matrix of unstabilized soil.

Stabilization does not affect all the aggregate.

A stabilized matrix covers the composite aggregations of sand and clay.

Effectiveness and proportions

The greatest effect is obtained by compression in the moist state. In the plastic state 50% more cement is required to achieve the same effect. The greatest compressive strength is obtained with gravels and sands rather than with silts or clay.

For soil, the required quantities of cement depend on its grain size distribution and structure, and the way that it is used.

Good results can be obtained with between 6 and 12%. Some soils require only 3% while the same proportion in others behave less well than if no cement at all had been used. Generally speaking at least 6% of cement is required in order to obtain satisfactory results. The compressive strength remains highly dependent on the quantity of cement used.

In comparable local conditions and for the same wall thickness (15cm), a stabilized earth brick will not necessarily be more economical than a concrete block. A preliminary study of the comparative costs is recommended.

Parameters of effectiveness

Soil Nearly all soils can be cement stabilized. The best results are obtained with sandy soils.

Organic matter This is generally recognized as being harmful, particularly if it contains nucleic acid, tartaric acid, or glucose. The effect is to slow the setting of the cement, and lower its strength. As a general rule an organic matter content greater than 1% represents a hazard, and soil containing more than 2% must not be used.

Sulphates These have very harmful side effects, calcium sulphate (anhydrite and gypsum) in particular, and are often encountered. They result in the destruction of the hardened cement from the inside of the cement-soil; an increase in the sensitivity of the clay to moisture. For soils with a sulphate content of more than 2 to 3% a special study is indispensable.

Oxides and metallic hydroxides Basically these are iron and aluminium oxides. They rarely exceed 5% of the soil, and thus have little effect.

Water In principle water containing organic matter and salts will not be used as these may result in efflorescence. Water with a high sulphate content may also have harmful effects.

Effects

Dry density This is less for soils which compact well. It increases for soils which compact less well.

Dry and wet compressive strength The influence of the cement on this parameter is a function of the dry density, and the voids ratio: $e = (\gamma S - \gamma d)/\gamma d$, of PI, LL and of M (proportion of elements with a diameter < 0.4mm in %) according to studies by the EIER at Ougadougou (Burkina Faso).

Tensile strength This varies from one fifth to one tenth of the compressive strength.

Dimensional stability Cement stabilization reduces the magnitude of dry shrinkage and swell when wet.

Erosion Improvement of the soil's resistance to rain erosion particularly when the soil contains large grains.

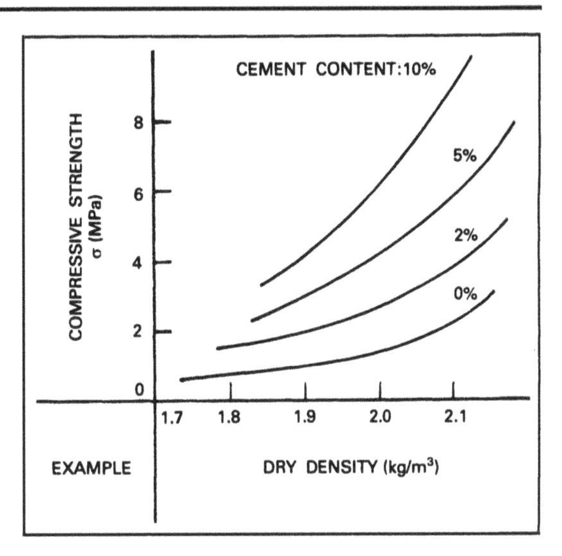

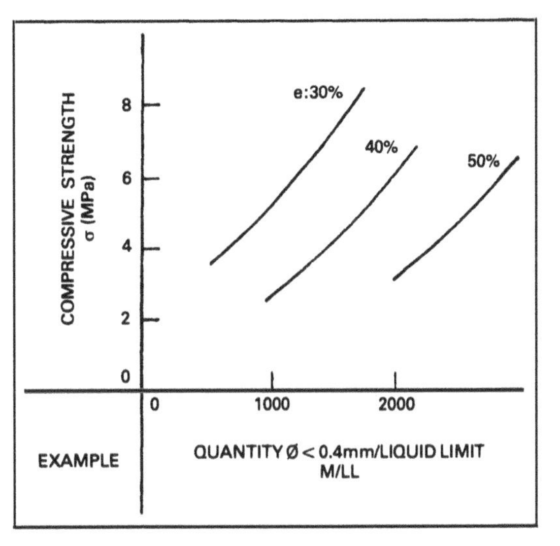

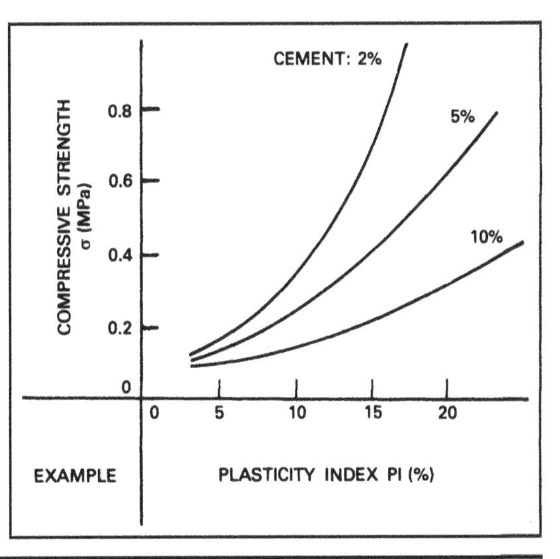

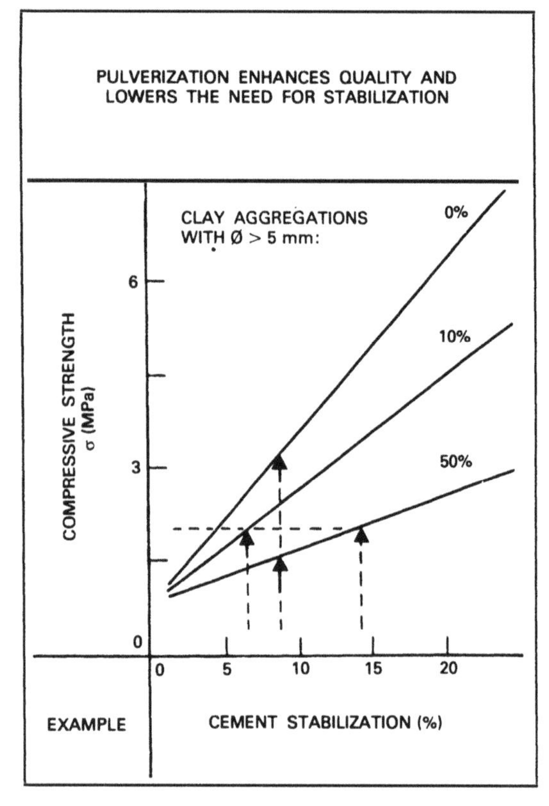

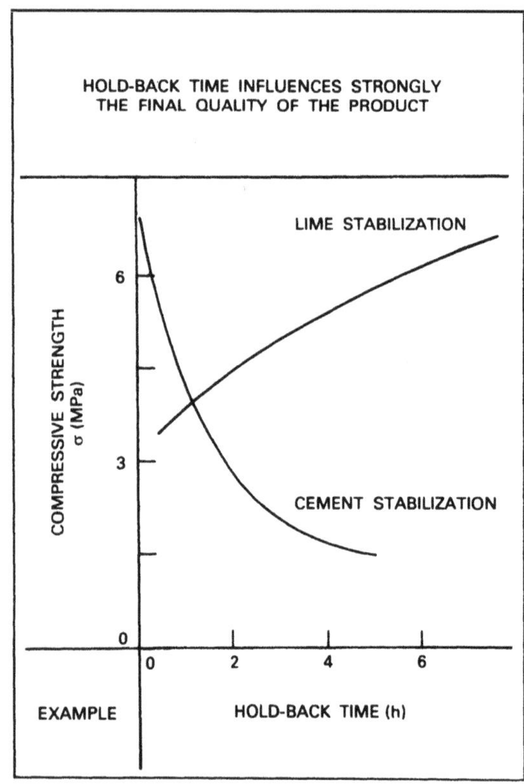

Cements

Ordinary Portland cement or cements of the same class are very suitable. There is no point in using high-strength cements, as these produce no particular improvement and are moreover very expensive. Higher-strength cements spoil very easily, making them unsuitable for use on work sites far from the factory. Preference should therefore go to Portland cements of the classes 250 or 350. Cements containing other materials such as slags, fly ash and pozzolanas can also be used, although these cements will only be available close to steel plants and power stations, and similar localities. In contrast cements with high contents of other materials should not be used because of their sensitivity when curing. These include iron Portland cement, blast-furnace cement, mixed metallurgical cements and slag clinker cements.

Additives

Certain products, added in small quantities to the soil cement during mixing, may improve the characteristics of the finished product.

1. Some organic products (amine acetate, melamine, aniline) and certain inorganic products (ferrous chloride) reduce the sensitivity of some soils to water.

2. Lime (2%) can reduce the harmful effects of organic matter, as does calcium chloride (0.3 to 2%), which also speeds setting. The lime also serves to modify the plasticity of the soil and to limit the formation of nodules.

3. Soda-based additives increase the reactivity of the soil and can induce cementation reactions supplementary to that of the cement with the soil particles. NaOH (sodium hydroxide) can be added in a proportion of 20 to 40g per litre of the mixing water, while between 0.5 and 1.1% of $NaSO_4$ can be added, or 1% of Na_2CO_3; or 1% Na_2SiO_2.

4. Between 2 and 4% of bitumen added as emulsion or cut-back makes the soil-cement waterproof.

Implementation

1. Crushing

Good cement stabilization demands thorough mixing of the components. The fine elements must not be allowed to form nodules with a size of more than 10mm. The presence of 50% nodules > 5mm can cut compressive strength by half.

2. Mixing

Good distribution of the cement and uniformity of the material is provided by mixing. It is important to have a dry soil if the best mixing conditions are to be attained. In wet climates this may make preliminary drying of the soil necessary. Grinding may accelerate drying and help to break up lumps. The water required for the mix should only be added at very end of the mixing process, after the very necessary dry mixing phase.

3. Moulding

The material should be compacted immediately after mixing, before the cement starts to set, and with a controlled moisture content which will be close to the OMC. A 4% difference in moisture content, either more or less, can make a significant difference to the quality of the material. As a general rule, soil with a high clay content ought to be compacted slightly moister than the OMC, while sandy soils should be compacted slightly drier than the OMC.

4. Drying

The strength of a soil cement increases with age.

A minimum curing period of 14 days is absolutely essential, although 28 days is better. During this period the material should be kept in a moist environment, sheltered from the sun and protected from the wind, in order to prevent excessively rapid surface drying, which may cause shrinkage cracking. The materials will be allowed to dry when compacted, and moistened by spraying or covered with a plastic sheet, which allows temperatures to rise and giving an RH close to 100%. The longer this moist-drying cure is allowed to go on the greater will be the strength of the material.

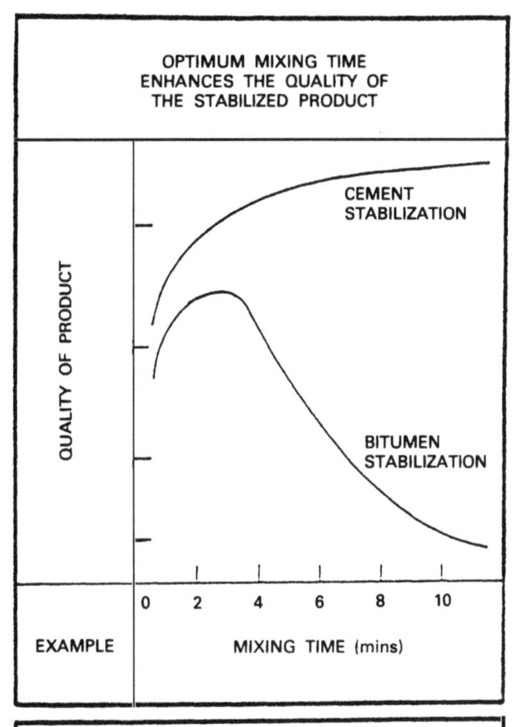

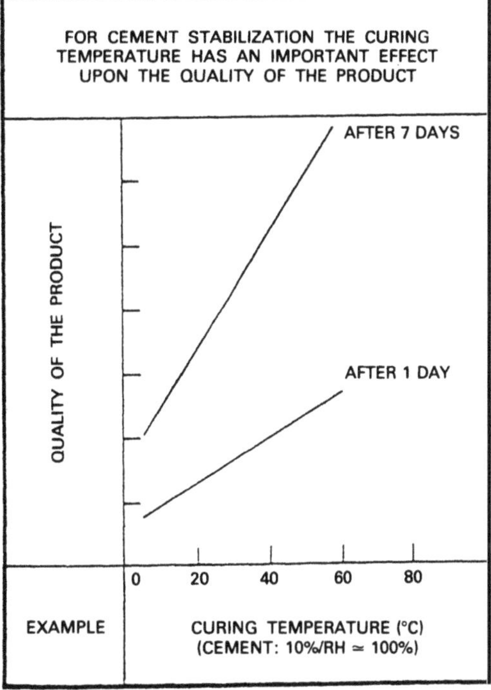

Historical background

It appears that the systematic use of lime for stabilizing soils is fairly recent, and was pioneered in the 1920s in the USA. Since then millions of m^2 of roads have been constructed in lime-stabilized soil and immense experience has been gained with them. The construction of the Dallas–Fort Worth airport, covering some $70km^2$, in 1974 is one of the most spectacular applications of this technique. Indeed more than 300,000 tonnes of lime were used for the stabilization works.

Lime was also, and still is, used for constructing buildings and there is an increasing interest in lime stabilization in this field.

Mechanisms

The theory of lime stabilization suggests five basic mechanisms.

1. Water absorption Quicklime undergoes a hydration reaction in the presence of water or in moist soil. This reaction is strongly exothermic with the release of about 300kcal for every kg of quicklime.

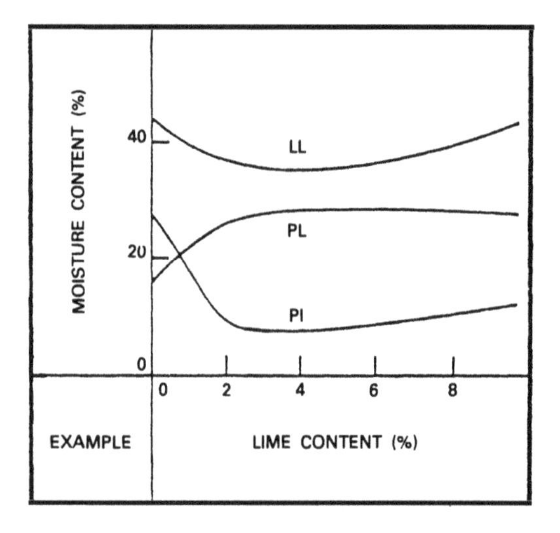

2. Cation exchange When lime is added to a moistened soil the latter is flooded with calcium ions. Cation exchange then takes place, with calcium ions replacing exchangeable cations in the soil compounds, such as magnesium, sodium, potassium, and hydrogen. The volume of this exchange depends on the quantity of exchangeable cations present in the overall cation exchange capacity of the soil.

3. Flocculation and aggregation As a result of the cationic exchange and the increase in the quantity of electrolytes in the pore water, the soil grains flocculate and tend to accrete. The size of the accretions in the fine fraction increases. Both grain size distribution and structure are altered.

4. Carbonation The lime added to the soil reacts with carbon dioxide from the air to form weak carbonated cements. This reaction uses part of the lime available for pozzolanic reactions.

5. Pozzolanic reaction This is by far the most important reaction involved in lime stabilization. The strength of the material results largely from the dissolution of clay minerals in an alkaline environment produced by the lime and the recombination of the silica and alumina in the clays with the calcium to form complex aluminium and calcium silicates and thus cementing the grains together. The lime must be added to the soil in sufficient quantities in order to proceed and maintain a high pH, which is necessary for the dissolution of the clay minerals for long enough to allow an effective stabilization reaction.

Effectiveness and proportions

When 1% of quicklime is added to the soil the exothermic hydration reaction dries the soil, removing between 0.5 and 1% of water. The addition of 2 to 3% of quicklime immediately provokes a reduction of the plasticity of the soil and the breaking of lumps. This reaction is called the fixing point of the lime. For ordinary stabilization work between 6 and 12% is used, similar to the amounts required for cement stabilization, the difference being that with lime there is an optimum quantity for each soil.

Sophisticated industrial procedures make use of high pressures and steam treatment in an autoclave with the proportion of lime rising to as much as 20%. The products obtains are similar to those obtained in the silica-lime industry. Lime stabilization is particularly well suited to pressure moulding procedures.

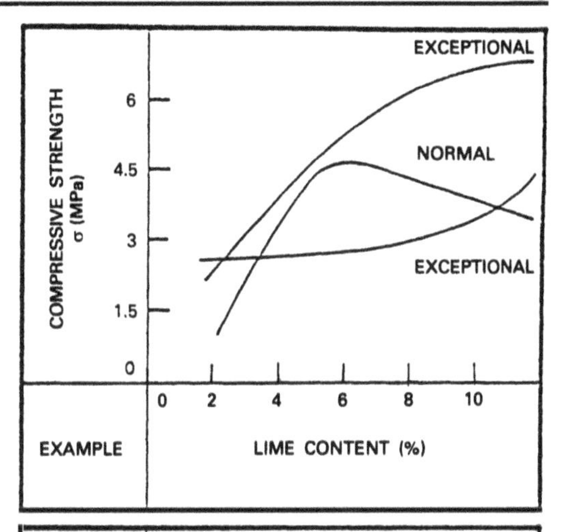

Parameters of effectiveness

Soil The soil must contain a significant clay fraction. Results vary depending on the nature of the clay minerals and are best with those with high contents of alumina silicates, silicas, and ferrous hydroxides. Natural pozzolanas react quickly and well with lime.

Organic matter These can block ionic exchange in clayey soils without blocking the pozzolanic reaction. Soils containing up to 20% organic matter can be stabilized with lime, but care is essential.

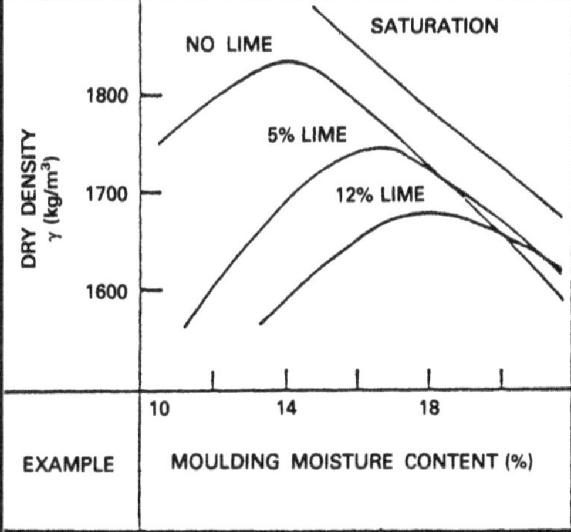

Sulphates When dry, calcium sulphates are less dangerous than magnesium sulphates. When moist, all sulphates are harmful.

Effects

Dry density For a given compression, lime reduces the vδ max. and raises the OMC because of the flocculation.

Compressive strength The optimum proportion of lime should be determined by preliminary tests. Compressive strength tends to increase with the age of the product. Values of between 2 and 5MPa can easily be obtained and when industrial procedures are employed values of between 20 and 40MPa can be expected.

Tensile strength This is highly influenced by the quantity and quality of clays contained in the soil, which reacts with the lime.

Dimensional stability Just one to two per cent of lime can reduce shrinkage of 10% to 1% and eliminate swell.

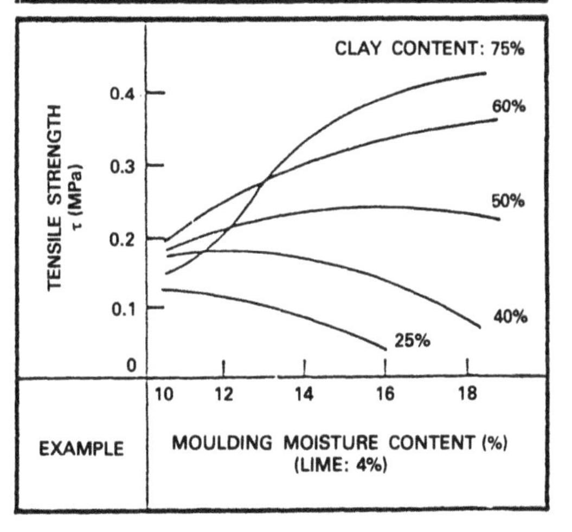

Limes

1. Non-hydraulic limes

These are produced by burning very pure limestone and represent the main source of lime for use in stabilization.

Quicklime (CaO) Quicklime is produced directly by burning the stone in kilns. The delicate conditions of storage and maintenance can limit its use. Quicklime is extremely hygroscopic (i.e. it attracts water) and must be protected from moisture. It is a caustic material and must be handled with great care. It becomes very hot in the hydration stage (up to 150°C). Weight for weight it is more effective than slaked lime, because it can supply greater quantities of calcium ions. In moist soils it can absorb the water required for its hydration.

Slaked lime $(CaOH)_2$ Slaked lime is obtained by hydrating quicklime. Widely used for stabilization, it does not have the drawbacks of quicklime. Fat slaked limes do not have to be very finely crushed to be effective. Industrial qualities contain between 90 and 99% of 'active lime' while craft-produced lime may only contain between 70 and 75%, with the rest being unburnt or excessively burnt materials. The proportions used for stabilization must be adapted in consequence.

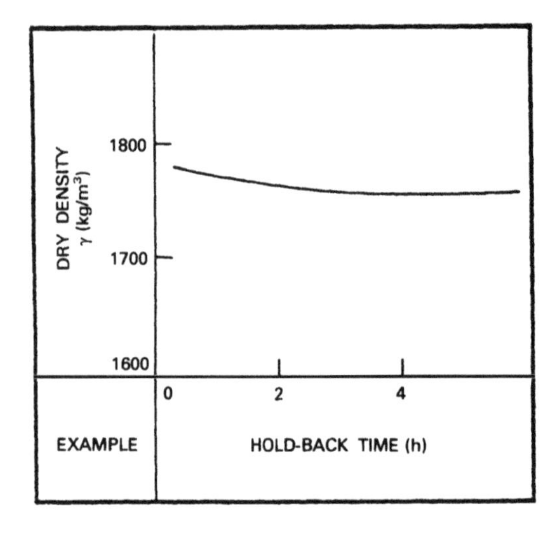

2. Hydraulic limes

These are similar to the cements. Their use should not be considered unless other qualities of lime are not available. Natural hydraulic (XHN) limes are more effective stabilizers than artificial hydraulic limes (XHA).

3. Agricultural limes

These are used to improve agricultural soils and usually have no stabilizing effect at all.

Additives

Some additives mixed with limes in small quantities can have special effects.

1. To increase the reactivity of the soil:

- Caustic soda: NaOH
- Sodium sulphate: Na_2SO_4
- Metasilicate of sodium: $Na_2SiO_3(9H_2O)$
- Sodium carbonate: Na_2CO_3
- Sodium aluminate: $NaAlO_2$.

The quantity used varies from 0.25 to 2 mol.g per litre of the water used for compaction.

2. To increase compressive strength:

- Portland cement in quantities of up to 100% of the lime.

3. To increase the effectiveness of stabilization for sandy silt soils and reduce the swell due to the slaked lime:

- Magnesium sulphate ($MgSO_4$) added at a rate of about a quarter of the weight of the lime.

4. To make the soil waterproof:

- Potassium sulphate (K_2SO_4)
- Bitumen products
- Other water repellents.

Implementation

1. Crushing

This operation is important and must be carried out with the greatest care. The finer the clay is crushed, the more active the lime will be in attacking the clay. The operation may be difficult because clay is highly cohesive.

Very moist soil can be dried and broken with quicklime. Stabilization will be effective if at least 50% of the aggregated clay is crushed to a diameter of less than 5mm.

2. Mixing

The mixing must be very carefully carried out in order to ensure the intimate mingling of the soil and the lime. For very plastic soil, the process should be carried out in 2 stages, with one or two days between them. This gives the lime a chance to loosen the lumps. This 2 stage procedure may nevertheless reduce the effect of the lime on strength. The homogeneity of the mix can be checked by observing the uniformity of its colour. No trail of lime not incorporated in the soil should be visible.

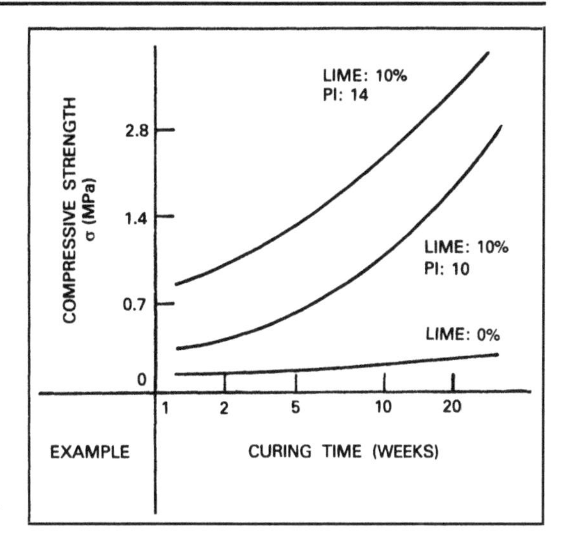

3. Hold-back time

If the moist method is used, it is advantageous to allow the mixture to rest after mixing. A wait of at least 2 hours should be allowed for quantities of lime above the fixing point; although a period of 8 to 16 hours is preferable. The effect on dry density is negligible but greater strengths can be achieved. If the plastic approach is used, everything is to be gained by allowing the mixture of soil and quicklime or slaked lime to rest for several weeks. This is particularly the case for renderings which become greasier and more adhesive.

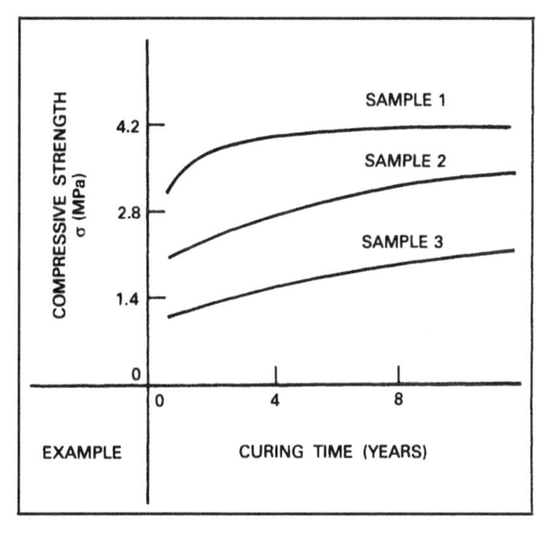

4. Compression

Dry density is very sensitive to the degree of compaction, especially for high proportions of lime. The moisture content will be close to the optimum, on the moist side, and an adequate hold-back time have been allowed (longer for higher quantities). The exothermic reaction set off by the quicklime consumes close to 1% of the moisture content per % of quicklime added. The moisture content will thus be corrected as the OMC is approached during the second mixing stage.

5. Drying cure

An increase in compressive strength can be obtained if the curing period is extended. This phenomenon lasts several weeks and persists for months, and is even better in a warm and humid environment. Lime-stabilized products can be very advantageously exposed to high temperatures (± 60°C). Curing in the sun under a sheet of plastic or in a tunnel built in corrugated iron makes it possible to achieve such high temperatures and relative humidities. Research carried out at the University of Denmark has shown that very good products can be obtained by drying for 24 hours in an autoclave at 60–97°C with an RH of 100%.

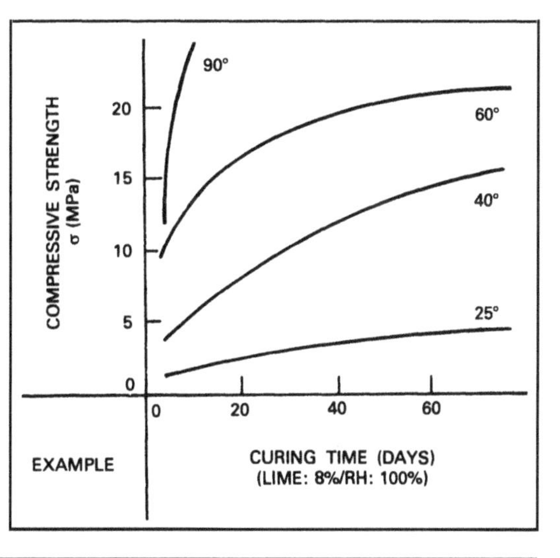

Bitumen and asphalt: terminology

Bitumen is often associated with the hydrocarbon pavement of roads. However, it should not be used confused with asphalt, tars, and so on.

The Sanskrit term for bitumen is *jatu krit*, meaning 'generator of pitch', by analogy with the resins of certain conifers. In Latin the equivalent term is *pix-tumens* meaning 'exuding pitch', i.e. exuded from the earth's crust. Later on 'bitumen' was adopted by most West European languages. Originally the word bitumen was used for a naturally occurring material made up of a 'mixture of hydro-carbons of high molecular weight, soluble in carbon disulphide and which may contain minerals in varying quantities'. In contrast the term 'asphalt' comes from the Greek adjective 'asphales' meaning durable and is used to designate 'a sedimentary calcareous rock, impregnated with 8 to 10% natural hydrocarbon'.

Nowadays bitumen is used to refer to a product consisting of at least 40% of heavy hydrocarbons and filler. The term asphalt refers to product containing less than 20% hydrocarbons, the rest being filler, sands or gravel. A word of warning should be given about the American use of the word 'asphalt' and is used to refer what Europeans call 'bitumen', while to an American 'bitumen' means any black binder and is applied equally to bitumen resulting from petroleum distillation and to coal tar. Sometimes soils which are naturally impregnated with bitumen can be found: this being the case for the bitumen sands of Nigeria.

If it is to be used, the bitumen must be:

- heated;
- or mixed with solvents, resulting in 'cut-back';
- or dispersed in water as an emulsion.

It is the two latter techniques which are used for stabilization.

Historical background

The use of bitumen as a stabilizer is very ancient. The Greek historian Herodotus describes how it was used in Babylon in the 5th century BC for making mortar to lay unbaked moulded bricks. Even so the use of bitumen throughout the cours of history has been limited. Indeed bitumen as suc was first produced on an industrial scale only a fev decades ago, in the 1940s in the USA.

Stabilized bricks have been sold under the name o Bitudobe or Asphadobe. Civil engineers have learned to use the technique in road construction. In Algeria, for example, close to 28,000km of road has been constructed using this technique. Nowadays, in the USA, bitumen stabilized adobe is as widely used as in Central and South America. Recent attempts to transfer the technique to Africa have not met with success, even in oil producing nations. Bitumen, which only a few years ago was regarded as a miracle product capable of finally solving all stabilization problems, is being used less and less because of the cost of oil products.

Mechanisms

Cut-back and bituminous emulsions come in the form of microscopic droplets in suspension in a solvent or in water. The stabilizer is mixed in the soil, and when the water or solvent evaporates the droplets of bitumen spread out to form very thin strong films which adhere and cover the soil. Bitumen improves the water-resisting properties of the soil (less absorption by clays) and can improve the cohesion of soils with non-cohesive soils, by acting as binder.

Effectiveness and proportions

In order to obtain uniform distribution of the bitumen throughout the soil, it is better to use a technique making use of large quantities of water. The adobe technique is thus the most suitable. Normally 2 to 3% bitumen is added, but this can rise to 8%.

Proportions vary in accordance with the grain size distribution of the soil because bitumen stabilization involves the coating of the specific surface of the grains. The values given here are for the bitumen prior to being diluted in a watery suspension or by a solvent. The bitumen has only a very slight effect on the colour of the material and has no typical odour once the stabilized products have dried.

Parameters of effectiveness

Soil Bitumen stabilization is at its most effective with sandy or silty soils. It is not suitable for fine soils in dry regions, where the pH and salt content of the soil may be high.

Organic matter and sulphates Their presence in the soil hinders the efficiency of bitumen stabilization as their adhesion to the grains prevents the adhesion of bitumen. Acid organic matter (e.g. forest soils) are very harmful. The neutral and basic organic matter found in arid and semi-arid regions are not particularly harmful.

Salts Mineral salts are very harmful. They can be neutralized by adding 1% of cement. When bitumen stabilization is carried out on an industrial scale salt contents of more than 0.2% are not accepted, but sometimes up to 6% of sodium chloride (NaCl) can be accepted.

Effects

Dry density Bitumen brings about a drop in density and increases the OMC (water plus bitumen).

Compressive strength In the dry state this increases with the proportion of bitumen to a certain threshold value after which it falls sharply, once the ideal level of coating has been achieved. In the moist state, strength rises steadily with the quantity of bitumen, independently of dry strength.

Absorption Absorption is a function of the moisture content during mixing and falls to very low levels after a certain threshold value. It is advisable to determine this value. After drying for several days, water absorption is stable in time.

Swell This is a function of the liquids content during mixing. The greater the degree of mixing in the liquid state, the less swell there will be.

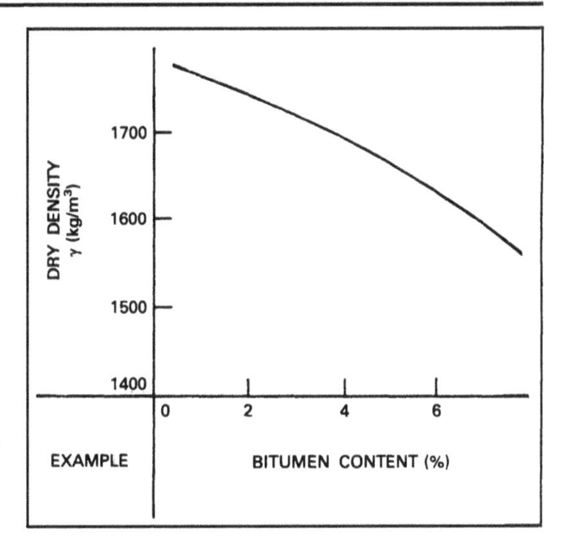

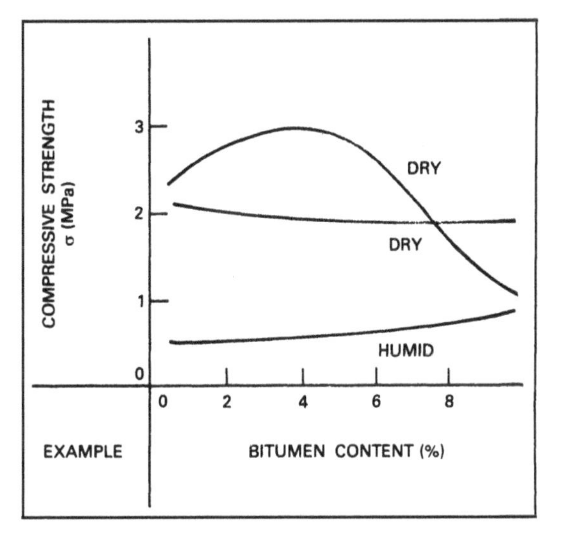

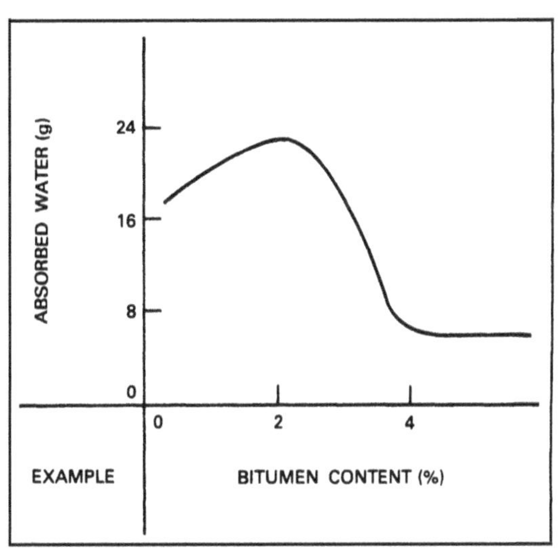

Bitumens

When bitumen is used for stabilization it is generally in the form of cut-back or emulsion. Both forms break down on drying, and this break-down can be slow or fast. Rapid break-down is suitable for temperate climates and slow break-down is better for hot climates.

1. Cut-back

This refers to bitumens to which a volatile solvent has been added making them less viscous. Solvents can be diesel-oil, kerosene and naphta. Some of them dry slowly, some moderately quickly, while others dry rapidly. They cannot be used in the rain and are flammable. Their viscosity is indicated by an index number: 0 = very fluid, 3 = viscous. RC 250 is a cut-back which is widely used in the USA nowadays.

2. Emulsions

Bitumen (55 to 65%) is dispersed in water with the help of an emulsifying agent (1 to 2%). This agent moreover keeps the bitumen in suspension. There are two sorts of emulsion.

- Anionic: unusual and not suitable for all aggregates; used mostly in Europe;
- Cationic: more widespread and compatible with virtually all soils, used in particular in the USA.

Emulsions are usually very fluid and mix easily with soil which is already moist.

They are less stable than cut-back, and there is a danger of the water and bitumen bond breaking down (separation).

SS 1 h is an emulsion recommended for hot regions in Africa.

Additives

These have various supplementary effects and be used to:
1. Neutralize salts: cement, 1 to 2% by weight of the soil;
2. Flocculate the soil: lime, 1 to 2% by weight of the soil;
3. Improve the coating of the grains with bitumen: quaternary amines, use 0.6% by weight of cut-back or 0.01% by weight of the soil;
4. Improve the adhesion of the bitumen to the grains: quaternary amines, use 0.03% by weight of the soil;
5. Increase the rigidity of the bitumen film: waxes, use 0.07% by weight of the soil or 1% by weight of the bitumen;
6. Raise the dry and wet compressive strength: phosphoric anhydride (P_2O_5), use 2% by weight of the soil.

CRACKING OR DRYING	CUT-BACK			EMULSIONS		
	EUROPE	USA (ASTM)	BITUMEN CONTENT (%)	ANIONIC (ASTM)	CATIONIC (ASTM)	VISCOSITY
SLOW	SC 0		45–50			
	SC 1	SC 70	55–61	SS 1	CSS 1	FLUID
	SC 2	SC 250	63–70	SS 1 h	CSS 1 h	VISCOUS
	SC 3		70–75			
MEDIUM	MC 0		61–65			
	MC 1	MC 70	68–72	MS 2	CMS 2	FLUID
	MC 2	MC 250	73–77	MS 2 h	CMS 2 h	VISCOUS
	MC 3		79–82			
RAPID	RC 0		62–65			
	RC 1	MC 70	70–73	RS 1	CRS 1	FLUID
	RC 2	MC 250	74–78	RS 2	CRS 2	VISCOUS

Implementation

1. Mixing

The effectiveness of the bitumen stabilization depends very largely on this operation. Too much mixing can increase water absorption after drying because of the premature break-down of the emulsion. If the mixing is carried out in the liquid or plastic state (adobe, cob, mortar, or rendering), no problems of this sort are encountered.

On the other hand, if the soil is going to be compacted, the mixing will be carried out at OMC. When soils are already moist care must be taken not to add excessive quantities of stabilizer (water and bitumen). Wet strength and impermeability may be less good.

When low proportions of bitumen are used (e.g. 2%), it is preferable to add the bitumen to a small quantity of soil, and then to mix this small quantity with the remainder of the soil. This applies particularly to cut-back. Emulsions should be diluted in the mixing water.

2. Hold-back time

When bituminous stabilizers with slow or moderately quick break-down time are used, it is possible to wait between mixing and moulding. When products with rapid break-down times are used, operations should follow one another without delay.

3. Compaction

Two to four MPa is enough and leaves the material a fairly porous structure in order to facilitate the evaporation of volatile solvents while ensuring good dry density. When turning out from the moulds the bitumen acts a release product, the blocks have an attractive appearance with sharp arrises.

4. Curing

Cut-back and emulsion which breaks down rapidly both shorten drying times. It is preferable to allow bitumen-stabilized material to cure in dry air rather than in a moist environment. Compressive strengths are related to the quantity of bitumen used and the duration of the drying period. These two parameters should be determined in advance by means of tests to find out what the optimal values are. The loss of volatiles is greater for longer curing periods and higher temperatures, and this has a beneficial effect on absorption and expansion. Above 40°C, however, no further improvement is noted.

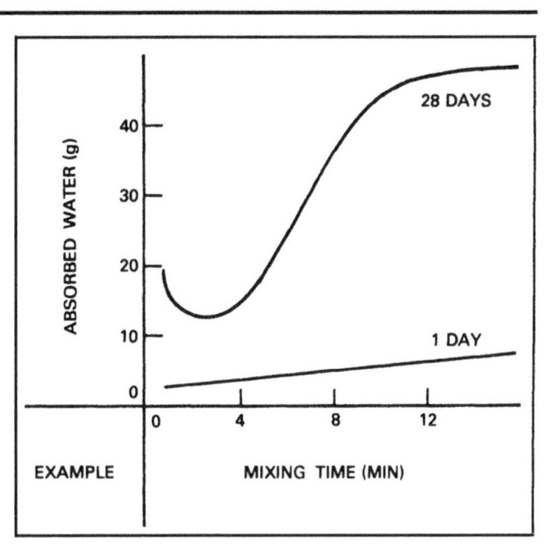

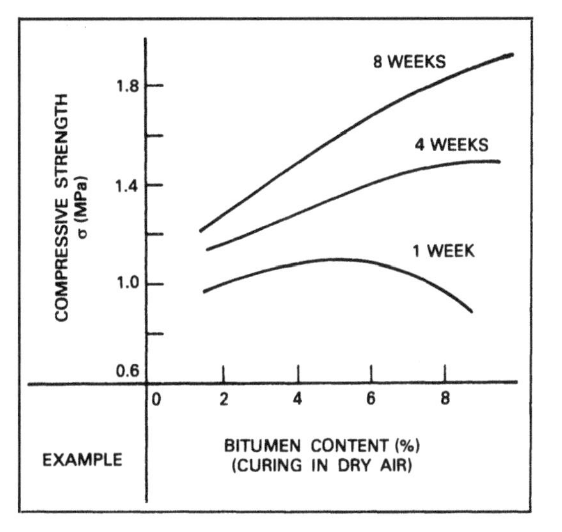

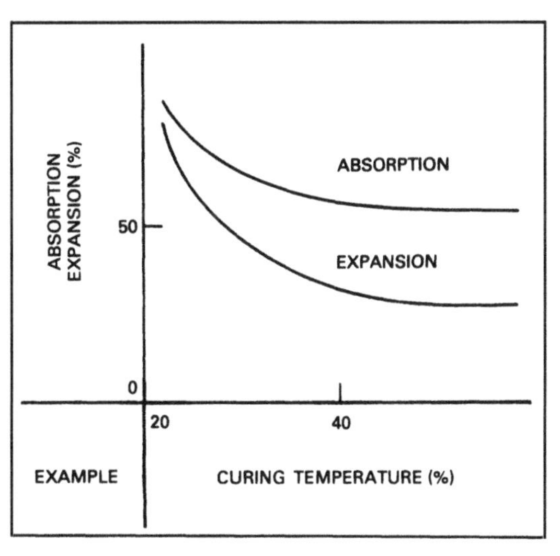

A great deal of recent research work has been concentrated on chemical stabilization by means of chemical resins, particularly in the civil engineering field. The object of this research has been to achieve an increase in load-bearing capacity while at the same time reducing the weight of the stabilized earth courses. The highest shear strength combined with the greater elasticity of the wearing courses has been sought. These objectives, which correspond to civil engineering requirements, are not necessarily applicable to the construction of buildings, except for some horizontal surfaces; such as footways, stabilized slabs and similar. Quite extraordinary results have been obtained with resin stabilization. The considerable extra cost compared with ordinary stabilization procedures, however, remains the greatest barrier to its widespread use.

Benefits

Vigorous action, rapid setting, and easy incorporation into the soil as viscosity is comparable to that of water. Very moist soils can be solidified.

Drawbacks

High cost, sophisticated production technology, available only in the industrialized countries, Quantities required large compared to conventional stabilizers, products are toxic, difficult handling requiring the use of catalysts, sensitive to water, and service life uncertain, products are biodegradable.

Mechanisms and principles

These resins are made up of long chain molecules resulting from the linking (polymerization) of certain chemical agents (monomers and polymers). They can be used in two different ways:

- Monomers are added to the soil at the same time as the catalyst: reaction between the soil and the monomers as well as polymerization is immediate. This is the case with abietic resins, for example.
- Polymer is formed in advance by synthetic or natural means and then added to the soil as a solid, solution, or emulsion.

The resins act in different ways: as a flocculant, as a dispersant or as an acid. Nevertheless the majority of products act to render the soil impermeable, while the more sophisticated can improve the cohesion of the soil.

1. Processed natural products

Gum arabic Obtained from the acacia tree. Its ability to provide impermeabilization is low because it is soluble in water. It acts primarily as a flocculant, helps to increase dry compression strength, and slows capillary absorption of water by acting on the kinetics of this phenomenon.

Palmo-copal Copal is a resin obtained from certain tropical trees. Palmo-copal is a solution of copal obtained by pyrogenation from palm oil. The amount required varies from 3 to 8% for sandy soils. Another variety, manilla copal, is the only copal resin which has an impermeabilizing capacity.

Wallaba resin Water repellent.

Rosin Obtained during the distillation of turpentine essences from oily pine resins. Soluble in organic solvents and in aqueous alkaline solutions. Rosin forms a gel after reacting with certain metallic salts (iron and aluminium). Reduces the water absorption of soils.

Vinsol Also obtained in the production of turpentine. It is used in acid soils at critical control rates (± 1%). Water repellent; improves cohesion but does not affect compressive strength.

Lignin A by-product of the paper industry. It is a sort of alkaline resinous liquor with an impermeabilizing capacity. It is soluble and can become insoluble when reacted with chrome. Chromolignin is unfortunately an expensive product.

Molasses Sugar aldehydes from dehydrated molasses can be polymerized at high temperatures with phenolic catalysts. The resinous material obtained has characteristics similar to that of a naturally occurring asphalt and synthetic resins.

Ethyl cellulose A synthetic resin which has been tested but without satisfactory results.

Carboxymethyl cellulose A non-ionic stabilizer with a coagulating action and is water soluble.

Shellac Confers excellent strength on sandy soils but the stabilized material does not stand up well to water.

2. Furfural-based resins

Furfural is a toxic aldehyde found in grain alcohols and in the following materials: rice hulls, peanut shells, cotton seed, bagasse, maize cobs and stalks, olive pits. It is present in percentages ranging from 10 to 20%.

Furfural aniline A resin formed from furfural and aniline, which is a cyclic amine derived from benzene, and is nowadays obtained from coal. Mix 70% of aniline with 30% of furfural. The product is highly toxic and between 2 to 6% is adequate for soil stabilization. It makes grains water repellent by ionic exchange and cements them together by polymerization.

Furfuryl alcohol This is an organic compound derived from furfural. It polymerizes in the presence of certain catalysts giving rise to polymers with excellent mechanical properties. Improves both the dry and wet mechanical strengths, and slows water absorption.

Resorcinol-furfural This product is used in an aqueous solution catalysed by soda in a basic milieu. It is a toxic product and is often very expensive.

Furfural-urea and phenol-furfural Stabilizers which have been tested alone or mixed with furfural aniline: disappointing results.

3. Formaldehyde-based resins

Formaldehyde is a volatile liquid obtained by the oxidization of methyl alcohol.

Resorcinol-formaldehyde This is a mixture of resorcinol, antiseptic phenol derived from benzene and formaldehyde. This yields a resin which acts by cementation and as a water repellent. Reduces water absorption.

Phenol-formaldehyde Results known to be satisfactory.

Formaldehyde-urea The action of this compound is similar to furfurol aniline.

Calcium-sulphamate-formaldehyde The results obtained have not been satisfactory.

Melamine-formaldehyde Good results for dry strength but a 50% loss of strength in the moist state.

4. Resins based on acrylic compounds

Calcium acrylate This water-soluble resin forms an insoluble gel with soil which is rubbery or stiff depending on the moisture content.

Acrylic nitrite This resin is used for grouting. Cements and waterproofs. Acrylic amine nitrite, which is another derived compound has a similar action.

Polyacrylamines Cationic polymer.

5. Urea-based resins

Urea-formol

Methyl urea

Dimethylol urea Gives the best known results for urea-based resins although wet strengths remain low.

6. Polyvinyl-based resins

Polyvinylic alcohol Non-ionic stabilizer which forms strong flexible films by the evaporation of the aqueous solution although it is soluble in water. Must be combined with natural oils or water repellents in order to be effective.

Polyvinyl acetate This is the best product for sandy soils because it improves their cohesiveness. The product is completely destroyed if immersed in water.

7. Other products

Aluminium compounds

Epoxy resins

Phenol-formalin

Polyurethane resin

The term 'natural product' covers a wide range of stabilizers of animal, vegetable and mineral origin. In our discussion of vegetable products we will concentrate on those which are obtained directly from vegetation or are subject to a simple preparation process rather than those which are the result of involved processing, more like the synthetic products, such as agricultural wastes. Accurate scientific information about these natural stabilizers, whose use is largely dependent on traditional skill, is rare and research is virtually non-existent. Nevertheless in recent years, various research laboratories have started to take a closer interest in investigating these products.

The attraction of these stabilizers is that they are locally available. Even so they are rarely available in large enough quantities to be used on anything more than the individual level. On the other hand because of their rarity and the fact that they produced by hand their market value is higher than industrial products, such as gum arabic. Often these products have a social utility, and could be used for agriculture or even in food, so that their use as stabilizers is difficult to countenance.

These stabilizers are often only effective in very specific conditions of preparation and environment.

Even so the products have the advantage, for example with respect to water resistance, of slowing the dynamics of deterioration. The water absorption of a treated wall will be slower than that of an untreated wall allowing deterioration to be remedied (between two rainy seasons for example); while an untreated wall would deteriorate very rapidly.

It should be remembered that natural products are not as effective as industrially manufactured products such as cement, lime, bitumen, and so on.

1. Mineral products

These products are used to correct the grain size distribution of soils. For example sands can be added to clays and vice versa. Sometimes very specific soils are selected in order to obtain very particular effects. A good example is bentonite, which is added in small quantities to the soil being treated. This smectic clay has powerful degreasing properties and expands in the presence of water, thus preventing the passage of water.

Some soils (volcanic sands in particular) have natural pozzolanic properties. They can be used for stabilizing clayey soils but often require the addition of at least 30% of lime or cement before becoming an effective binder which in turn will stabilize clay when added at a rate of about 8%.

2. Animal products

These are only very rarely used for stabilizing walls or solid elements of a building. They tend to be reserved for the stabilization of renderings.

Excrement All sorts of excrement are used. Cowpats are undoubtedly the most widely employed, although better used as a manure or fuel. This particular excrement has in the final analysis only a very limited effect on water resistance and reduces compressive strength. Other traditions make use of horse or camel dung, or of pigeon droppings. The action of these sorts of excrement is probably due to the presence of fibre (mixed straw), phosphoric acid and potassium. The use of animal urine is also known. Horse urine when used to replace the mixing water (e.g. for daub) effectively eliminates cracking and gives a marked improvement in the ability of the soil to stand up to erosion. Surprisingly good results can be obtained when it is combined with lime.

Animal blood The use of bull's blood is known from Roman times. When combined with lime or polyphenols stabilization with bull's blood is effective. The blood must be fresh and not be in powder form.

Animal fur and hair Animal hair and fur play much the same role as some vegetable fibres. Their use is generally reserved for stabilizing renderings.

Casein Proteinic casein (middle fraction of the protides of milk) is sometime used in stabilization in the form of whey combined with bull's blood. Certain milk powders have been tried and gave good results. 'Poulh's soap' is also used. This is diluted casein mixed with brick dust and beaten like a paste, prior to being to added to the soil.

Lime Lime can be prepared from shells or coral. This is still done in some countries such as Somalia and Senegal.

Animal glues These can be used for stabilization, particularly for renderings. Animal glues are produced from horn, bone, hooves, and hides.

Termite hills Termites secrete an active substance, which appears to be a non-ionic cellulose polymer of the polysaccharide type. Termite hills stand up well to rain and their soil can be mixed with another for the production of blocks. The substance has been synthesized by research workers in South Africa, but costs three times as much as cement.

Oils and fats Fish oils and animal fats can serve as waterproofing agents. The stearates contained in animal fats play the same role.

3. Vegetable products

Ashes Hardwood ash is rich in calcium carbonate and has stabilizing properties but is not always suitable for soils which may be suited to lime stabilization. Classic proportions suggest the addition of 5 to 10% ash. They improve the dry compressive strength but do not reduce sensitivity to water.

Vegetable oils and fats If they are to be effective vegetable oils must dry quickly so that they harden upon contact with the air and are insoluble in water. The use of castor oil is highly effective, but it is extremely expensive because it also used in aviation. Coconut, cotton, and linseed oils are also used. Kapok oil prepared first by roasting kapok seeds, turning them into a flour which can then be transformed into a paste (20 to 25 litres of water to 10kg of powder) can be effective. This depends on the quality of the seeds and the roasting process, which increases the yield, as well as the length of time it takes to prepare the paste (6 hours boiling). There is also palmitic acid which is obtained from saponified palm oil precipitated by 25% HCl. About 1kg of palmitic acid is obtained per kg of palm soap in solution.

Palmitic acid mixed with lime gives calcium palmitate, which is used in stabilizing renderings.

Shea (karite) oil or butter is also used for renderings. This product is also used for soap production but is increasingly scarce for construction applications.

Tannins Tannins often act as dispersants and improve the coating of grains by clays. They are also good compaction acid (break-up of lumps) and they reduce permeability. The amount of tannin required varies from a small percentage of the mixing water for the most active products, to the total replacement of the mixing water when tannin decoctions are used. The most commonly known are tannin from the bark of néré, oak, chestnut, scorpioid acacia.

Humic acid or polyphenols These are derived from lignin, and form hard stable compounds, particularly in ferrallitic soils.

Sap and latexes The juice of banana leaves precipitated with lime improves erosion resistance and slows water absorption. The latex of certain trees, such as euphorbia, reduces permeability slightly.

The same applies to hevea rubber, and concentrated sisal juice in the form of organic glue. Latexes mix with acid soils (coagulation), but mix better with basic soils.

These stabilizers are industrial or synthetic products, and may even be industrial wastes. Others in this group may be natural products which require sophisticated processing. These products are currently the object of laboratory research. They are not very satifactory from the economic point of view and their effectiveness is often doubtful. Some of the products mentioned have long been known, while others have been abandoned. In general they are not widely used.

1. Acids

Using acids always entails some degree of risk. Each type of acid involves a specific reaction. They modify the pH of the soils in which they are incorporated, resulting in flocculation, the effects of which are often reversible. Some acids act as catalysts to form insoluble phosphates. Hydrochloric acid (HCl) and nitric acid (HNO_3) result in moderate stabilization. Hydrofluoric acid (HF) is very effective in all soils except those with high aluminium contents, inducing a reaction which brings about the formation of insoluble and strong silica fluorides. The effectiveness of sulphuric acid (H_2SO_4) is doubtful. If phosphoric acid (P_2O_5) is incorporated a hydration reaction is set off with the formation of phosphoric anhydride (H_3PO_4) which reacts with clayey minerals, and the creation of an insoluble gel of aluminium and iron phosphates, which cement the grains together.

2. Sodas

Sodas induce cementation by reacting with minerals which produce insoluble silicates and aluminates. Caustic soda acts as a dispersant by degrading the minerals by alkaline attack.

The product reacts vigorously both with lateritic soils and with soils with a high aluminium content. The best strengths are obtained when an adequate curing period for the material is allowed. Caustic soda is not suitable for soils with high montmorillonite contents. The following are also known to be used:

- barium hydroxide: $Ba(OH)_2$, $8(H_20)$;
- calcium hydroxide: $Ca(OH)_2$;
- potassium hydroxide: KOH, $\frac{1}{2}$ H_2O;
- lithium hydroxide: $LiOH,H_2O$.

3. Salts

Salts acting on soils induce colloidal reactions, alter the characteristics of the water and lead to flocculation. By increasing the attraction between fine soil grains, salts help to create larger particles. This flocculation reaction leads to a reduction in density and an increase in OMC, in permeability and strength but also to a reduction in plasticity. Salts act on pore water and reduce the loss of water from the soil, slowing evaporation, and reducing water absorption. However, the effectiveness of treating soil with salt depend on the magnitude of the moisture movements in the stabilized material. The treatment is not always lasting as the salts can be leached out and dissolved when the material is moistened again. The quantity of soil required lies in the region of 0.5 to 3%. Four main salts are used:

- Sodium chloride: flocculant and aid to compaction; effective in non-saline soils;
- Calcium chloride: impermeabilizing agent;
- Ferric chloride: powerful coagulant and flocculant;
- Aluminium chloride: electrolytic coagulant, electrochemical consolidation of soil.

Salts should never be used in conjunction with cement.

4. Quaternary amine derivatives

Some cationic quaternary amine compounds are used alone or occasionally as secondary additives to cement or bitumens in concentrations of 5 to 10% of the cationic exchange capacity of the clay fraction. They act as binders and water repellents. They require a sophisticated production process and they are difficult to mix with the soil when only low concentrations are involved. These products are expensive and are not readily available. The most effective quaternary amine derivatives are the aromatic or aliphatic amines, and the amine salts.

These products are often effective in quantities as low as 0.5%. They form a water-repellent film around the grains, which because of its tensio-active properties reduce capillary water absorption. These treatments are particularly suitable where capillary rise presents problems and may be suitable for foundations which are constantly exposed to moisture. The products may lose their effectiveness if they are immersed in water or if they are totally dried out over an extended period.

5. Silicates

Sodium silicate is fairly cheap and is available in many parts of the world. The quantity usually used is 5% and it has proved to be attractive for stabilizing sandy soils, clayey and silty sandy soils, sands rich in limonite (some laterites) in arid regions, and in general soils which lack cohesion. Sodium silicate is not suitable for clay soils. Sodium silicate also acts as an impermeabilizing agent, particularly where a surface treatment of the materials is required. A curing period of at least seven days is required if the effectiveness of the treatment is to be assured. The product is highly soluble but can be rendered insoluble by allowing it to react with slaked lime. Some sodium silicate can be dissolved in water and is then known as 'waterglass'. Other silicates may be used such as potassium silicate and calcium silicate.

6. Stearates

Stearates are salts or esters of the stearic acid contained in animal fats. They act as impermeabilizing agents. Aluminium and magnesium stearates as well as zinc stearate may be suitable.

7. Paraffins

Paraffins are a mixture of solid saturated hydrocarbons characterized by their inertness in the presence of chemical agents. They can be used as a compaction agent but must first be dissolved in a fatty medium.

8. Waxes

Industrial waxes can be used as an aid to compaction. They are often added to other stabilizers.

9. Latexes

Industrial or synthetic latex dissolved in water and added at the rate of 3 to 15% can give good results. These products are binders and impermeabilizing agents.

10. Synthetic adhesives

Synthetic glues with one or two components.

11. Soaps

The use of between 0.1 and 0.2% of ionic detergent has no effect on strength but reduces water sensitivity by about 25%.

12. Industrial wastes

Certain industrial wastes can be used for stabilizing soil.

Sump oil It has no lasting effect because it is washed off by rain. It is an impermeabilizing agent.

Blast furnace slag These are silica slags which can approach Portland cement in composition. Some qualities of fly ash have no effect at all.

Lignin and lignosulphates These are by-products of the wood industry. Soluble in water, they can be rendered insoluble by mixing with chrome salts (potassium or sodium bichromate) resulting in a thick gel, known as chromolignin. Good impermeabilizing agents, but expensive.

Molasses A product of the sugar industry. It improves compressive strength and reduces capillarity. A quantity of 5% is suitable for sandy and silty soils. Add lime for clayey soils.

Pozzolanas If these are to be effective they must be used with lime.

Other products Plastified sulphur, sulphonates and siliconates (water repellents).

13. Plaster

Plaster or calcium sulphate is an attractive stabilizer for sandy soils which lack cohesion. It is not recommended for clayey soils. Plaster by itself gives good results when used in quantities which do not exceed 15%. There is a risk of premature setting, i.e. prior to moulding, and when small quantities are prepared problems may be encountered. Plaster can be combined with lime (1:1) but not with cement, as this can have disastrous results. A plaster-lime mix may be suitable for clayey soils which would not stand up well to water, if stabilized with plaster alone. When used for sandy soils moulded into adobes at least 5 to 10% of plaster is required, and this may rise to 20%. The wet strength of plaster-stabilized materials is equal to or lower than that of non-stabilized materials. Fruitless attempts at stabilizing materials with calcium anhydrite have been made.

The products mentioned below are commercially available. Many of these products are presented by the makers or their commercial agents as being 'miracle' products. It cannot be said too often how important it is to systematically verify the real performance of a product and to determine how well-founded the sales arguments are which are put forward in publicity, oral presentations, and in laboratory reports.

The majority of these commercial products are based on known industrial products and act in the same way. Nevertheless, manufacturers do not always describe their product in precise detail, and keep their formula secret. It will be noted that the majority of these products are not patented. This is because they are already public property. To give an example a product which is 90% sulphuric acid will be described as 'a liquid catalyst, soluble in water and inducing ion exchange'. Consequently it appears to be essential to insist that salesmen identify their product exactly and, if necessary, obtain the advice of specialists (e.g. from chemical engineers). This distrust does not necessarily lead to the systematic rejection of these products, as some them are effective, albeit for clearly specified applications.

Many of the products on the market have been formulated for use in road stabilization. A large number of these were developed in a military context in recent wars, with a view to achieving the quickest possible stabilization of impassable roads or for constructing landing areas and heliports on marshy land in the space of just a few hours. The service life of such applications was often not guaranteed to be anything more than a few months. The original intention of these products should therefore not be forgotten. The demands imposed by the temporary stabilization of roads and the conditions of use are very different from those demanded by the construction of permanent buildings. Some of these products, which are highly effective on roads, rapidly lose their effectiveness when used to stabilize walls. Laboratory reports must be interpreted in the light of these applications in the road construction field. Moreover a large number of tests and trials are carried out on samples prepared by the makers of the products themselves. This practice of course does makes it impossible for the laboratory to guarantee the product itself even though the results obtained on the material can be guaranteed.

The quantities of these products when used for stabilization are extremely low, being often of the order of 1% or 0.1% and even 0.01%. The difficulty of achieving uniform mixing can thus be appreciated.

For example when one tonne of soil must be processed the use of one kg of the product requires very thorough mixing and professional attention to the work.

Price studies carried out by manufacturers usually indicate a price clearly lower than that of cement stabilization, which is usually used as a reference figure. However, as soon as the suggested quantity proves not to be effective and it becomes necessary to increase the quantity required, there is a distinct danger of exceeding the budget. Furthermore the real sale price of these integrated products often includes an excessive profit margin, especially if one remembers that these products are only moderately effective and that they are the most common industrial chemicals. The prices are often set with respect to a ceiling of acceptability compared with conventional stabilizers. It is not unusual to observe that products with identical formulas can be purchased directly from suppliers of industrial chemicals for prices which may be 20 to 50% less.

The use of these products may prove to be satisfactory, but it is essential to carry out thorough preliminary tests and studies.

COMMERCIAL PRODUCTS			
PRODUCT	MANUFACTURER OR DISTRIBUTOR	COUNTRY	REMARKS
STABILIZERS			
CONSERVEX	CONSOLID AG	Switzerland-France	Impermeabilizing agent to be used with CONSOLID 444
CONSOLID 444	CONSOLID AG	Switzerland-France	Compacting aid to be used with CONSERVEX
MUREXIN SB-86	FORSTER & HAENDEL	Austria	Probably of the same nature as the CONSOLID products
MUREXIN SB-99	FORSTER & HAENDEL	Austria	Probably of the same nature as the CONSOLID products
UNIVEST	HÜLST	Germany	Liquid polymer
RENDER ADDITIVES			
ACROPOL	REX CANDY	Australia	–
DARAWELD-C	–	United States	High polymer resin emulsion
SUPERIOR ADDITIVE 200	EL REY	United States	Acrylic modifier for renders
UNI 719	F.I.T.	Australia	Alkali metal alkyl silicone compounds
WATERPROOFERS			
700 S	SICOF	France	Epoxy resin
ADOBE PROTECTOR	ELREY	United States	Water repellent
DARAWELD-C	–	United States	High polymer resin emulsion
DYNASYLAN FH	DYNAMIT NOBEL	France	Silicon resin, silane
PROTIDRAL	SARL CYBEO	France	Silicone resin, siloxane
REPELLIN S-101	PIDILITE IND. LTD	India	To be sprayed or brushed
ROCAGIL AL 6	RHONE POULENC	France	Acrylic resin, organic polymer
SD 104	RHONE POULENC	France	Acrylic resin, organic polymer
SOIL SEAL	–	United States	Latex acrylic balanced polymers spray mixed with water

AGRA. *Recherche Terre*. Grenoble, AGRA, 1983.
Alexander, M.L. *et al. Relative Stabilizing Effects of Various Limes on Clayey Soils*. In HRR, 1972.
Coad, J.R. *Parpaings de Terre Stabilisée à la Chaux*. In Bâtiment International, Paris, CIB, 1979.
Cytryn, *S. Soil Construction*. Jerusalem, the Weizman Science Press of Israel, 1957.
Doat, P. *et al. Construire en Terre*. Paris, Editions Alternatives et Parallèles, 1979.
Doyen, A. 'Objectifs et Méchanismes de la Stabilisation des Limons à Chaux Vive'. In *La Technique Routière*, Brussels, CRR, 1969.
Dunlap, W.A. *Soil Analysis for Earthen Buildings*. 2nd regional conference on earthen building materials, Tucson, University of Arizona, 1982.
El Fadil, A.A. *Stabilised Soil Blocks for Low-cost Housing in Sudan*. Hatfield Polytechnic, 1982.
Ephoevi-Ga, F. 'La protection des murs en banco', in *Bulletin d'Information*, Cacavelli, CCL, 1978.
Eurin, P.; Rubaud, M. *Etude Exploratoire de Quelques Techniques de Stabilisation Chimique de la Terre*. Grenoble, CSTB, 1983.
France, S. *et al. Traitement des Sols à l'Anhydrite en Vue de la Construction de Parpaings de Terre Stabilisée*. Douai, Ecole des ingénieurs des Mines, 1978.
Gallaway, B.M.; Buchanan, S.J. *Lime Stabilization of Clay Soil*. Agricultural and mechanical college of Texas, 1951.
Geatec. *Etude d'une Terre Crue Renforcée à la Résine Furanique*. Venelles, Geatec, 1982.
Grésillon, J.M. 'Etude de l'aptitude des sols à la stabilisation au ciment. Application à la construction'. in *Annales de l'ITBTP*, Paris, ITBTP, 1978.
Habibaghi, K.; Nostaghel, N. 'Methods of improving low-cost construction materials against earthquake'. In *New Horizons in Construction Conference*, Envo publishing company.
Hammond, A.A. 'Prolongation de la durée de vie des constructions en terre sous les tropiques'. In *Bâtiment Build International*, Paris, CSTB, 1973.
Herzog, A.; Mitchell, J.K. 'Reactions accompanying the stabilization of clay with cement'. 42nd annual meeting of the HRB, Washington, HRB, 1963.
Houben, H. *Technologie du Beton de Terre Stabilisé pour l'Habitat*. Sidi Bel Abbes, CPR, 1974.
Ingles, O.G. 'Advances in soil stabilization'. In *Pure and Applied Chemistry Revue*, 1968.
Ingles, O.G.; Lee, I.K. *Compaction of Coarse Grained Sediments*. Amsterdam, G.V. Chilingarian and K.H. Wolf, 1975.
Ingles, O.G.; Metcalf, J.B. *Soil Stabilization*. Sydney, Butterworths, 1972.
Lilley, A.A.; Williams, R.I.T. Cement-stabilized materials in Great Britain. In *Highway Research Record*, Washington, HRB, 1973.
Markus, T.A. *et al. Stabilised Soil*. Glasgow, University of Strathclyde, 1979.
Markus, T.A. 'Soil stabilization by synthetic resins'. In *Modern Plastics*, New York, 1955.
Martin, R. 'Etude du renforcment de la terre à l'aide de fibres végétales'. In *Colloque Construction en Terre* 1984, Vaulx-en-Velin, ENTPE, 1984.
Mitchell, J.K.; El Jack, S.A. 'The fabric of soil cement and its formation'. In *Fourteenth national conference on clays and clay minerals*.
Patty, R.L. 'Soil and mixtures for earth walls'. In *Agricultural Engineering*, Saint Joseph, ASAE, 1942.
Ringsholt, T.; Hansen, T.C. 'Lateritic soil as a raw material for building blocks'. In *American Ceramic Society Bulletin*, 1978.
Rocha Pitta M. *Uma proposta para o estabelecimento de um método de dosagem de solo-cimento para uso na construção de moradias*. São Paulo, Associação Brasileira de Cimento Portland, 1979.
Seed, H.B. *et al.* 'The strength of compacted cohesive soils', *ASCE research conference on the shear strength of cohesive soils*, Denver, University of Colorado, 1960.
Stulz, R. *Appropriate building materials*. St. Gallen, SKAT, 1981.
The Asphalt Institute. *The Asphalt Handbook*. Maryland, The Asphalt Institute, 1975.
Uzomaka, O.J. Performance characteristics of plain and reinforced soil blocks. In *The International Conference on Materials of Construction for Developing Countries*, Bangkok, AIT, 1978.
Williams, W. 'Construction of homes using on-site materials'. In *International Journal IAHS*, New York, Pergamon Press, 1980.

The regional base of the Ministry of Hydraulics, Niger, built using compressed earth blocks (architect: Josep Esteve, CRATerre-EAG)

5. SOIL SUITABILITY

Immense experience has been acquired with building in earth in recent times. It is, however, far from being complete. The criteria for the suitability of a soil currently in use are far from final and should not be too literally interpreted.

The majority of design charts used have been borrowed from road engineering techniques and these are fairly suitable for earth construction. Many suitability criteria, however, were drawn up on a regional scale and are therefore not always universal in application. The best course is to draw on their general approach while adapting them to local conditions. They should be used primarily for their qualitative information.

Interpretation should be as flexible as possible as should be their amendment, taking into account the ranges of values, which can to some extent be enlarged and still provide good results. Even so only experienced personnel who can appreciate the consequences of their use should be allowed to interpret suitability criteria.

In stabilization, for example, it is possible to depart to some extent from the ideal conditions described by the suitability criteria, but at the same time the dangers involved in doing so must be appreciated. When a well-defined soil is used regularly, experience and know-how will confirm the accuracy of the values considered (e.g. proportions). It should, however, never be forgotten that the economy of the final product depends basically on choosing a new soil. For large-scale projects preliminary comparative tests can be carried out on test walls built using different methods and make it possible to find the most suitable solution. Apart from these, indicator and (if necessary) laboratory tests are essential if the greatest efficiency is to be obtained. Similarly the suggested suitability criteria must be regarded as a starting point. They cannot be considered as binding recommendations and even less as standards.

Can this soil be used for construction?

There is no direct reply which can be reasonably made to this question. It is preferable to adopt a gradual approach by asking a series of questions of the type:

– What do you intend to build? An outside wall? A single-storey dwelling? A building with several floors? And so on.
– Where do you intend to build? In an earthquake region? In a dry or a wet region, etc.?
– What will it be doing? A bearing or non-bearing wall? An inside or outside wall? An arch, vault, or dome? A terrace? Rendered or not? Will there be any form of protection? And similar questions.
– What means are available? Can it be stabilized or not? Can the soil be improved? And so forth.

A reply to the question 'Can this soil be used for construction?' is both impossible – if a too direct reply is given – and disconcertingly easy – if care is taken in replying. In general all soil with good cohesion can be used for construction, but it is advisable to make sure that all the means for using it are available. Another consideration becomes important where the suitability of the soil for a particular building technique is being questioned, or, in the reverse case, where the building technique is being questioned in the light of the available soil. Should the soil be changed, if it does not suit the technique, or should it be improved so that it is suitable? And again should another technique be chosen, if it is not suited to the available soil, or should the technique be modified so that it is suitable? Suitability criteria and reference nomograms will guide choice but their use remains problematical. They should not be interpreted too strictly and are best used only by competent persons. Thus when theoretical decisions are taken, tests must be made to check on practical performance.

ENGINEERING PROPERTIES OF THE VARIOUS SOIL COMPONENTS

++ VERY HIGH; + HIGH; M MEDIUM; – LOW; –– VERY LOW

GROUP DESIGNATION	MEAN SIZE	PERMEABILITY DRY	PERMEABILITY WET	VOLUME STABILITY	PLASTICITY AND COHESION	COMPACTIBILITY AT OMC	DURABILITY (SPRINKLING)	ABRASIVENESS
SILTS		–	+	++	––	M	+	M
VERY FINE SAND	1 μ	––	++	++	––	+	++	++
MICA	1 μ	M	+	++	–	––	––	–
CARBONATE	ANY	M	M	++	–	++	++	–
SULPHATE	> 1 μ	M	M	++	–	+	–	M
ALLOPHANE	ANY	M	++	M	++	++	–	–
KAOLIN	≃ 1 μ	–	–	+	M	–	+	–
ILLITE	≃ 0.1 μ	––	––	–	+	M	M	––
MONTMORILLONITE	≤ 0.01 μ	––	––	––	++	––	+	––
CHLORITE	≃ 0.1 μ	–	–	–	M	M	M	–
ORGANIC MATTER	ANY	++	++	+	M	––	––	–

SOIL		SHRINKAGE AND SWELLING	SENSITIVITY TO FROST ACTION	BULK DENSITY AT OMC (kg/m^3)	VOIDS RATIO (ρs = 2700 kg/cm^3)	COMPRESSIVE STRENGTH DRY	GENERAL SUITABILITY (WITHOUT STABILIZATION)
GW	Clean gravel Well graded	ALMOST NONE	ALMOST NONE	> 2000	< 0.35		NOT SUITABLE. FINE SOIL SHOULD BE ADDED
GP	Clean gravel Poorly graded	ALMOST NONE	ALMOST NONE	> 1840	< 0.45		NOT SUITABLE. FINE SOIL SHOULD BE ADDED
GM	Silty gravel	ALMOST NONE	SLIGHT TO MEDIUM	> 1760	< 0.50		SUITABLE, BUT LACKS COHESION. ERODES EASILY. ADD FINE SOIL
GC	Clayey gravel	VERY SLIGHT	SLIGHT TO MEDIUM	> 1920	< 0.40		SUITABLE. SOMETIMES FINE SOILS SHOULD BE ADDED
SW	Clean sand Well graded	ALMOST NONE	ALMOST NONE	> 1920	< 0.40		NOT SUITABLE. FINE SOIL SHOULD BE ADDED
SP	Clean sand Poorly graded	ALMOST NONE	ALMOST NONE	> 1600	< 0.70		NOT SUITABLE. FINE SOIL SHOULD BE ADDED
SM	Silty sand	ALMOST NONE	SLIGHT TO HIGH	> 1600	< 0.70		SUITABLE, BUT LACKS COHESION. ERODES EASILY. ADD FINE SOIL
SC	Clayey sand	SLIGHT TO MEDIUM	SLIGHT TO HIGH	> 1700	< 0.60		SUITABLE. SOMETIMES FINE SOIL SHOULD BE ADDED
CL	Low-plasticity clay	MEDIUM TO HIGH	SLIGHT TO HIGH	> 1520	< 0.80	SLIGHT TO HIGH	SOMETIMES SUITABLE. SANDY SOIL SHOULD BE ADDED
ML	Low-plasticity silt	SLIGHT TO HIGH	MEDIUM TO VERY HIGH	> 1600	< 0.70	VERY SLIGHT	SUITABLE, BUT EVENTUALLY LACKS COHESION
OL	Organic silt and clays with low plasticity	MEDIUM TO HIGH	MEDIUM TO HIGH	> 1440	< 0.90		NOT SUITABLE. SOMETIMES ACCEPTABLE
CH	Highly plastic clay	HIGH	VERY SLIGHT	> 1440	< 0.90	MEDIUM TO VERY HIGH	RARELY SUITABLE. SANDY SOIL SHOULD BE ADDED
MH	Highly plastic silt	HIGH	MEDIUM TO HIGH	> 1600	< 0.70	VERY SLIGHT TO MEDIUM	VERY RARELY SUITABLE
OH	Highly plastic organic silt and clay	HIGH	VERY HIGH	> 1600	< 0.70	MEDIUM TO HIGH	NOT SUITABLE
Pt	Peat and other highly organic soils	VERY HIGH	SLIGHT				SUITABLE AS SOD

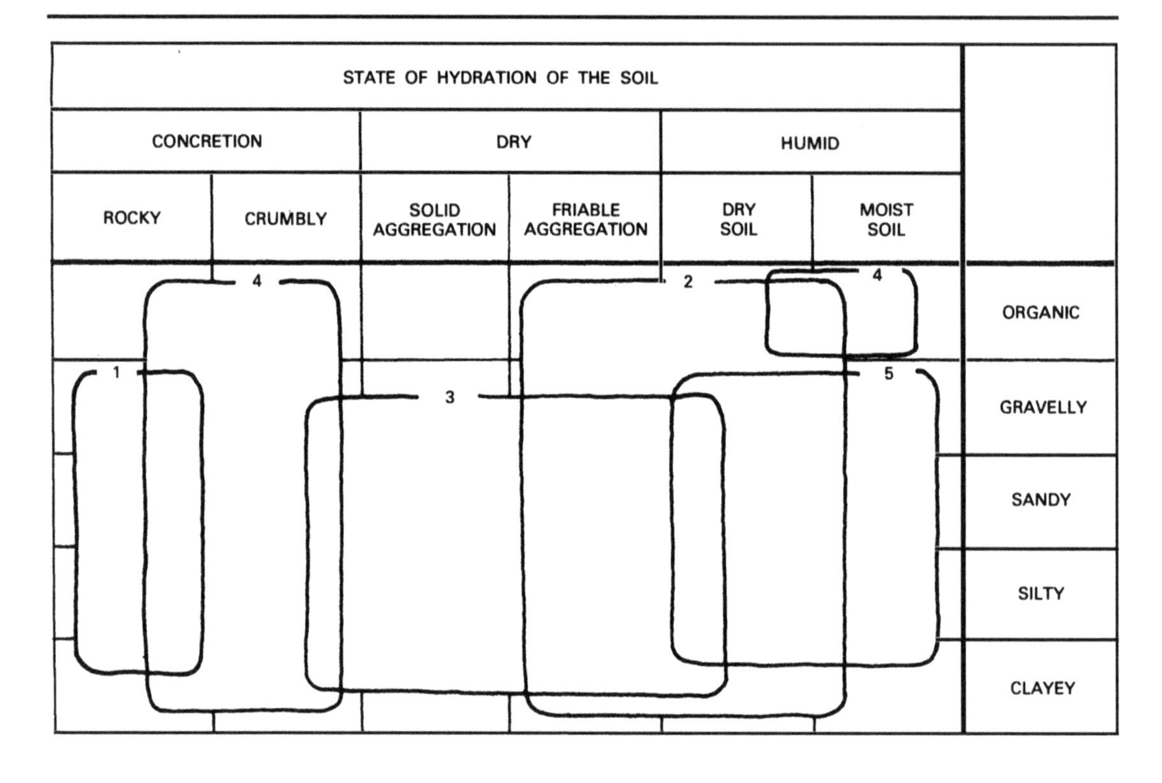

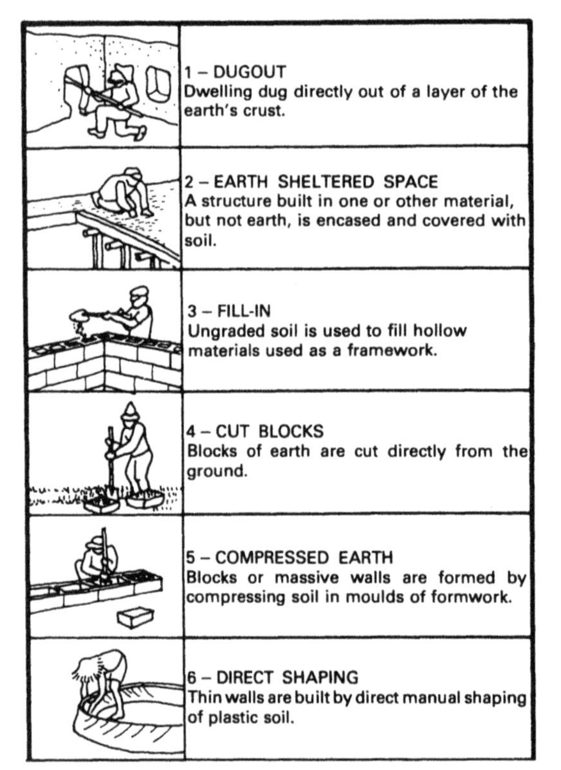

1 – DUGOUT
Dwelling dug directly out of a layer of the earth's crust.

2 – EARTH SHELTERED SPACE
A structure built in one or other material, but not earth, is encased and covered with soil.

3 – FILL-IN
Ungraded soil is used to fill hollow materials used as a framework.

4 – CUT BLOCKS
Blocks of earth are cut directly from the ground.

5 – COMPRESSED EARTH
Blocks or massive walls are formed by compressing soil in moulds of formwork.

6 – DIRECT SHAPING
Thin walls are built by direct manual shaping of plastic soil.

Rocky concretion Monolithic agglomerations of coarse material; compact and heavy soil which is difficult to cut.

Crumbly concretion Agglomerations of crumbly or decomposed material, including peat and sod, which is easy to cut.

Solid aggregation Perfectly dry soil in large, solid lumps.

Friable aggregation Absolutely dry soil in powder form.

Dry soil Soil characterized by a naturally low humidity (4 to 10%); it is dry rather than moist to the touch.

Moist soil Soil unmistakably moist to the touch (8 to 18%), but cannot be shaped because of its lack of plasticity.

* Values for moisture content are only indicative and vary greatly according to soil type.

	STATE OF HYDRATION OF THE SOIL					
	PLASTIC		SOFT		LIQUID	
	SOLID PASTE	SEMI-SOLID PASTE	SEMI-SOFT PASTE	SOFT PASTE	MUD	SLURRY
ORGANIC						
GRAVELLY			7		10	
SANDY		9	8			
SILTY	6			12		11
CLAYEY						

Solid paste An earth ball (moisture content 15 to 30%) which flattens only slightly when dropped from a height of one metre is formed by powerful kneading with the fingers.

Semi-solid paste Only slight finger pressure is sufficient to form an earth ball (moisture content 15 to 30%) which flattens slightly but does not disintegrate when dropped from a height of one metre.

Semi-soft paste With this very homogenous material it is very easy to shape an earth ball which is neither markedly sticky nor soiling (moisture content 15 to 30%) that flattens markedly without disintegrating when dropped from a height of one metre.

Soft paste This kind of soil is so adhesive and soiling (moisture content 20 to 35%) that it is extremely difficult, if not impossible, to make balls from it.

Mud This kind of soil is saturated with water and forms a viscous, more or less liquid mass.

Slurry This consists of a suspension of clayey earth in water and constitutes a highly liquid fluid binder.

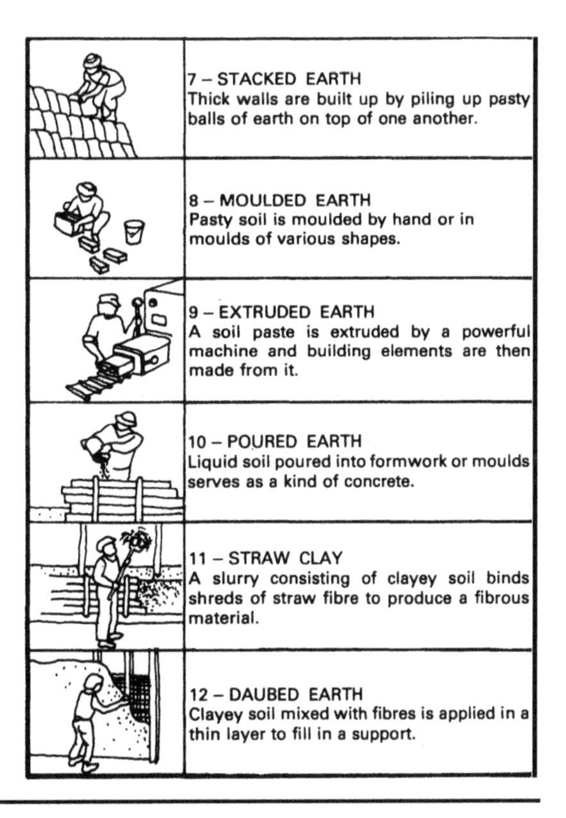
7 – STACKED EARTH
Thick walls are built up by piling up pasty balls of earth on top of one another.

8 – MOULDED EARTH
Pasty soil is moulded by hand or in moulds of various shapes.

9 – EXTRUDED EARTH
A soil paste is extruded by a powerful machine and building elements are then made from it.

10 – POURED EARTH
Liquid soil poured into formwork or moulds serves as a kind of concrete.

11 – STRAW CLAY
A slurry consisting of clayey soil binds shreds of straw fibre to produce a fibrous material.

12 – DAUBED EARTH
Clayey soil mixed with fibres is applied in a thin layer to fill in a support.

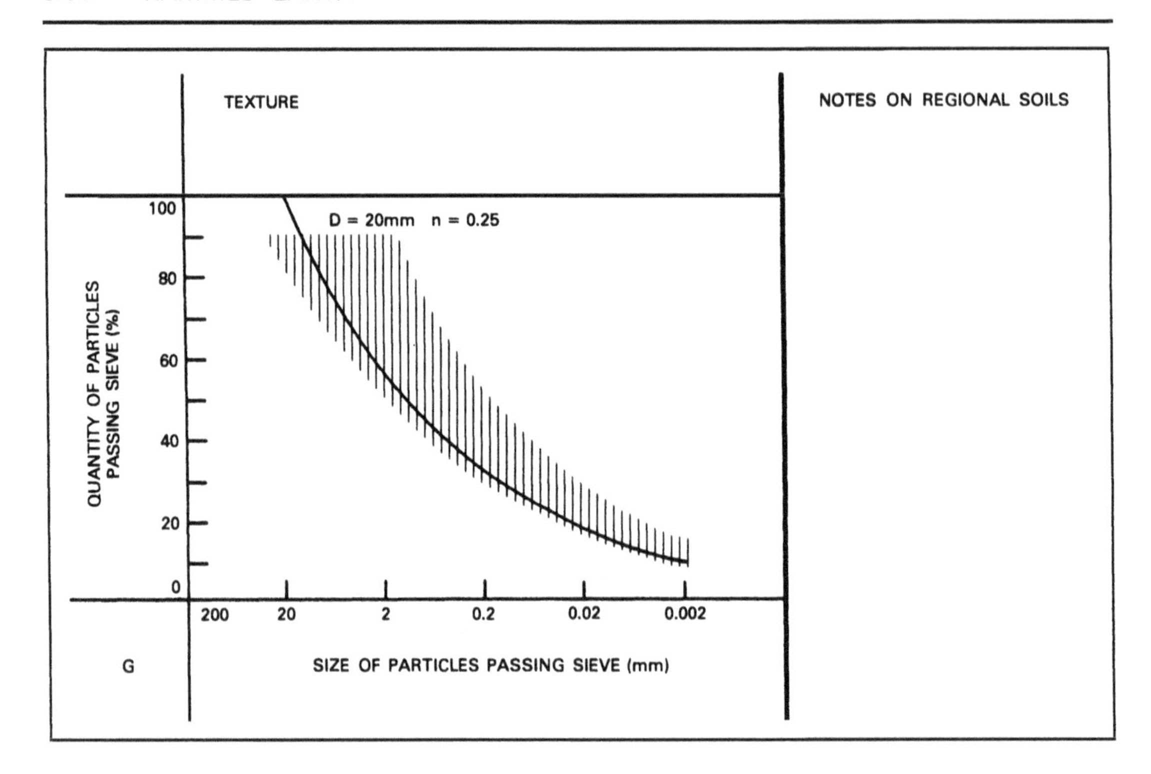

The limits of the zones recommended are approximate. The tolerances permitted vary considerably. Present knowledge does not justify the application of narrow limits. It is admitted that many soils which fail, for one reason or another, to comply with the requirements have been found satisfactory in practice;

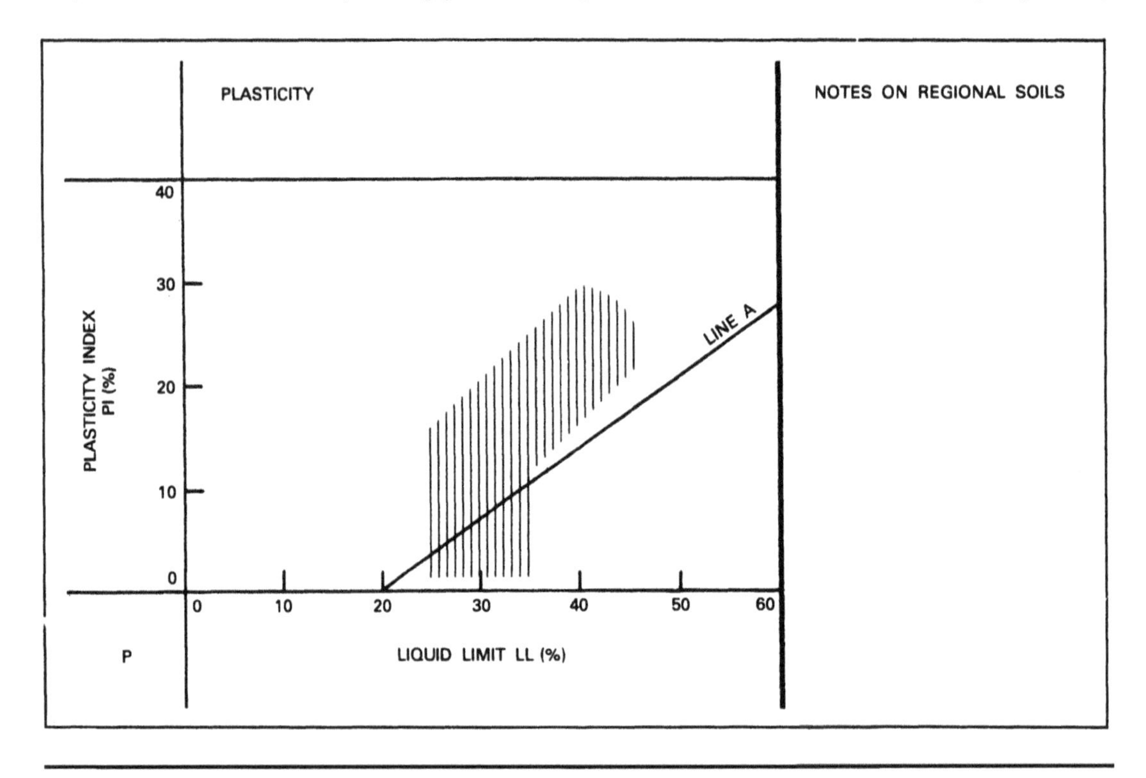

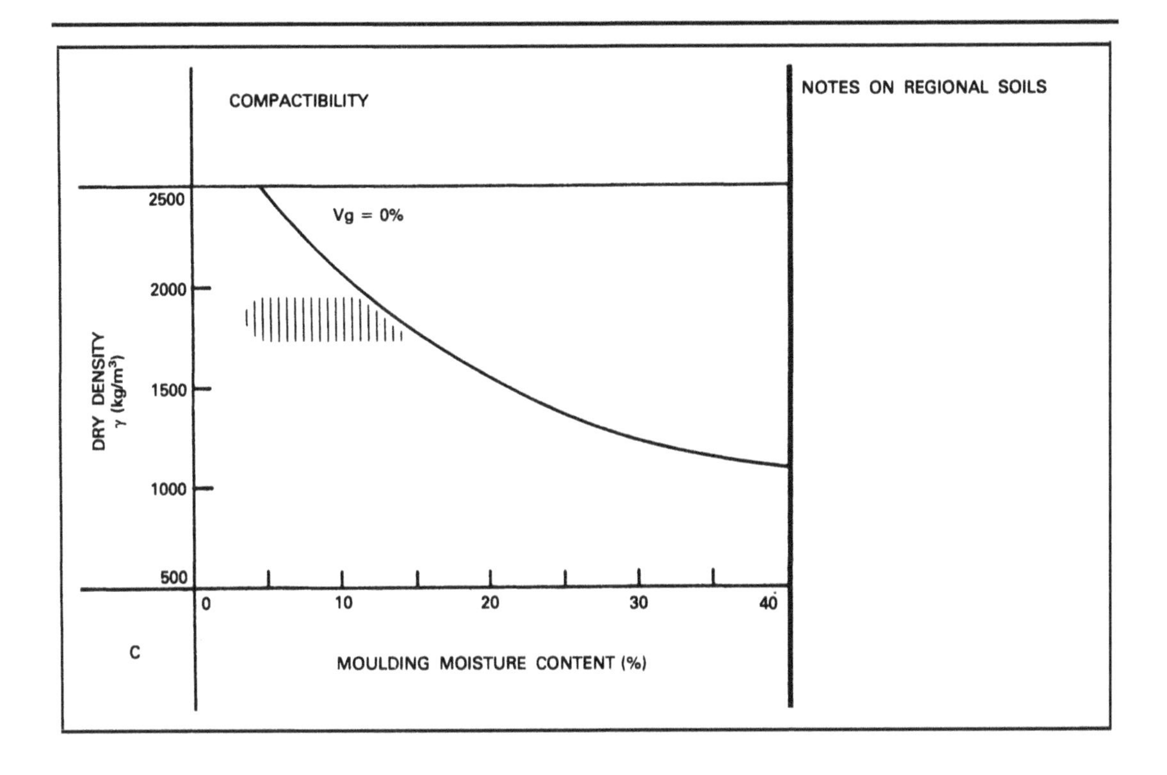

all that is claimed for the recommendation is that materials which comply with it are more likely to be satisfactory than those which do not. The zones are intended to provide guidance and are not intended to be applied as a rigid specification.

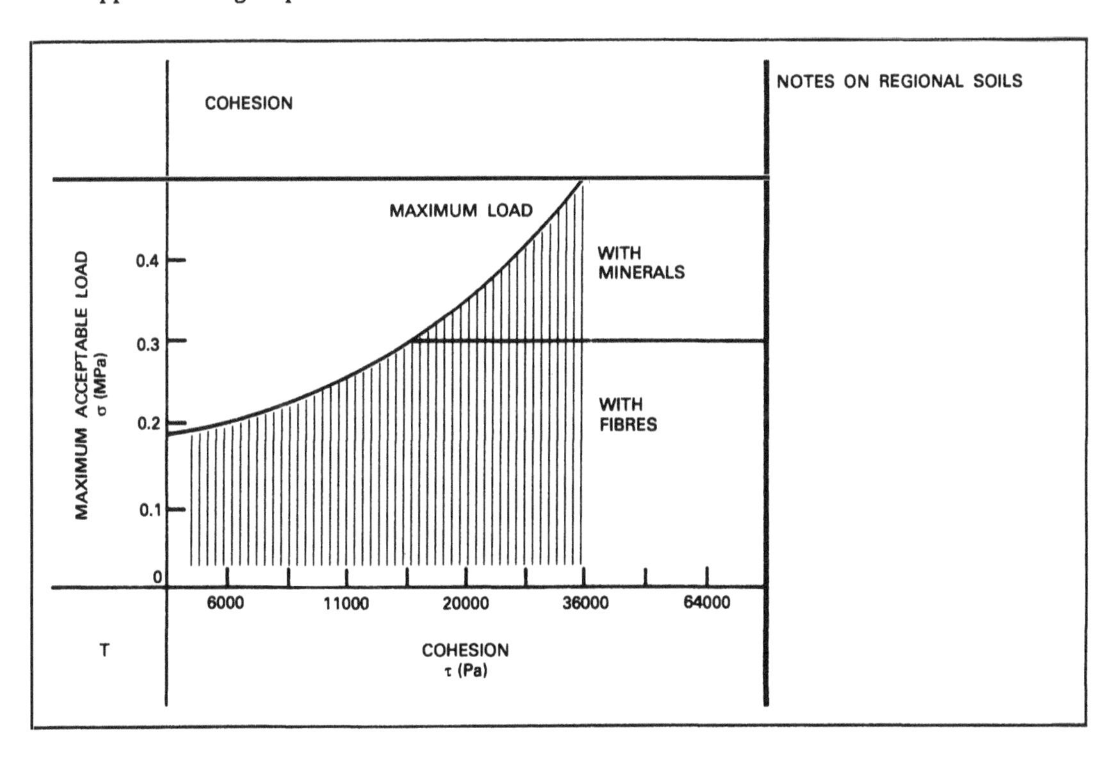

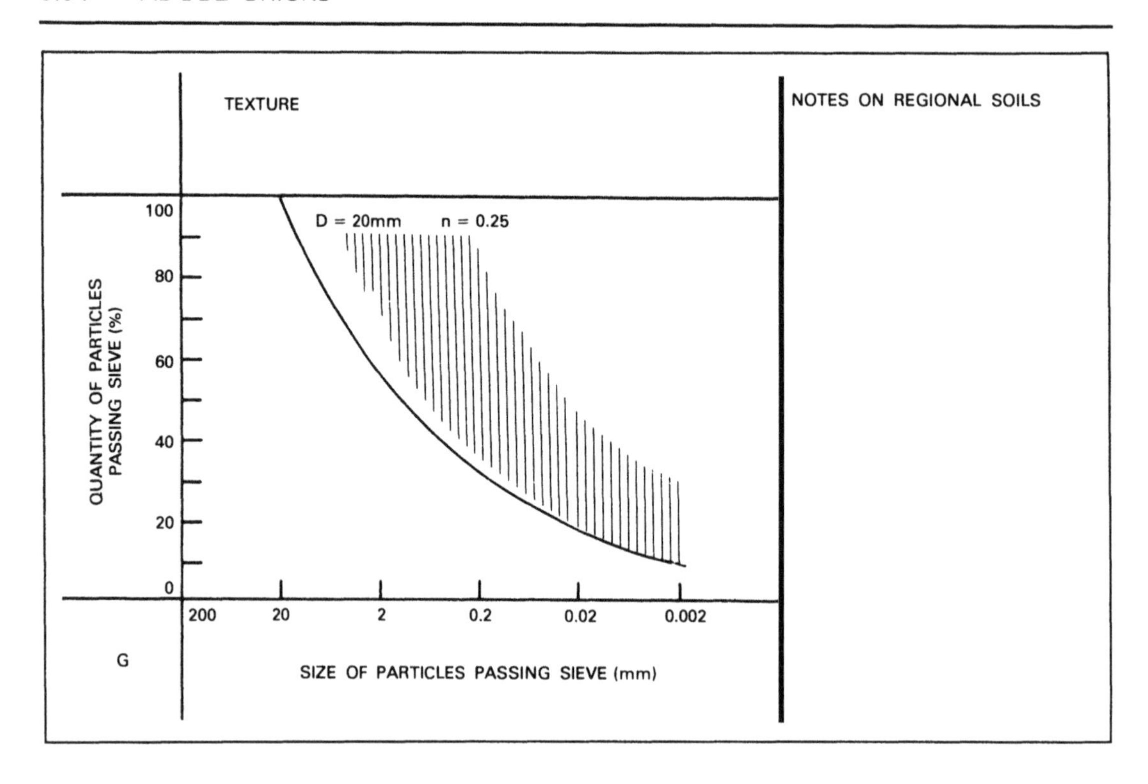

The limits of the zones recommended are approximate. The tolerances permitted vary considerably. Present knowledge does not justify the application of narrow limits. It is admitted that many soils which fail, for one reason or another, to comply with the requirements have been found satisfactory in practice;

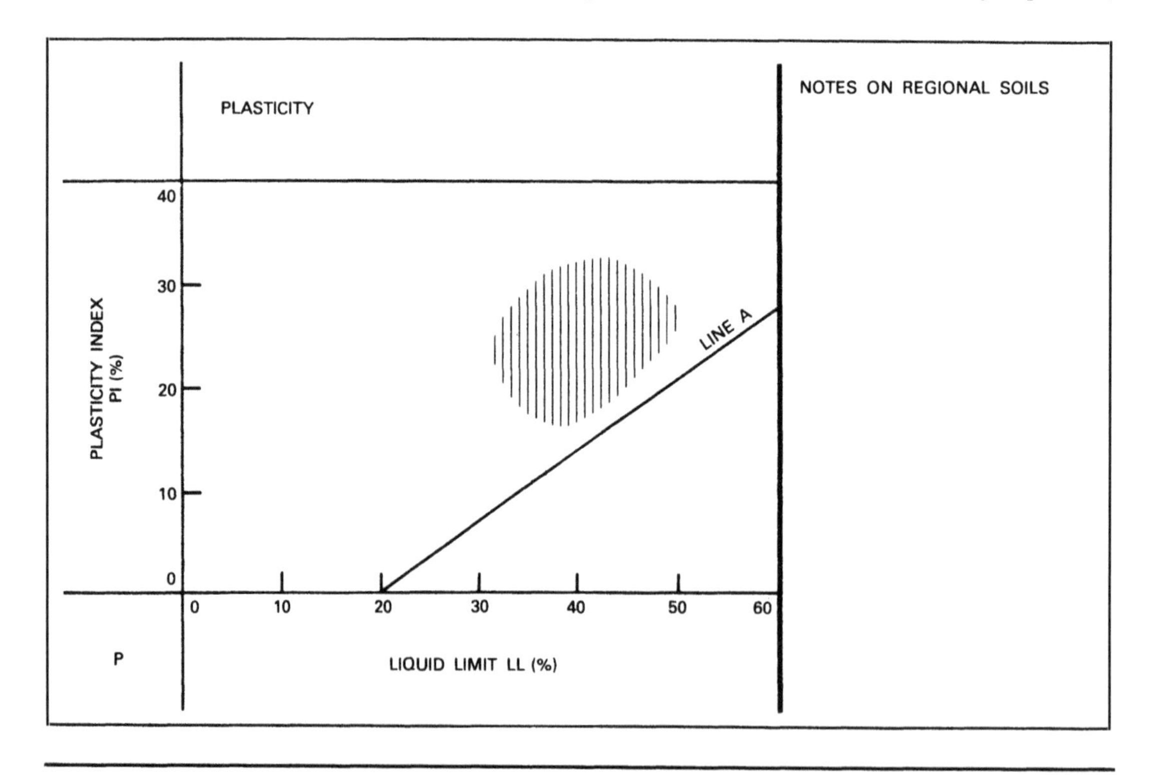

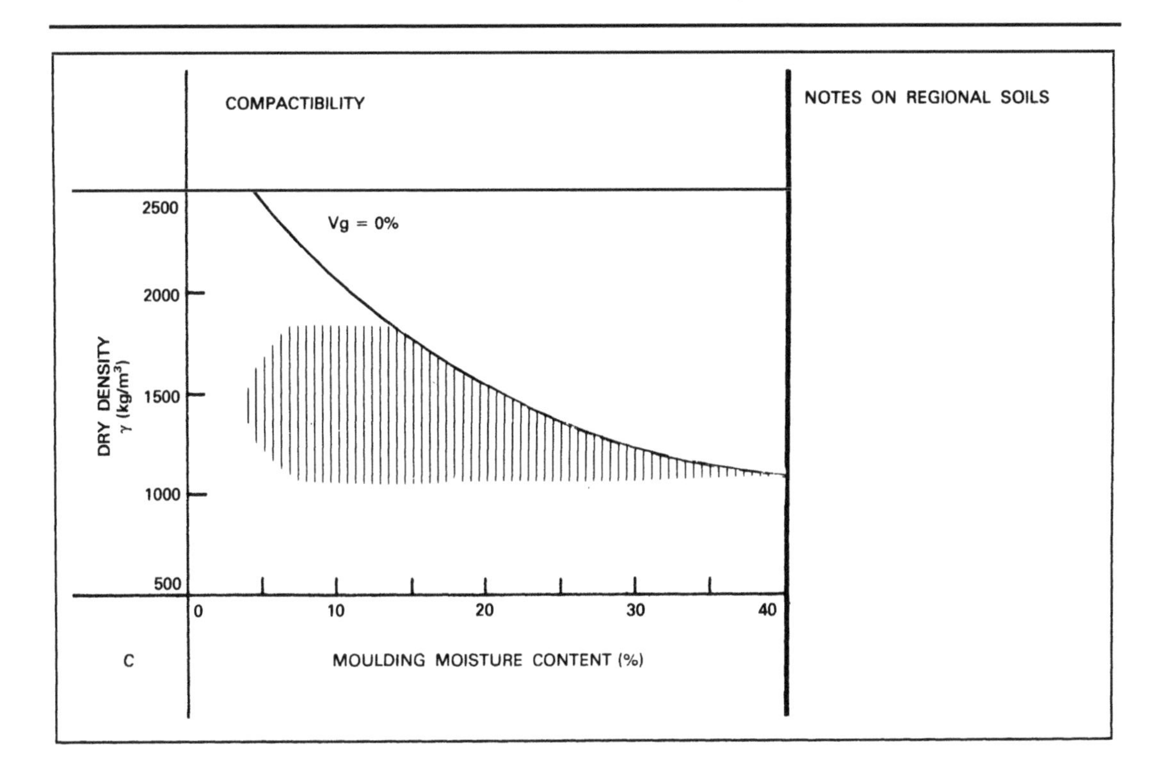

all that is claimed for the recommendation is that materials which comply with it are more likely to be satisfactory than those which do not. The zones are intended to provide guidance and are not intended to be applied as a rigid specification.

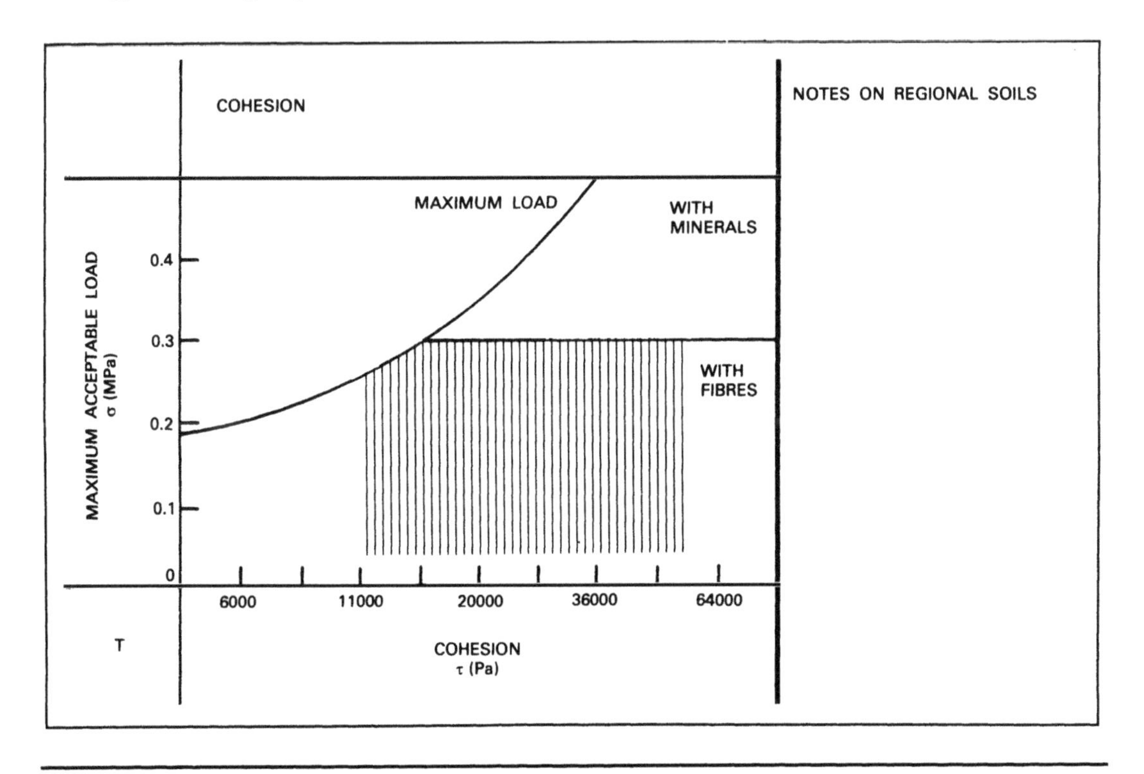

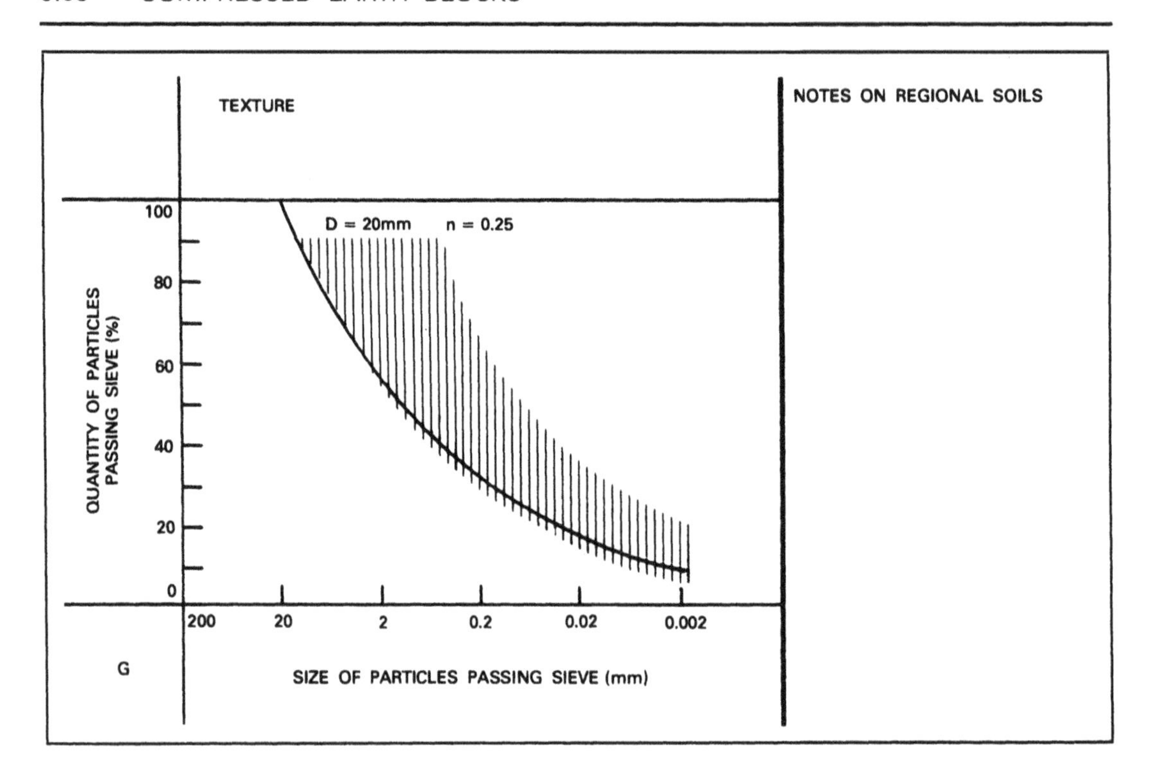

The limits of the zones recommended are approximate. The tolerances permitted vary considerably. Present knowledge does not justify the application of narrow limits. It is admitted that many soils which fail, for one reason or another, to comply with the requirements have been found satisfactory in practice;

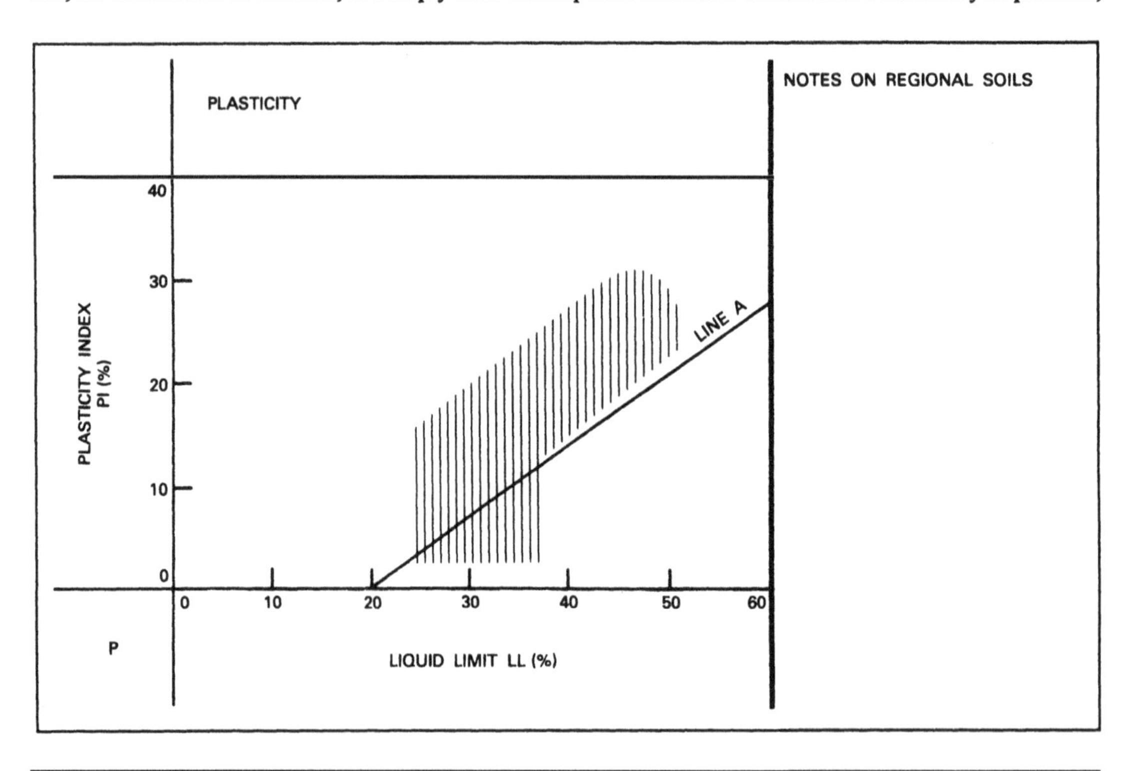

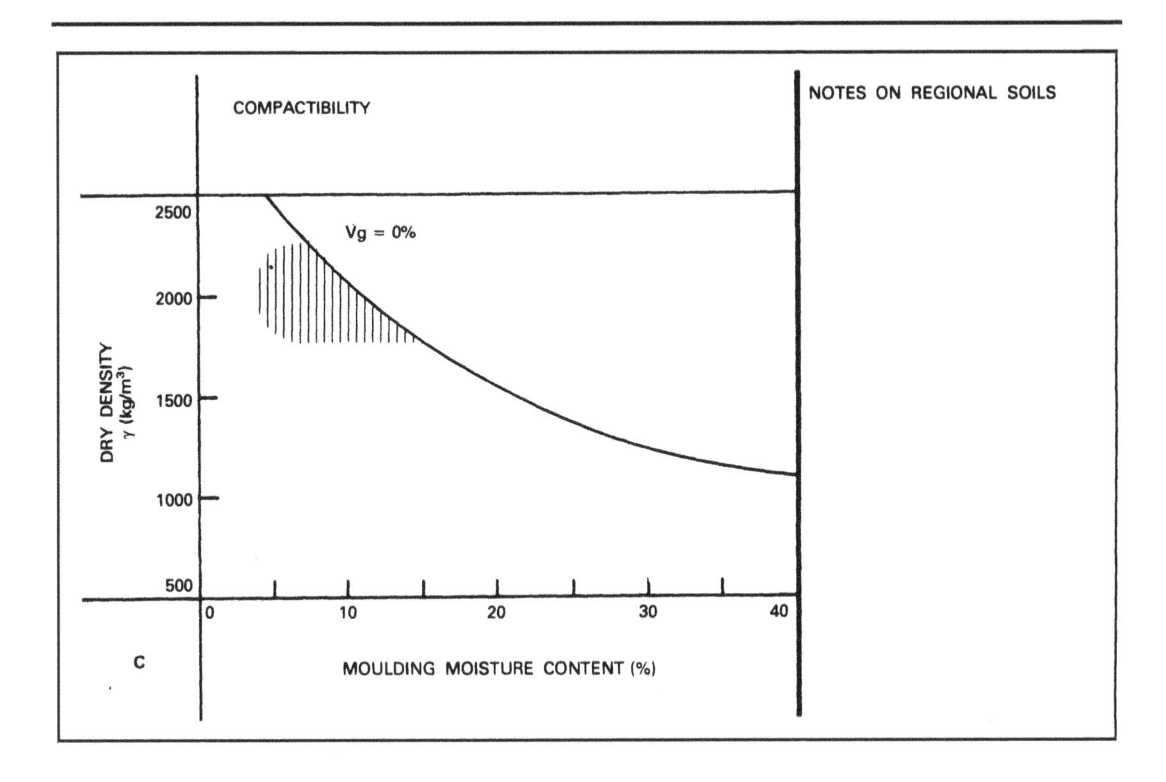

all that is claimed for the recommendation is that materials which comply with it are more likely to be satisfactory than those which do not. The zones are intended to provide guidance and are not intended to be applied as a rigid specification.

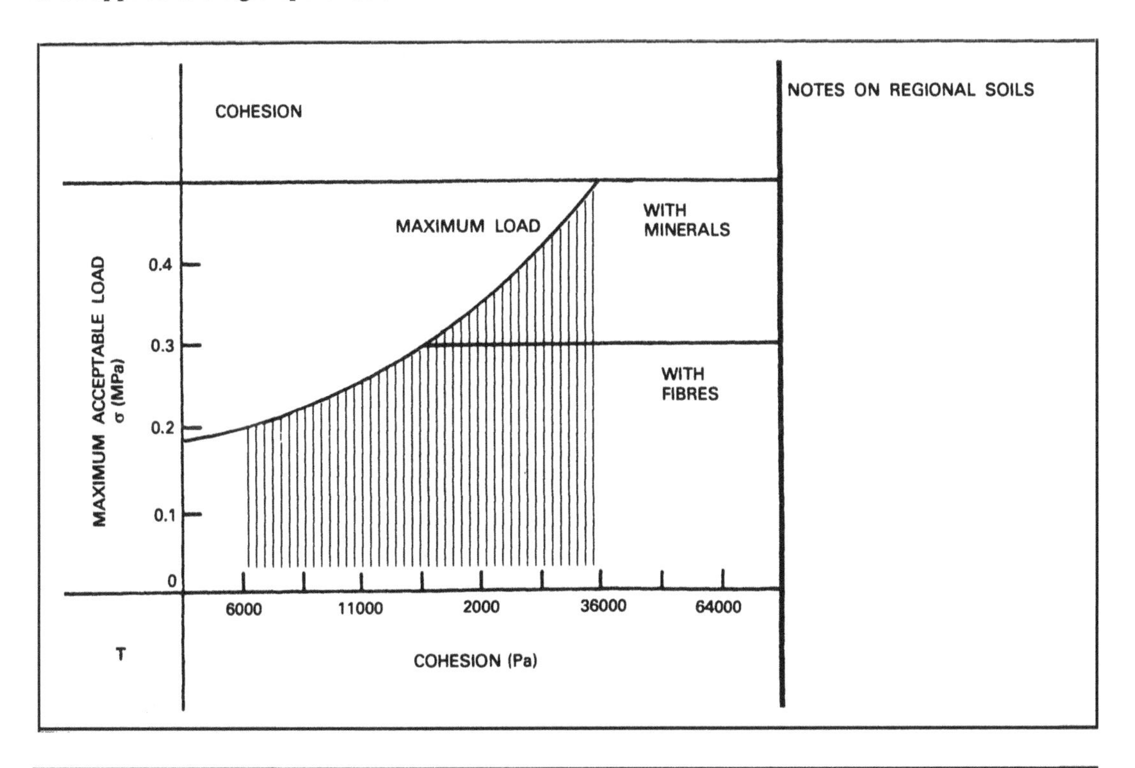

When the properties of a soil are not entirely satisfactory, it may be possible to use stabilization in order to bring about an improvement. The knowledge of the general suitability of soils and a certain skill in identifying soils will serve as a guide to decisions on stabilization.

The general rule which says that 'cement and bitumen is good for sandy soils and lime is good for clayey soils' is perfectly applicable but ignores many other stabilization methods. It is nonetheless true that the main stabilization methods make use of compaction, fibre, aggregate, cement, lime, or bitumen. Even so there are many other methods and products, but either their efficiency is less, or they are applicable to fewer soils. Furthermore these other methods may be extremely expensive and thus be ruled out for economic reasons.

Decision-making may make use of three main procedures:

1. by referring to nomograms. These are based essentially on information from the road engineering field. Care must be taken to interpret them correctly.
2. by performing direct tests. Measurements of shrinkage or pH, for example, which make it possible to judge suitability directly and what proportion of stabilizer is required.
3. by carrying out all the required tests on samples or on sample bricks.

In general it is better to try for a good result with a minimum of stabilizer, than trying for the best possible result. It should be remembered that laboratory conditions differ from field conditions, and that these may require a 150% increase in the amount of stabilizer.

ASTM SOIL CLASSIFICATION			CEMENT STABILIZATION	LIME	BITUMEN
GRANULAR SOILS	A1	A-1-a	o		
		A-1-b	o		
	A2	A-2-4	o		
		A-2-5	o		
		A-2-6	o	o	
		A-2-7	o	o	
	A3		o		o
FINE SOILS	A4		o	o	
	A5		o	o	
	A6		o	o	
	A7	A-7-6	o	o	
		A-7-5	o		

BITUMEN	LIME	CEMENT STABILIZATION	USCS SOIL CLASSIFICATION	
		o	GW	GRANULAR SOILS
		o	GP	
	o	o	GM	
	o	o	GC	
o		o	SW	
o		o	SP	
o		o	SM	
o		o	SC	
	o	o	CL	FINE SOILS
		o	ML	
		o	OL	
	o	o	CH	
		o	MH	
		o	OH	

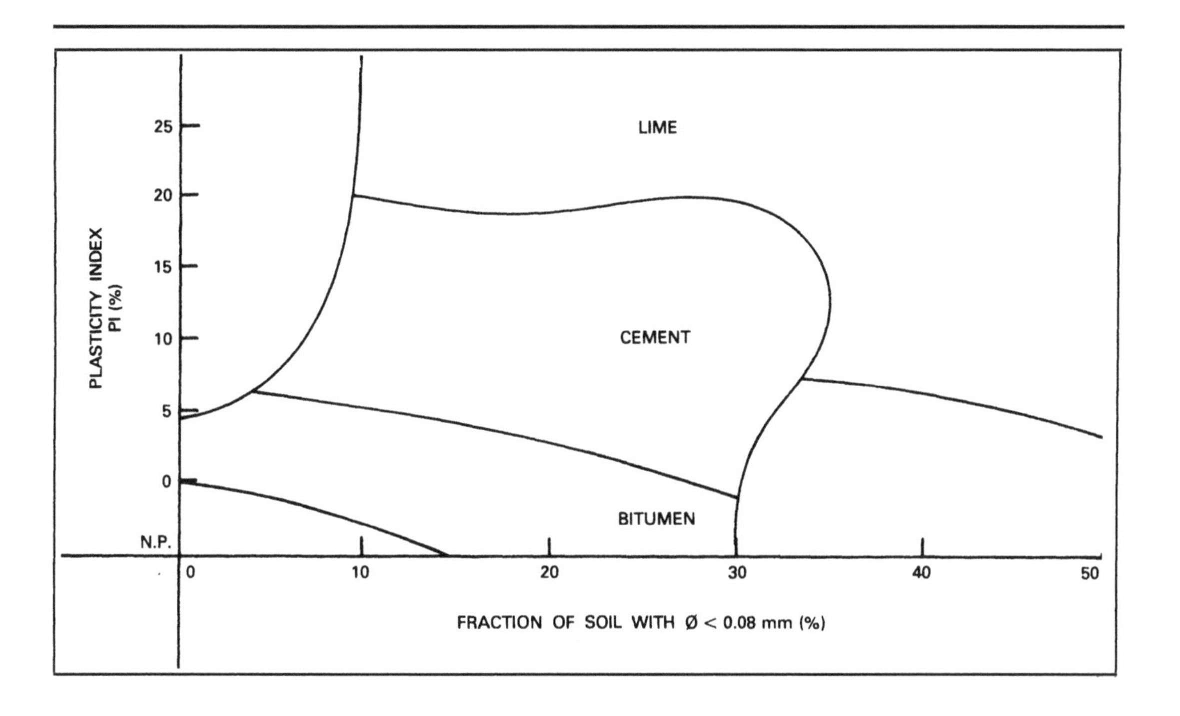

CEMENT	LIME	BITUMEN	POLYMERIC ORGANIC	MECHANICAL	DOMINANT SOIL COMPONENT	RECOMMENDED STABILIZER	REASONS
o		o		o	COARSE SANDS	CLAY LOAM	FOR MECHANICAL STABILITY
o		o	o	o	FINE SANDS	CEMENT BITUMEN	FOR DENSITY AND COHESION FOR COHESION
o			o	o	COARSE SILTS		
o	o		o	o	FINE SILTS		
o	o		o	o	COARSE CLAYS		
o	o				FINE CLAYS		
					ALLOPHANES	LIME LIME + GYPSUM	FOR POZZOLANIC STRENGTH AND DENSIFICATION
					KAOLIN	CEMENT LIME	FOR EARLY STRENGTH, WORKABILITY AND LATER STRENGTH
					ILLITES	CEMENT LIME	FOR EARLY STRENGTH FOR WORKABILITY AND LATER STRENGTH
					MONTMORIL-	LIME LONITES	FOR WORKABILITY AND EARLY STRENGTH
					CHLORITES		

Usage criteria applicable to soil stabilized with fibre and minerals exist. These were drawn up in the 1940s as the result of prolonged laboratory research on a very large number of samples. These laboratory observations were subsequently enriched by a mass of practical construction experience. Since then these criteria have been successfully applied to thousands of projects. Nevertheless, these criteria were established in Germany, and are above all applicable to the soils of that region, which are of a loess-based silt sort. It is possible that these criteria can be be adapted to other soil types, but only after extensive verification.

These criteria for fibre-stabilized or mineral-stabilized soils are accompanied by the maximum compression rates to which the materials may be subjected and at which they can be used in all safety. It may thus be observed that fibre-stabilized soils should not be worked with at above 0.3MPa and that mineral-stabilized soils have a limit of 0.5MPa

These are the maximum compression rates. Certain soils may not be able to achieve these rates, so the values given by the performance curve should not be interpreted as permissible values but as values which should in no case be exceeded.

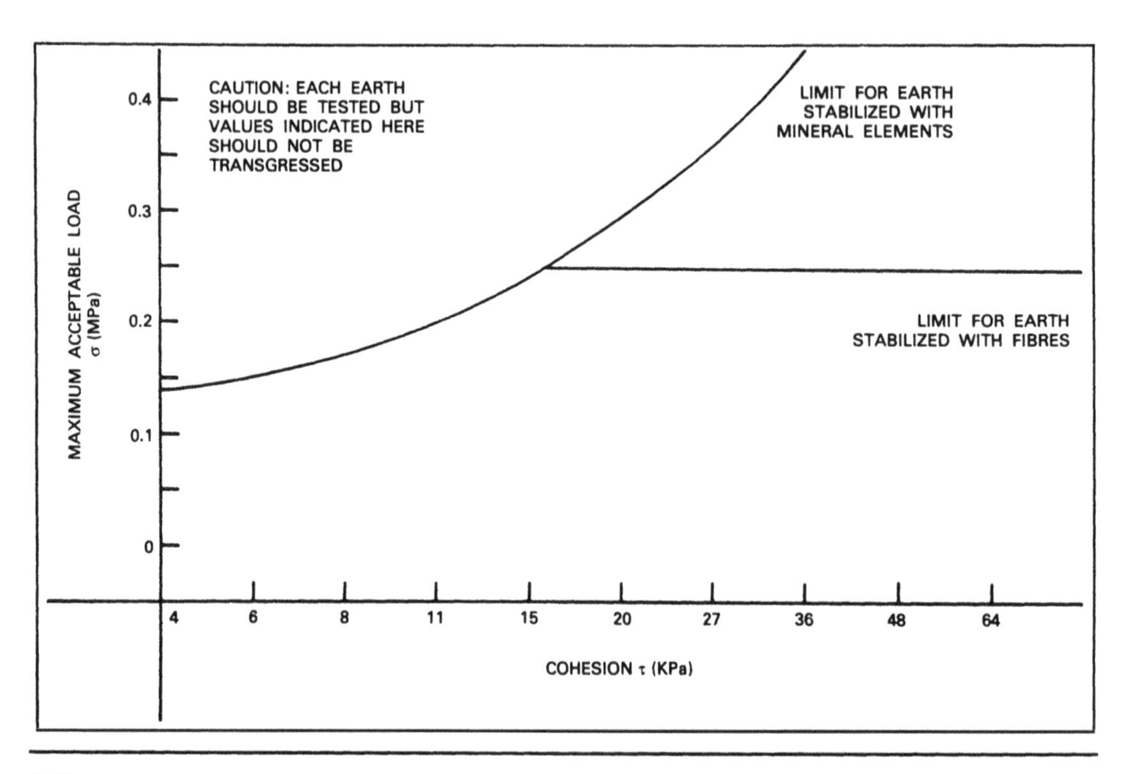

RESULTS OF COHESION TEST	STABILIZATION WITH FIBRES (kg/m^3 LOOSE EARTH)					STABILIZATION WITH MINERAL ELEMENTS (VOLUME)			
T	RAMMED EARTH	ADOBE	COMPRESSED BLOCKS	COB	STRAW CLAY	RAMMED EARTH	COMPRESSED BLOCKS	POURED EARTH ($700kg/m^3$)	POURED EARTH ($300kg/m^3$)
KPa	FIBRE LENGTH 5–10cm	FIBRE LENGTH 8–12cm	FIBRE LENGTH 4–12cm	FIBRE LENGTH 30–40cm	FIBRE LENGTH 30–40cm	ø max = 60mm	ø max = 20mm	kg CINDERS/m^3 LOOSE EARTH	kg PUMICE/m^3 LOOSE EARTH
4–6	*			–	45–70	o*		–	–
6–8	*		4	–	45–70	o*		–	–
8–11	*		4	(20)	45–70	o*		(125)	(60)
11–15	4–5*	3–5	4–5	22–23	45–70	1:5–1:4*	1:5–1:4	200	90
15–20	6–8*	6–8	6–8	24–25	50–70	1:4–1:3.5	1:4–1:3.5	350	150
20–27	8–11	9–11	8–10	25–26	60–70	1:3–1:2	1:3–1:2	500	225
27–36	10–14	12–14	10–12	26–28	70	1:2–1:1.5	1:2–1:1.5	700	300
36–48	(14) oo	15	(12) oo	* *	80	(1:1.5) oo	* *	1000	450
48–64	* *	* *	* *	* *	90	* *	* *	1400	600
> 64	* *	* *	* *	* *	90	* *	* *	1400	600

* RENDERING NECESSARY
** ADD SAND AND TEST AGAIN
o ADD CLAY TO MORTAR
∞ EFFECTUATE PRELIMINARY TESTS
( ) EVENTUALLY
– NOT RECOMMENDED

Nearly all soils, except those which have an excessive content of organic material, can be treated with cement and thus undergo a sharp improvement in their properties. Salt-rich soils are also difficult to stabilize with cement; even so, an increase in the proportion of cement can often bring good results. Soils which have a large clay fraction mix only with difficulty and require large quantities of cement. When the mixing process is very closely controlled under laboratory conditions good results can be achieved with clayey soils. In practice, however, cement is not used for stabilizing clay when the liquid limit is higher than 50 and the clay content is higher than 30%. Preliminary treatment of these extremely clayey soils with hydrated lime may improve the likelihood of obtaining good results with cement added at later date. Numerous tests give indications regarding the suitability and proportion of cement.

Abrasion test The proportion of cement should reduce material losses to 3% after 50 cycles, which is an excellent performance.

Erosion test The proportion of cement should reduce the mean depth of holes to 15mm – an excellent performance for this extremely severe test.

Wetting-drying An optimum proportion should reduce material losses to 10% – an excellent performance for this extremely severe test.

Freeze-thaw An optimal proportion should reduce material losses to 10%: an excellent performance for this extremely severe test.

Shrinkage (based on the Alcock test).

Linear shrinkage (mm)	Cement : soil (vol.)
less than 15	1 : 18
from 15 to 30	1 : 16
from 30 to 45	1 : 14
from 45 to 60	1 : 12

These values are applicable to soils compressed to to a maximum of 4MPa. The quantity of cement can be reduced to less than 30% for soils compressed to 10MPa.

Organic matter When the pH > 7 (alkaline or basic): calcareous soils, brown alkaline soils, and some gley soils can be stabilized with 10% cement: rates of between 1 and 2% of organic matter are in general not a problem.

When the pH < 7 (acid): gley soils can be successfully stabilized with 10% cement if the content of organic matter is less than 1%. Podsols and acidic brown soils can sometimes be stabilized with success if they contain less than 1% of organic matter. If anomalies are found to exist, preliminary treatment with calcium chloride (1 to 2%) may bring about a certain improvement.

Note: The following figures apply to compressed earth blocks.

AASHO SOIL CLASS	USCS SOIL CLASS	CEMENT REQUIREMENTS FOR VARIOUS SOILS* BY VOLUME (%)	BY WEIGHT (%)	WT % FOR PROCTOR TEST	WT % FOR FREEZE-THAW TEST
A–1–a	GW, GP, GM, SW, SP, SM	5–7	3–5	5	3–5–7
A–1–b	GM, GP, SM, SP	7–9	5–8	6	4–6–8
A–2	GM, GC, SM, SC	7–10	5–9	7	5–7–9
A–3	SP	8–12	7–11	9	7–9–11
A–4	CL, ML	8–12	7–12	10	8–10–12
A–5	ML, MH, CH	8–12	8–13	10	8–10–12
A–6	CL, CH	10–14	9–15	12	10–12–14
A–7	OH, MH, CH	10–14	10–16	13	11–13–15

* For most A horizon soils the cement should be increased 4 percentage points, if the soil is dark grey to grey, and 6 percentage points if the soil is black.

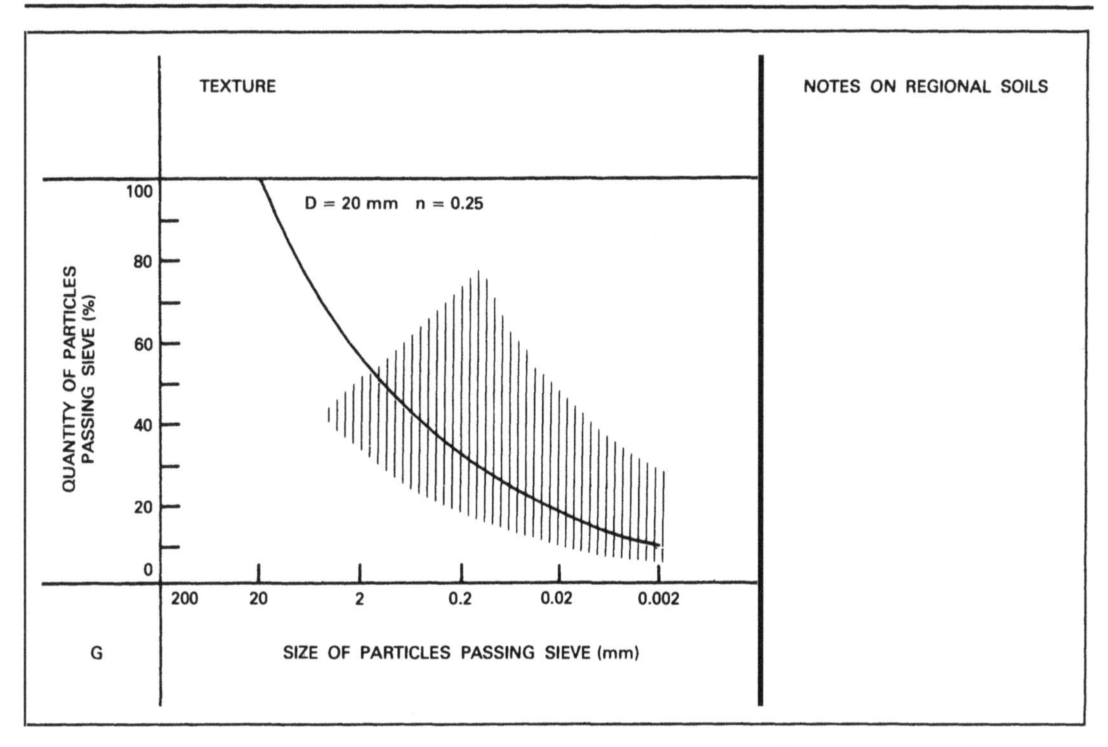

The limits of the zones recommended are approximate. The tolerances permitted vary considerably. Present knowledge does not justify the application of narrow limits. It is admitted that many soils which fail, for one reason or another, to comply with the requirements have been found satisfactory in practice; all that is claimed for the recommendation is that materials which comply with it are more likely to be satisfactory than those which do not. The zones are intended to provide guidance and are not intended to be applied as a rigid specification.

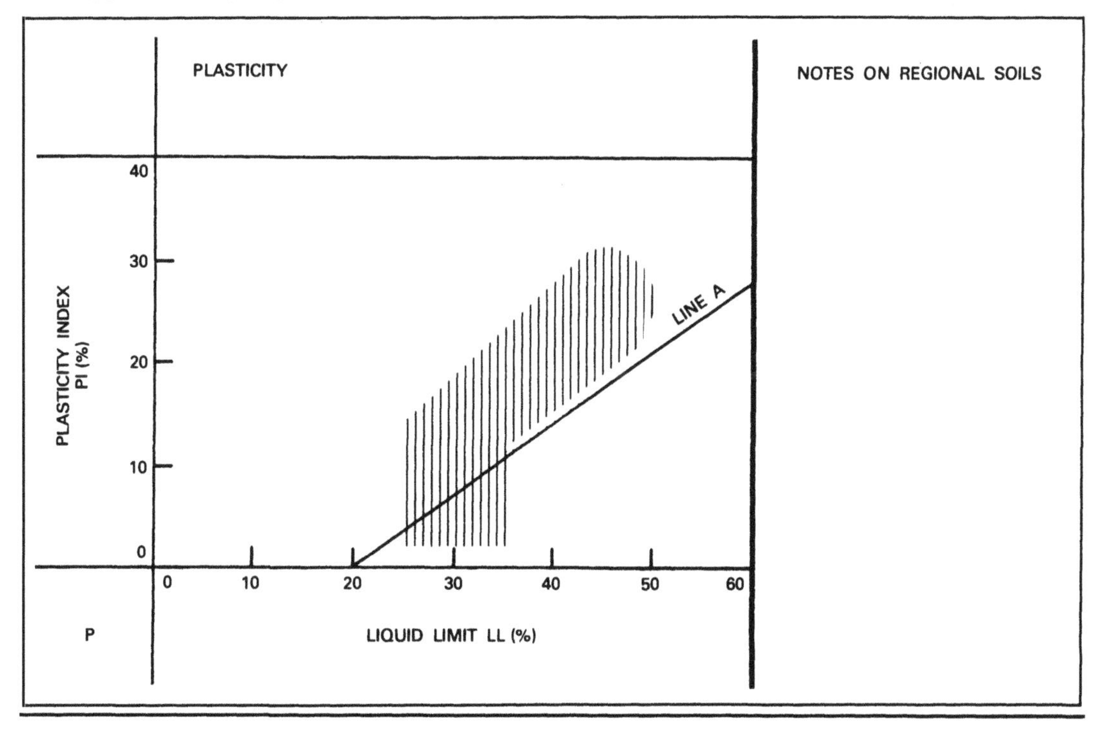

Lime has only a very limited effect on soils with a high organic matter content (content higher than 20%) and on soils short of clay. It can be more effective than cement on clay-sand soils and especially on very clayey soils. The effects of lime are thus highly dependent on the nature of the soils involved but a comparison with the effects of cement can, in many cases, be attempted. It has been observed that lime reacts far more quickly with montmorillonite clays than with the kaolinites, reducing the plasticity of the montmorillonites and having only a slight effect on the plasticity of the kaolinites.

Water content has a significant effect on clay soils which can be stabilized with lime, particularly in the pulverization and compaction stage. Natural pozzolanas react particularly well with lime.

For the rest we may note that the proportions of lime quoted are for industrial quality lime containing between 90 and 99% of quick lime.

For lime produced by less sophisticated methods, which may contain only 60% of quick lime (the rest being made up of unfired or over-fired components), the proportion must be increased. Below the two main methods of improving the performance of soil with lime are summarized:

1. Modification of the soil: the lime is added until a setting point is reached. This operation reduces the plasticity of the soil and improving its flocculation.

2. Soil stability: the proportions are much higher. Reference nomograms on the suitability of soils and the proportion of lime must be interpreted with a great deal of reserve.

Tests should be carried out only after allowing a curing period of 3 months.

Abrasion test The proportion of lime should reduce material lost to 3% after 50 cycles – an excellent performance.

Erosion test The proportion of lime should reduce the mean depth of holes to 15mm – an excellent performance for this extremely severe test.

Wetting-drying The proportion should reduce material losses to 10% – an excellent performance for this excessively severe test.

Freeze-thaw The proportion of lime should reduce material losses to 10%: an excellent performance for this excessively severe test.

Compressive strength The reactivity of soils containing lime was determined by Thompson in 1964. The soils are stabilized with the optimal proportion of lime; the increase in compression strength after curing for 7 days at 23°C is defined as the reactivity of the soil with the lime:

Group	Increase (MPa)	Reactivity
1	0.1	non-reactive
2	0.1 to 0.35	non-reactive
3	0.35 to 0.7	reactive
4	0.7 to 1.05	reactive
5	1.05 and more	reactive

This makes it possible to decide quickly whether further testing is justified.

Note: The following figures apply to compressed earth blocks.

SOIL CLASS		LIME REQUIREMENTS FOR MODIFICATION (WT %)		LIME REQUIREMENTS FOR STABILIZATION (WT %)	
		SLAKED	UNSLAKED	SLAKED	UNSLAKED
WELL GRADED CLAY GRAVELS		1–3		3 AND OVER	
SAND		NOT REC		NOT REC	
SANDY CLAY		NOT REC		5 AND OVER	
SILTY CLAY		1–3		2–4	
HEAVY CLAY		1–3		3–8	
VERY HEAVY CLAY		1–3		3–8	
ORGANIC SOIL		NOT REC		NOT REC	
GC, GM–GC	(A–2–6, A–2–7)			2–4	2–3
CL	(A–6, A–7–6)			5–10	3–8
CH	(A–6, A–7–6)			3–8	3–6

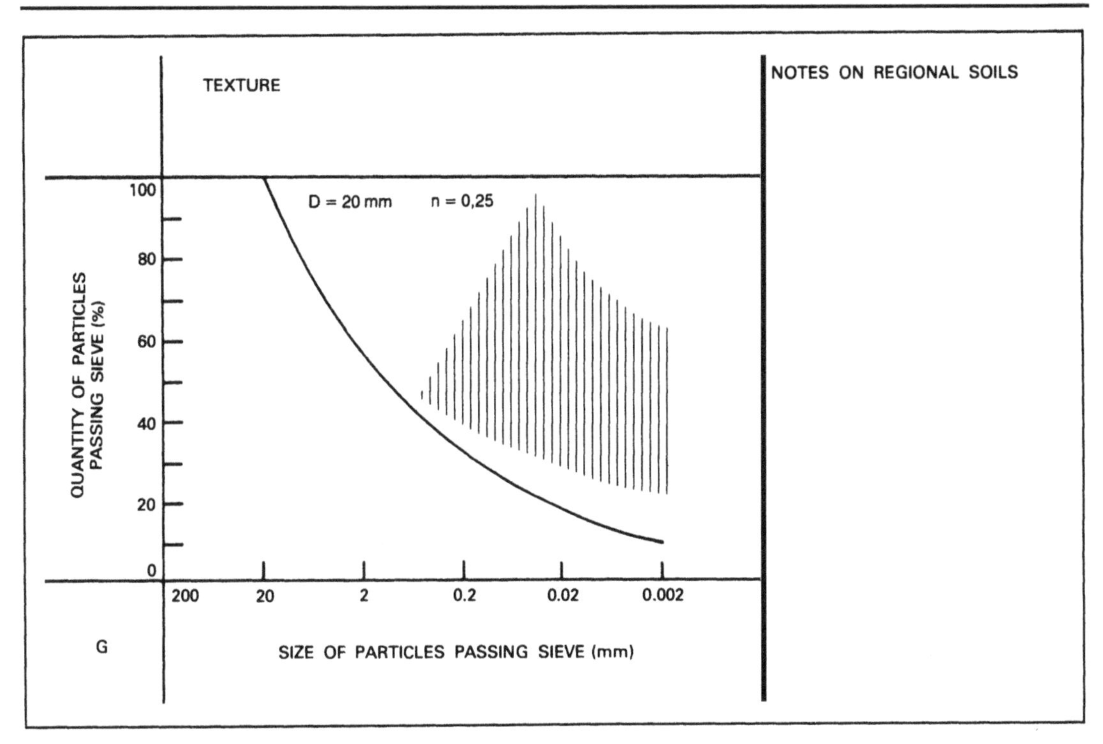

The limits of the zones recommended are approximate. The tolerances permitted vary considerably. Present knowledge does not justify the application of narrow limits. It is admitted that many soils which fail, for one reason or another, to comply with the requirements have been found satisfactory in practice; all that is claimed for the recommendation is that materials which comply with it are more likely to be satisfactory than those which do not. The zones are intended to provide guidance and are not intended to be applied as a rigid specification.

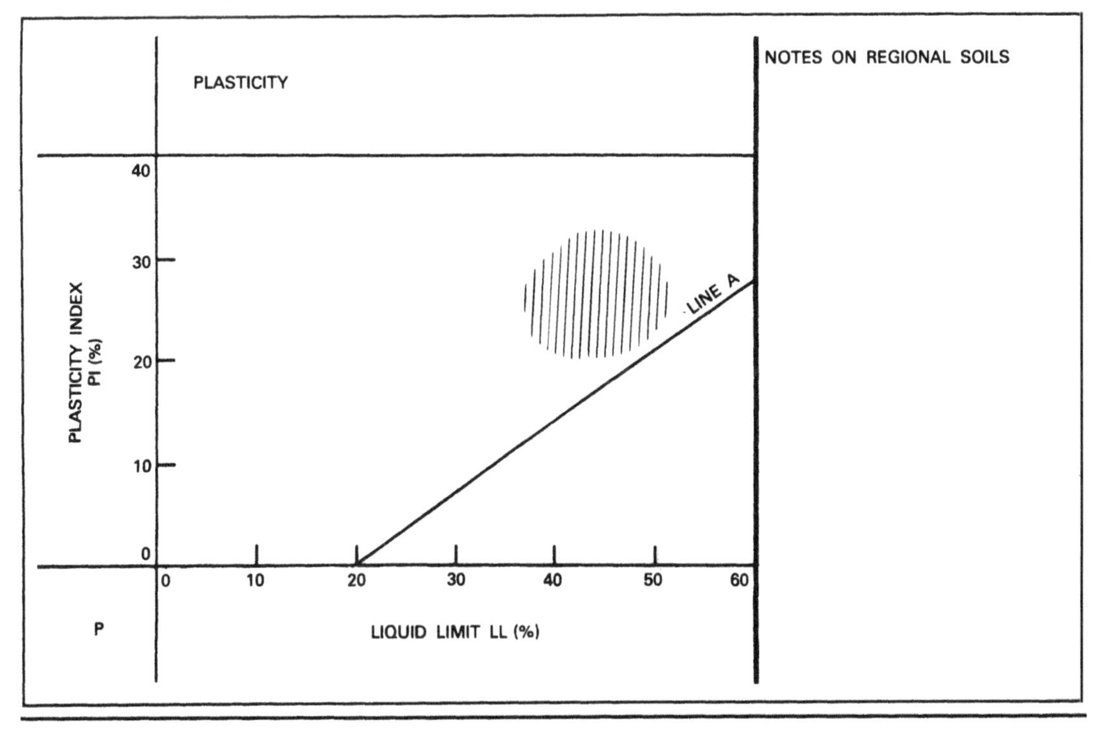

Although clayey soils have been successfully treated with cut-back or hydrocarbon emulsions, stabilization with hydrocarbons is more suitable for sandy soils or sandy-gravel soils, for soils lacking in cohesion or when an impermeable finish is particularly desired. With extremely clean sandy soils, the low adhesion of the bitumen to the surface of the siliceous particles can lead to the separation of the bitumen under the action of water, with the result that the stabilizing effect of the bitumen on the soil is considerably reduced. Moist soils are in general not suitable for bitumen stabilization, because of the difficulty of mixing the hydrocarbon with the soil.

Soluble salts Their presence in a soil is likely to lead to the deterioration of the soil as a result of the successive hydration and dehydration of the soil. Salts moreover have a tendency to cause efflorescence. Furthermore, in the presence of a stabilizer such as bitumen, they can be very harmful to the binding films between the bitumen and the clays. The presence of salts in a soil which can be stabilized with a bitumen should preferably not exceed 0.25%.

Proportioning The IIHT (California, USA) makes recommendations for adobe which involves carrying out tests by progressively increasing the bitumen content as follows:

Cut-back: 2%, 3%, 4%, 5%
Emulsion: 3%, 4%, 5%, 6%.

Each test is carried out on 3 to 4 samples which are tested for compression and bending strength, and for erosion resistance, until satisfactory results are obtained. The IIHT also points out that excessively clayey soils requiring more than 3% of cut-back or 6% of emulsion are not suitable for making adobes because of their marked shrinkage. For emulsions the following figures may be given:

High sand content soils: 4 to 6%
Low sand content soils: 7 to 12%
Clayey soils: 13 to 20%

The percentage is for the hydrocarbon itself and not for the suspending liquid.

Choosing the bitumen

Fraction with grains of Ø < 0.08 mm	Moist soil with more than 5% water	Dry soil with less than 5% water
0 to 5%	SS – 1h (CSS – 1h)	CNS – 2h (SS – 1h*)
5 to 15%	SS-1, SS-1h (CSS-1, CSS-1h)	CMS – 2h (SS – 1h*, SS – 1h*)
15 to 25%	SS – 1h (CSS – 1h)	CMS – 2h

* The soil must be wetted in advance.

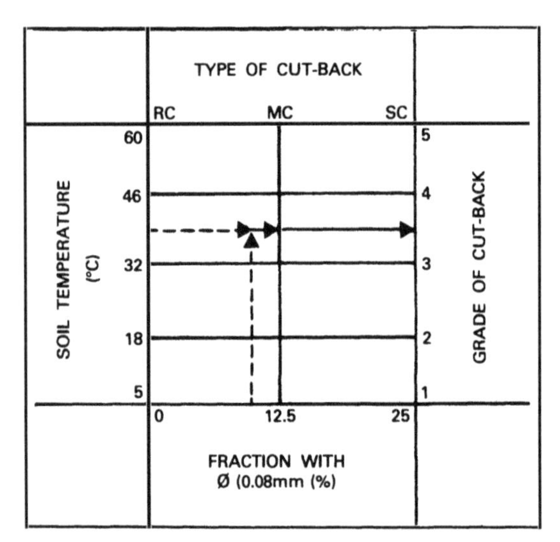

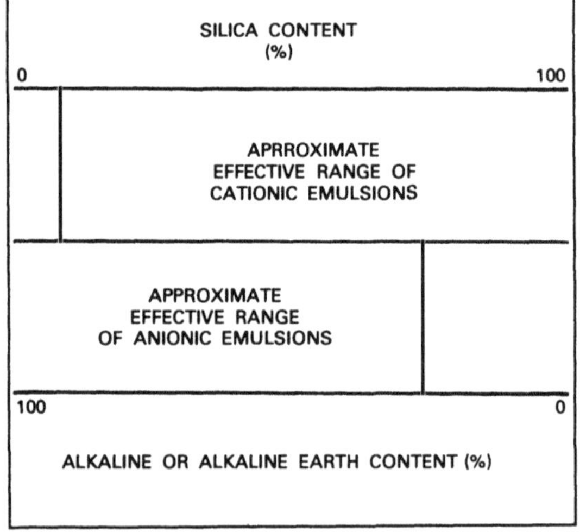

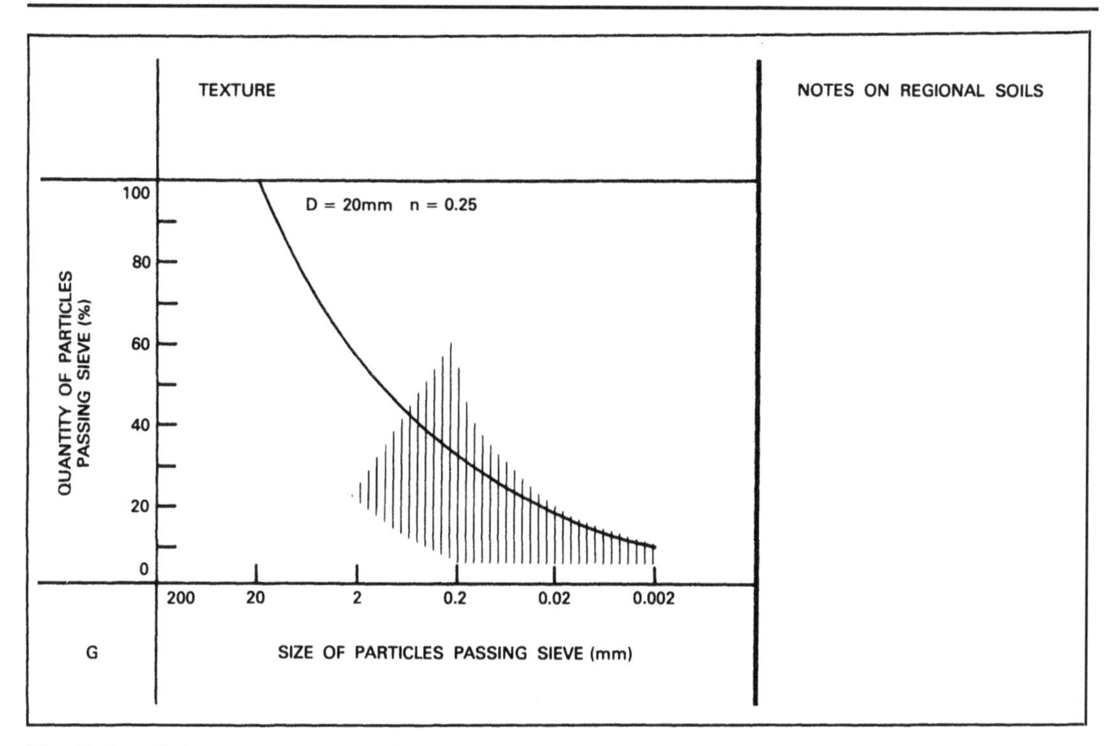

The limits of the zones recommended are approximate. The tolerances permitted vary considerably. Present knowledge does not justify the application of narrow limits. It is admitted that many soils which fail, for one reason or another, to comply with the requirements have been found satisfactory in practice; all that is claimed for the recommendation is that materials which comply with it are more likely to be satisfactory than those which do not. The zones are intended to provide guidance and are not intended to be applied as a rigid specification.

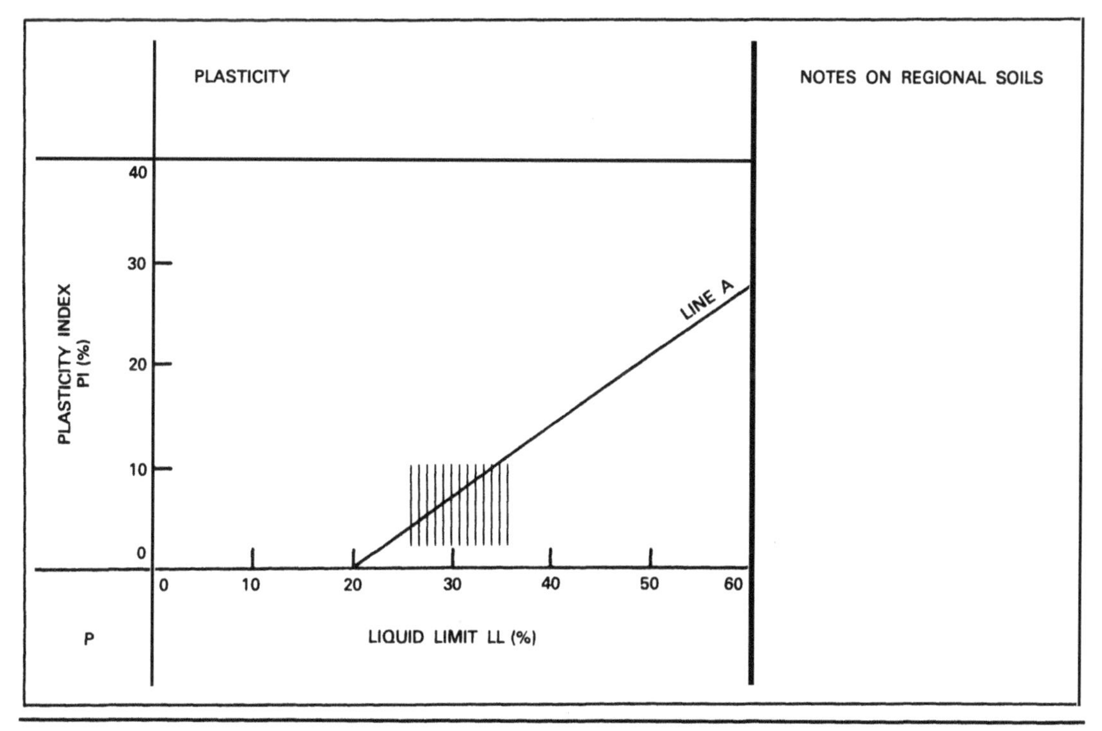

CINVA. *Le Béton de Terre Stabilisé, Son Emploi dans la Construction*. New York, UN, 1964.
Doyen, A. 'Objectifs et mécanismes de la stabilisation des limons à la chaux vive'. In *La technique routière*, Brussels, CRR, 1969.
Dunlap, W.A. 'Soil analysis for earthen buildings'. *2nd Regional Conference on Earthen Building Materials*, Tucson, University of Arizona, 1982.
El Fadil, A.A. *Stabilised Soil Blocks for Low-cost Housing in Sudan*. Hatfield Polytechnic, 1982.
Ingles, O.G.; Metcalf, J.B. *Soil Stabilization*. Sydney, Butterworths, 1972.
Kahane, J. *Local Materials, a self-builders manual*. London, Publications Distribution, 1978.
Niemeyer, R. *Der Lehmbau und seine Praktische Anwendung*. Grebenstein, Oko, 1982.
Road Research Laboratory DSIR. *Soil Mechanics for Road Engineers*. London, HMSO, 1958.
Somker, H. *Traitement des Sols au Ciment et Béton Maigre dans la Construction Routière Européenne*. Paris, Laboratoire Central des Ponts et Chaussées, 1972.
Stulz, R. *Appropriate Building Materials*. St. Gallen, SKAT, 1981.
Webb, D.J.T. Priv. com. Garston, 1984.

Rammed earth and adobe house, Peru, with adobe blocks drying in the foreground (Theo Schilderman, IT)

6. TESTS

Soil can be subjected to numerous tests, the majority of which are not standardized, or even regulated. From the engineering and scientific point of view, it is always interesting to submit a soil and earth construction materials to the widest possible of series of analyses, tests and trials. It should, however, not be forgotten that the main object is to use the material for construction and not to carry out as many analyses and tests as possible. Analysis and trial procedures will therefore be kept to the minimum required to ensure that the soil will behave well in the finished structures.

With this in mind, and with the acquisition of know-how and experience, careful observation of environments built in earth and the lessons drawn from them make it possible to shorten painstaking analysis procedures, especially as these can consume large amounts of time and money.

When in doubt, though, carrying out tests is strongly recommended.

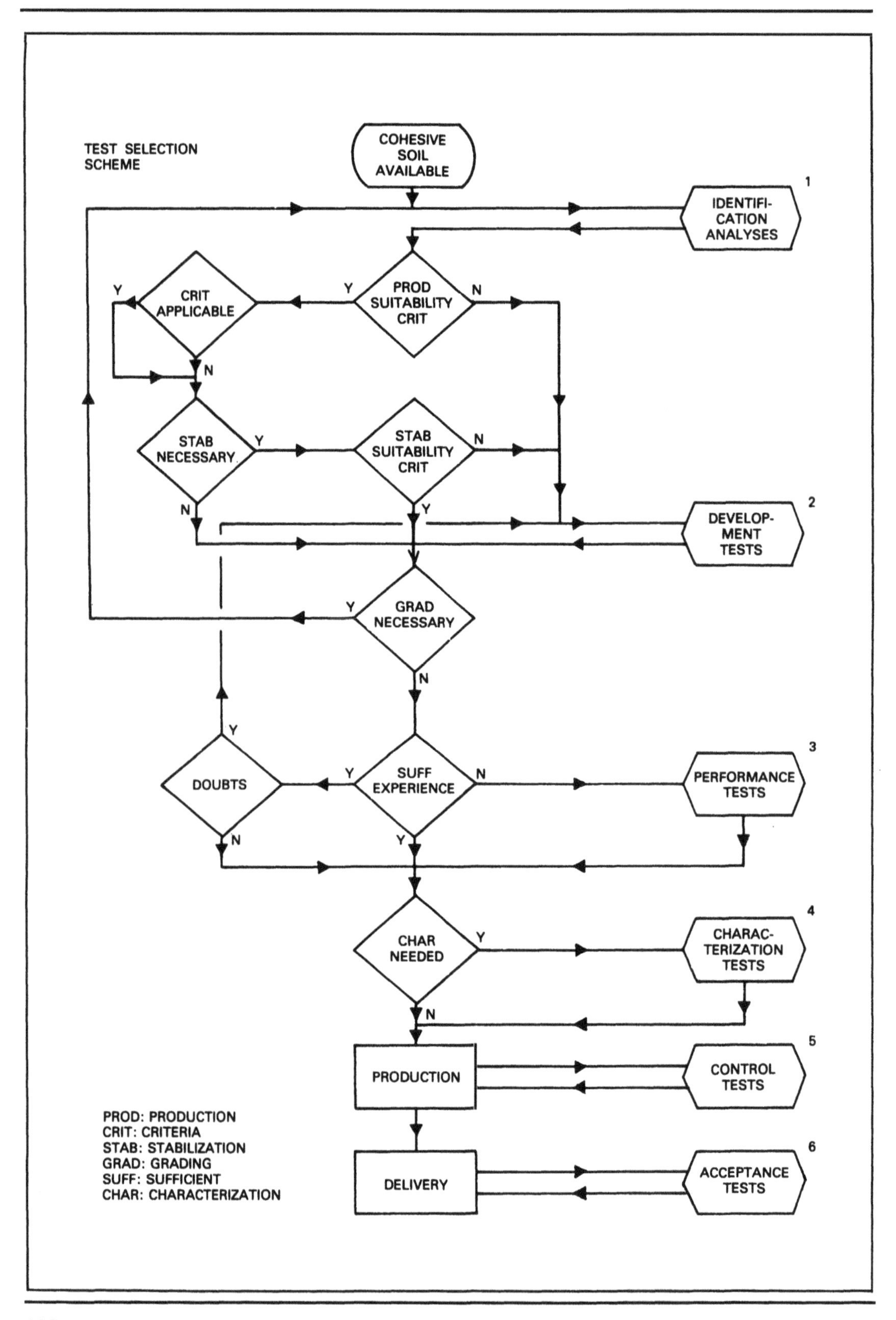
TEST SELECTION SCHEME
COHESIVE SOIL AVAILABLE
1
IDENTIFI- CATION ANALYSES
Y
CRIT APPLICABLE
Y
PROD SUITABILITY CRIT
N
N
STAB NECESSARY
Y
STAB SUITABILITY CRIT
N
N
Y
2
DEVELOP- MENT TESTS
Y
GRAD NECESSARY
N
Y
3
DOUBTS
Y
SUFF EXPERIENCE
N
PERFORMANCE TESTS
N
Y
4
CHAR NEEDED
Y
CHARAC- TERIZATION TESTS
N
5
PRODUCTION
CONTROL TESTS
6
DELIVERY
ACCEPTANCE TESTS
PROD: PRODUCTION
CRIT: CRITERIA
STAB: STABILIZATION
GRAD: GRADING
SUFF: SUFFICIENT
CHAR: CHARACTERIZATION

Earth is a material which can be analysed and tested by means of six basic test series. The application of each component of this screening process is, however, not always necessary. It is, indeed, possible, depending on the soil being analysed, the ease with which it can be characterized, the conditions under which the work is carried out, and the experience of the user or builder, to ignore one or other test series. All the analyses, tests and trials described below can be carried out with sophisticated laboratory apparatus as well as with smaller and lighter equipment suitable for on-site use.

It will be noticed that many of the individual tests described are common to several test series. It is moreover not necessary to repeat these tests, although the results obtained must be differently interpreted according to the series in which they are included. To take an example, in the tests grouped under the heading 'Development Tests' the object of the freeze-thaw test is to show that the material analysed (stabilized compressed brick) can withstand the required minimum of 12 test cycles, as required by the most frequently applied standards. For the 'Performance Tests' on the other hand, the material is tested to see, for example, if it can stand up to 17 freeze-thaw cycles. Passing this test shows that the material is of a high quality and suitable for use in very severe conditions. At present very few countries have developed standards for analyses and tests specifically suited to soil. Use is thus usually made of tests originating in other disciplines, such as concrete construction materials, road pavements, etc. These standards are not necessarily suited to earth. It is thus advisable when adopting these standards to be flexible about their application. Similarly the results should also be regarded with some degree of tolerance as it is generally acknowledged that they are extremely severe, as is the case for example with the durability tests, and rarely correspond to the realities of their use, except in specific circumstances. A few examples with respect to compressive strength illustrate this very well. The same applies to many of the tests listed here.

Example

The compressive strength test must be carried out at a moment when the quality of the material is representative of the normal state of the material. This is why the ultimate compressive strength of a sample cement-stabilized soil cube is measured after a period of 28 days, as this curing period is long enough to allow it to be assumed that the cube has reached or is approaching its ultimate strength. In fact this period is suitable for concrete of cement blocks, which reach 80 to 90% of their ultimate strength within 28 days. Cement-stabilized soil, however, reaches only 60 to 70% of its ultimate strength in the same period. If the soil is stabilized with lime this period is by no means long enough, the cube must be allowed to cure by at least another 3 weeks if incorrect comparisons and interpretations are to be avoided.

On the other hand, it is perfectly acceptable to carry out compressive tests on CINVA module soil bricks (29.5 × 14 × 8.8cm) by crushing them flat. Some standards permit this procedure. The advantages of crushing the material flat are that the test is representative of the load exerted on the brick in the wall, that ordinary production samples can be used, and that it is not necessary to place two bricks on top of one another bound by a mortar as is usually the case in standard versions of this test. Taking the Belgian Standard NBN B/24/201 on this subject, it can be seen that CINVA bricks have a height to short side ratio of 8.8/14 = 0.63, which is higher than the limit coefficient of 0.55.

The results obtained can therefore be compared to those obtained for other masonry elements and even with those obtained on 20cm cubes of the same material. For compressed material measures of compressive strength must be carried out in the same direction as the material was compressed in. Indeed if the measurements are carried out at right angles to that direction the values obtained may be between 25 and 45% lower, totally falsifying reality.

Identification analyses

The object of these analyses is to determine the characteristic of the basic materials with a view to obtaining greater insight into the behaviour of the finished product.

Once these characteristics are known the possible uses of the material can be envisaged, by making using of tables, nomograms, and rules governing decision-making. The identification analyses are as follows:

- Visual Identification
- Tests guided by sensory perception
- Natural Moisture Content
- Grain Size Distribution
- Sedimentation
- Sand Equivalence
- Limit of Liquidity
- Limit of Plasticity
- Limit of Absorption
- Limit of Shrinkage
- Bulk Shrinkage
- Linear Shrinkage
- Proctor Tests
- Bulk Density
- Apparent Bulk Density
- Wet Tensile Test
- Wet Shear Test
- Water Content after Drying.
- Dry Colour and Wet Colour
- Dissolution in Water
- Mineralogical Identification
- Specific Area
- Emerson Test
- Pfefferkorn Tests
- Quantity of organic material and humus
- Nature of organic material and humus
- Iron Oxide Content
- Magnesium Oxide Content
- Calcium Oxide Content
- Carbonate Content
- Sulphate Content
- Soluble Salts Content
- Insoluble Salts Content
- Loss on Ignition
- pH

These lists are merely indicative and not intended to be exhaustive.

Development tests

These tests are used to ensure that a good construction material is obtained, after the basic materials have been identified. The tests determine the parameters which must be respected when the product is mixed and produced.

The parameters also serve as a reference for when carrying out the analyses and tests prior to approving the products. The development tests are as follows:

- Grain Size Distribution
- Sedimentation
- Sand Equivalent
- Limit of Liquidity
- Limit of Plasticity
- Bulk Shrinkage
- Linear Shrinkage
- Proctor Tests
- Bulk Density
- Apparent Bulk Density
- Minimum Wet Weight
- Minimum Dry Weight
- Voids Ratio
- Water Content after Drying
- Degree of Pulverization
- Penetration Tests
- Wet and Dry Compressive Strength
- Tensile Strength
- Bending Strength
- Shear Strength
- Poisson's Ratio
- Young's Modulus (Elasticity)
- Swell
- Dry Shrinkage
- Thermal Expansion and Thermal Shock
- Permeability
- Water Absorption
- Frost Susceptibility
- Efflorescence
- Erosion
- Abrasion
- Fire Resistance
- Compatibility with Mortars
- Compatibility with Renderings
- etc.

Wetting and drying test

The procedure described corresponds to that laid down by ASTM D 599 and AASHO T 135.

After a period of storage of 7 days in an extremely moist atmosphere, the samples are totally immersed in water at the same temperature as the laboratory for a period of 5 hours. After this period the samples are removed from the water and dried in an oven or cabinet at a temperature of 71°C for a period of 42 hours. After drying the samples are to be withdrawn from the oven and the brushed one by one. The brushing is to be carried out with a brush composed of metallic fibres and serves to remove all the fragments of the material affected by the wetting and drying cycles. The brushing is to be firm and is to be carried out on each surface of the samples, in both directions (e.g. top to bottom), for a total of 18 to 25 brush strokes.

The force applied during brushing is to be of the order of 1.5kg. The procedure described above constitutes one 48-hour wetting-drying cycle. The samples are then immersed once again in the water and subjected to a further wetting-drying cycle. The procedure is repeated for 12 cycles.

If the test must be interrupted (e.g. weekends), the samples are to be stored in the oven or cabinet. After twelve test cycles the samples are to be dried at 110°C until a constant dry weight is obtained. The weight loss with respect to the initial weight is then calculated. When the samples have been stabilized with lime the wetting-drying tests are performed after a curing period of one month. This test is considered as being extremely severe.

Erosion test

This test creates a standard artificial rain on the face of the sample block assumed to be exposed to the rain. Spraying is provided by a pump which maintains a constant pressure of 0.14MPa connected to a 10cm diameter sprinkler rose or shower head at a distance of 20cm from the block under test. The pressure of the spray is monitored by a manometer. The spray is held for a period of 2 hours at right angles to the block.

Subsequently the depth of the holes in the block are measured. The depth of the 18 deepest holes in each block is then averaged and is recorded as Pmm. The results of this test are no more than indicative. The appearance of mild erosion or pitting on a block of stabilized soil should not necessarily be interpreted unfavourably.

Freeze – thaw test

The procedure described below is the same that described by ASTM D 560 and AASHO T 156.

After storing in an extremely moist environment for a period of 7 days, the samples are placed on an absorbent material which has been saturated with water, and then put into a refrigerator, at a constant temperature of no more than −23°, for a period of 24 hours and then removed. The samples are thawed in a moist environment (RH = 100%) at a temperature of 21°C for a period 23 hours and then removed. During this thaw phase the samples must absorb water by capillarity (from the absorbent material). The samples are then brushed in accordance with the procedure used in the wetting-drying test. The test is continued for 12 freeze-thaw cycles, the sample being returned to the absorbent material between each cycle. Some samples made of silty or clayey material may scale, particularly after the sixth test cycle. Care will be taken to eliminate this scale so that brushing is not hindered. If the test must be interrupted (e.g. at the weekend), the samples should be stored in the refrigerator.

After 12 test cycles, the samples are dried in an oven at a temperature of 110°C until a constant weight is obtained. The weight loss with respect to the original weight is then calculated. This test is considered as being extremely severe.

Abrasion test

The blocks are tested dry. A metal brush weighted to 6kg is used to scrub the face of the block exposed to the rain (according to the way the masonry is bonded). A single back and forth motion of the brush is regarded as one cycle of abrasion. Brushing is continued for fifty cycles. The measurement consists of weighing the material detached by the brushing. The dry weight of this material is recorded per square cm of brushed area in order to obtain a test result which is independent of the shape and size of the block.

Blocks stabilized with cement or bitumen are tested after a curing period of 28 days, while those stabilized with lime are tested after a curing period of 3 months.

Performance tests

The purpose of these tests is to check the performance observed in the laboratory for the materials, by testing them under simulated conditions of use or in structural systems. This is a matter of testing the behaviour of walls or other structural elements as they would be in a structure. The most practical performance tests are the following:

- Dry and wet compressive strength. This test is realized by centric loading, and by eccentric loading when, for instance, a wall is tested.
- Wet and dry tensile strength
- Wet and dry bending strength
- Transverse pressure strength
- Lateral pressure strength
- Impact strength (soft body)
- Resistance to seismic tremors
- Loading of vaults, domes, and arches
- Loading of beams and lintels
- Buckling
- Flow
- Poisson's Ratio
- Young's Modulus (Elasticity)
- Swell and Dry Shrinkage
- Passage of Water
- Capillary Rise
- Water Erosion
- Wind Erosion
- Freeze-thaw
- Thermal Expansion
- Adhesion of Mortar to Blocks
- Adhesion of Renderings to Walls

These lists are merely indicative and not intended to be exhaustive.

Characterization tests

The purpose of these tests is to determine certain physical properties of the building materials. Some of these properties make it possible to calculate the thermal performance of the building, for example, or to evaluate the ageing performance of the building structures, or to make statements about the likely comfort, and general safety. The main characterization tests are listed below:

- Compatibility with Renderings
- Compatibility with Mortars
- Fire Resistance
- Coefficient of Conductivity
- Specific Heat
- Thermal Buffering Coefficient
- Thermal Storage Coefficient
- Thermal Effusion and Diffusion
- Thermal Shrinkage
- Susceptibility to Frost
- Water Absorption
- Permeability
- Capillarity
- Dry Shrinkage
- Bulk Shrinkage
- Linear Shrinkage
- Water Content after Drying
- Bulk Density
- Colour
- Surface Texture
- Radiation and Nuclear Protection
- etc.

Compressive strength test

The conventional presses found in test laboratories are suitable for the dry compression of material samples. If, for example, the tester wishes to crush stabilized compressed bricks of the high performance CINVA type he should use a press capable of testing bricks up to at least of 10MPa, or a press which can achieve 400KN. In the field it is possible to construct small presses using small steel beams or even with a lorry jack. This site equipment should, if possible, be fitted with a manometer so that the force applied to the sample can be read off directly. If no manometer is available a dynamometric comparator may be used. However, this sort of apparatus is somewhat fragile and use must be made of small samples. Small lever presses in metal or wood can also be made.

The force applied by such lever presses cannot be made very high, which means that the samples must be specially prepared (cylinder or cubes with a side of 5cm). The preparation of samples of rammed earth, cob, or adobe does not pose many problems, but when compressed blocks require testing they must be sawn in order to obtain a small enough sample. Unfortunately structural damage can be expected. Another solution is to make special samples but there is a danger that they are not representative of the compressed bricks under test.

Tensile strength test

This can be measured using a procedure developed by the I.I.H.T in California (USA). A sample soil brick is placed (on one of its large faces) on 2 tubes with a diameter of 2.5cm which have been set 20cm apart and perpendicular to the longitudinal axis of the brick. Another identical tube is placed on the upper surface of the brick parallel to the short side. This tube is surmounted by a balanced plate which is then carefully loaded with bricks at a rate of 250kg per minute, until the brick under test breaks. The intention is not to learn the exact strength but to see that it exceeds a certain predetermined threshold. The tensile strength in MPa is given by:

$$\tau = \frac{0.15 \times 20\text{cm} \times (\text{load in kg})}{(\text{width of block in cm}) \times (\text{thickness in cm})^2}$$

Control tests

Following on the identification and development tests, control tests serve to check the quality of the basic materials used in production and their conformity with the requirements and properties determined in the laboratory. The type of analyses and the frequency of tests are determined by the partners in the operation, though in practice a daily test is carried out either when the soil is delivered to the site or the brickyard, or when a major sample is taken from the soil borrow. These analyses are thus carried out with site equipment either at the borrow, or the site, or at the place where the materials are produced. The results are carefully recorded and submitted for the approval of the partners in the operations: the test laboratory, the site manager, the owner, the contractor. The main control tests are the following:

- Visual Inspection
- Grain Size Distribution
- Sand Equivalent
- Quantity of Organic Matter
- Degree of Pulverization
- Water Content
- Uniformity of the Mix
- Penetration Tests
- Weight

These lists are merely indicative and not intended to be exhaustive.

Acceptance tests

These tests are carried out when the construction materials are being produced or when they are received, either in the brickyards, or on the site. Their purpose is to determine the quality of the production and its conformity with the requirements and properties determined in the analytic and test laboratories. For blocks and bricks five samples are usually taken every 1000 bricks when production is started, and one sample every 1000 bricks when controlled production is underway.

In small brickyards which produce fewer than 1000 bricks per day the practice is to take two samples per day. For techniques such as rammed earth and cob it is usual to take two samples per day when the site is opened and one per day later on.

These tests are carried out on site using mobile apparatus. The results are carefully recorded and submitted for the approval of the partners in the operations: the test laboratory, the site manager, the owner, and the contractor. In the event of a legal dispute or doubts about the actual quality of the material, samples are sent to a laboratory which makes up an official test report. The main acceptance tests are the following:

- Colour when Wet and Dry
- Efflorescence
- Uniformity of the Material
- Content of Stabilizer
- Dry Weight
- Apparent Bulk Density
- Dimensions
- External Appearance
- Dry Shrinkage
- Compressive Strength
- Tensile Strength
- etc.

Pocket penetrometer

This small and extremely handy device makes it possible to check the density of bricks when they are actually being produced. Each brick is penetrated at least five times. The area of penetration is 3 to 5mm^2 and the depth is about 5mm. These penetration tests give a general idea and the results are compared to a predetermined threshold of acceptability for the materials.

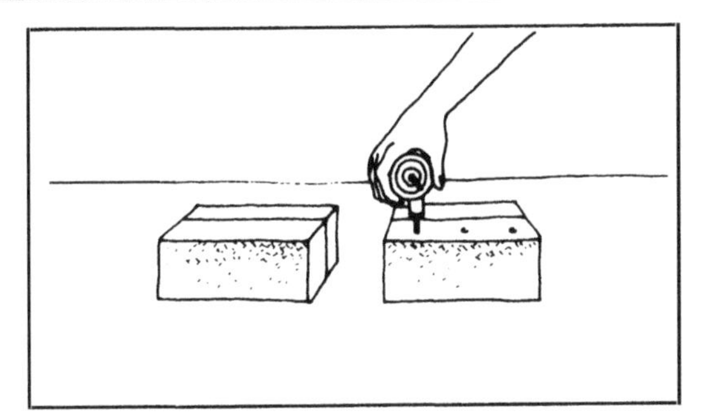

Pendulum scleroscope

This is a cunning device which makes a non-destructive check on the quality of the materials, usually on a finished structure such as a wall. The pendular scleroscope measures the compressive strength of the soil in MPa. When working with earth structures, a scleroscope suitable for low-strength materials should be used: from 5 to 8 MPa.

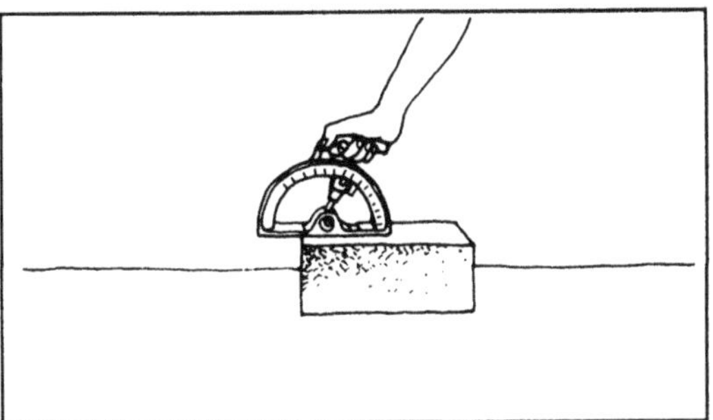

Optimal moisture content test

A simple test of optimal moisture content on a batch of soil destined for compression can be carried out as follows. Take a handful of the soil, compress it firmly in the fist, and then allow it to drop onto a hard and flat surface from a height of about 1.1m. The moisture content is right if the ball of earth breaks into 4 or 5 lumps. If the ball flattens without disintegrating, the moisture content is too high. If the ball breaks into lots of small pieces, the soil is too dry.

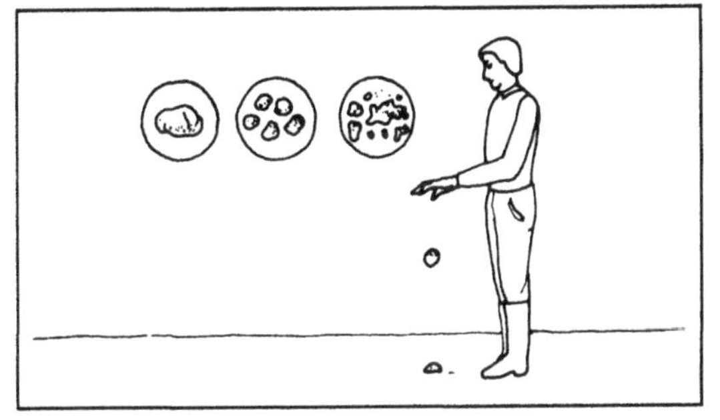

Penetration test

A probe is pushed in the soil sample or sample brick from a distance of between 5 and 10cm. The strength of the material is estimated in relation to the penetrating force of the probe and the effective hardness of the material. A predetermined threshold of acceptability is taken as reference. The test suffers from its subjectivity.

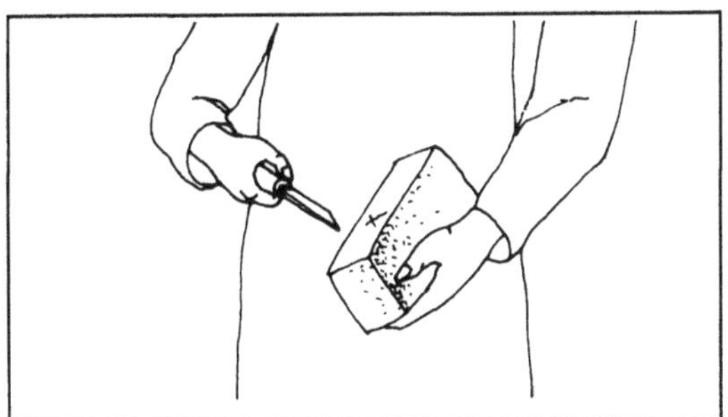

Impact test

Two samples of cement-stabilized brick or two samples of the material are held perpendicularly with respect to one another and are banged against one another with increasing force between each bang. The hardness of the material is gauged in relation to the sound generated by the bangs.

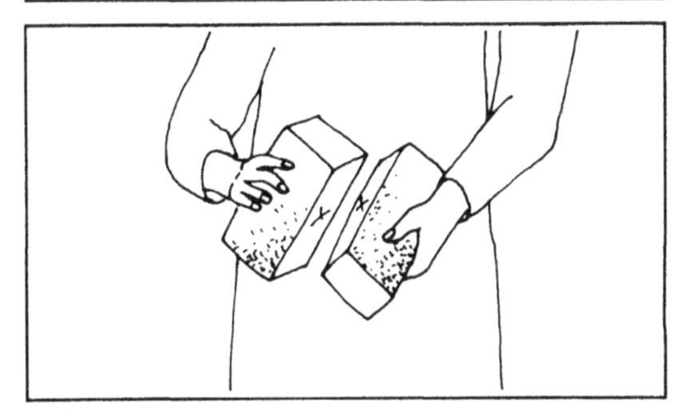

The analyses, tests, and trials for earth construction can be carried out with the standard apparatus of laboratories equipped for civil engineering, soil science, or even building. These tests require very little special equipment. The main materials, tools and products discussed are useful but not all essential. A well-equipped laboratory for this work can be had for an investment in the region of US$50,000. This investment can, however, be reduced to US$5000, and this would be enough to obtain the right basic equipment. This investment can be reduced even further by adopting simple and practical analysis procedures, on-site tests and trials.

Prospecting

- Soil Auger, 80mm diameter
- Auger Extension Rods, 1 metre long
- Torsion Lever
- Three Tonne Hand Jack
- Tarpaulin
- Shovel
- Spade
- Pick
- Geologists Hammer
- Comparative Chart for Soil Colours

Grain Size Distribution

- Complete Set of Standard Sieves, maximum diameter 60cm
- Sieve Bottom
- Sieve Cover
- Sieve Stand
- 500ml Washing Bottle

Sedimentation

- Manual Agitator for 1000ml Sample Jar
- 995–1050g/l Hydrometer

Proctor

- Proctor Mould with Stand
- Levelling Scraper
- Compacting Tamper

Atterberg

- Complete Casagrande Apparatus
- Smooth and Rough Cups
- Flexible Straight Spatula, 150mm
- Casagrande Grooving Tool
- A.S.T.M. Grooving Tool
- Marble Slab, 45 × 30 × 3cm
- Cat's Tongue Trowel, 12cm long
- Plasticity Gauge, Diameter 3mm

Cohesion

- Mould
- Tension Rings
- Bracket
- Sand (for ballast)

Strength

- 400 KN Press
- Table for Cleaving Tests
- Tensioning Machine
- 50 litre Planetary Mixer
- Cube Moulds
- Cylindrical Moulds
- Pocket Needle Penetrometer
- 0.5–8MPa Scleroscope

Shrinkage

- Alcock Shrink Mould
- Shrinkage Cup, diameter 50mm, 415mm

Durability

- Refrigerator, –30°C
- Wire Brush
- 10cm Diameter Sprinkler Rose
- Manometer
- Pump
- Sprinkler Nozzle

Chemicals

- Hydrochloric Acid
- Nitric Acid
- Oxalic Acid
- Barium Chloride
- Potassium Chromate
- Lime
- Distilled Water
- Sodium Hexametaphosphate
- Universal pH Indicator Paper
- Universal pH Indicator Ethanol
- Milk of Lime
- Phenolphtalein
- Potassium Hydroxide
- Methyl Red
- Sodium Hydroxide

Miscellaneous Apparatus

- Portable pH metre
- Hygrometer
- Contact Thermometer
- Atmospheric Thermometer –10/+80°C
- Penetration Thermometer
- Microscope
- Calibrated Magnifying Glass
- 20kg Balance
- 10kg Hanging Balance for Field Use
- 1g to 7kg Balance
- 0.1g to 500g Balance
- 0.001g to 100mg Balance
- Chronometer 1/5 secs
- Timer
- Alarm Clock
- Dial Comparator 1/100
- Sliding Stand
- 120°C Drying Cabinet

Miscellaneous Equipment

- 2 litre Glass Flasks, diameter 12 to 15cm
- Flask Covers
- Pans
- 20 litre Polyethylene Barrel
- Burette
- 70cm of Rubber Tubing, diameter 0.5cm
- Pyrex Recipients, 400ml and 2000ml
- Test-tubes 1000ml
- Test-tubes 500ml
- Test-tubes 100ml
- Rubber Corks for Test-tubes
- Funnel, diameter 140mm
- 10 litre Polyethylene Sacks
- 10 litre Jute Sacks
- 12 litre Containers
- Heating Plate
- 2 litre Mortar
- Hard Pestle
- Rubber Pestle
- Labels

Miscellaneous Tools

- Asbestos Gloves
- Small Scoop
- 150mm Flexible Straight Spatula
- 16mm Square Trowel
- Rigid Straight Spatula
- Stainless Steel Ladles, One Large, One Small
- Stainless Steel Laboratory Spoon
- 960g Hammer, 42 × 36
- 300mm Chisel
- Knife
- Ruler
- Folding Rule
- Pincers
- Other Small Tools

AGRA. *Recommandations pour la Conception des Bâtiments du Village Terre.* Grenoble, AGRA, 1982.
Bertram, G.E.; La Baugh WM.C. *Soil Tests.* Washington, American Road Builders' Association, 1964.
CRET. *Maisons en Terre.* Paris, CRET, 1956.
Doat, P. *et al. Construire en Terre.* Paris, Editions Alternatives et Parallèles, 1979.
Grésillon, J.M. 'Etude de l'aptitude des sols à la stabilisation au ciment. Application à la construction'. In *Annales de l'ITBTP*, Paris, ITBTP, 1978.
Hernández Ruiz, L.E.; Márquez Luna, J.A. *Cartilla de Pruebas de Campo para Selección* de *Tierras en la Fabricación de Adobes.* México, CONESCAL, 1983.
International Institute of Housing Technology. *The Manufacture of Asphalt Emulsion Stabilized Soil Bricks and Brick Maker's Manual.* Fresno, IIHT, 1972.
PCA. *Soil-cement Laboratory Handbook.* Stokie, PCA, 1971.
Sulzer, H.D. Priv. com. Zürich, 1984.
Verlarde Gonzalez, J.M. *La Tierra Estabilizada y su Utilización en la Producción de Componentes para la Construcción.* Panamá, Universidad de Panamá, 1980.

Control tower and arrival/departure hall made of compressed earth blocks, on the island of Mayotte near Madagascar where 10,000 buildings have been constructed with compressed earth blocks. (architect: Pascal Rollet/GEP-HUOC, Direction de l'equipement) (Thierry Joffroy, CRATerre-EAG)

7. CHARACTERISTICS

One of the main problems of earth construction appears to be the lack of standard criteria which would make it possible to come to an accurate evaluation of the finished material.

This absence has a number of harmful consequences as it has not only a negative influence on potential owners and decision-makers, but also on financial backers, because when considering investment in earth structures they have no guarantee of the technical quality of the buildings, particularly with respect to their durability beyond the period of the loan.

At long last, however, this lack of descriptions of the characteristics of the material and total absence of standards is slowly being remedied.

In recent years numerous research efforts and experiments have made it possible to make a fairly good characterization of soil as a building material. Furthermore several countries have drawn up valuable standards.

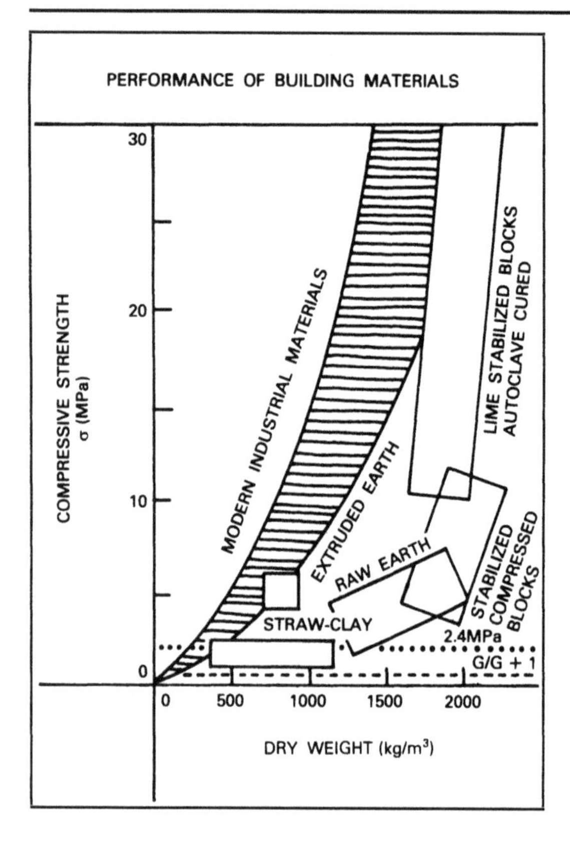

COMPRESSIVE STRENGTH	
1 STRENGTH REQUIRED BY DOWNWARD LOADING	0.1MPa
– single-storey dwellings	
2 SAFETY COEFFICIENT	×3
– variation in production quality – variation in construction quality – accidental excess loading	
3 REDUCTION COEFFICIENT	×4
– nature of the material – strength of the mortar – buckling of the wall – method of loading	
4 SATURATION COEFFICIENT	×2
– ratio of wet strength to dry strength	
	TOTAL × 24
σ 28 = 2.4MPa	

Whenever the subject of the quality of soil as a construction material is brought up, it is almost immediately followed by the subject of compressive strength. It is generally agreed that earth is a heavy material with a low compressive strength. Numerous soil materials fall within the same order of compressive strength and behave in the same way as low-strength concrete. Soil can, in fact, be considered as a kind of lean concrete. Pursuing the comparison with other materials, we may say:

– The lowest quality soil products have little strength for a much higher specific gravity than conventional mineral materials.

– Over a wide range of strengths, soil materials are comparable to conventional mineral materials.

– Soil is capable of extremely high performance. Research conducted at the ICAM in Lille, France has established that autoclaved lime-stabilized soil products (250°C at 16atm) have a compressive strength of 90MPa. Strength jumps to 200MPa when the material is processed in a drying oven (350°C). Indeed it is possible to obtain extremely high strengths by means of a number of processes (including compression, stabilization, and oven-drying). The question arises, however, of whether such performances are necessary. With single-storey dwellings or two-storey dwellings the downward thrust is of the order 0.1 to 0.2MPa. It is thus pointless to use materials with strengths approaching 10MPa and more.

Nevertheless, 0.1MPa is not enough as there are other problems apart from the simple performance of a building brick or a structure. Thus a safety coefficient of between 20 and 30 is considered desirable.

A safety factor of between 2 and 2.5MPa is generally regarded as being large enough to satisfy the requirements of most modern standards, and, when brick production is carefully monitored, 1 to 1.5MPa can be regarded as an absolute minimum which can guarantee adequate strength, particularly with respect to handling.

The second requirement of soil used for construction purpose is that it can stand up to water. Conventional tests used on other materials (wetting and drying, spraying, total or partial immersion, and freezing and thawing) have been adopted for soil without modification. Consequently the results obtained in the laboratory do not always agree with those observed in actual practice. These results are for isolated samples and not for walls or full-scale structures.

In 1974 in Burkina Faso the EIER developed some comparative data on theoretical tests and practical performance. The results clearly show that the erosion test for unrendered walls in stabilized soil is 'very severe compared with the behaviour of control walls exposed to the vagaries of the weather over a three year period'. The erosion of an exposed wall is insignificant compared with the erosion observed in the test. The durability of earth as a construction material is its best quality. This can be expressed in terms of resistance to the weather (rain, frost, wind) and the use to which it is put (behaviour of the inhabitants, animals, and so on). Theoretical tests do not allow for the complexities of the situation. A simple sun-baked adobe does not stand up to these laboratory tests; after immersion it may have a mean compressive strength of 0.5 to 1.0MPa, or disintegrate completely, eliciting a harsh verdict: unserviceable material. However, the same bricks are used to construct multi-storey buildings in the Yemen and reservoirs in Iran. On the other hand, the theoretical durability of many modern materials is far higher than what is observed in reality. Estimates of the durability of a material cannot be based on a simple extrapolation from theoretical tests alone; the performance of the structure must also be taken into account. In this respect, the centuries-old buildings which are still standing in countries around the world bear witness to the lasting qualities of soil as a construction material.

In the present context, however, an attempt must be made to quantify the properties of soil. While figures may not be of overriding importance for small structures, they serve as a point of departure for decision-making bodies, financial institutions, insurance companies, architects and contractors. The construction industry would be well-advised to refrain from hasty interpretations based on the theoretical and scientific data which fail to take full account of reality.

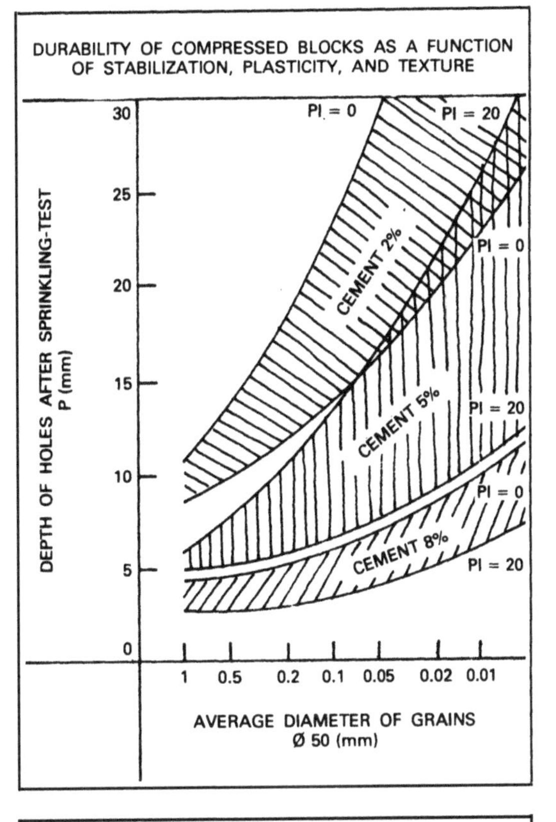

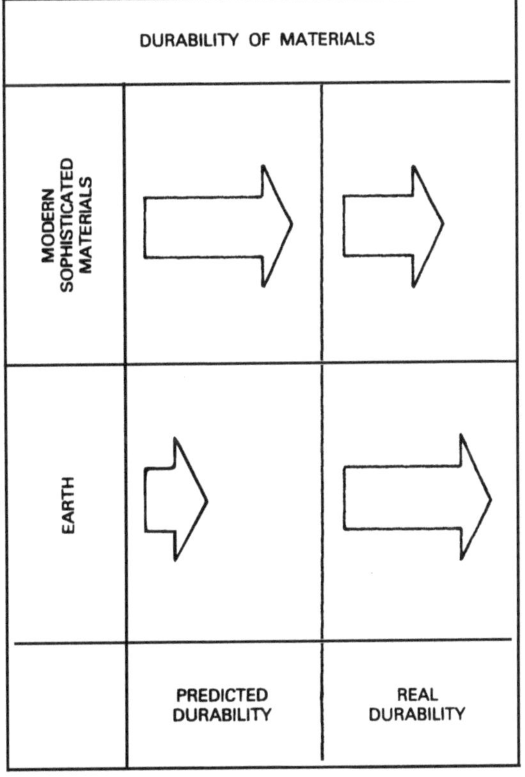

PROPERTIES	S	U	CLASSES A	B	C	D
28 DAY COMPRESSIVE STRENGTH * (+40% after 1 year, +50% after 2 years)	σ 28	MPa	> 12	5 12	2 5	ABOUT 2
28 DAY WET COMPRESSIVE STRENGTH (24 h in water)	σ 28	MPa	> 2	1 2	0.5 1	0 0.5
28 DAY TENSILE STRENGTH (Brazilian method)	τ 28	MPa	> 2	1 2	0.5 1	0 0.5
28 DAY TENSILE STRENGTH (on a core)	τ 28	MPa	> 2	1 2	0.5 1	0 0.5
28 DAY BENDING TEST	τ 28	MPa	> 2	1 2	0.5 1	ABOUT 0.5
28 DAY SHEAR TEST	τ 28	MPa	> 2	1 2	0.5 1	ABOUT 0.5
POISSON'S RATIO	μ		0 0.15	0.15 0.35	0.35 0.50	> 0.5
YOUNG'S MODULUS	E	MPa		700 7000		
APPARENT BULK DENSITY	γ	kg/m^3	> 2200	1700 2200	1200 1700	>1200
UNIFORMITY OF DIMENSIONS (individual finished products)			EXCELL.	GOOD	AVERAGE	POOR

The values given above are the result of research conducted in laboratories by recognized authorities. They give an idea of what can be reasonably expected of a product made in accordance with the rules of the art. The fact that no figures are given indicates that insufficient data are available.

I COMPRESSED BLOCKS			II ADOBE		III RAMMED EARTH		IV EXTR. BRICKS	V DAUB	VI STRAW CLAY	VII BURNED BRICKS	VIII CEMENT BLOCKS
1 RAW	2 STA	3 STA	4 RAW	5 STA	6 RAW	7 STA	8 STA	9 RAW	10 STA	11	12
D *	C *	A *	D	C	D *	C *	B				
D	A	A	D		D	A	A				
	B					B					
C					C						
C					C						
D					D						
	B					B					
	B			B		B					
B	B	A	C	C	B	B	D				
B	B	A	C	C	D	D	A				

1. Compressed at 2MPa
2. Stabilized with 8% cement
   Compressed at 2–4MPa
3. Stabilized laterite with 12 to 19% lime
   Compressed at 30MPa
   Autoclaved at an RH of 95% and at 90°C
4. Sand-moulding method
5. Stabilized with 5 to 9% bitumen emulsion
6. Compressed to 90–95% Proctor Standard
7. As above, stabilized with 8% cement
8. Hollow products weighing 1100kg/m^3
9. Double-face on wattle
10. 600 to 800kg/m^3
11. Indicate values in your region
12. Indicate values in your region

PROPERTIES	S	U	CLASSES A	B	C	D
RESISTANCE TO TANGENTIAL IMPACT BY A SOFT BODY (initial height of a 27kg sandbag vertically suspended from a point above the wall)		m	> 3	2 3	1 2	< 1
RESISTANCE TO CRUSHING BY AN ECCENTRIC LOAD (reduction coefficient for walls with a forward height to thickness ratio of 7 to 8; 30cm wall)	R		> 0.50	0.40 0.50	0.30 0.40	0.20 0.30
BENDING STRENGTH (uniform horizontal pressure – wind)		MPa	0.5×10^{-3} 0.6×10^{-3}	0.4×10^{-3} 0.5×10^{-3}	0.3×10^{-3} 0.4×10^{-3}	0.2×10^{-3} 0.3×10^{-3}
RESISTANCE TO LOCALIZED HORIZONTAL THRUST (pressure due to a 2.5cm diameter disc) – walls h = 2.5m, L = 1.20m, b = 30cm		N	> 4 500			
COEFFICIENT OF THERMAL EXPANSION		mm/m°C	< 0.010	0.010 0.015		

The values given above are the result of research conducted in laboratories by recognized authorities. The give an idea of what can be reasonably expected of a product made in accordance with the rules of the art.

I COMPRESSED BLOCKS			II ADOBE		III RAMMED EARTH		IV EXTR. BRICKS	V DAUB	VI STRAW CLAY	VII BURNED BRICKS	VIII CEMENT BLOCKS
1 RAW	2 STA	3 STA	4 RAW	5 STA	6 RAW	7 STA	8 STA	9 RAW	10 STA	11	12
	B		B	C	B	C					
	A	A	B	D	C	A					
	A		D	C	D	A					
	A		A	A	A	A					
						B					

1. Compressed at 2MPa
2. Stabilized with 8% cement
   Compressed at 2–4MPa
3. Stabilized laterite with 12 to 19% lime
   Compressed at 30MPa
   Autoclaved at an RH of 95% and at 90°C
4. Sand-moulding method
5. Stabilized with 5 to 9% bitumen emulsion
6. Compressed to 90–95% Proctor Standard
7. As above, stabilized with 8% cement
8. Hollow products weighing 1100kg/m^3
9. Double-face on wattle
10. 600 to 800kg/m^3
11. Indicate values in your region
12. Indicate values in your region

PROPERTIES	S	U	CLASSES			
			A	B	C	D
SWELL (immersion until saturated)		mm/m	0 0.5	0.5 1	1 2	> 2
POTENTIAL SHRINKAGE (artificial drying)		mm/m	0 1	1 2	2 5	> 5
SHRINKAGE DUE TO DRYING		mm/m	> 0.2	0.2 1	1 2	> 1
PERMEABILITY		mm/sec		1.10^{-5}		
WATER ABSORPTION OF THE SURFACE TO BE RENDERED		% WEIGHT	0 5	5 10	10 20	>20
TOTAL ABSORPTION		kg/m^3	0 7.5	5 10	10 20	>20
FROST SUSCEPTIBILITY			NIL	LOW	AVERAGE	HIGH
SUSCEPTIBILITY TO EFFLORESCENCE			VERY LOW	LOW	AVERAGE	HIGH
DURABILITY UPON EXPOSURE TO WEATHER (wall only – no protection)			EXCELL.	GOOD	AVERAGE	POOR

The values given above are the result of research conducted in laboratories by recognized authorities. They give an idea of what can be reasonably expected of a product made in accordance with the rules of the art.

I COMPRESSED BLOCKS			II ADOBE		III RAMMED EARTH		IV EXTR. BRICKS	V DAUB	VI STRAW CLAY	VII BURNED BRICKS	VIII CEMENT BLOCKS
1 RAW	2 STA	3 STA	4 RAW	5 STA	6 RAW	7 STA	8 STA	9 RAW	10 STA	11	12
							C/D				
							C				
B	B		B	B	C	C					
	B					B					
				A			A				
	C	A				C	C				
D	B	A	D	B	C	B	A/B				
B	B	A			B	B					
D	B	A	D	B	C	A	B				

1. Compressed at 2MPa
2. Stabilized with 8% cement
   Compressed at 2–4 MPa
3. Stabilized laterite with 12 to 19% lime
   Compressed at 30MPa
   Autoclaved at an RH of 95% and at 90°C
4. Sand-moulding method
5. Stabilized with 5 to 9% bitumen emulsion
6. Compressed to 90–95% Proctor Standard
7. As above, stabilized with 8% cement
8. Hollow products weighing 1100kg/m^3
9. Double-face on wattle
10. 600 to 800kg/m^3
11. Indicate values in your region
12. Indicate values in your region

PROPERTIES	S	U	CLASSES A	B	C	D
SPECIFIC HEAT	C	KJ/kg	1.00 0.85	ABOUT 0.85	0.65 0.85	< 0.65
COEFFICIENT OF CONDUCTIVITY (depends largely on the apparent density – see that heading)	λ	W/m°C	0.23 0.46	0.46 0.81	0.81 0.93	0.93 1.04
DAMPING COEFFICIENT (40cm walls)	m	%	< 5	5 10	10 30	> 30
LAG TIME COEFFICIENT (40cm walls)	d	h	> 12	10 12	5 10	< 5
COEFFICIENT OF ACOUSTIC ATTENUATION (40cm walls at 500Hz)		dB	> 60	50	40	30
COEFFICIENT OF ACOUSTIC ATTENUATION (20cm walls at 500Hz)		dB	> 6	50	40	30
FIRE RESISTANCE			EXCELL.	GOOD	AVERAGE	POOR
FLAMMABILITY			VERY POOR	POOR	AVERAGE	GOOD
SPEED OF FLAME SPREAD			VERY SLOW	SLOW	AVERAGE	FAST

The values given above are the result of research conducted in laboratories by recognized authorities. They give an idea of what can be reasonably expected of a product made in accordance with the rules of the art.

I COMPRESSED BLOCKS			II ADOBE		III RAMMED EARTH		IV EXTR. BRICKS	V DAUB	VI STRAW CLAY	VII BURNED BRICKS	VIII CEMENT BLOCKS
1 RAW	2 STA	3 STA	4 RAW	5 STA	6 RAW	7 STA	8 STA	9 RAW	10 STA	11	12
B	C		B	B	B	C					
C	C	D	B	B	C	C					
B	B	B			B	B					
B	B	B			B	B					
B	B	B			B	B					
C	C	C			C	C					
B	B	B	B		A	A					
					A	A					
					A	A					

1. Compressed at 2MPa
2. Stabilized with 8% cement
   Compressed at 2–4MPa
3. Stabilized laterite with 12 to 19% lime
   Compressed at 30MPa
   Autoclaved at an RH of 95% and at 90°C
4. Sand-moulding method
5. Stabilized with 5 to 9% bitumen emulsion
6. Compressed to 90–95% Proctor Standard
7. As above, stabilized with 8% cement
8. Hollow products weighing 1100kg/m^3
9. Double-face on wattle
10. 600 to 800kg/m^3
11. Indicate values in your region
12. Indicate values in your region

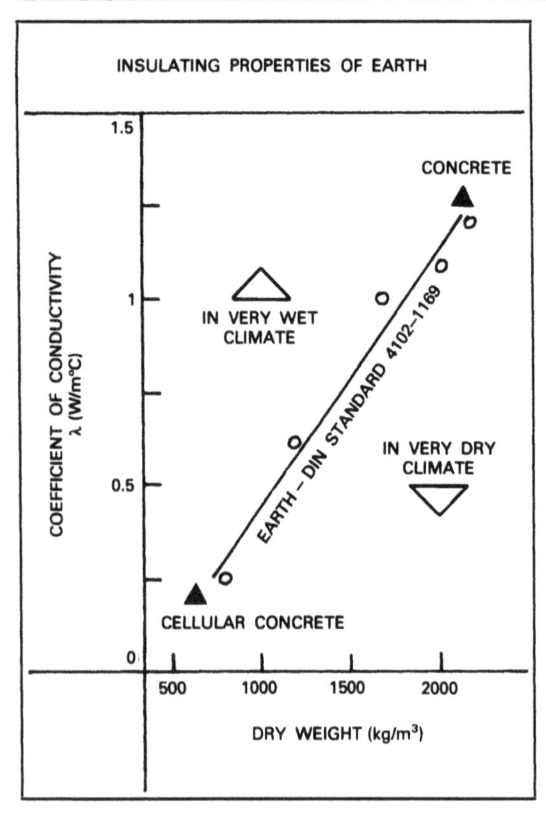

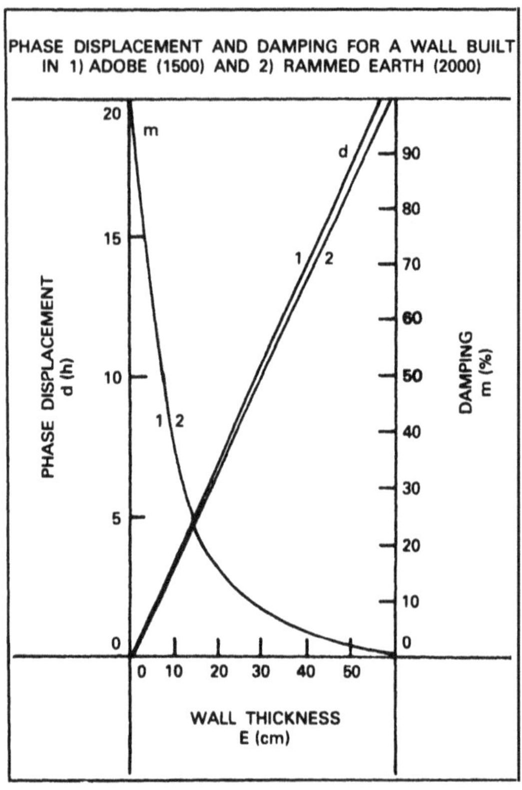

Although the thermal inertia of earth dwellings is widely recognized, this property of the material does not constitute a solution to all thermal problems. Soil simply does not possess all the legendary properties which have been claimed for it. The material is not particularly insulating, and its thermal capacity is far below that of solid concrete, as the following values indicate: for equal volumes, concrete: 590 (Wh/m^{3}°C); rammed earth 510 (Wh/m^{3}°C); and adobe 380 (Wh/m^{3}°C).

It is thus important to understand the thermophysical properties of soil in order to make the most of it. Current literature reveals a lack of data. The older specialist literature reports on isolated experiments carried out by various research centres under very diverse conditions, which hinders comparison. Research programmes currently underway in a number of countries will make it possible to describe the thermal properties of soil more accurately. In the period before new data becomes available curves can be constructed and information about reasonably stable properties can be disseminated on the basis of the older literature. For example, the relationship between the coefficient of thermal conductivity λ (W/m°C) and the bulk density of the material (kg/m^{3}) is very close to that of other mineral materials. Nevertheless variations in the density of the material used must be considered. This is typically the case with rammed earth and clay-straw techniques, and is dependent on the force with which the ramming was carried out. The curve for λ shown here was established for a mean moisture content in equilibrium in a temperate climate of 2.5% by weight. This is the least favourable mean. Values for λ can be doubled for an increase in moisture content of between 1 and 7%. This variation in moisture content is typical for unstabilized earth walls from summer to winter in temperate climates and in regions lying between a dry area such as the Sahel and a wet region such as the tropical coasts. The same value for λ can therefore not be used in all climates for calculation purposes. Empirical observations indicate that a 1% variation in water content results in 15% change in λ. Finally, according to the literature the specific heat of soil is between 800 and 1000 (J/kg°K).

These values for dry materials may increase with residual moisture content. An understanding of the dynamic behaviour of earth walls can be attained by studying the properties of thermal effusion and diffusion of the material in question. These properties are furthermore very similar to those in other mineral materials. Effusion, which considers the rate of propagation of a heat wave from one surface through the bulk of the wall and its restitution by the opposite surface, determines the suitability of the material for thermal storage. Even though soil has a lower thermal capacity than other heavier materials, its ability to store heat, which logically speaking should be lower, is nevertheless excellent. Soil benefits from a latent inertia related to its absorption capacity. The slowness with which water migrates through earth walls enhances their storage capacity over the long annual cycle. Diffusion is a measure of the lag time and damping of a heat wave as it is propagated through the wall. With its low diffusion, soil has the advantage of possessing a significant ability to damp and delay thermal variations and external thermal inflows. This property is particularly valuable in regions characterized by highly variable climatic and atmospheric conditions. The greatest asset of soil is the ease with which its specific gravity can be altered during construction. This is particularly true of the clay-straw technique, where bulk density can be made to vary from 300 to 1300kg/m^3. This ability makes it possible to change storage, damping, lag time wave displacement and insulating capacities as required by manipulating the weight and the thickness of the walls, and thus satisfying the conditions imposed by thermal considerations. Leaving myths aside, the thermophysical properties of earth can be considered as favourable, but nevertheless calculations must allow for variations due to the hygrothermic behaviour of the material.

WALL THICKNESS NEEDED TO OBTAIN A DAMPING EFFECT EQUAL TO THAT OF A 36.5cm BURNED BRICK WALL (COOLING TIME 82 HOURS)

UNRENDERED MATERIAL	DRY WEIGHT (kg/m^3)	WALL THICKNESS (cm)
CLAY-STRAW	300	28
CLAY-STRAW	400	27
WOOD-FIBRE CEMENT	400	17
CLAY-STRAW	600	28
CELLULAR CONCRETE	600	30
FIBRE PANELS	600	17
CLAY-STRAW	800	29
HOLLOW BRICK	800	35
CLAY-STRAW	1000	31
CLAY-STRAW	1200	35
ADOBE	1400	35
ADOBE	1600	36
BURNED BRICK	1800	36.5
RAMMED EARTH	1800	39
HEAVY CONCRETE	2400	52

INSULATION OBTAINED WITH A WALL THICK ENOUGH TO OBTAIN THE DAMPING EFFECT OF A 36.5cm BURNED BRICK WALL

UNRENDERED MATERIAL	DRY WEIGHT (kg/m^3)	$1/\lambda$ (m^2°C/W)
CLAY-STRAW	300	2.80
CLAY-STRAW	400	2.25
WOOD-FIBRE CEMENT	400	1.83
CLAY-STRAW	600	1.65
CELLULAR CONCRETE	600	1.58
FIBRE PANELS	600	1.31
CLAY-STRAW	800	1.16
HOLLOW BRICK	800	1.06
CLAY-STRAW	1000	0.86
CLAY-STRAW	1200	0.74
ADOBE	1400	0.59
ADOBE	1600	0.49
BURNED BRICK	1800	0.45
RAMMED EARTH	1800	0.43
HEAVY CONCRETE	2400	0.24

It is often mistakenly thought that there are no standards for earth construction. There have, however, been several attempts to establish standards. The known publications on the subject, which emanate from various parts of the world, bring together enough data to serve as a basis for the widest possible range of contexts and conditions of use of the material and for the greatest possible variety of national requirements. Though standards have little significance for small-scale projects such as individual dwellings, this is certainly not the case for more ambitious projects, which are becoming increasingly common. Reference manuals are nowadays indispensable in certain countries where technical inspections are very strict. In the majority of the developing countries, the quality of earth construction is often guaranteed by local know-how or the competence of technical assistance, but may also be guaranteed by official bodies.

Standards are, however, required by the decision-makers, the financiers, and the builders. Existing publications usually contain technical recommendations or practical tips, which do not cover the entire field but are limited to adobe, compressed blocks, or rammed earth.

The subjects dealt with mainly concern structural and thermal behaviour and test methods, but overlook a whole series of important points relating to production and construction. Several attitudes can be discerned among standardizers:

- The material is wilfully passed over and its importance minimized. It does not appear in the manuals.
- Earth construction is deemed to be adequately known and that it is enough to study existing examples. It is thus not necessary to standardize. This attitude is assumed sufficient to handle problems arising in connection with decision-making bodies, financial institutions and inspecting organizations.
- Over-strict standards are insisted upon, which could endanger the development of soil technology.
- Some standardizers consider their work as being necessary for the advancement of earth technology. In this case the manuals are highly didactic and are constantly being updated with reference to actual practice. These codes influence practice, which in turns influences the codes.

The absence of standards can no longer be invoked as an obstacle to the development of earth as a construction material. Numerous specialist organizations are capable of compiling codes of good practice which would serve as standardizing works for those responsible for earth construction projects. Below we list some good standardizing works, without attempting to be exhaustive. The pages of this manual deal with much of the information contained in these texts.

France During the period of reconstruction after World War II, three documents of an official nature were published:

- REEF DTC 2001 *Béton de terre et béton de terre stabilisé*, 1945
- REEF DTC 2101 *Constructions en béton de terre*, 1945.
- REEF DTC 2102 *Béton de terre stabilisé aux liants hydrauliques*, 1945.

A special set of specifications were prepared for the Village Terre de l'Isle d'Abeau project (72 housing units). This official document served as a reference document for the financial backers, insurers, the site manager, architect, contractors, and the inspecting body:

- 'Recommandations pour la conception des bâtiments du Village Terre – Plan Construction, 1982'.

The thermal characteristic of the soil can be found in the publications of the C.S.T.B. (Paris):

- N° 215. Cahier 1682. 198.

Germany This was one of the first countries in the world to draw up standards. The first DIN standards were published in 1944. Between 1944 and 1956, a series of standards and recommendations came into effect. These publications, which were deemed to be outmoded, were withdrawn in 1971.

- DIN 18951, Blatt 1 Lehmbauten (lehmbauordnung) Vorschrifter für die Ausführung/Blatt 2 dsgl. Erlaüterung, 1951. This document is accompanied by official commentaries:
- Lehmbauordnung nebst bäuerlichen Siedlungs richtlinien-Hölsher Wambsganz Dittus, 1948.
- DIN 18952, preliminary standard, Blatt 1 Baulehm Begriffe, Arten, 1956/Blatt 2 Prüfung von Baulehm, 1956.
- DIN 18952, draft, Lehm als Baustoff, 1951.
- DIN 18953, preliminary standard, Blatt 1 Baulehm Lehmbauteile Verwendung von Baulehm, 1956/Blatt 2 dsgl. gemauerte Lehmwände, 1956/ Blatt 3 dsgl. gestampfte Lehmwände, 1956/Blatt 4 dsgl. gewellerte Lehmwände, 1956/Blatt 5 dsgl. Leichtlehmwände in Gerippebauten, 1956/Blatt 6 dsgl. Lehmfusshöden, 1956.
- DIN 18953, draft, Lehmbau, Eigenschaften, Bauarten, Anwendungsbereich, 1951.

– DIN 18954, preliminary standard, Ausführung von Lehmbauten Richtlinien, 1956.
– DIN 18955, preliminary standard, Baulehm, Lehmbauteile, Feuchtigkeitsschutz, 1956.
– DIN 18956, preliminary standard, Putz auf Lehmbauteile, 1956
– DIN 18957, preliminary standard, Lehmschindeldach, 1956.

Apart from these standards covering rammed earth, compressed blocks, adobe, clay-straw and daub techniques, the German literature abounds in extremely detailed practical construction manuals. The thermal characteristics of soil still form part of standard: DIN 4108, 1981

India This is another of the few countries to have prepared official standards:

– 'Specification for soil-cement blocks used in general building construction, IS 1725', Indian Standards Institute, New Delhi, 1960.

Ivory Coast This country possesses a standardizing publication:

– Recommandations pour la conception et l'execution de bâtiments économiques en geobéton. LBTP, 1980

Peru Peru has published official standards on anti-seismic adobe structures:

– 'Norma de diseño seismo resistente, construcciones de Adobe y bloque estabilizado'. RNC. Reglamento Nacional de Construcciones. Resolucion Ministerial nº 159–77/UC-110º, 1977.
– Itintec 331.201. Diciembre, 1978, 9 p. 'Elementos de suelo sin cocer: adobe estabilizado con asfalto para muros: Requisitos'.
– Itintec 331.202. Diciembre, 1978, 8 p. 'Elementos de suelo sin cocer: adobe estabilizado con asfalto para muros: Metodos de ensayo'.
– Itintec 331.203. Diciembre, 1978, 4 p. 'Elementos de suelo sin cocer: adobe estabilizado con asfalto para muros: muestra y recepcion'.

United Nations Two extremely valuable codes of good practice have been published by the United Nations:

– 58/II/H/4, Manual on stabilized soil construction for housing. Fitzmaurice, R., 1958.
– 64.IV.6, Soil-cement, its use in building. 1964

As part of the Cissin project at Ouagadougou, Burkina Faso, the PNUD prepared certain basic standards:

– Cahier des charges dressé pour le project Cissin, 1973.

USA As long ago as 1941 the National Bureau of Standards was carrying out studies of the properties of adobe, compacted blocks, and rammed earth. The following paper was published:

– Report BMS 78 – Structural, Heat-Transfer and Water Permeability Properties of Five Earth-Wall Constructions, 1941.

Two years later, the United States Department of Interior Office of Indian Affairs published codes of good practice and technical specifications:

– *Earth Brick Construction*, U.S. Office of Indian Affairs, 1943.

Nowadays adobe construction practice has been integrated into national construction codes and standards.

– *Uniform Building Code Standards* – Section 24–15, 2403 Unburned Clay Masonry Units and Standards.
– *Methods of sampling and testing. Unburned Clay Masonry Units*. Recommended standards of the International Conference of Building Officials, 1973.
– Uniform Building Section 2405 – Unburned Clay Masonry, 1973.

Several states have contributed improvements to these national standards, including Arizona, New Mexico, California, Nevada, Utah, Colorado, and Texas. In 1983, certain states published regulations for rammed earth construction, compressed blocks, and even for sod.

Other countries have published basic texts which, however, have no official character; these include Tanzania, Ghana, Mozambique, Morocco, Tunisia, Kenya, Zimbabwe, Mexico, Brazil, Australia, the USSR, Turkey and Costa Rica.

International The International Union of Test and Research Laboratories for Materials and Construction (RILEM) and the International Council for Building Research Studies and Documentation (CIB) have established a new technical commission on earth construction in 1987. This commission (RILEM/CIB: TC 153-W90 'Compressed Earth Block Technology') will for instance elaborate recommendations and technical specifications for earth construction. Without doubt, these texts will be adopted as norms and standards by many countries.

AGRA. *Recommandations pour la Conception des Bâtiments du Village Terre*. Grenoble, AGRA, 1982.
Bernard, P.A. 'L'inertie, Facteur d'Economie de Chauffage'. In *Moniteur BTP*, Paris, 1979.
Diaz Pedregal, P. 'Les caractéristiques thermiques de la terre: du mythe à la réalité'. In *Revue de l'Habitat Social*, Paris, Union Nationale HLM, 1981.
Grésillon, J.M. 'Etude de l'aptitude des sols à la stabilisation au ciment. Application à la construction'. In *Annales de l'ITBTP*, Paris, ITBTP, 1978.
Mukerji, K.; Bahlmann, H. *Laterit zum Bauen*. Stamberg, IFT, 1978.
Penicaud, H. 'Caractéristiques hygrothermiques du matériau terre'. In *Colloque Actualité de la Construction en Terre en France*, Lyon, PCH, 1982.
Pollet, H. *Un Nouveau Matériau de Construction Obtenu par la Réaction Argilo-calcaire*. Paris, RILEM, 1970.
Simonnet, J. *Recommandations pour la Conception et l'Exécution de Bâtiments en Géobéton*. Abidjan, LBTP, 1979.
Vollhard, F. *Leichtlehmbau*. Karlsruhe, CF Müller GmbH, 1983.
Wittemore, H. *et al. Building Materials and Structures*. Washington, US Printing Office, 1941.

Compressed earth block demonstration house in Sri Lanka (Theo Schilderman, IT)

8. CONSTRUCTION METHODS

The possible ways in which earth can be used as a construction material are very numerous. For the sake of simplicity, a dozen or so fundamentally different building methods can be identified. Even so there are close to a hundred variations on these basic themes.

Among the most widely known and practical construction methods are rammed earth in formwork, bricks moulded in raw earth and baked by the sun or 'adobe', and compressed earth blocks, which are produced in presses.

There are, however, other construction methods, such as cob and daub, which continue to be widely used as they were in the past. The largest and grandest earth buildings were not necessarily built in rammed earth, adobe, or compressed earth blocks.

The twelve construction methods described here have all been used to construct dwellings ranging from modest huts to splendid palaces, settlements going from small villages to imperial cities.

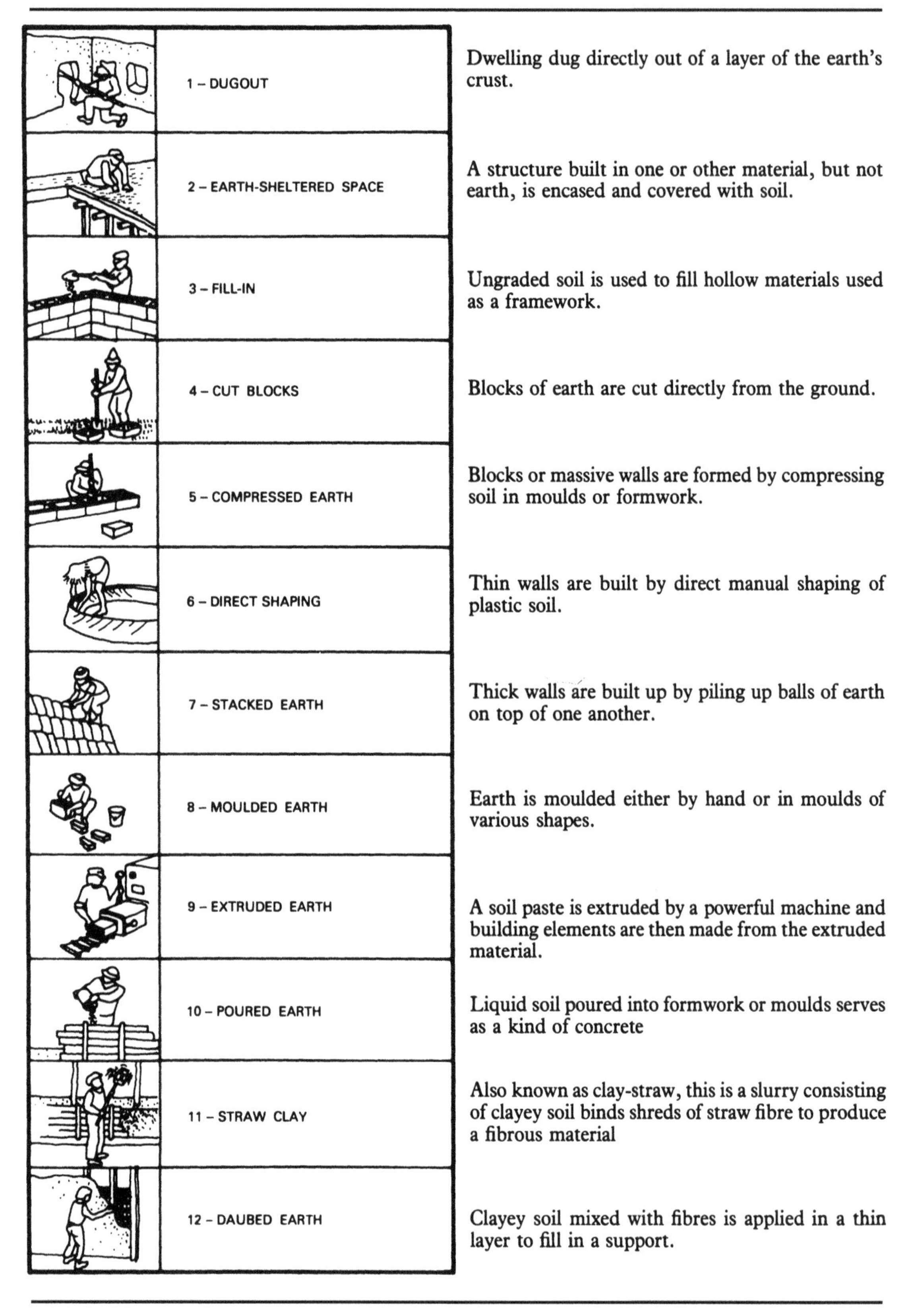

1 – DUGOUT	Dwelling dug directly out of a layer of the earth's crust.
2 – EARTH-SHELTERED SPACE	A structure built in one or other material, but not earth, is encased and covered with soil.
3 – FILL-IN	Ungraded soil is used to fill hollow materials used as a framework.
4 – CUT BLOCKS	Blocks of earth are cut directly from the ground.
5 – COMPRESSED EARTH	Blocks or massive walls are formed by compressing soil in moulds or formwork.
6 – DIRECT SHAPING	Thin walls are built by direct manual shaping of plastic soil.
7 – STACKED EARTH	Thick walls are built up by piling up balls of earth on top of one another.
8 – MOULDED EARTH	Earth is moulded either by hand or in moulds of various shapes.
9 – EXTRUDED EARTH	A soil paste is extruded by a powerful machine and building elements are then made from the extruded material.
10 – POURED EARTH	Liquid soil poured into formwork or moulds serves as a kind of concrete
11 – STRAW CLAY	Also known as clay-straw, this is a slurry consisting of clayey soil binds shreds of straw fibre to produce a fibrous material
12 – DAUBED EARTH	Clayey soil mixed with fibres is applied in a thin layer to fill in a support.

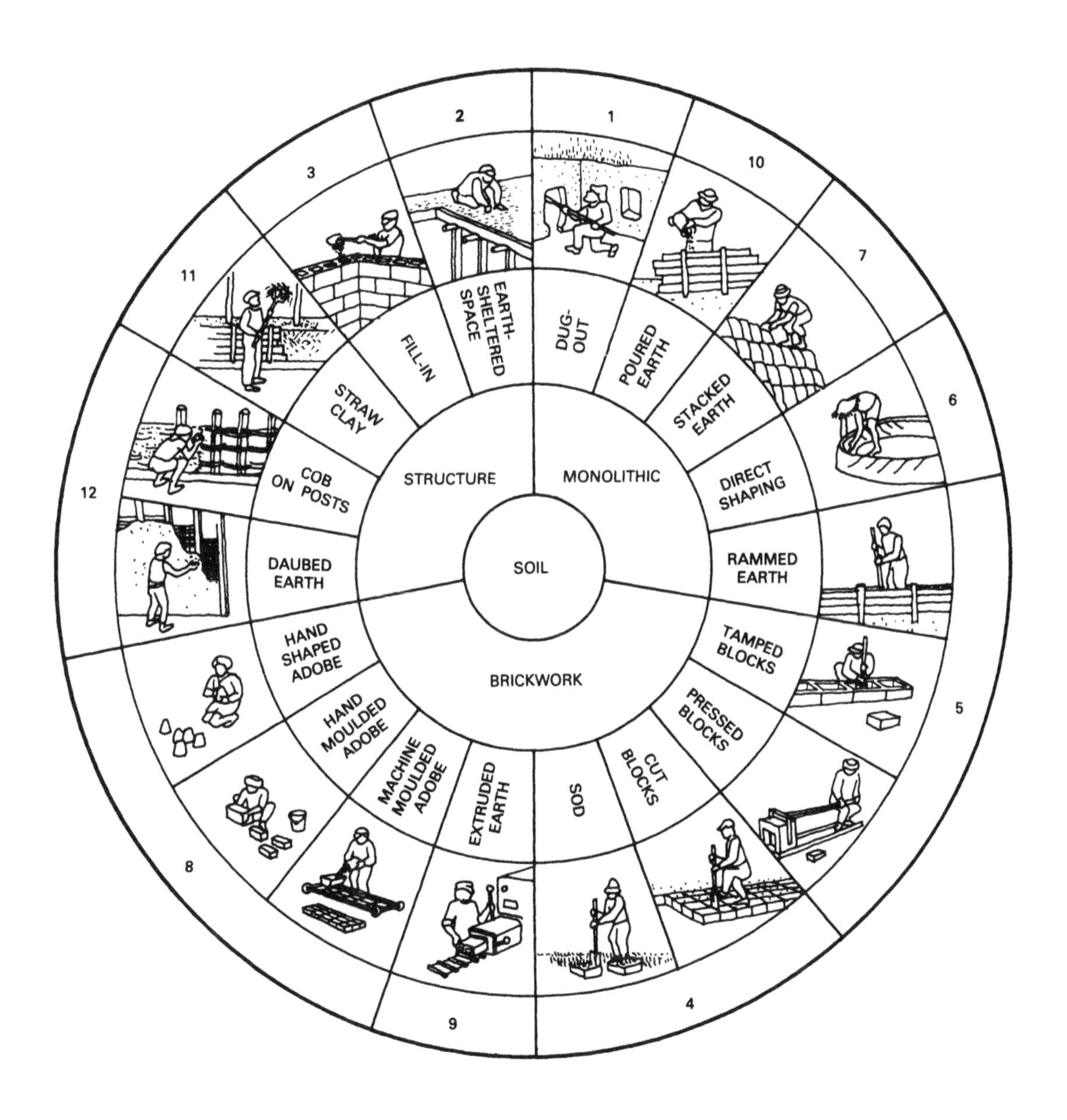
SOIL
MONOLITHIC
STRUCTURE
BRICKWORK
DUG-OUT
POURED EARTH
STACKED EARTH
DIRECT SHAPING
RAMMED EARTH
TAMPED BLOCKS
PRESSED BLOCKS
CUT BLOCKS
SOD
EXTRUDED EARTH
MACHINE MOULDED ADOBE
HAND MOULDED ADOBE
HAND SHAPED ADOBE
DAUBED EARTH
COB ON POSTS
STRAW CLAY
FILL-IN
EARTH-SHELTERED SPACE
1
2
3
4
5
6
7
8
9
10
11
12

A close relationship emerges between the worldwide distribution of dugout architecture and factors of climate and soil type. In the majority of cases, sites are dug out in soft soil, tuff, loess, or porous lava in areas with hot dry climates. Similarly the underground dwellings in loess of China and Tunisia are found in desert and steppe. The greatest concentration of such structures can be found in the countries of the Mediterranean region of Europe (i.e. Spain and Italy), Turkey and the countries of North Africa (i.e. Morocco, Algeria, Tunisia and Libya).

Horizontal dugouts

Construction methods reached the point at which it was possible to artificially dig out the earth horizontally. Using these methods, relatively complex underground dwellings were built. Generally, the interior of such dwellings was inspired by other structures in the same area. In fact, these interiors could be considered to be mirror images of one another. In many cases it can be seen that the dugout is camouflaged by a sophisticated lean-to structure, by a simple porch roof, or by a covered extension. The façade also contributes to the camouflage effect, and this for the reason that it is constructed using dug-out material the visual impact of which evokes the traditional local architecture. An isolated dugout dwelling built for temporary residence or as a shelter is highly exceptional. Historically, the permanent settlement of fertile land lying near mountainous areas capable of easy defence provided the basis for the establishment of entire communities in villages of dugouts. Laid out in rows on one or several levels and in such a way as to draw the maximum benefit from the sun, these dwellings are connected to one another by a complex network of paths and stairways.

Many dugout dwellings have a very similar layout with a daytime area consisting of a living room and kitchen directly illuminated by means of openings made in the façade and a night-time area consisting of rooms illuminated by a second source of light. In the more spacious dwellings an inner central area ensures access to each of the rooms. These inner areas are limited by the walls left behind after the completion of the excavation works. Dugout volumes and areas are not subject to the same restrictions as those of standard construction materials. Underground dwellings are so to speak sculpted with ease from the earth mass. Moreover it is not rare to observe highly irregular shapes among them with prominent curves. The areas referred to above are usually oblong or more or less quadrangular. If the dwelling contains more than one storey, the floors are often constructed using a layer of harder and more cohesive material. Vaults, or dropped arches, and domed ceilings confer stability upon the system. Underground

The dugout home offers protection from the daily heat and smooths the difference between night-time and daytime temperatures due to the thermal buffer effect of the earth mass. Apart from defensive reasons, the lack of building techniques and materials (quarrying, transport, and long and costly construction) must have led people to live underground. Nowadays, urban concentration, the wastage of space and energy, has thrust a new form of troglodytism upon us. Skyscraper development is yielding to underground developments of often alarming dimensions.

dwellings may also assume the form of traditional wooden structures using posts, beams and joists, and boards. This approach makes it possible for far larger volumes to be freely divided up as required.

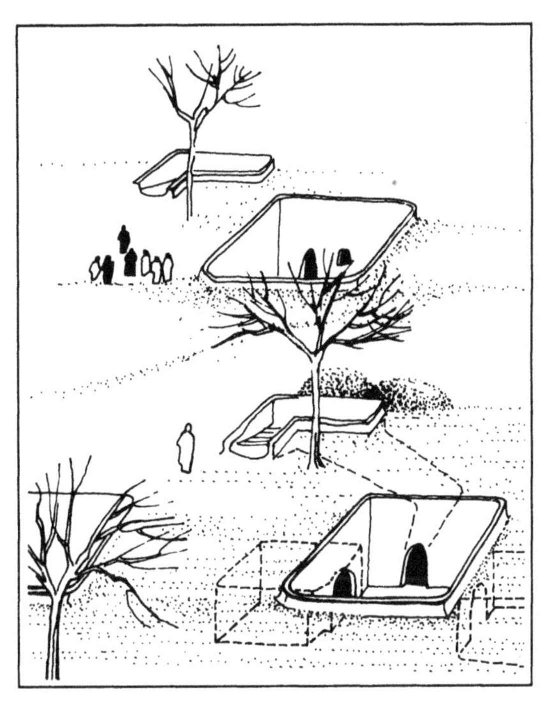

Vertical dugouts

In flat areas such as plateaus or plains underground dwellings are dug out vertically. Generally a ramp or stairway of gentle gradient leads to the surface from the bottom of the hole, which penetrates the soil to a greater or lesser extent. Vertical dugouts can be seen among the Matmata of Tunisia as well as in the loess regions of the Chinese provinces of Hunnan, Shansi, Shensi and Kansu which together have a total population of close to ten million people.

In China dugout dwellings are systematically planned with thousands of dwellings and workshops, small industries, schools, offices and hotels laid out on a gridiron pattern. In Loyang, the dugouts are nine to ten metres deep and are built around a rectangular patio approximately 200 to 250 square metres in area. The living quarters are organized around this patio. The rooms are dug so that they run outwards from the vertical faces of the patio hole. They are vaulted and low, as soil has very poor bending characteristics. Dwellings constructed in such a way are at one and the same time sheltered from prevailing winds, ventilated by the patio and extraordinarily resistant to the seasonal extremes of weather. The dugout dwelling suffers from two major problems: erosion and the accumulation of rainwater. Driving rain and wind progressively wear down the outer walls which must be regularly resurfaced. During the rainy season, water which runs down into the patio is collected in wells, which constitute a valuable reserve of water.

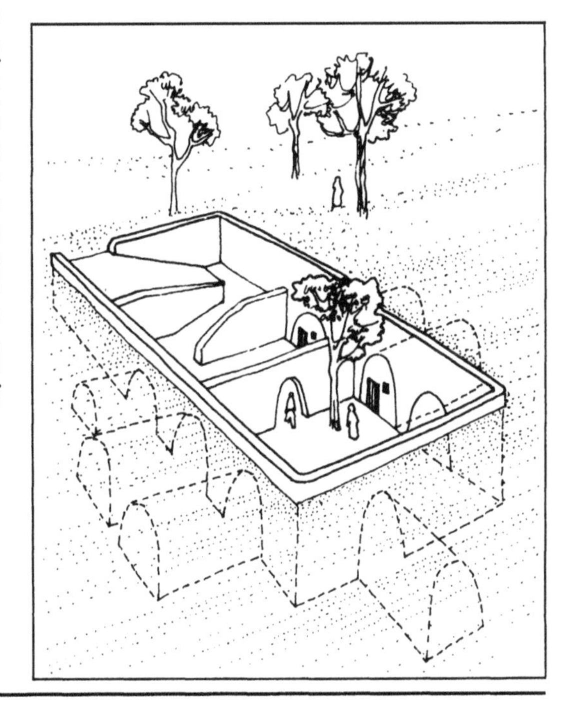

There are numerous examples of underground architecture throughout the world and it is commonly believed that they belong to a bygone age. That some people can nowadays again consider it desirable to live in dwellings sheltered by earth provokes wonderment. Conventional construction methods are not compatible with the concept, whereas modern technology and present-day standards and notions of comfort favour it. It was only in a period of abundant fossil energy, now also past history, that man could conceive of an architecture which did not take local climatic factors into consideration. The energy crisis has forced the necessity upon us of finding efficient and cost-effective solutions to such problems. The hyperinsulation provided by a thick layer of earth heaped onto a structure built with other materials constitutes a barrier to undesirable heat or heat loss. Quite apart from the considerable energy-saving

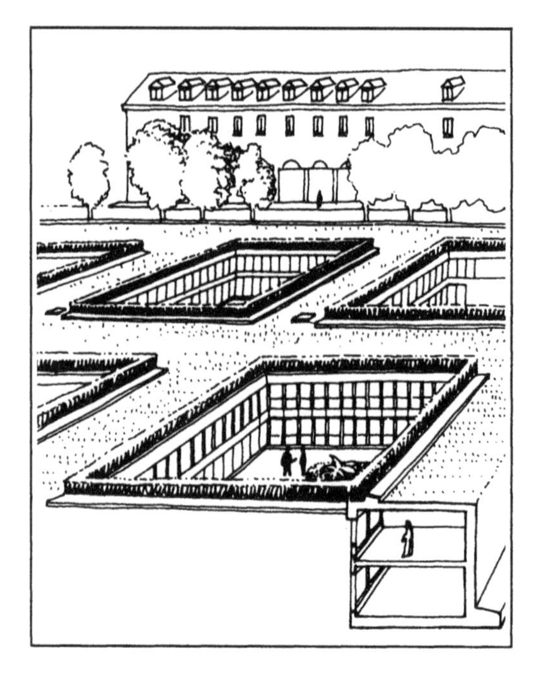

Underground dwellings

The structure of an underground building is not subject to any particular specifications and a great variety of materials other than earth may be used. Earth does not form an integral part of the structure. It is only used to shelter it. The Paris headquarters of UNESCO offer a good example of the use of backfilled earth.

In contrast to the above, earth can be used structurally in the formwork of underground buildings. The 'reinforced earth' method of the architect Henri Vidal is a good example. In this method linear metal reinforcement is placed in a horizontal layer and covered over with layers of highly compacted earth. The solid mass of earth thus obtained is uniform throughout its entire height and to a depth equal to the length of the reinforcement. The facing consists of a flexible skin or of concrete 'shells'. The Vidal method, tested in public works projects, has found architectural applications in various countries including France, Spain and the United States. Very steep sites with slopes as high as a 50% can now be built upon.

Earth-sheltered dwellings

The earth-sheltering dwelling is a very old concept as the circular dwellings at the Banpo site in China constructed about 4000 years ago attest to. Comfort in a variety of weather conditions, more than defence considerations, was the basic premise of these dwellings.

This ancient method of heat insulation is once again being used. The most recent bio-climatic concepts in architecture lend themselves to it.

The insulating earth can be used to protect the roof alone. This principle is just as valid in hot as in cold climates. Several different types of African dwellings (for example, in Tanzania, Ethiopia, Burkina Faso and Niger) are sheltered with a thick layer of beaten, clayey earth. In Iceland and Norway

achieved by the new architecture, the latter forces a new attitude toward the landscape upon us. The fact that the stage of trials and models has given way to numerous actual structures confirms the growing popularity of underground dwellings. In the same vein, international conferences bringing together researchers and specialized builders are regularly held on the subject. It is a fact moreover that a number of universities and research centres have set up high-level scientific and technical programmes for the study of this subject. A distinction must be made here between underground structures and earth-sheltered structures. In the first case an excavated pit is filled with the structure which is then covered with soil. In the second, the structure is built at ground level and then sheltered wholly or in part with earth.

sod is used to preserve the roofs of traditional wooden dwellings. Recent research conducted at the GHK in Kassel, Germany, into the concept of insulating roofs with earth tends to confirm the benefits of the technique. A layer of expanded clay sheltered by 40cm of grass-covered earth reduces heat loss by between 50 and 90%. In addition, a thick layer of earth considerably enhances the soundproofing of a dwelling.

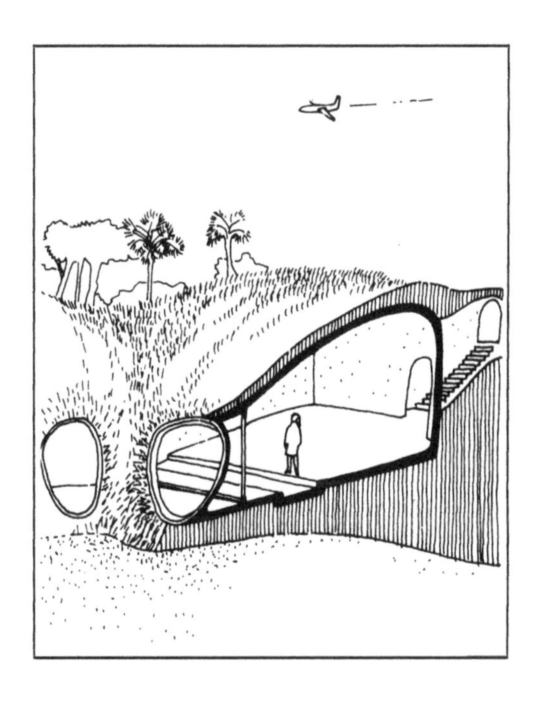

Structures, whether they be underground or earth-sheltered, must be damp-proof.

Dampness can be caused by any of the following:

- the absorption of water from a poorly drained surface;
- temporary water pressure on the walls;
- variation in underground water levels and capillary rise;
- the transmission of water vapour through walls.

Dampness can be prevented by providing drainage, protection and effective damp-proofing. There are numerous techniques available: filter membranes and drainage around the sides, vapour barriers, watertight dressings and paint, reinforced plastic membranes, etc.

Underground structures must be able to withstand the vertical pressure due to the mass of earth covering the roof, as well as the horizontal pressure caused by the soil pressing against the buried walls. Temporary, lateral pressure caused by water and frost and upward pressure caused by fluctuations in underground water levels constitute loads of variable magnitude. Massive supporting walls, cantilever walls, reinforced masonry, reinforced earth, hull-shaped, arched and dome-shaped structures are some of the solutions that have been adopted by builders of underground dwellings.

Dry earth can be used to fill a whole range of hollow ready-made building materials which can then be used in the actual fabric of a structure. When the materials used for the structure are light and not very durable in themselves, filling them with earth gives them weight and renders them stable. In cases in which structures built using hollow construction materials are stable to begin with, the added earth enhances thermal and acoustic insulation. The application of those techniques to the construction of emergency dwellings for disaster victims of is well known. In the majority of cases in which these techniques have been applied, it was for the purpose of building temporary dwellings, and they offer the advantages of economizing rare materials, rapid execution, and low cost.

Bags

Jute bags are filled with dry earth and then piled on top of one another. These are staggered in their respective successive rows like the bond in brickwork. Arched openings are reinforced by steel bars while horizontally and vertically positioned posts strengthen the walls.

Tubes

Cotton tubes filled with volcanic sand are soaked in whitewash, placed on top of one another and inserted between a rigid bamboo reinforcing structure. The two sides of each wall are white-washed so that the wall can breathe. One such project was completed in Guatemala in 1978.

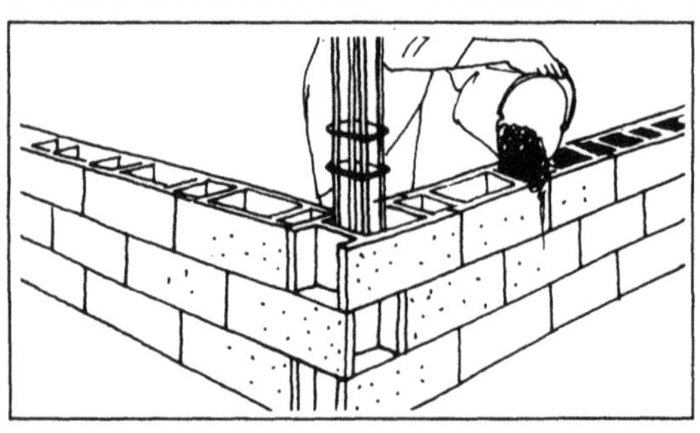

Blocks

Large hollow cement breeze blocks are filled with earth. The corner units are filled with reinforcement and concrete, constituting a rigid self-coffering structure. This method was tested in the construction of the agricultural village of Abadla in Algeria in 1975 (2).

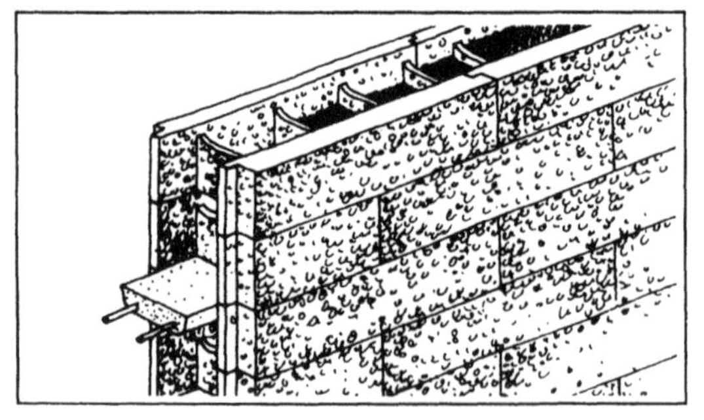

Insulating units

Hollow blocks made of cork are filled with earth. This system combines mass and thermal insulation. Horizontal and vertical bonding in reinforced concrete is cast in the hollows of the structure. This method, which is still in the experimental stage, was devised by the Studie group in Corsica in 1981.

Boxes

In 1971 NASA patented a system using folded boxes of waxed cardboard. The idea was for the boxes to be filled with earth and used to construct walls on the moon. The ingenious folding system divides the boxes into four compartments. When folded the flaps project from the flat upper surface and serve to lock the boxes in position.

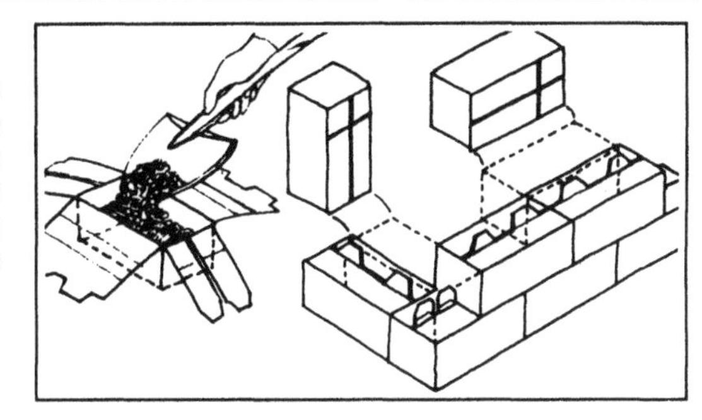

Containers and refuse

Discarded reusable material such as barrels, drums, etc. can serve as construction units. Tyres filled with earth and piled on top of one another result in heavy, insulated walls. These are dressed with earth on the outside and plastered on the inside. Dwellings built by self-help builders are constructed in this fashion.

Hollow walls

The so-called in-fill technique is very old. Walls built using stone or brick are filled with earth. Numerous structures dating from Roman times as well as cathedrals and the Great Wall of China are examples of this technique. It was employed recently in the construction of the village of Zeralda in Algeria.

Textiles

In this method sheets of canvas hung from frames driven into the ground are filled with earth. Units of up to 10 metres long can be produced in series and rapidly set up. The taut side bulges at its base and constitutes a structure which is highly resistant to earthquakes. The technique was perfected at the GHK in Kassel in Germany.

Latticeworks

Wooden latticeworks or a wooden framework surfaced with chicken wire are filled with earth. This system unites simplicity with speed, but must be protected. The technique was employed on Mayotte Island in the Comores Archipelago in 1982.

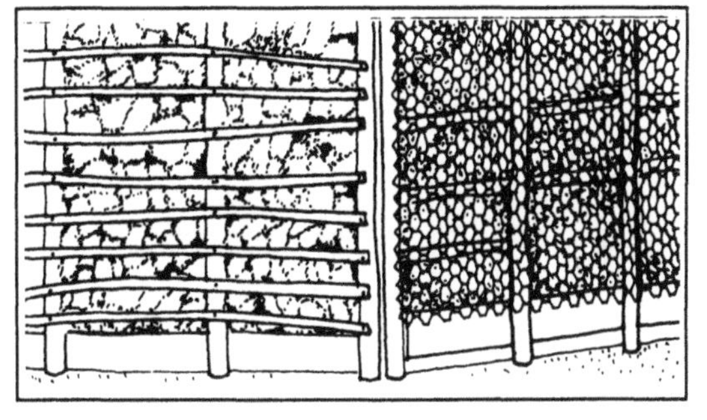

Surface soil, whether organic or mineral in content, which is naturally cohesive can be cut without any preliminary steps. The removal of blocks of earth from a layer of soil is a very ancient practice. Organic earth containing plant material and coherent because of the abundance of roots was cut into sods. The technique was practised throughout the ages in the Scandinavian countries as well as in Asia and America. Soil in arid regions rich in concretions containing carbonates (a chemical source of cohesion) is suited

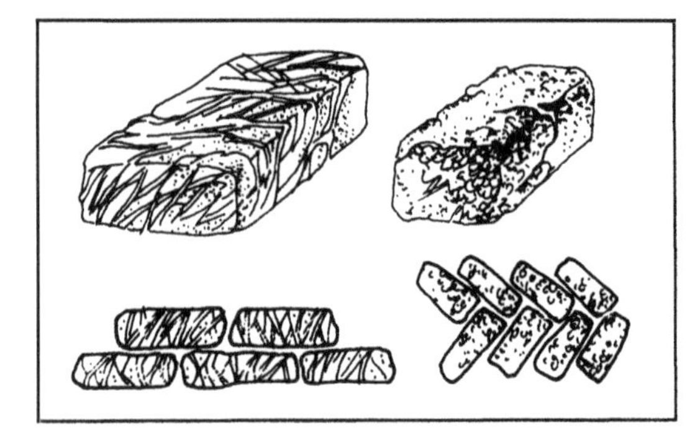

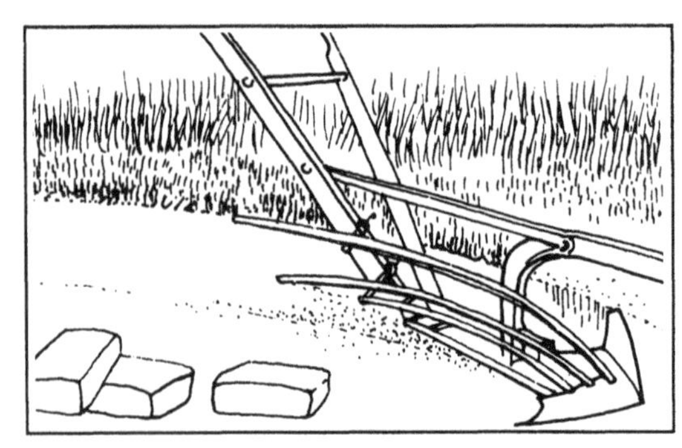

Blocks of earth and grass roots

Blocks of earth used for the purpose of construction are referred to by various terms: for example, 'sod' in England and the United States, 'turf' in Ireland and 'terrone' in Central and South America (this section will use the word 'sod'). If the blocks are cut from a single layer of earth containing plant material, the sod is removed in a single cut, or in a number of slices depending on the thickness of the stratum. The American use of this technique deserves comment. The European immigrants that settled the states of Missouri and Nebraska found themselves lacking a source of wood and stone. There was a plentiful supply of clay but the scarcity of fuel excluded the possibility of baking it. To build their farms, the settlers were inspired by the local Indian dwellings, i.e. the round huts peculiar to the Omaha and Pawnee tribes, and they took up the practice of cutting sod. It was difficult to obtain uniformly shaped pieces of earth using a spade. The situation improved with the development of a specially adapted plough called the 'grasshopper' which featured a right-angled blade able to cut uniformly shaped chunks delicately without breaking them. The blocks of earth are placed in the structure (with the grass-covered surface facing inwards) just as they would be in a conventional brickwork. Sod walls and structures suffer from settlement and they rarely have foundations. It was customary for constructors to provide empty spaces at right angles to the doors and windows which were gradually closed by the settling process. The first rectangular dwellings were simple in design, and as time went on, they became more elaborate, describing L and T shapes. Some of these began to take on the appearance of multi-level middle-class homes. They were particularly well insulated and resistant to erosion. Furthermore, the clearing of land overgrown with grass freed it for cultivation. At present this technique is rarely used. It should be pointed out, however, that 'terrone' is officially approved by the construction standards of the state of New Mexico in the United States.

to this technique. Only elementary tools (spade, pick and quarry tools) are required for the removal of chunks and blocks. Advances have been made in this area only recently. The production process involved is minimal. There is no pulverization, mixing, or compaction. The blocks or chunks are ready for use as soon as they are cut. The cut block technique, which has gradually fallen into disuse, could be due for a comeback.

Cut blocks

Some soils have a cohesion and hardness which permits blocks of earth suitable for construction purposes to be cut from them. Such soils are characterized in particular by a high carbonate content or they are the product of lateritic induration. The use of blocks of earth cut out of the ground is familiar in a number of countries. Such blocks are known variously as 'tepetate' in Mexico, 'caliche' in the United States, 'mergel' in the Low Countries, 'marl' in England and 'tuf' in most of the Mediterranean countries. The soils in question are most often the result of the consolidation of disaggregated parent rock. One type of indurated laterite for example, the plinthite common in Burkina Faso, possesses the property of reacting on contact with air. Although they are friable and disaggregate in water, when they are removed from the ground plinthite blocks become as hard as rock and resistant to water after a few months' exposure to the air. The traditional method of making these blocks is similar to the one used for stone. One labourer using only picks, chisels, wedges and saws can turn out fifty big blocks a day. By about the early 1960s in Libya, the removal of plinthite blocks was mechanized with powered rotary saws being used. Fragile in water, plinthite cannot be cut with diamond saws. Special saws fitted with tungsten carbide inserts were used. These are highly resistant to abrasion and shocks from part of the blocks that have already hardened. Guides enable these saws to cut both horizontally and vertically through the thickness of the plinthite blocks, greatly increasing productivity. Recently in Mexico the idea of building schools using 'tepetate' has resurfaced in the CONESCAL programme. The state of Texas is once again considering using caliche. These materials are so solid and durable that interest in them has revived. Specially designed equipment for extracting them remains to be developed.

Naturally permeable and porous, earth is extremely vulnerable to water. Once water penetrates the voids of this material, it causes irreparable damage. Vulnerability to water can be decreased and strength increased by reducing the voids ratio. This improvement is achieved by compressing or compacting the earth thus raising its bulk density. The technology of compressed earth has been thoroughly investigated and is widely applied, as a result of the attention it has received in the past from engineers and scientists. Technically advanced and uncompromisingly modern, compression is very highly rated and appears to be in for a promising future.

The mechanisms of compression

Static compression: a dead weight exerts static pressure on the surface of the soil. The effect below the surface is limited as a result of the soil's internal friction.

Dynamic compression: impact or vibration.

When an object falls onto the surface of the earth, the impact produces a shock wave and pressure which sets the particles in motion. The effect deep below the surface is enhanced. The force of the impact of an object which has fallen from a height of 50cm gives rise to a pressure fifty times as great as the static pressure that the same object would produce.

Vibrating machines produce a series of rapid impacts on the soil at a rate of 500 to 5000 vibrations per minute. The pressure is propagated by waves to beneath the surface. The motion imparted to the particles temporarily overcomes internal friction and rearranges the grains. In this fashion maximum density can be achieved. While vibration compression is swift and equipment need be less heavy, the plant must be more sophisticated than that required for static compression.

1. Static compression

In practice the static compression of an earth mass is effected through the application of a force. This is produced by means of a hydraulic or mechanical press. Kept in a mould, the soil is compressed between two plates that slowly approach one another. The compression drives the air out of the soil. The densities thus achieved are of the order of 1700 to 2300kg/m^3. The static press is only suitable for the production of small building units (bricks, in-fill, etc.) and cannot be adapted to the construction of walls or monolithic paving. Depending on the type of press used, production ranges from 300 to 200,000 bricks a day and the cost of such a piece of equipment from US$800 to $2,000,000.

2. Dynamic compression by impact

The manual production of bricks by compressing soil in wooden or steel moulds with tampers is a very old technique. It requires painstaking work and the yield is low. In addition, it is difficult to control the thickness of the blocks. Several attempts at mechanization have been made with presses using heavy hinged lids. These were motorized and made for very laborious handling. They are no longer on the market because of the problems presented by their slowness of operation, the control of the thickness and the density of the blocks. In contrast, dynamic compression by means of impact is perfectly suited to the construction of walls made in monolithic compressed earth. The universally known *pisé* involves ramming earth in formwork. The final product of this process is very solid and durable as universally attested to by an architectural legacy as rich as it is varied and frequently of an astonishing quality. A similar technique was and continues to be used for making beaten earth paving.

3. Dynamic compression by vibration

Raising the bulk density of soil by vibration is very widely used in roadworks. It is also used to produce blocks made of compacted earth. The production units in Burundi and the Sudan that were operating in the mid 1950s are a case in point. The advantage of vibratory compression is that large blocks (50 × 20 × 20cm) can be produced at a significant saving in energy compared with the results which would be achieved if the process were carried out using static compression. Hollow blocks can also be produced by using vibratory compression. The principle has been applied to rammed earth by the GHK in Kassel in Germany. This takes the form of a small electrical vibrator containing an eccentric rotating weight run by electricity which travels back and forth in the formwork. The movement of the machine away from and back to its starting point is controlled by a switch. This system represents a considerable saving in manpower. It was used to construct a home in Brazil in 1982. Compression by means of vibration is particularly well suited to making floors. The range of such machines available on today's market is very extensive.

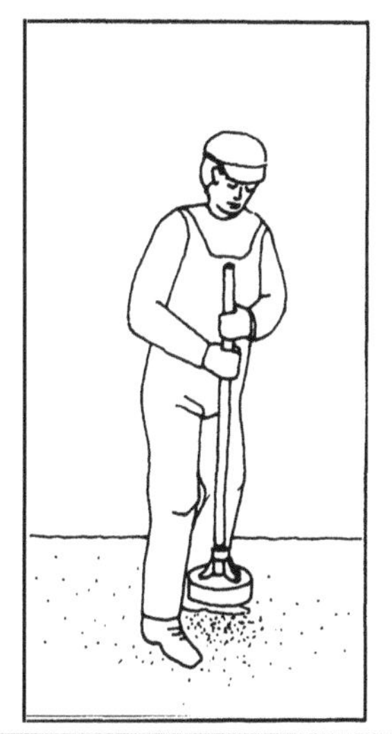

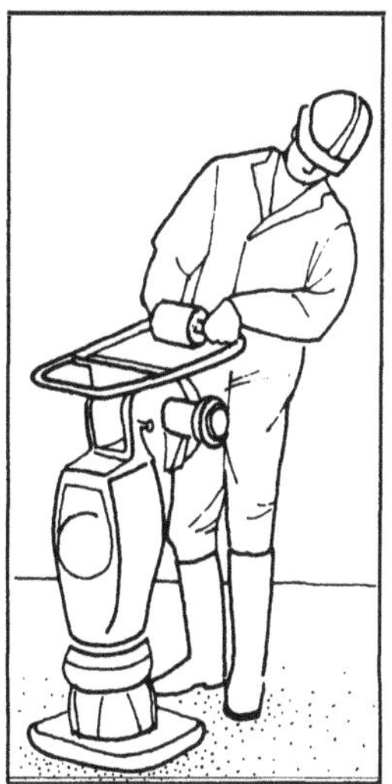

Direct shaping makes use of plastic earth. This is an essential precondition, but makes it possible to model forms without the use of a mould or formwork. In black Africa a large proportion of the dwellings in the Sahel and equatorial regions have been constructed in this fashion despite the extremes of the wet season which really puts the durability of a structure to the test. The architecture there is often breathtakingly beautiful. The quality and the preparation of the soil, and the appropriate consistency of the soil, are known only to the builders. This kind of know-how escapes scientific control. The principal

Rolls

Earth which is to be directly shaped by hand is often worked beforehand. Such preparation consists of the following consecutive stages: hydration, kneading and drying until the ideal plastic consistency is achieved. The material is sometimes enhanced by the addition of plant or animal substances or organic refuse. The earth is then shaped into sausages which are interlaced obliquely and then smoothed, or are dovetailed together, the actions of the builder being similar to those of the potter. The walls constructed are only from five to seven centimetres thick and hardly much more at the base. The world-famous *cases-obus* or shell dwellings of the Mousgoum of North Cameroon use this technique, and to this day their wholly original architecture continues to evoke amazement.

Balls

The use of large earth balls makes it possible to construct thicker walls. The material is laid down in several courses, the human hand being the main tool. Each layer must dry before the next can be put down. In order to ensure greater stability from the material, successive layers are made to overlap one another and fit into one another in the form of overlaying edges. The walls cannot bear much weight. Thicker layers are provided to support the heavy roof of timbers and heavy straw covered with kneaded clay. The farms of the Lobi of North Ghana and the fortress-homes of the Somba in North Benin are good examples of the technique.

Coils

The material can be prepared and used in a great many ways. Generous quantities of vegetable fibre, usually straw, may be added to the earth. Small sheaves of long twisted stalks of straw are dressed with plastic clayey mud. Thin walls (5cm thick) are built up by shaping actual coils of earth. This technique, examples of which can be found in Mexico, is used to construct roofs whose form is reminiscent of pottery kilns. These structures last a surprisingly long time.

advantages of the technique are as follows: great architectural variety, a limited manpower requirement, the low cost of the structures, the use of elementary tools, a good ground for applying a finishing rendering (i.e. the roughing of the underlying support and the insertion of hard chips in the wall). On the other hand, drying and control over shrinkage cracking as well as the poor mechanical performance of the material present problems. Apart from carefully tabulating its styles of architecture, direct shaping has not been studied in any systematic fashion which would allow its advantages and its drawbacks to some extent to be overcome.

Although the technique is no longer used in Europe, where it is known as 'bauge' in France and 'cob' in Great Britain, cob construction continues to be practised in many developing countries. Found in simple rural dwellings in Afghanistan as well as in the cities of Yemen, where towering multi-storey buildings can be seen, cob architecture can be both modest and spectacular. In this technique clayey earth in the form of rather soft cohesive paste is used. A degreasing agent is added to improve its cohesiveness and tensile strength. This agent is usually straw or chaff, though certain soils are at times

Stacked earth balls

In France some examples of the application of this abandoned technique can be found in Brittany in the Bassin de Rennes and the Vendée. The term 'bourrine' in French, which is used to refer to the type of dwelling found in the Marais of the Vendée, comes from the verb 'bourriner' which means 'to thatch'. To construct walls using the cob technique, earth was taken from the slopes of the raised paths that stretched across the marshes ('Marais'). This earth was very fat and of the silt-clay type. Spread over the ground in circular heaps 1.5 to 2m in diameter to a depth of 10cm, it was thoroughly moistened and covered with stalks, straw and reeds and on occasion twigs of heather. These vegetable fibres limit the considerable shrinkage of the material. The heap is then trodden so as to obtain a homogenous mixture. Another layer of earth and fibres is added and the operation repeated on a mass of material almost one metre thick. The mixture is then left to rest for a day. The following day the pasty mixture is cut with a spade into more or less uniform pieces known as 'bigots'. A labourer uses a fork to serve the bricklayer and the 'bigots' are stacked and compressed by stamping with the feet. After a few days of drying, the surface of the material is worked using a small spade with a sharp triangular blade. As soon as this is completed, the outer surface of the wall is scored with a small rake and encrusted with chippings, which serve as a basis for surface rendering. After two weeks of drying, the next batch of earth is applied. A similar technique was once practised in Germany; draught animals trampled the mixture, and the balls of cob, which could be large, were lightly tamped with wooden mallet. The thatched dwellings of the Vendée, like the houses built in Germany, are relatively long narrow structures, small in overall area and built all on one floor. The construction of the walls is a major part of the work. The cob structures once built in England in Devon seem to have been more elaborate, in order to permit building multi-storey houses. Provided with 'good shoes' (the foundations) and a 'good hat' (the eaves), as a Devon proverb put it, cob-built houses can last for many years.

Pitched earth balls

In North Yemen cob construction has been perpetuated by extensive technical know-how. The architecture has been characterized by superimposed layers of earth which are raised at the corners of the buildings. The soil is taken from the site of the building itself, and straw added to it. The mixture is moistened and trod by human feet in shallow pits. The mixture is left for two days before being shaped into large balls. These are thrown up to the bricklayer who stands on a finished part of the wall under construction and forms the new layer of cob by throwing the pasty balls forcefully and rhythmically onto the wall. As soon as this has been done the mass of earth is beaten with a wooden mallet and then smoothed by hand. The technique practised in Yemen has the advantage of speedy execution but the long drying stage delays the completion of the building. The problems of erosion and durability can be solved by adequately protecting the material (i.e. rendering). The necessary know-how, which still exists in quite a number of countries, could be used to set up modern low-cost building programmes.

very gravelly and approach those used for rammed earth structures. The material is first kneaded and then shaped into large balls, which are either piled or forcefully thrown onto the wall. In this way very thick walls (40 to 200cm) consisting of several layers of monolithic appearance can be built. The many construction techniques using cob are not very well known. The cracking of the material upon drying, which can be avoided, has preoccupied engineers, discouraging them from reintroducing the technique. The merits and potential of cob have been sufficiently well established to justify an attempt at reviving it.

Sun-dried clay brick is without doubt one of the oldest materials used to construct human dwellings. It is usually referred to by the Spanish word 'adobe', derived from the Arabic 'ottob', which in turn is related to 'thobe', the Egyptian word for sun-dried brick. Adobes are made using thick malleable mud often with added straw. They are shaped by hand in moulds or by machines. They can be found in numerous shapes. In view of the great diversity in the mode of production, ranging from the craftsman who produces from 100 to 150 bricks a day to the completely automated factory capable of turning out several thousand bricks a day, adobe is an extremely versatile material and can be adapted to the

Pyramidical bricks

The oldest sun-dried brick currently known to man was found at the site of Jericho in the Jordan Valley. It dates from 8000 years BC and was shaped by hand to resemble an elongated loaf with the fingerprints of the craftsman who made it still visible on it. There is a certain amount of archaeological evidence that the shape of moulded sun-dried bricks has changed considerably through the ages. The earliest known bricks were conical (discovered in Peru) and cylindrical. Later bricks were hemispherical and humpbacked (Middle East). Cube and parallelepiped-shaped bricks are of quite recent origin. At present conical and pyramidical bricks are still frequently used in West Africa. In Togo and the north of Nigeria these, which are known as 'tubali', are made without a mould and with a mixture of earth and straw which also serves as the mortar. Tubali are laid next to one another in thick walls with a large quantity of mortar. In Niger urban houses are made of bricks shaped by hand. These houses are often multi-storey affairs, as for example in Zinder and Kano, and are lavishly decorated with geometric motifs and interlacing arabesques.

Cylindrical bricks

Sun-dried cylindrical bricks were used to construct homes in Germany after the First World War. This shape was introduced in the area of the village of Dünner by a missionary who had returned to his native country after having spent all his life in Central Africa – an early sort of North-South exchange. At the time Germany had undertaken a massive reconstruction project and was thus interested in developing ways of using local materials. Immediately after they were moulded, clay cylinders were used to fill in the wooden frames of the walls. Erecting the frame and immediately filling it in offered protection from the vagaries of the weather.

Because the clay cylinders, which pressed hard against one another, were adhesive, it was possible to dispense with the use of mortar. Shrinkage of the walls was prevented by reinforcing the courses of the bricks with small branches. The visible surfaces of the walls were pierced with even series of holes so as to secure the rendering. The 'Dünner' method was disseminated throughout many areas of Germany and within a few years it had resulted in the construction of nearly 350 separate dwellings.

Parallelepiped bricks

Of all the various shapes of bricks the parallelepiped is the best known. These bricks, which are of varying size (from 0.20 × 0.11 × 0.05cm up to 0.60 × 0.30 × 0.10cm) and weight (from 2kg up to 30kg), are usually shaped in wooden or steel moulds containing one or two spaces using the slop or sand-moulded technique. In the case of the slop-moulded technique the clay is quite soft and the mould is dipped in water to facilitate its subsequent removal. The brick is made on the ground itself on a drying surface covered with straw, sawdust or sand. The mould is filled several times, and the soil compressed so as to drive out the air. A short, sharp vertical jerk is necessary to effect the removal of the brick from the mould. Bricks made in this fashion are rarely solid. In Egypt an experienced craftsman can produce 2000 small bricks a day. In the case of the sand-moulded technique the soil is usually plastic. The bottom of the mould is full of holes and as in the previous case it is dipped in water, but this time it is sprinkled with sand. The process is carried out over a table. The mould is filled only once, the earth is forcefully ejected and the excess removed by scraping. The mould is then turned over and shaken so that the brick can be removed. Bricks made in this fashion are generally of good quality and it is possible to produce from 400 to 600 of them per man-day.

widest possible range of socio-economic circumstances. Adobe construction is still widely practised in most developing countries but is regarded as a symbol of poverty. However, adobe has experienced a resurgence of popularity during the last ten years in the industrialized countries. In the Southwest of the United States, for example, international conferences have been held to bring together research workers and people from the practical side of adobe construction, and adobe has been made subject to building standards. Although sun-dried brick does not appear destined for any extraordinary technical advances, it will undoubtedly continue to play a major role in earth construction in the coming years.

The earth extrusion techniques have long been adopted by the brick industry. In the technical literature the origin of the machines presently used in most of the countries known for traditional skill in brickmaking is traced to the last century. The principle of the technique remains unchanged, while the machines themselves have been slowly improved. The earth used is very clayey, large aggregate having been removed, and it is processed in the form of a semi-solid paste. Fed into the machine, the material is extruded

Adobe

Extrusion can be adapted to the processing of sun-dried earth products such as adobe. The elimination of burning makes it necessary to check the aggregate content of the soil which must be less clayey and sandier. Sun-dried material is more abrasive than burned material and machines must be able to stand up to it. The fact that the soil is less adhesive in this case implies that the extrusion is a less power-consuming process. Adobes stabilized with asphalt used to be made by extrusion in the United States in the 1940s and 1950s. In India this type of process continues to be used to this day. Heavy expensive machines are generally used for this, but small mobile units mounted on trucks and combining a shovel excavator, a grader, a mixer machine, and the die were used in the United States in 1950. Small machines fitted out with vertical drums were used in Europe at the beginning of the century, as were animal-powered systems, which are currently being reconsidered.

Rolls

The concept of an extruder, combining a vertical mixer and a barrel in the bottom of the drum which operates horizontally, has been readopted by researchers at the GHK in Kassel in Germany. The GHK extruder was equipped with an electric motor and a helical cutter, which made it possible to extrude earth with a high sand content. The machine can produce a 1.5 metre long strip with a section of 8 × 16cm per minute or 360 bricks an hour. The relative elasticity of earth strips makes it possible to construct walls with flowing shapes, as wells as domes and arches, without using mortar. This technique, which is reminiscent of direct shaping, was tested in Kassel in 1982 but remains experimental. Cracking and shrinkage present problems which must still be resolved.

Extruded hollow bricks

The energy crisis spurred producers to consider ways of economizing the heat energy consumed in baking bricks. At Rennes in Brittany the INSA has experimented with producing extruded earth bricks using the so-called Stargil process. The CTTB has launched a similar product. The bricks are produced using all the equipment of a brickworks except for the kiln. After the earth is stabilized with cement (approximately 15%), and plasticizers (molasses) added to it, it is extruded through the barrel which turns out conventional bricks. In comparison with burned materials, there is an overall energy saving of 58% and plants reduce their investment outlay by 30%. Because of their smoothness it has been necessary to develop a special rendering with a glue and plaster base.

Loaves of earth

In this technique cylindrical mud loaves of earth are extruded by a small mobile extruder. The loaves laid in the structure are used immediately upon being extruded. They stay in place for the sole reason that they are plastic and adhesive. A wooden frame is filled in with the units and the resulting structure is particularly resistant to earthquakes. The technique was perfected in Dünner, Germany after the Second World War. In fact it represents a mechanized version of the manual process for producing earth cylinders.

through a specially designed barrel and comes out in the form of a continuous earth roll. Once on a delivery belt the paste roll is cut by a steel wire into standard lengths. This effective technique, which is still used in the ceramics industry, offers the advantage of being capable of being adapted to the production of a great variety of materials including plain or hollow bricks, in-fill, tiles, piping and drainpipe.

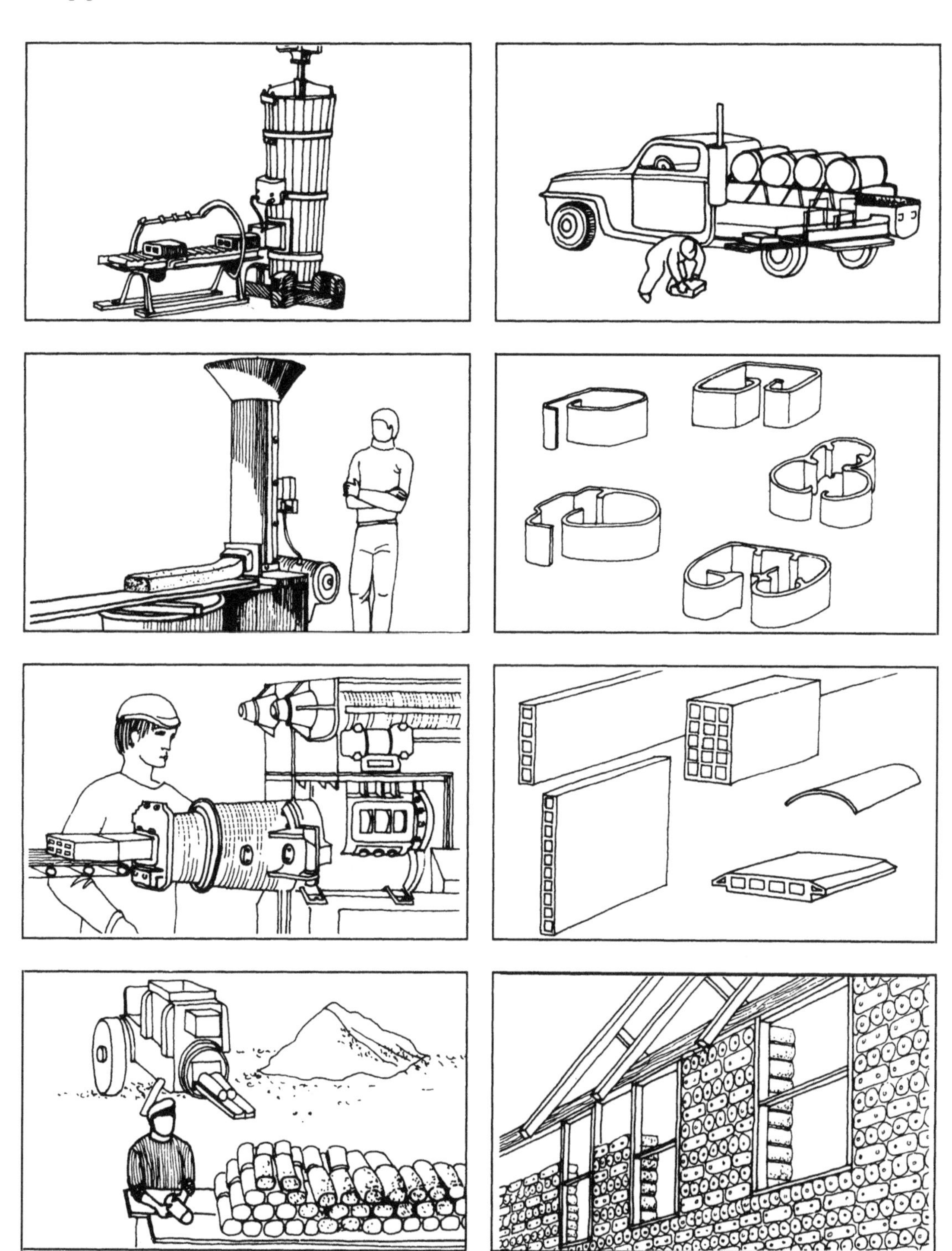

Soil in the form of liquid mud but containing fairly sandy aggregates, even to the point of being gravelly, can perform the same function as lean concrete. Poured earth offers several advantages such as easy preparation with a minimum energy requirement, great ease of use and a wide range of applications ranging from the prefabrication of units to the construction of monolithic walls or the production of

Bricks

Poured earth can be used to prefabricate bricks. This process is currently used for small-scale production. The mud can be prepared in large ditches by means of a frontloader. The frontloader then carries the material in its bucket to moulds set up on the moulding area. Mobile machines equipped with hoppers which tip the mud into a mould of the same capacity as the machine can produce 25 bricks or more at a time. Production units utilizing such machines can produce up to 20,000 bricks a day. Another method consists of pouring the earth into a large mould (3 × 3m for example) and then cutting the bricks by means of a cutting wire or else mechanically, using discs. All these methods are used today in the Southwest of the United States. Poured earth technology requires a great deal of know-how with respect to the evaluation of the consistency of the material. Bricks moulded with earth whose liquid content is too high, for example, collapse on being removed from the mould, and bricks moulded with earth whose liquid content is too low stick to the mould and will be damaged by the time they are removed from the mould.

Poured earth walls

It is possible to cast bricks *in situ*. Here bricks are cast in crenellations which are then filled up. The problem of shrinkage is overcome as a result of the small size of the crenellation. Small light modular crenellations are used which can be rapidly assembled and disassembled. This construction method makes it necessary to proceed spirally around the structure. At present the technique is practised by a number of building contractors in the United States. The earth can be poured or pumped just like concrete.

Monolithic walls can be reinforced against earthquakes with steel or bamboo reinforcement. The technique was used in Brazil by the PCA in 1950. The Adriano Jorge hospital in Manaus, which occupies an area of 10,800m^2, was built in this fashion. There are significant cracking problems. A third technique was developed by the engineer David in the Ivory Coast in 1980. It consists of prefabricating stabilized earth with aggregates of varying diameter. After drying, these aggregates are used as they normally would be to make a concrete bound by a stabilized earth mortar. The technique makes use of ordinary concrete formwork.

Paving

The poured earth technique is ideally suited to making of paving. Here too know-how is indispensable, if unsightly cracking is to be avoided. The supporting soil must be properly prepared and drained. A sand or set stone foundation is required. Poured earth paving can be decorated using geometric motifs or else it can be given the appearance of slabs, but in that case contraction joints are a necessity. A multi-coloured surface, after the fashion of the Indians of the southwest United States, is another possibility. The lifetime of poured earth paving can be enhanced through the use of surface hardeners or waterproofing agents such as asphalt or linseed oil which gives excellent results. This technique is suited to both indoor and outdoor paving. Cement can be used. The technique referred to as the plastic soil–cement technique is used for making paths and even irrigation and drainage canals. In the United States special machines have been developed for this technique.

paving. Nevertheless the technique is rarely used as the material suffers severe shrinkage upon drying. The latter problem can be solved by means of stabilization, by dividing slabs up into smaller units, or by sealing cracks at a later date if there is no risk of structural damage. Furthermore, the poured earth technique can be adapted to the entire range of tools used in concrete technology including the concrete pump. This technology may be destined for future development.

In this method (also known as clay straw) the function of the earth is to bind straw stalks together. Any kind of straw will do: wheat, barley, rye and winter barley can all serve, as well as other kinds of fibres such as hay and heather. The straw is well carded; 15 to 40cm is the optimal length for the stalks. If they are any longer there is a risk of nests being formed. Large grains are removed from the clayey earth. This is then dispersed in water in drums or barrels by means of paddles or a screw mixer until a greasy slurry is obtained. The straw is then sprinkled with the clay slurry and stirred with a fork. The mixture maintains a very straw-like appearance. The conventional formula calls for 70kg of straw to 600kg earth for an average density of 700kg/m^3 to 1200kg/m^3 (with λ ranging from 0.17 to 0.47λ[W/M°C]) depending on the quality of the earth, the viscosity of the slurry and the degree of compression of the mixture in the formwork. The lightly compressed straw clay is used to fill in wooden frames of thickness ranging from 15 to 30cm. There is no observable horizontal shrinkage. Slight vertical settlement sometimes results from a compression of the material. The preparation and actual use of straw clay require little special skill and makes use of only ordinary tools. It is thus accessible to the greatest possible number of people. In view of the ease with which it is put to use and the advantage of long lifetime and resistance to intemperate weather that it offers, there is no doubt that straw clay is a technique with a future.

Construction elements

Straw clay can be adapted to the prefabrication of various building elements. These are in fact bricks that are laid using clay mortar. Small bricks dry rapidly and it is possible to throw up structures quite quickly using larger bricks. The walls are in all cases light. As a result of various experiments, floor blocks (70 × 30 × 10cm) have been built and reinforced with concrete. These floors can support a load of 200kg with a deflection of 0.5mm. Small arches (50 × 50 × 10cm) 4cm thick at the key easily support the weight of three persons. Sections of wall can also be prefabricated (i.e. wooden frames filled with straw clay.) These vary from 1 to 4m^2 in area and are quickly installed. The entire house can be prefabricated.

Walls

Outer walls are commonly 20 to 30cm thick. Inner partitions are thinner, being about 12cm thick. The structure is built using either sawn boards of standard cross-section or else rough non-standardized timber. Depending on the load due to the roof, vertical posts are spaced 1 to 2m apart. Their function is to maintain the formwork. Openings are included in the fabric of the frame. Horizontal lintels nailed to the frame brace the sections filled with straw clay. The walls must dry for several months before rendering can be undertaken. The texture conferred on the walls by the straw helps to secure the finish. Usually one-storey houses are built using this method, but with certain adaptations to the structure, multi-storey houses are possible.

Paving and floors

Paving Straw clay is applied in a layer 6 to 10cm thick on top of a set stone foundation affording protection from rising capillary damp. Tubular straw is selected so as to trap a maximum of air, which acts as an insulating agent. Conventional flooring materials should be used.

Floors The same procedure as for walls is followed but the work is carried out horizontally. Straw clay covers over a support consisting of laths wedged between planks. Another technique permits the underside of a floor to appear as the ceiling of the storey below. A layer of earth or lime takes care of the finish. A wooden floor is unnecessary since a straw clay floor can support more than 500kg/m^2.

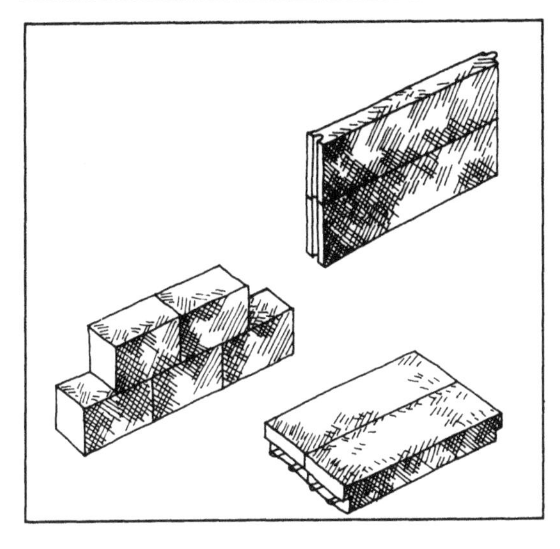

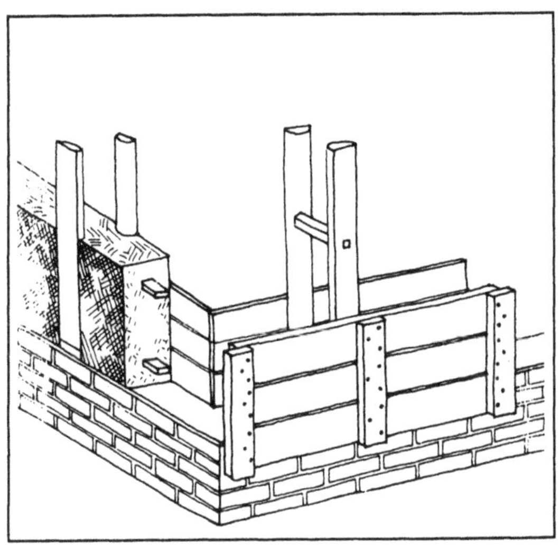

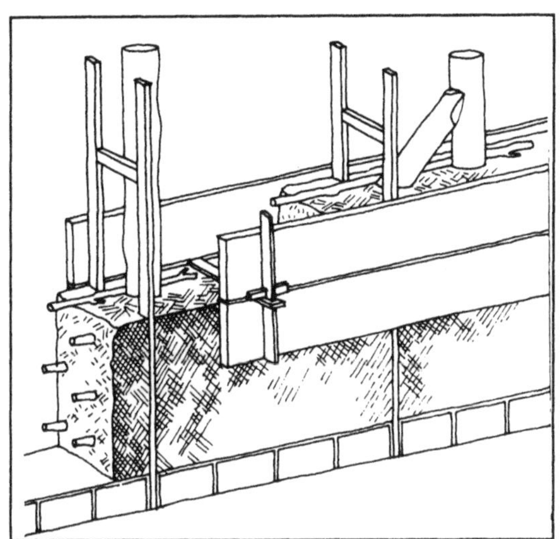

This technique is commonly known as daub and continues to be used in many areas of the world. A weight-bearing structure, generally wooden (a frame), is filled in with daub which covers a lattice consisting of wooden poles tied or nailed together, a wickerwork or plaited straw support. The earth is very clayey and mixed with straw or other vegetable fibres. This technique is very poorly rated as it is regarded as having a limited life. Nevertheless, a daub dwelling that is properly planned and adequately protected against capillary rise, the caprices of the weather and rodents and termites, can have a surprisingly long life, leaving aside its high resistance to earthquakes. Europe's heritage of rural architecture is rich in structures more than one hundred years old. A section of Versailles as well as part of the Faculty of Medicine in Lima, Peru was built using the daub technique. Daub construction is once again being practised in the industrialized countries, and builders are emerging which are specialized in it. In addition, training programmes for would-bc practitioners have been organized. This economic technique is suitable for the construction of dwellings for a very large number of people. The quality and lifetime of such dwellings depend on the degree of care that goes into the planning and execution stages.

Cob on posts

This is the most widely used technique in tropical countries. The load-bearing frame and the wattle panel are made up entirely of plant materials. These include wood, bamboo, palm trees, mock elephant grass, split cane and lianas. Filling in with earth is completed within a few days and is done during the dry season. The walls are quite thick – from 10 to 15cm. The technique is very simple and economic and accessible to a very great number of self-help builders. A good foundation and a good roof assure the survival of this type of structure for well beyond fifty years, and this is even true of regions where rainfall is heavy and rodents and termites are remorselessly active.

Straw bottles

This technique has remained largely confined to Germany. Earth balls are used to fill a kind of basket made of plaited straw which is then coated with earth. As soon as this has been done, each straw and earth 'bottle' is installed in a wooden frame. The units stay in position next to one another owing to the adhesive action of the earth. A large quantity of straw is required to carry out this technique, thus has the advantage of being highly insulating. The material is reminiscent of the straw-clay technique discussed above.

'Reels' of earth

In this technique long stalks of straw coated with clayey earth are wound around 'spindles' of wood. The resulting elements are stored until they are half dry and then fitted in a wooden load-bearing structure. The earth surfaces, which are in mutual contact with one another, cause the reels to adhere to one another. There were several variations on this technique, which was widely used in Europe. The French word 'fusée' (spindle) is the term applied to the technique used in Anjou, France. Although it provides a high degree of insulation and is easy to carry out, it is hardly used any longer.

Wattle and daub

This is undoubtedly one of the oldest known construction techniques, and it is still one of the most widely used in the world today. The load-bearing structures and wattles that serve as a support for the filled-in daub vary in design but the construction principle is the same everywhere. With a view to extending the life of the daub the earth is mixed in certain areas with horse urine instead of water: the results are impressive. Daub is currently the object of a great deal of research. The ININVI, in Peru, for example, has conducted experiments with easy-to-assemble prefabricated panels. Efforts are going on as well to rediscover exactly what went into the 'finishing touch' of former craftsmen.

Blown earth

Daub has remained an essentially manual technique. Recently, in the context of a large housing project in the Ivory Coast, there have been trials involving the blowing of earth onto wooden, bamboo and expanded metal support structures. The technique requires the use of high-pressure rendering pumps. The main difficulties revolve around achieving the right consistency of earth – earth that is too muddy is subject to major shrinkage problems – and the obstruction of the jets of the pumps by plant fibres which accumulate in knots and balls. These difficulties have as yet not been resolved and they point to the necessity of modifying the plant that is currently used for blowing earth renderings.

Adobe News. *Adobe Codes from around the Southwest.* Albuquerque, Adobe News, 1982.
AGRA. *Recommandations pour la Conception des Bâtiments du Village Terre.* Grenoble, AGRA, 1982.
An. 'Adobe brick stabilized with asphalt'. In *Engineer News Record*, 1948.
An. *L'habitat Traditionnel Voltaïque.* Ouagadougou, Ministère du plan et des travaux publics, 1968.
Barnes and associates. 'Ablobe debuts in Tucson'. In *Adobe Today*, Albuquerque, Adobe News, 1982.
Bourdier, J.P. 'Houses of Upper Volta'. In *Mimar*, Singapore, Mimar, 1982.
Cervantes, M.A. *Les Trésors de l'Ancien Mexique.* National museum of anthropology. Barcelona, Geocolor, 1978.
Charneau, H.; Trebbi, J.C. *Maisons Creuses, Maisons Enterrées*, Editions Alternatives, 1981.
Chesi, G. *Les Derniers Africains.* Paris, Arthaud, 1977.
Conescal. *Cartilla de Autoconstrucción para Escuelas Rurales.* México, Conescal, 1978.
CRATerre. 'Casa de tierra'. In *Minka*, Huankayo, Grupo Talpuy, 1982.
Davco. 'Procédé de constructions industrialisées et composants'. Priv. com. Paris, 1982.
Dayre, M. *et al.* 'Les blocs de terre compressés. Elaboration d'un savoir faire approprié'. In *Colloque L'habitat économique dans les PED*, Paris, Presses Ponts et Chaussées, 1983.
Denyer, S. *African Traditional Architecture.* New York, Africana, 1978.
Department of Economic and Social Affairs. *The Development Potential of Dimension Stone*, New York, UN, 1976.
Dethier, J. *Des Architectures de Terre.* Paris, CCI, 1981.
Diamant Boart. *Catalog*, Brussels, 1979.
Doat, P. *et al. Construire en Terre.* Paris, Editions Alternatives et Parallèles, 1979.
Dubach, W. *Yemen Arab Republic, A study of traditional forms of habitation and types of settlement.* Zürich, Dubach, 1977.
Fauth, W. *Der praktische Lehmbau.* Singen-Hohentwiel, Weber, 1948.
Fitch, J.M.; Branch, D.P. 'Primitive architecture and climate'. In *Scientific American*, 1960.
Gardi, R. *Maisons Africaines.* Paris-Bruxelles, Elsevier Séquoia, 1974.
Gossé, M.H. 'Algérie: Abadla, villages agricoles au Sahara'. In *A+*, 1975.
Guidoni, E. *Primitive Architecture.* New York, Harry N. Abrams, 1975.
Heufinger Von Waldegg, E. *Die Ziegel und Röhenbrennerei.* Leipzig, Theodor Thomas, 1891.
Houben, H. *Technologie du Béton de Terre Stabilisé pour l'Habitat.* Sidi Bel Abbes, CPR, 1974.
Kemmerer, J.B. 'Adobe goes modern'. In *Popular mechanics*, 1951.
Lavau, J. 'Le durcissement chimique des latérites pour le bâtiment'. Priv. com. St Quentin, 1982.
Mariotti, M. *Les Pierres Naturelles dans la Construction.* Paris, CEBTP, 1981.
Markus, T.A. *et al. Stabilised Soil.* Glasgow, University of Strathclyde, 1979.
Massuh, H.; Ferrero, A. 'El centro experimental de la vivienda economica'. In *Colloque L'habitat Economique dans les PED*, Paris, Presses Ponts et Chaussées, 1983.
Miller, T. *et al. Lehmbaufibel.* Weimar, Forschungsgemeinschaften Hochschule, 1947.
Minke, G. *Alternatives Bauen.* Kassel, Gesamthochschul-Bibliothek, 1980.
Minke, G. 'Low-cost housing. Appropriate construction techniques with loam, sand and plant stabilized earth'. In *Colloque L'habitat Economique dans les PED*, Paris, Presses Ponts et Chaussées, 1983.
NASA. 'Foldable patterns form construction blocks'. In *NASA Tech Brief*, NASA, 1971.
Palafitte jeunesse. *Minimôme Découvre la Terre.* Grenoble, Palafitte Jeunesse, 1975.
Pellegrini. *Catalog*, 1979.
Pollack, E.; Richter, E. *Technik des Lehmbaues.* Berlin, Verlag Technik, 1952.
Reuter, K. Lehmstakbau als Beispiel wirtschaftlichen Heimstättenbaues. In *Die Volkswohnung*, 1923.
Riedter. *Catalog*, 1983.
Rock. *Catalog*, 1983.
Schöttler, W. 'Das Dünner Lehmbauverfahren'. In *Natur Bauweisen*, Berlin, 1948.
Schultz, M. *et al. Les Bâtisseurs de Rêve.* Paris, Chêne-Hachette, 1980.
Schuyt, E. *Adobe Bricks in New Mexico.* Socorro, New Mexico Bureau of Mines and Mineral Resources, 1982.
Studie. *Résumé de proposition village terre de l'Isle d'Abeau.* Priv. com. 1981.
Svare, T.I. 'Stabilized soil blocks'. In *BRU data sheet*, Dar-Es-Salaam, BRU, 1974.
The Underground Space Center. *Earth Sheltered Housing Design.* New York, Van Nostrand Rheinhold company, 1979.
Turbosol. *Catalog*, 1983.
Vallery-Radot, N. 'Des Toits d'Herbe Sage'. In *La maison*, 1980.
Vibromax. *Catalog*, 1983.
Vidal, H. *et al.* 'Architerre habitat-paysage'. In *Annales ITBTP*, Paris, 1981.
Webb, D.J.T. 'Stabilized soil construction in Kenya'. In *Colloque L'habitat économique dans les PED*, Paris, Presses Ponts et Chaussées, 1983.
Welsch, R.L. *Sod Walls.* Broken Bow, Purcells Inc., 1968.
Yurchenko, P.G. 'Methods of construction and of heat insulation in the Ukraine'. In *RIBA journal*, London, 1945.

Modern private house built with compressed earth blocks, adobe blocks and rammed earth in Morocco (architect: Elie Mouyal) (Thierry Joffroy, CRATerre-EAG)

9. PRODUCTION METHODS

The variety of production methods rivals that of the construction methods.

Three of the main production procedures are well-known and have attained a degree of sophistication similar to that of industrially produced building materials. These methods are rammed earth, adobe, and compressed blocks.

Unbaked earth can therefore no longer be dismissed as a backward material suitable only for use in 'underdeveloped regions'.

The gulf between craft and industrial production methods has been bridged. Researchers, craftsmen, and builders using earth are no longer condemned to work with a material which is devoid any potential for industrial development.

Development potential

Present day earth construction employs production techniques ranging from the most rudimentary and artisanal to the most sophisticated, relying on industrial, mechanized and even automated processes. The upper end of this range has little to learn from other modern production processes for other construction materials, including even the most advanced. Some turnkey stabilized-earth plants represent an investment of close to US$1,000,000. The tendency to industrialization began approximately twenty-five years ago with among other things the production of compressed blocks. If today this trend has reached its technical fruition, it doesn't automatically follow that it is essential or even desirable for every construction project. Even so this tendency is now a reality which must be taken account of. Earth technology is no longer a matter of purely artisanal production or third world techniques which offer no potential for development. The passage from artisanal to industrial production, although technically possible, must obviously be justified by the parameters governing each particular case such as development policy, socio-economic and cultural considerations, economic and technical base, investment, work procedures, etc. For example in craft production, five people can produce from 500 adobes (in West Africa) to 2500 adobes (in Egypt and Iran) per day, with no investment outlay being required.

In industrial production in the USA five workers can produce up to 20,000 adobes per day, but with an investment outlay of the order of US$300,000. Despite the current infatuation with mechanized systems, these are frequently unable to achieve qualitative and quantitative targets. The following remark by Joe Tibbets in the American magazine *Earth Builder* is worth quoting: 'A machine is no better than the soil you put into it'. It is not the machine that guarantees the quality of production but rather how production is organized and the skill of the operators. The real return of a mechanized production process is often only a tenth of what is commercially claimed for it, and a product that looks beautiful at a demonstration may be a sorry sight after a few years. The fact that earth construction is practised throughout the world today dictates that an in-depth study of the tools of production is undertaken at all levels of the production process right up to the building and maintenance stages. Here we will limit ourselves to the three most frequently used of the twelve techniques discussed: compressed block, rammed earth and adobe.

Production cycle

Whatever the production methods, manual or mechanical, the operations involved are almost identical. This is particularly true for the excavation, transport, preliminary drying, and storage of the raw material, and the crushing and screening, proportioning and mixing, and drying and storage of the end product. It is obvious, however, that the overall production scheme must be adapted to each technique. Certain processes may be left out, split into two or switched round. Reference to the following scheme will help the builder to estimate productivity and execution times.

T01 Excavation of the soil from a borrow.

T02 Drying by spreading in thin layers or small ventilated heaps or passing through a hot-air cyclone.

T03 Storage of a reserve of raw or prepared soil at the work site.

T04 Screening if the soil contains too many large stones, which have to be removed.

T05 Pulverizing to break up clayey aggregations.

T06 Sifting to eliminate undesirable elements after general preparation.

T07 Dry proportioning of the earth by weight and by volume with a view to mixing it with water and/or stabilizer.

T08 Dry mixing to maximize the effectiveness of powder stabilizer.

T09 Wet mixing this consists of adding water by spraying after dry mixing correctly or directly in the form of a liquid stabilizer.

T10 Reaction variable hold-back time depending on the nature of the stabilizer; very short for cement, longer for lime.

T11 Trituration should be done just before using the soil.

T12 Moulding the final shaping of the earth using various processes.

T13 Curing best carried out under the same conditions pertaining during the first drying stage.

T14 Spraying spray, if necessary, to moisten the stabilizer properly.

T15 Drying should be carried out for a sufficiently long period to ensure an acceptable product quality.

T16 Final storage stock of products for immediate use.

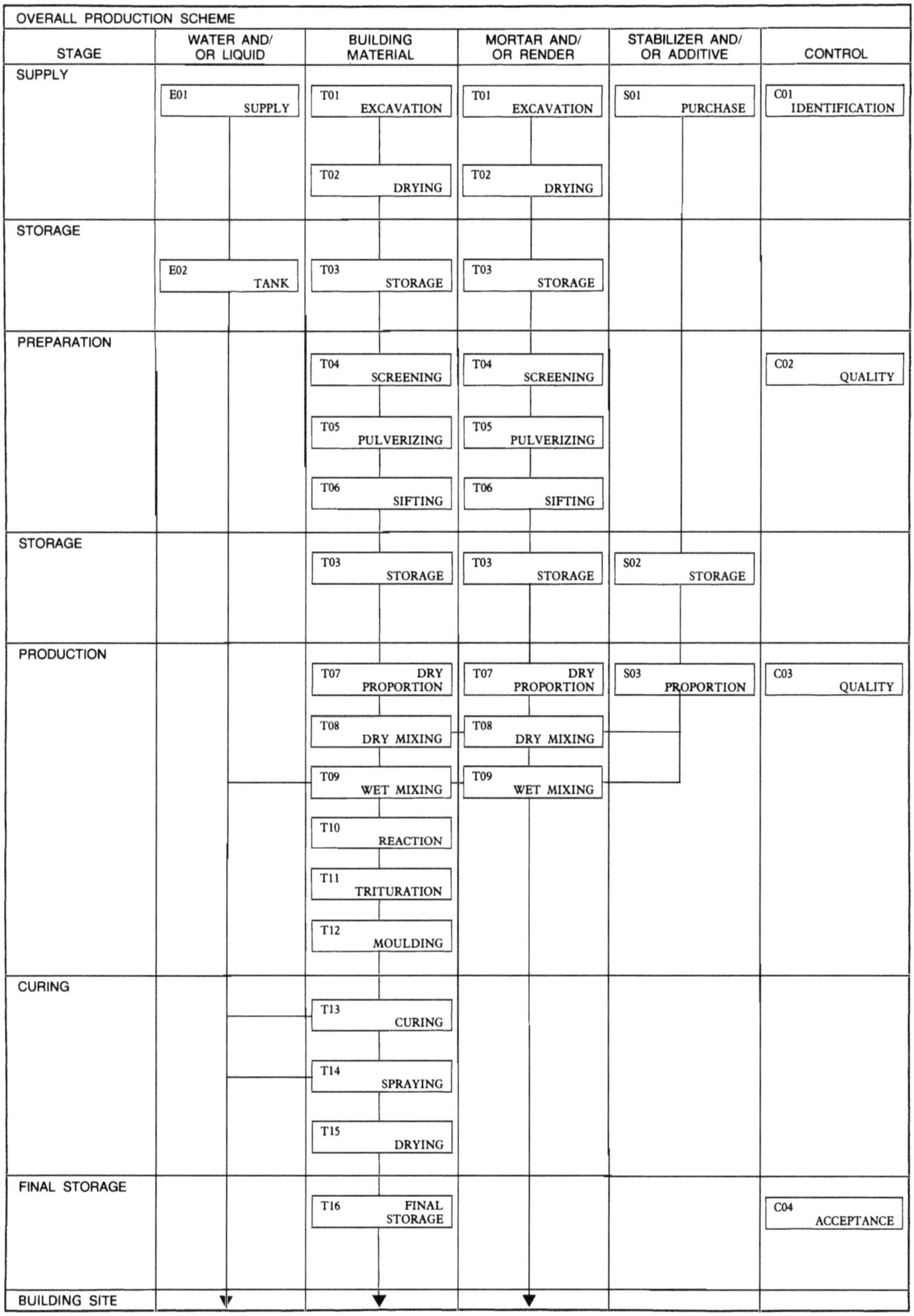

OVERALL PRODUCTION SCHEME

STAGE	WATER AND/ OR LIQUID	BUILDING MATERIAL	MORTAR AND/ OR RENDER	STABILIZER AND/ OR ADDITIVE	CONTROL
SUPPLY	E01 SUPPLY	T01 EXCAVATION T02 DRYING	T01 EXCAVATION T02 DRYING	S01 PURCHASE	C01 IDENTIFICATION
STORAGE	E02 TANK	T03 STORAGE	T03 STORAGE		
PREPARATION		T04 SCREENING T05 PULVERIZING T06 SIFTING	T04 SCREENING T05 PULVERIZING T06 SIFTING		C02 QUALITY
STORAGE		T03 STORAGE	T03 STORAGE	S02 STORAGE	
PRODUCTION		T07 DRY PROPORTION T08 DRY MIXING T09 WET MIXING T10 REACTION T11 TRITURATION T12 MOULDING	T07 DRY PROPORTION T08 DRY MIXING T09 WET MIXING	S03 PROPORTION	C03 QUALITY
CURING		T13 CURING T14 SPRAYING T15 DRYING			
FINAL STORAGE		T16 FINAL STORAGE			C04 ACCEPTANCE
BUILDING SITE	▼	▼	▼		

Excavation

The problems involved in excavating earth required for construction purposes are about the same for all earth construction techniques with the exception of dugout dwellings. Similar problems are encountered in connection with the excavation of the materials used in the ceramics industry, the cement and binder industries, stone quarrying and even agriculture and road building. These problems are dealt with very extensively in the specialist literature. Methods of excavation vary in accordance with several factors where geological and engineering problems and economic considerations are the overriding concerns. A number of problems must be tackled well before excavation is actually undertaken with a view to optimizing the work and streamlining its organization. From the geological and engineering viewpoints, for example, the following questions must be considered: What type of soil is required and at what depth? What area having what boundaries will be worked? What will be the angle of dip of the borrow? What preparatory work will be necessary as regards, for example, clearing away undergrowth, levelling vegetation, the removal of stones and the use of explosives? Should contingency plans be made for the storage of organic soil with a view to reusing it for agriculture or forestry? Is the soil horizon to be worked homogenous or mixed and stratified? Is the terrain adequately drained or must drainage of the borrow be planned for? Is the terrain sufficiently stable to guarantee the safety of the excavation personnel? Is it safe to use heavy machinery? What excavation plant is available – manual, artisanal, agricultural tools or mechanized, motorized plant? If machines are available, are they heavy or light, static or mobile? Is manual excavation to be preferred over mechanized or vice-versa? If the soil is not stratified, will the most suitable mode of excavation be horizontal and hence either manual or mechanized, or will it be vertical thus giving rise to several layers and necessitating the sinking of shafts and the use of explosives? How is the raw excavated material to be transported from the borrow? What is the most feasible overall economic strategy in the regional context? How can the exploitation of the deposit be managed as efficiently as possible from every conceivable viewpoint and the entire operation be optimized? What legal regulations are in force? How can the site be cleaned up properly? These are some of the many questions that must be asked before even thinking of picking up a shovel or putting oneself behind the controls of a scraper.

The example of Burkina Faso is instructive as regards the economic aspect of such an undertaking. There it was decided to excavate earth manually and transport it by means of donkey carts. In this way ten people were provided with work over a period of two years. If a powerful bulldozer had been used, the same job could have been completed in four days at the same cost, but robbing ten people of work.

EXTRACTION MODE	HYDROUS STATE OF THE EARTH	LIQUID	SOFT	PLASTIC	HUMID	DRY	CONCRETION
UNMOTORIZED	1 PICK AND SHOVEL		O	O	O	O	O
	2 HAND SCRAPER				O	O	
	3 POLE				O	O	O
	4 EXPLOSIVE				O	O	O
	etc.						
MOTORIZED	5 PNEUMATIC PICK				O	O	O
	6 MOTORIZED WHEELBARROW		O		O	O	
	7 EXCAVATOR TRACTOR	O	O	O	O	O	O
	8 SCRAPER				O	O	
	9 BULLDOZER			O	O	O	O
	10 GRADER			O	O	O	O
	11 EXCAVATOR			O	O	O	O
	12 PLOUGH			O	O	O	
	13 POWER CULTIVATOR				O		
	14 BUCKET CHAIN	O	O	O	O	O	
	etc.						

Transport

The engineering factors determining the mode of transport of the soil relate to the nature of the soil, its moisture content and the conditions under which the borrow is being worked. From the logistic point of view, transport problems should be taken into account at every stage in the overall production process, but such problems are not in fact specific to unburnt soil technology. They must also be faced when working quarries containing other materials as well as the later phases involved in the production of building components. Consequently transport must be considered both when working the borrow and later stages of provisioning the brick production area and conveying the material to the construction site, directly (e.g. rammed earth), or after the product has been manufactured (e.g. adobes). In order to plan transport effectively the following factors must be evaluated: the distances involved, the suitability of the ground for the construction of roads, access and whether an existing road network can be used or not. Transport around the edge of the site, the construction of approaches to the area being worked so as to provide transport with a maximum of manoeuvrability, and soil stability where traffic involving heavy plant is likely, are also problems that must be reckoned with. The safety of the workers whose job it is to excavate the material, of the transporters themselves, that of the production staff who work near the brickyards and of the workers at sites close to storage areas must also be considered. In the final analysis, the preceding problems represent technical choices about the appropriate transportation strategy and mode of transport. They must be jointly assessed in the light of the following factors: the geological and engineering restrictions on the operation of the borrow, technical and local economic limitations imposed by the production area and the construction site, and lastly, the overall economic considerations which will determine the optimal management of the production process. Next comes the choice of tools, which will be determined by their availability or non-availability as well as the cost of using them. Depending on the context, the most suitable mode of transport will be manual, draught animal, or by motor vehicle, in which case, the energy costs of transport must be taken into account. In cases where a large borrow site is close to a large-scale project, it may be feasible to build up a system of cable cars or telpher carriers, belt conveyors, mining cars on rails, etc. In most cases the transport problem is fundamental to the truly economic management of a project, and it frequently becomes necessary to reduce the strain on this link in the production chain to a minimum.

TRANSPORT MODE	HYDROUS STATE OF THE EARTH	LIQUID	SOFT	PLASTIC	HUMID	DRY	CONCRETION
MOBILE	1 BAG, BASKET, BUCKET	O	O	O	O	O	O
	2 WHEELBARROW	O	O	O	O	O	O
	3 MOTORIZED WHEELBARROW		O	O	O	O	O
	4 DUMPER, WAGGON		O	O	O	O	O
	5 CART		O	O	O	O	O
	6 LORRY, LIGHT MOTOR LORRY, PICK-UP		O	O	O	O	O
	7 BULLDOZER, FRONT LOADER			O	O	O	O
	etc.						
STATIONARY	8 PLASTER PUMP	O					
	9 CONCRETE PUMP	O	O				
	10 UNIVERSAL TRANSPORT PUMP					O	O
	11 CONVEYOR BELT				O	O	O
	12 ROPE WAY	O	O	O	O	O	O
	13 CRANE	O	O	O	O	O	O
	14 TIP WAGGON	O	O	O	O	O	O
	etc.						

Grading

For every earth construction technique there is a preferred grading for the material. Such grading requirements should be respected or approached as closely as possible in order to ensure the quality of the processed product or that used in construction. Upon excavation soil can present the two following major deficiencies.

1. Texture deficiency In this case the earth contains either too much or too little of a given grain fraction, e.g. too much clay and insufficient sand. This occurs, for example, when the earth contains too many large stones or too much plant material such as roots. The removal of coarse material (stones or roots) is fairly simple: removal by hand of stones, cleaning and screening, either manually or by machine. To remove an excessive amount of a finer fraction is a more delicate operation. It consists of screening to the required sieve size or screening several fractions separately and then reconstituting the soil; while clays must be leached out. Both are cumbersome procedures which hamper production and can push up costs. Care must also be taken that the undesirable grain fraction is properly separated, because if it is trapped in a layer of other material, the operation becomes difficult and leads to a structural problem caused by the decomposition of the soil into independent grain fractions. In contrast adding a missing grain fraction is a simple matter, and results in the improvement of the soil. It is carried out by mixing.

2. Structural deficiency It is sometimes desirable to break up, crush or pulverize a fragmented or consolidated soil. Pulverization is often indispensable in the case of compressed, cement or lime stabilized blocks and is largely responsible for ensuring the quality of the product. When a soil is well-composed into required distinct grain fractions and sieved, the grains, water and powder stabilizer will be evenly distributed and the product will be of excellent quality.

It is thus often necessary to carry out preliminary work on the texture and structure of the soil. This operation is either manual or involves tools. Machine designers have developed specific screening, crushing and pulverizing equipment for such work. However, depending on the conditions in the region, manpower or tools borrowed from another sector such as agriculture or public works may be able to serve the purpose.

GRADING MODE		HYDROUS STATE OF THE EARTH	LIQUID	SOFT	PLASTIC	HUMID	DRY	CONCRETION
TEXTURE	UNMOTORIZED	1 MANUAL SELECTION				O	O	
		2 MANUAL SCREENING				O	O	
		3 VIBRATING SCREEN				O	O	
		etc.						
	MOTORIZED	4 WASHING	O	O				
		5 VIBRATING SCREEN				O	O	
STRUCTURE	UNMOTORIZED	6 NATURAL BREAKDOWN				O	O	
		7 BREAK UP BY WATER		O		O	O	
		8 EARTHRAMMER					O	O
		9 ROLLER					O	O
		10 PULVERIZER OR CRUSHER				O	O	
		etc.						
	MOTORIZED	11 ROLLER				O	O	O
		12 PULVERIZER OR CRUSHER				O	O	O
		13 POWER CULTIVATOR				O	O	
		14 AGRICULTURAL CRUMBER				O	O	
		15 CRUSHING BY TRUCKS					O	O
		etc.						

Mixing

Few soils can be used straight from the ground. Usually a soil must be prepared in some way such as being crushed, sieved or pulverized before being mixed, which is the last stage prior to construction. The importance of mixing should not be underestimated, and is just as important as the prior operation of pulverization. It ensures the subsequent quality of the product and of the structure itself. It also guarantees that the structure is built economically in the sense that it optimizes the proportions of the materials and additives, if stabilization is carried out. Mixing can be between the ingredients of the same soil separated out into distinct grain fractions, between different soils (in the case of soil improvement), or between a soil of a particular type and other soils to which water, stabilizer, etc. have been added. On the other hand only dry materials and liquids, or liquids and liquids can be mixed. Mixing dry and moist materials poses certain problems. This is particularly the case when stabilization is carried out with powdered hydraulic binders, like cement and lime. In such cases, a preliminary dry mixing stage is essential. Water is only added afterwards by the progressive sprinkling of small quantities. The planning that must precede the mixing of soil is very different from that prior to mixing concrete, because while concrete is not cohesive, soil is, and hence with earth there is a risk of formation of lumps and crumbs that could reduce the strength of the material. It follows that the conventional concrete mixer is not up to the task. A longer mixing time and a sturdier piece of equipment fitted with a more powerful motor is required. Cleaning the equipment is in this case a more tedious process. Mixing can be carried out manually using a shovel or treading by foot, or mechanically using specialized machines or machines borrowed from other sectors such as agriculture, road works or the ceramics industry. The method of mixing adopted will depend on the quality of the soil, the building technique used, i.e. blocks, rammed earth or other, the desired quality of the product and the structure, the conditions under which production takes place in the context of the operation, the degree of skill available, limitations of price, energy and a social nature (organization of the work) – all of which must be optimized.

MIXING MODE		HYDROUS STATE OF THE EARTH	LIQUID	SOFT	PLASTIC	HUMID	DRY	CONCRETION
UNMOTORIZED	HUMANS	1 SHOVEL	○	○	○	○	○	
		2 TRAMPLING	○	○	○			
		3 RAMMING			○			
		4 ROTARY				○		
		etc.						
	ANIMALS	5 TRAMPLING	○	○	○			
MOTORIZED	VERTICAL SHAFT	6 PLANET WHEEL MIXER		○		○		
		7 VERTICAL MIXER			○	○		
		8 PUGGING MILLS	○	○	○			
		9 SCREW MIXERS	○	○				
		10 TURBOMIXER	○					
		etc.						
	HORIZONTAL SHAFT	11 PADDLE MIXER		○		○	○	
		12 LINEAR MIXER	○	○	○	○	○	
		13 POWER CULTIVATOR		○		○	○	
		etc.						

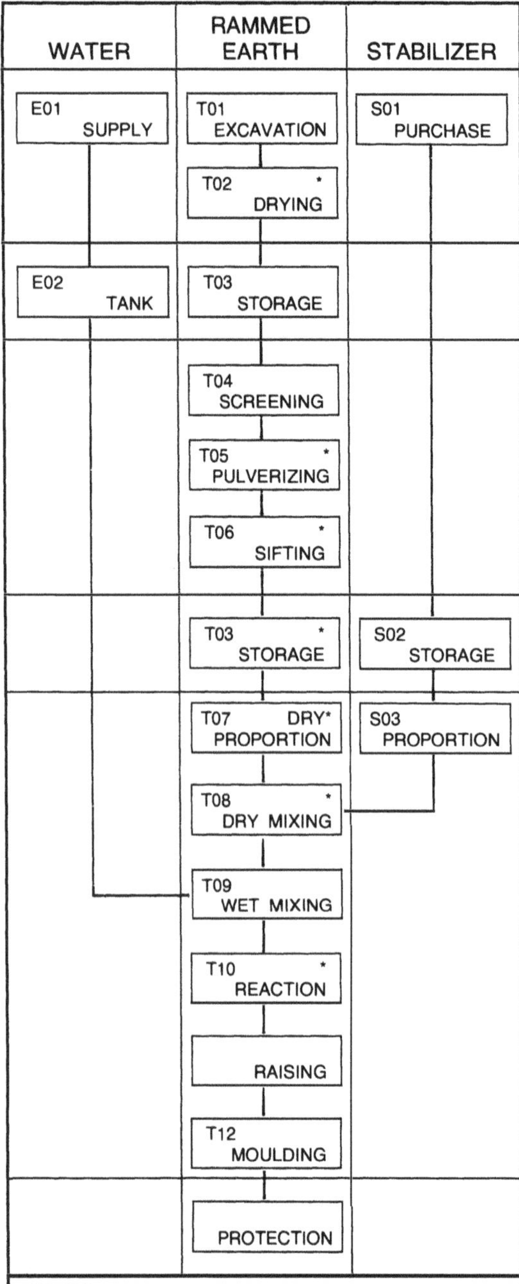

NOTES

– It should be borne in mind that the crew that produces the building material is the same crew that constructs the house.

– The production chart for stabilized rammed earth illustrated above clearly points up the complications caused by stabilization.

– Stabilization imposes a higher degree of organization on the production process sufficient to ensure that there is no slowup in the fabrication of the construction material or the construction itself.

* Stabilization makes these operations necessary.

1. Production

In theory the production of rammed earth is a fairly straightforward affair: earth tipped into a formwork is rammed. The ideal conditions of a high-quality earth and the right moisture content do not always coincide. An oversophisticated technology can invalidate the original simplicity that characterizes rammed earth construction. There are numerous technological variations depending on the part of the world in which the process is carried out.

2. Product

The appearance of rammed earth is highly variable. To simplify matters it can be stated that there are two classes of rammed earth: the gravelly variety – when the gravelly fraction of the soil is rammed at the outer face of the wall; and the very fine variety – when the gravels are rammed at the inner face of the wall. The visible particles of gravel are used to hold the rendering. The variation in the appearance of rammed earth also arises from the finish with or without pointing with mortar, for example.

3. Production season

In temperate regions no rammed earth construction should be carried out three months prior to and during the months in which frost can be expected. In wet climates the rainy season is avoided, and in a hot dry climate, the hottest months are to be avoided. The local economy, particularly if it is agricultural, may also dictate work priorities.

4. Work crews and productivity

The size of the work crews varies according to the nature of the construction site and numerous local factors. Crews consisting of five or six workers are suitable for small construction sites. Productivity will depend to a very great extent upon working conditions and the design of the building.

	MANUAL OR MECHANIZED	HIGHLY MECHANIZED
INVESTMENT ($)	200–6000	20,000
CREW (WORKERS)	6	5
- Preparation	2	1
- Transport	2	1
- Construction	2	3
OUTPUT (h/m^3)		
- Very good conditions	8–10	5
- Good conditions	15–20	9–10
- Bad conditions	25	15
- Very bad conditions	35	30

earth construction, builders must ensure that the earth to be used satisfies the selection criteria, and this in particular with respect to its texture and moisture content. This simplifies the entire production process, though this is not the case when other stages such as screening, pulverization and dry and wet mixing are necessary, for example for cement or lime stabilization. Too wide a departure from the limiting grain size distribution curves has, among other things, a very harmful effect on the cost of production as well as on productivity and on the quality of the product.

3. Pulverization

If earth is to be rammed, it must be pulverized. This also applies to excessively clayey earth containing hard lumps and to which a sandy fraction must be added. It is advisable to group the pulverizing, grating and mixing operations together. With regard to improving clayey soil with sand, processing the clayey fraction and the sandy component in the pulverizer in alternating fashion will result in a premix of reasonable quality. The mixture goes through the following sequence of further operations: transport, elevation and distribution within the formwork. The pulverizer must be a sturdy machine able to handle stony and sandy soil, and it must be able to project the earth a certain distance in order to ensure good aeration and proper premixing.

4. Mixing

Mixing is advisable when the soil requires homogenization or when it is desirable to add a stabilizer. The most suitable piece of equipment for this operation is a concrete mixer, but a motor cultivator also gives good results in most cases.

5. Transport

This is one of the major problems in rammed earth technology. Enormous quantities of earth are in fact required in the construction process. The material must be transported horizontally from the borrow to the construction site, and must be transported vertically as well to the required level. Traditionally, workers who build with rammed earth use manual labour to carry the soil in heavy or light baskets or other receptacles from the borrow to the construction site, the material then being raised by ladder or scaffold to where it is used. The same task can also be carried out in a highly efficient manner by means of hoists. Rendering projectors have been adapted to the same end. The adaptation was a complex affair as the material is not liquid. From a central position the earth can be pumped to anywhere within 40 to 50m to a height of 10m.

Experience shows that formwork is most effective when it is small and simply designed. It must be solid and stable in order to resist the pressure and vibrations resulting from the ramming (a minimum of 3000Pa). It must be easy to manage, i.e. light and easy to assemble and dismantle – plumb, fit and fastening must be good. Lastly and most importantly, the formwork must be perfectly capable of accommodating changes in height, length and thickness of the walls. Good design plays a major role in ensuring the productivity of rammed earth construction. A broad range of materials can be used for the formwork, such as wood in Morocco, logs in China, aluminium in France, steel in Algeria, and glass fibre.

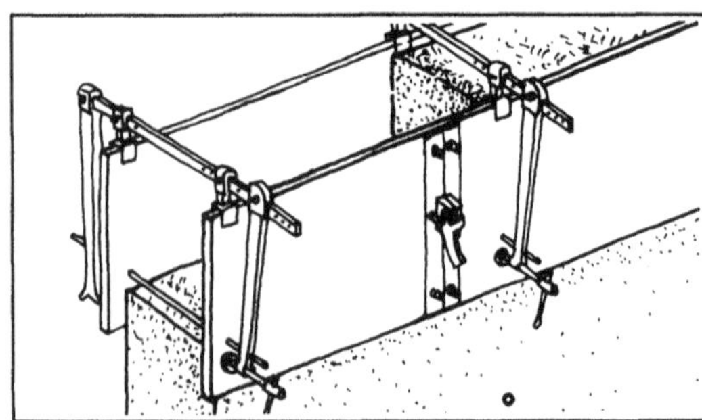

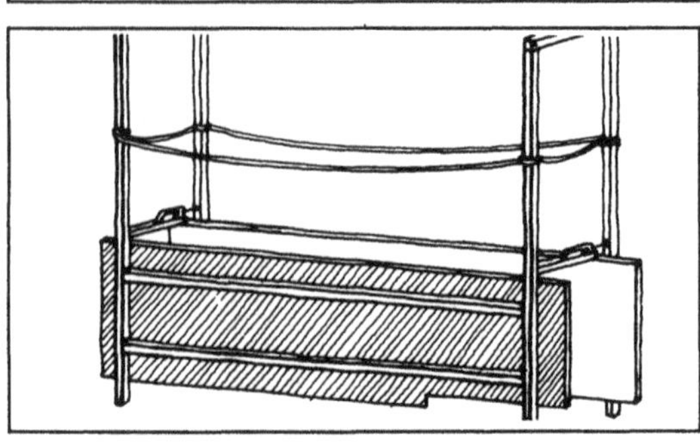

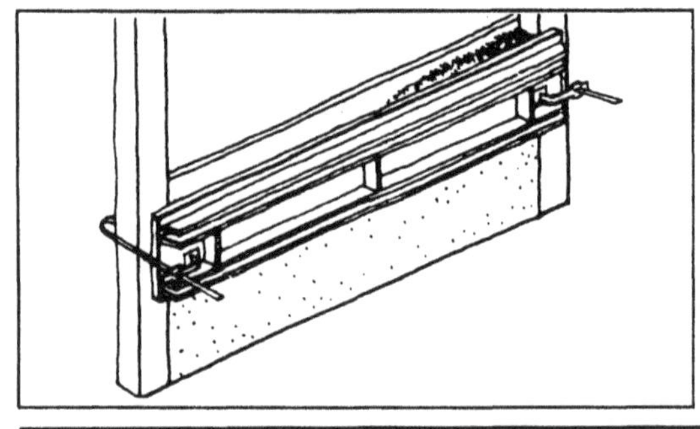

Keeping the formwork in position

Once the formwork has been set up and put in position, it must be securely fastened. Builders have developed various systems to this end.

1. Big clamps and large holes

The clamps are solid cross-pieces as thick as a rafter and provide a support for the boards of the formwork. The system is traditional in countries such as Peru and Morocco. With this type of clamp, large holes appear in the rammed earth when the forms are removed. Ideally the clamps should be slightly conical so that they can easily be removed without damaging the rammed earth. If they project from the boards of the formwork, planks serving as scaffolding can be put in position. The whole set-up is quite heavy and vulnerable. A sledge-hammer is required to remove the clamps.

2. Small clamps and small holes

This system is based on concrete formwork technology. Threaded steel rods, concrete bars or flat iron bars serve as clamps. Only small holes appear upon removal of the forms. The bars must be of a large enough cross-section to enable them to support the forms and avoid the shearing of the rammed earth. Sheathing in sections of plastic tubing affords a solution to this problem.

3. Without clamps or holes

This is the system of keeping the formwork in place which all builders dream of. Designers have settled on two solutions:

Independently gripping formwork The boards are kept in place by a tensioning device or screw-locking system or – yet another possibility – locking hydraulic jacks.

Formwork on posts To eliminate the risk of bulging walls, care must be taken that the distance between the posts is not too great. This system is widely used by the Ceped of Brazil.

Organization of formwork

The organization of the construction site, the plant and manpower available, execution times and the desired result, the architectural plans and the type of finish desired can all have a major effect on the choice of a formwork system.

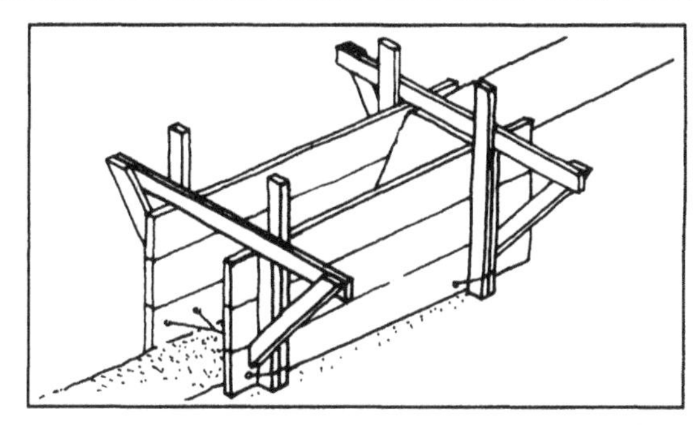

1. Formwork consisting of small units

Horizontally sliding formwork This system has been traditionally adopted for rammed earth construction. This type of formwork system was developed by craftsmen and differs widely. The system uses fastening systems which follow the principles elaborated above and has the following major advantages: lightness, manoeuvrability of the equipment, adaptability.

Vertically sliding formwork This system is ideally suited to the construction of rammed earth walls in piers. It facilitates and greatly accelerates the erection of a structure but the formwork must be carefully designed. The vertical elements holding the formwork in position can be formwork bottoms, construction posts or external frames.

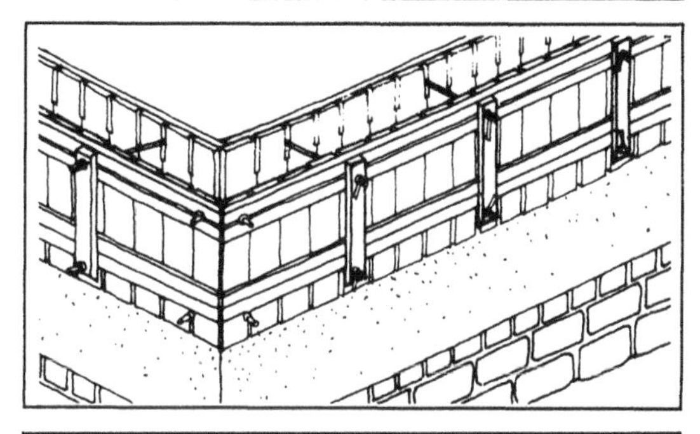

2. Integral formwork

Most attempts at using this concrete construction technology have produced rather disappointing results. The reliability of the system is highly dependent upon the architectural design, namely, on whether the latter features such characteristics as modular dimensions, full walls with independent bays and ultra-simple drawings.

Integral horizontal formwork A ring of formwork is moved vertically. Success demands that the elements be light and that assembly and disassembly be easy and rapid. The chief obstacles are the joints between the boards, horizontal alignment and the maintenance of plumb.

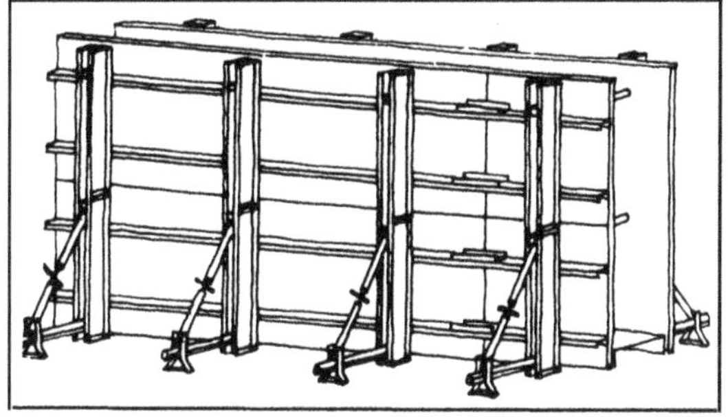

Integral vertical formwork This type of formwork lends itself to the construction of large piers, contained in formwork for their entire height. In order to facilitate ramming, only one side of the formwork is completely erected. The second is erected as the wall is constructed.

Integral – integral formwork The formwork for the building is put up at a single go. This system was adopted in Morocco in the BTS 67 project. Projects using this type of formwork should be small and offer easy access inside the formwork.

Moving formwork

The moving of formwork poses a serious problem for workers when they are perched at a height of 7m above the ground on a 40cm thick rammed-earth wall. The safety of the work is essential. As a general rule, the lightness and manoeuvrability helps to ensure safety. Means have been sought to avoid the total dismantling and reassembly of the formwork after moving.

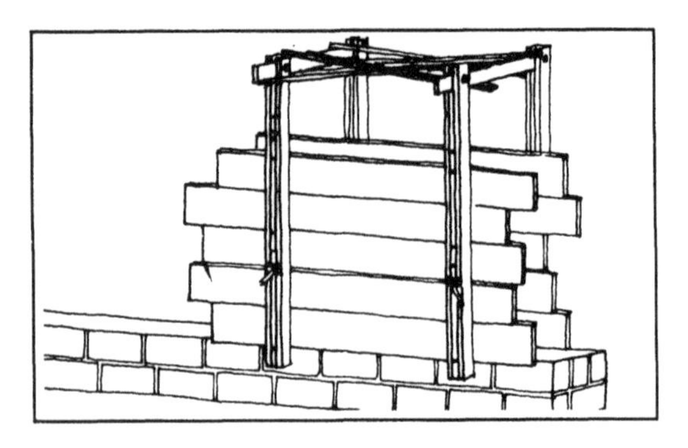

1. Gantry formwork

This technique is best suited to the construction of piers or wall sections. The formwork is light, consisting of simple planks, plywood panels or even billets which are kept in position by wooden supports driven into the ground and secured at the top. This type of system is used in Chinese rammed earth construction. GHK in Kassel has reintroduced the technique with a system involving a wooden hinged frame, fastened by threaded rods.

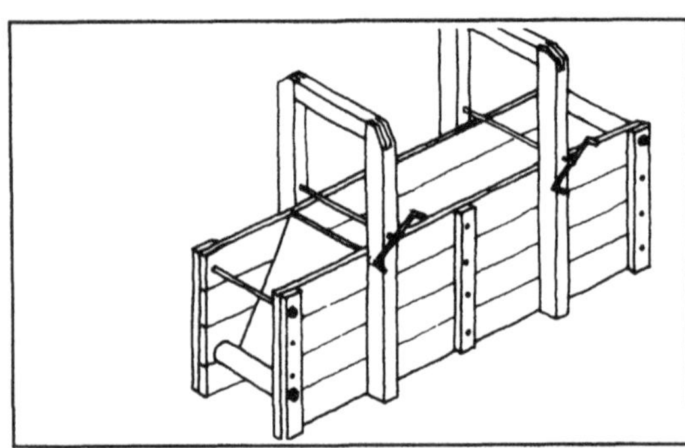

2. Formwork with rollers

The concept of a mobile travelling form based on the use of rollers was originated by the Australian G.F. Middleton, as early as 1952. The system is suitable for the construction of straight walls but requires stationary formwork for bays, corners and partition walls.

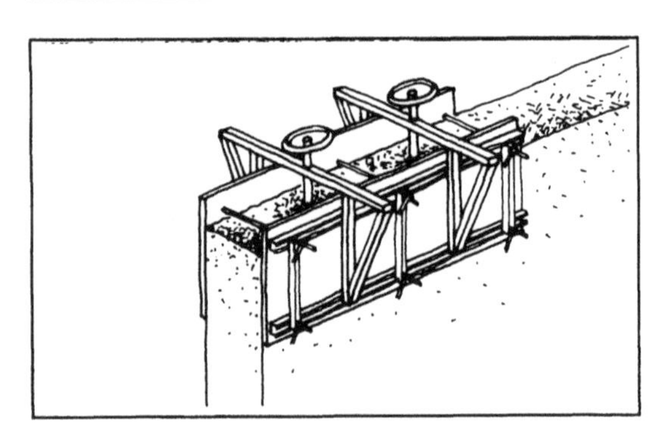

3. Sliding formwork

Various attempts have been made to adapt the sliding forms used in concrete construction. Up to the present time the various realizations are somewhat laborious affairs, though they work quite well.

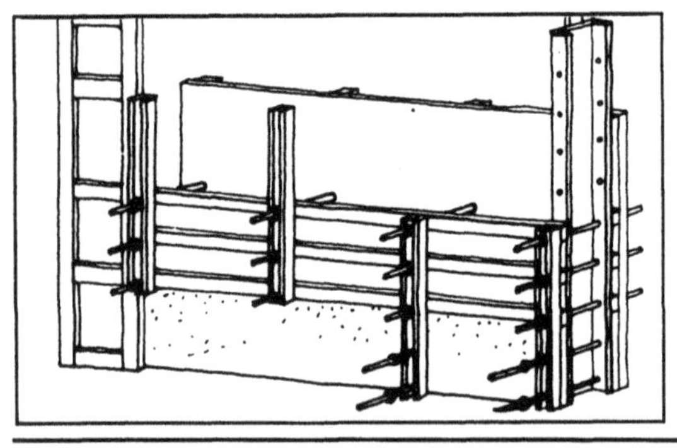

4. Leapfrogging formwork

A succession of forms is constituted by a vertical series of panels. The principle was revived by the Rammed Earth Institute International, Colorado, USA. Two sets of formwork panels are required. Panel number one is placed in position three and then in position five; panel number two is placed in position four and then position six. The system performs rather well but does not solve the problem of corners or the bonding of walls. Its primary advantage is that the lower panel serves as a support and reference for the one above.

Formwork shapes

Most formwork used, whether horizontal or vertical, is designed on the principle of straight panels which yield flat surfaces. Research work has been carried out to find ways of getting away from the limitations this imposes.

1. Plane and perpendicular surfaces

This is the most frequently encountered set-up, being used as often in traditional rammed earth construction as in modern rammed earth buildings regardless of the type of formwork used. The principle is simple and does not give rise to any complex problems.

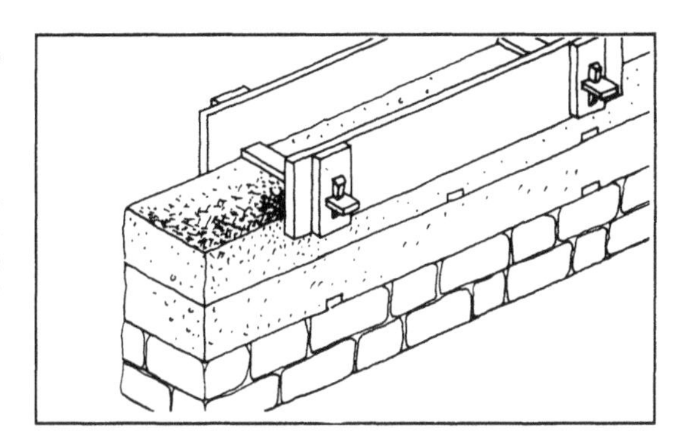

2. Batter

This formwork can be used to produce plane perpendicular or plane inclined surfaces otherwise known as the 'batter' of a wall. The slope is toward the centre of the wall and progresses as the wall rises. Only the outer face or both faces of the wall can be battered, and simple wooden packing pieces or wedges known as 'batter fixers' can be used to effect this design. The batter lightens the mass of the wall in proportion to its rise.

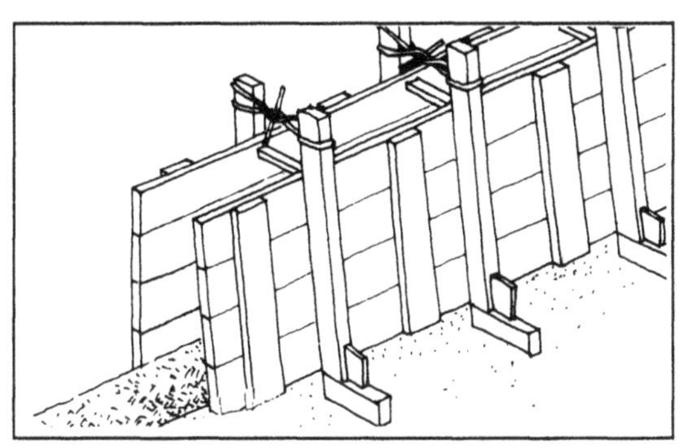

3. Curved surfaces

With certain types of formwork it is possible to make curved surfaces. This can be a matter of curving walls at corners or the construction of buildings with curved walls. The concept is well known in the Chinese tradition of rammed earth construction as illustrated in the 'round buildings' of the Hakkas of the Central Plateau. The development of such formwork was the subject of a recent project in the context of the Earth Village Project on the French Isle d'Abeau.

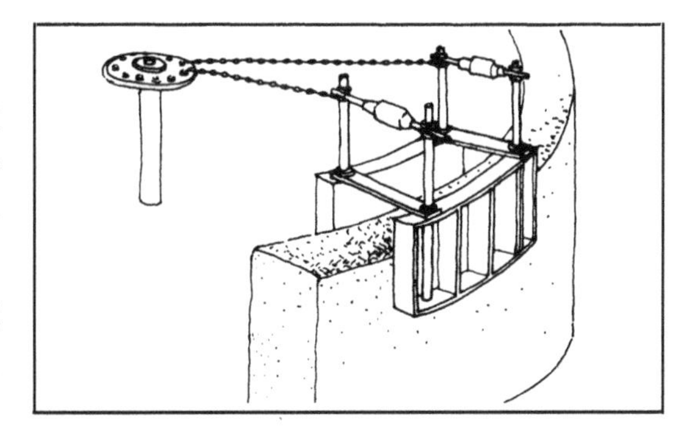

4. Composite face design

A formwork system using modular panels makes it possible to fashion surfaces composed of small vertical faces. The concept is suited to the production of diverse shapes including posts of geometrically variable cross-section. The design of the system is rather complex but the concept continues to prove practicable and should become simplified with time.

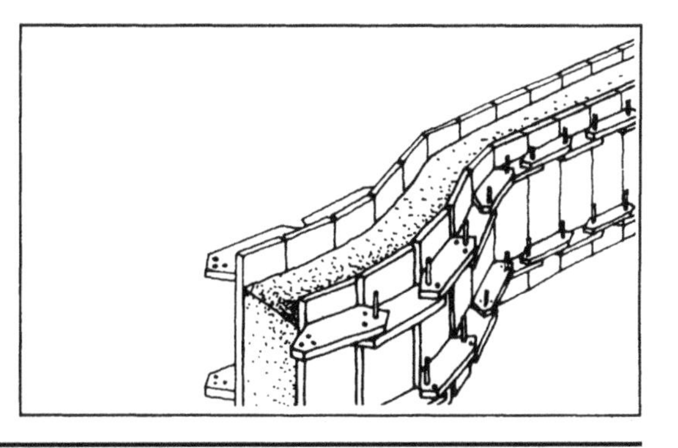

In rammed earth construction the building of corners between walls requires the use of special formwork. The play allowed for in the formwork for sections of perpendicular wall can prove to be inadequate if insufficient attention is paid to corners. These can be fashioned all of a piece or by the alternate, perpendicular overlapping of boards. The provision of chamfered edges reduces the erosion of outside corners. The 'T-system' used for bonding partition walls assumes the same principles as those applied to the corners.

1. Corner posts

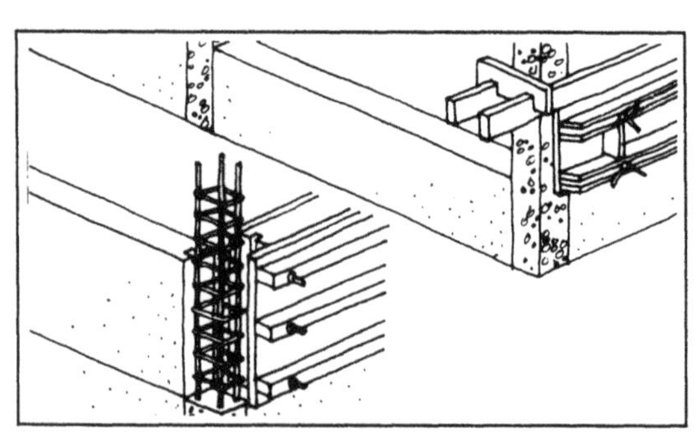

These can be constructed in concrete which can be poured either before or after erection of the rammed earth walls. Corners can be constructed in stone or brick masonry but should be toothed with (conventional) rammed earth.

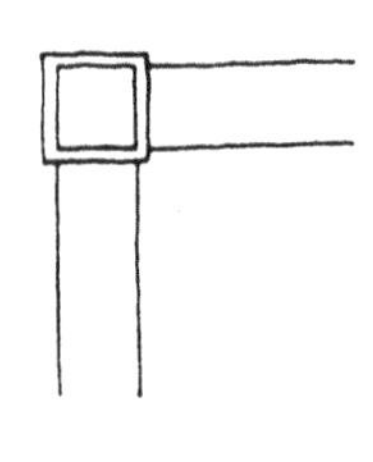

2. Formwork ends

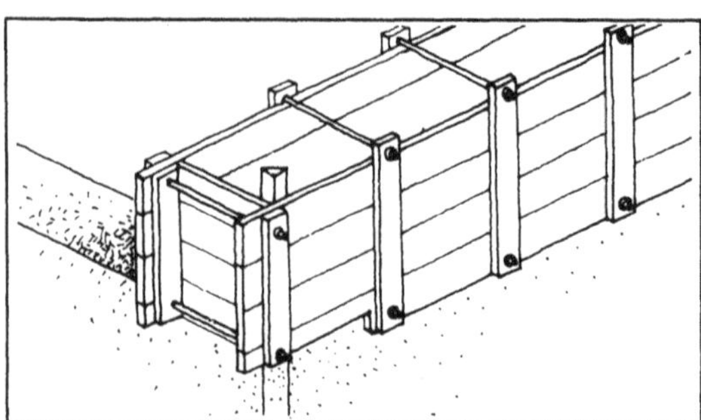

This is the system adopted in the Moroccan tradition of rammed earth construction. It is reminiscent of ordinary brick bonding. The corner results from the perpendicular overlapping of the rammed earth sections slabs in both directions.

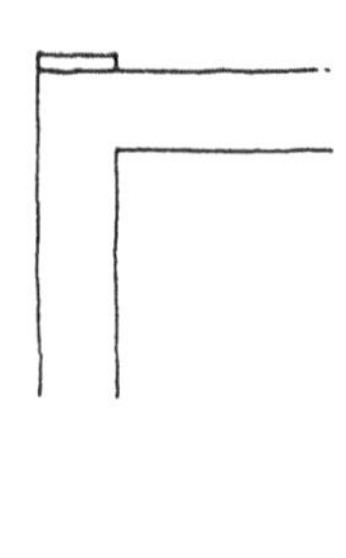

3. Non-modular formwork

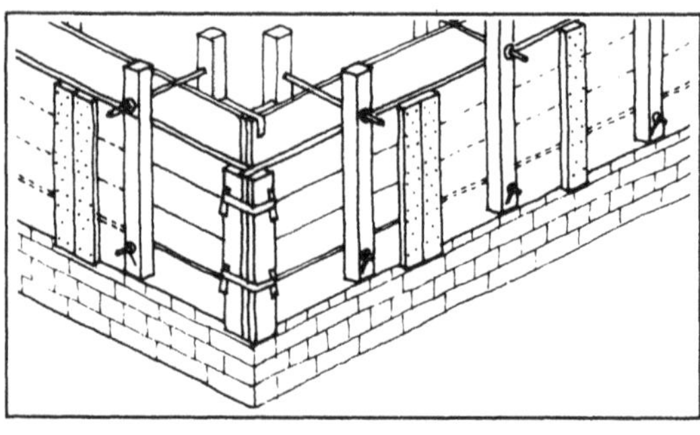

Each corner is constructed using a special element adapted to the particular conditions resulting from the use of non-modular formwork.

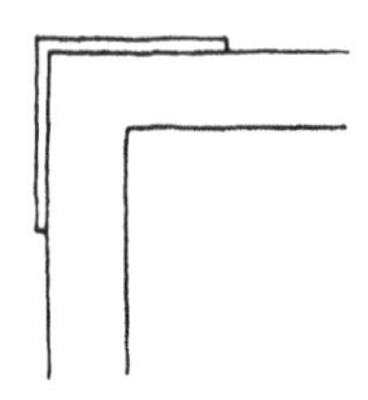

4. Modular formwork

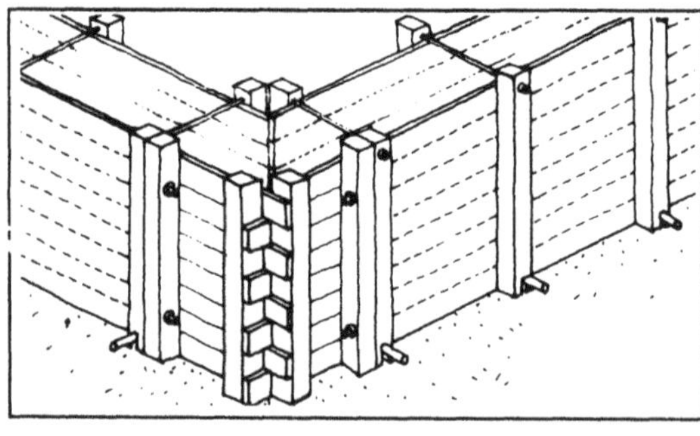

In this system corners are constructed as a single piece, coupling the two inner panels and using a modular formwork on the outside. The design and external dimension must be very precise.

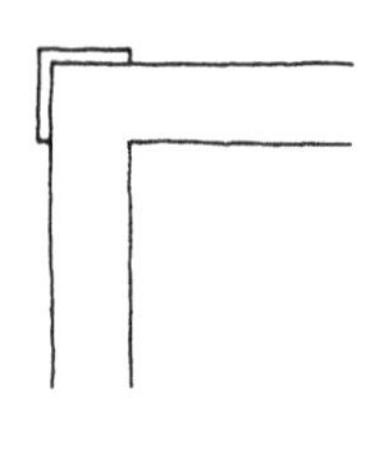

5. Integral corner

This system can accommodate the setting up of a formwork that can form an integral corner from the bottom all the way up to the top of the building under construction. In this way it takes care of the very tedious problems of plumb and adjustment.

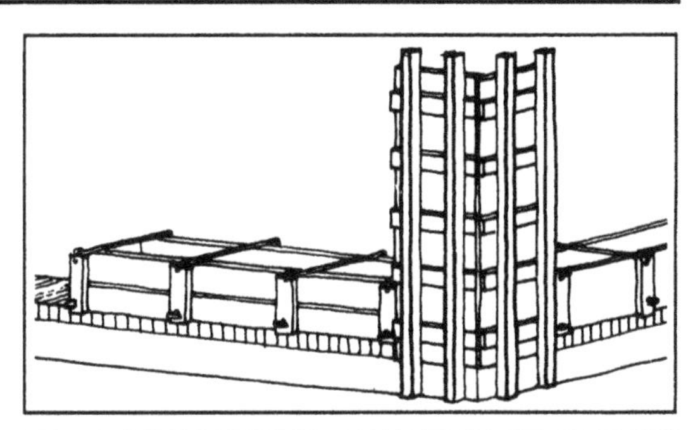

6. Symmetric formwork

The formwork for both inner and outer faces is modular and symmetric. This system solves the problem of adjusting the panels, but it does not altogether eliminate the risk of a corner separation crack.

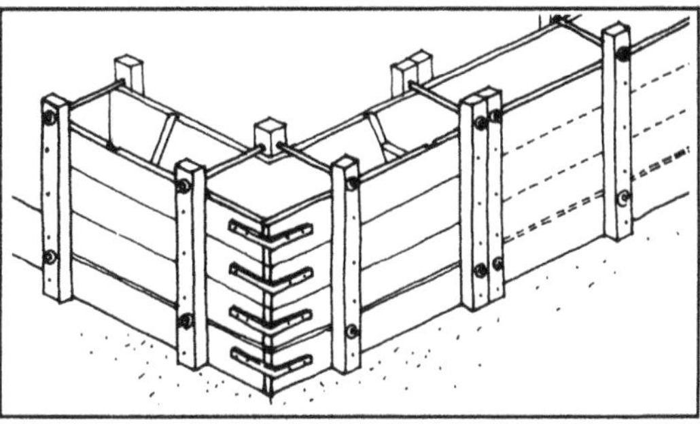

7. Assymetric formwork

This system is sounder than the completely symmetric corner, given that the forms can be inverted thus eliminate the danger of corner separation cracking.

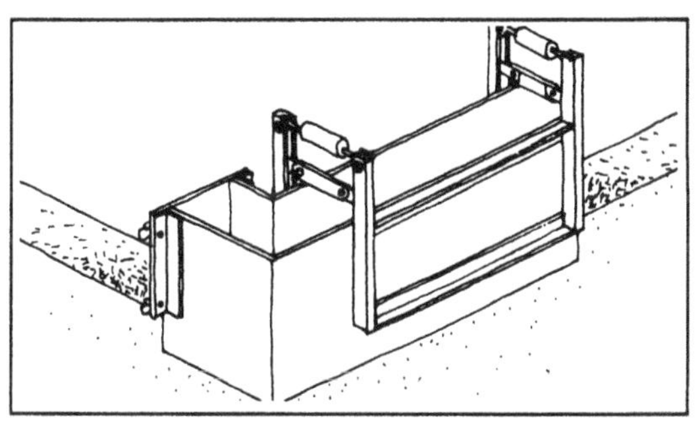

8. Variable formwork

The angle of the corner being formed can be varied by means of systems that incorporate regular or lift-off hinges. These systems are delicate and always pose the problem of gauging the fit of the panels.

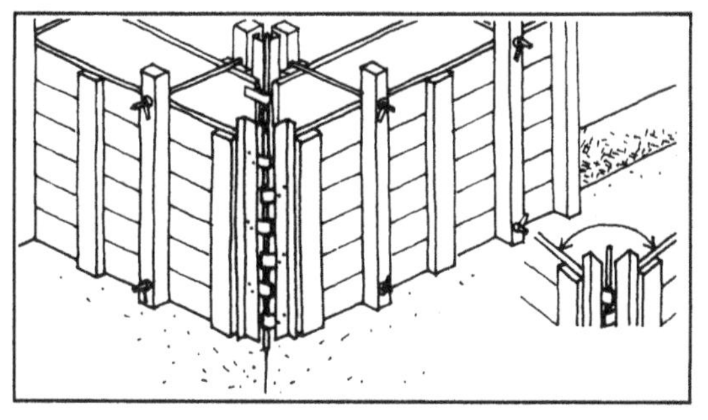

9. Rounded formwork

This corner requires special formwork, produced on the site, which can accommodate architectural features. The corresponding operation is very delicate, costly and difficult to carry out.

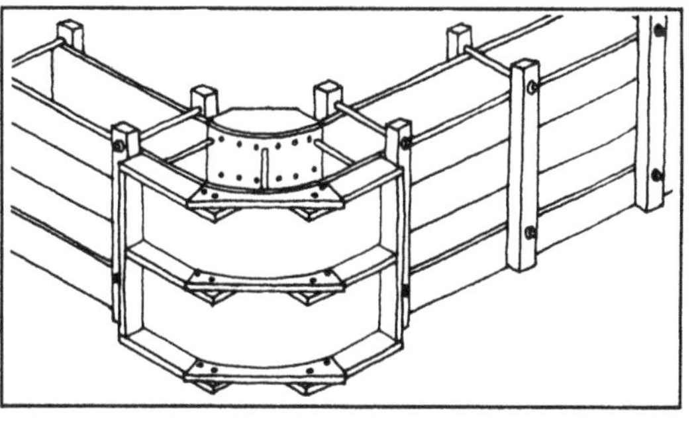

Conventional rammers

These are designed for manual ramming of the earth and consist of a mass of wood or metal fitted with a weighted handle. The diversity of the design of the tool as well as the vocabulary used to designate it throughout the world is very great. In certain countries several kinds of rammers are used in the same structure depending on the job at hand.

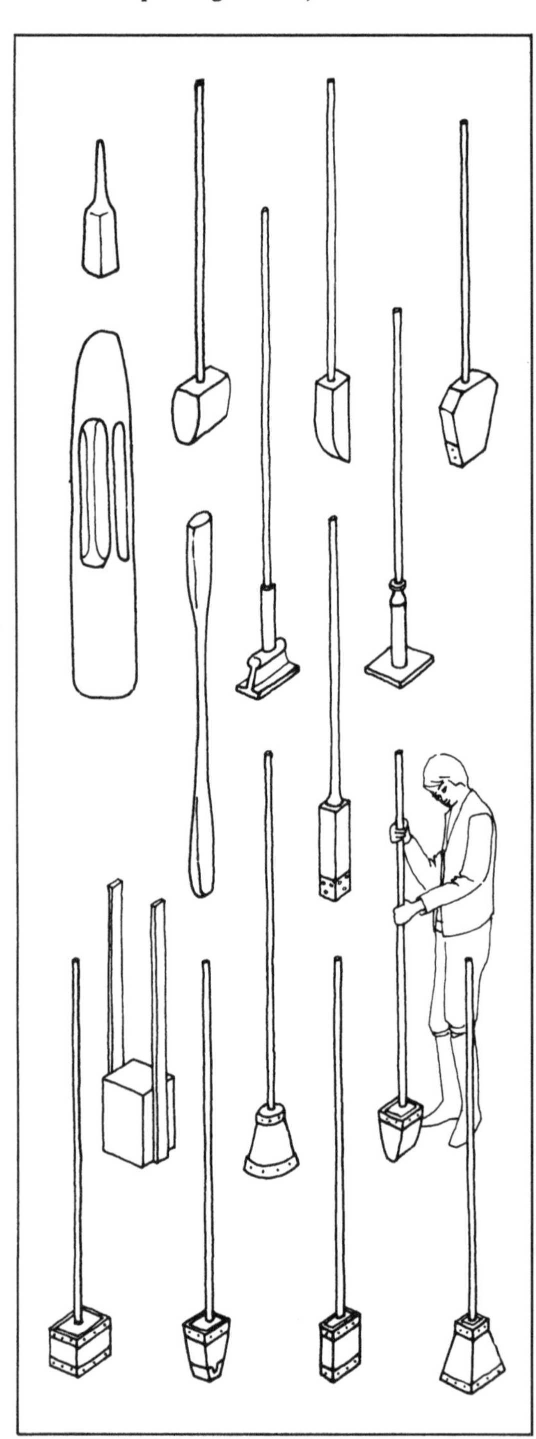

1. Parameters

The most important factors in the design of a conventional manual rammer are as follows: material, weight, area, shape, striking face, type of handle and size.

2. Striking faces

Rounded striking faces spare the formwork but are less effective in corners. Rounded edges are enough to prevent damage to formwork. A prism-shaped flat striking surface gives the best results as far as highly advanced research has been able to determine. A striking angle of 60° appears to give similar results, but effectiveness falls off quickly as the angle becomes sharper. Thus, for example, a striking angle of 45° could result in a 36% loss of effectiveness. Special conical shapes and wedge shapes permit access to difficult areas such as the angles in formwork or under clamps.

3. Striking area

This should preferably be in the region of 64cm^2. Research shows that an area of not greater than 225cm^2 ensures maximum effectiveness.

4. Striking head

This is usually made of wood or metal. The striking head of wooden rammers is protected by a metallic plate to reduce over-rapid wear. Metallic rammers are more solid and easier to handle as their heads are smaller. They are heavier and their striking face must large enough to prevent penetration into the formwork.

5. Weight

The recommended weight of a wooden or metallic rammer is between 5 and 9kg. This can vary depending on the size of the rammer and the strength of the person doing the ramming.

6. Handle

Rammers can have either a single or double handle made of either wood or metal. A hollow handle removes all restrictions about the way the rammer is weighted. The size of the handle is adjusted to the person doing the ramming and varies from 1.3 to 1.4m. Handles equipped with a sliding mass which strikes the earth a second time also exist.

Ramming is a slow and tedious task. Various machines can be used, most of them heavy. They exert a great deal of pressure on the formwork. Each machine has a different compaction energy, which is dependent on the ratio between the number of passes, the speed of the pass and the thickness of the layers of uncompacted earth. The cost of such machines is highly variable and ranges from US$3500 to $1500.

1. Impact ramming

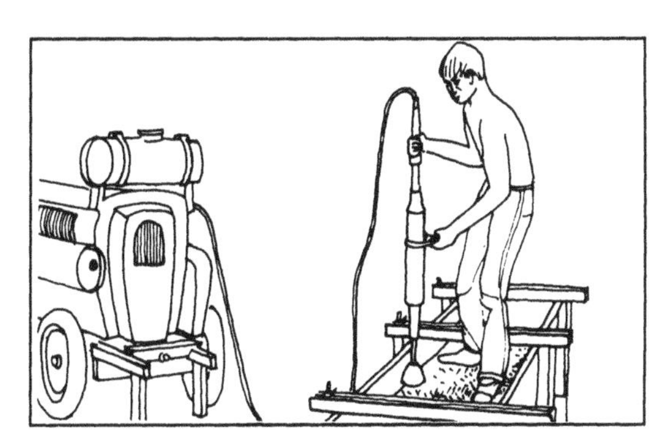

Pneumatic rammers These have been directly copied from the foundry industry where they are used to settle sand in moulds. Their mode of functioning imitates that of a manual rammer but they are capable of a much higher impact frequency (up to 700 strokes per minute). Of all the pneumatic rammers available, only the 'soil' rammers are effective, of which there are numerous commercial types (Atlas Copco, Ingersoll-Rand, Perret, etc.). Pneumatic rammers must be neither too heavy (15kg maximum) nor too powerful, as they might destabilize the formwork and cause the rammed earth to bulge or they might penetrate the earth. They should have a long stroke (approximately 20cm) and run on regulated air. They should be able to attain a pressure of 0.5MPa and barely beyond that. Compressors are very expensive, costing about US$5000. Notwithstanding the expense, however, ramming carried out by a pneumatic rammer is highly effective, and if the soil is of a high standard, the quality of the rammed earth is excellent.

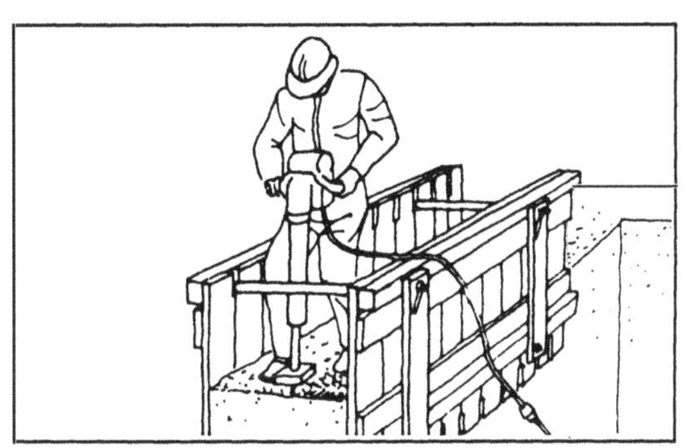

Pick hammers The idea of transforming a pick hammer by fitting it out with a special ramming plate has already been tried. These tools are, however, too powerful, and can set up resonance vibrations within walls thereby splitting the material.

2. Vibration ramming

Vibrating plates This method was developed by GHK in Kassel. In it, a motor with an eccentric rotating mass transmits vibrations to the plate, thus causing the machine to move. A switch enables the operator to determine the direction of this movement, and the machine then functions automatically. The ratio between the weight of the machine, the speed of operation and the vibration frequency is difficult to set.

Vibrating rammers Versions of these machines powered by combustion engines or electric motors are available on the market. They are heavy, cumbersome and expensive. Their use has been the object of numerous tests with very modest results and constructors are advised against using them.

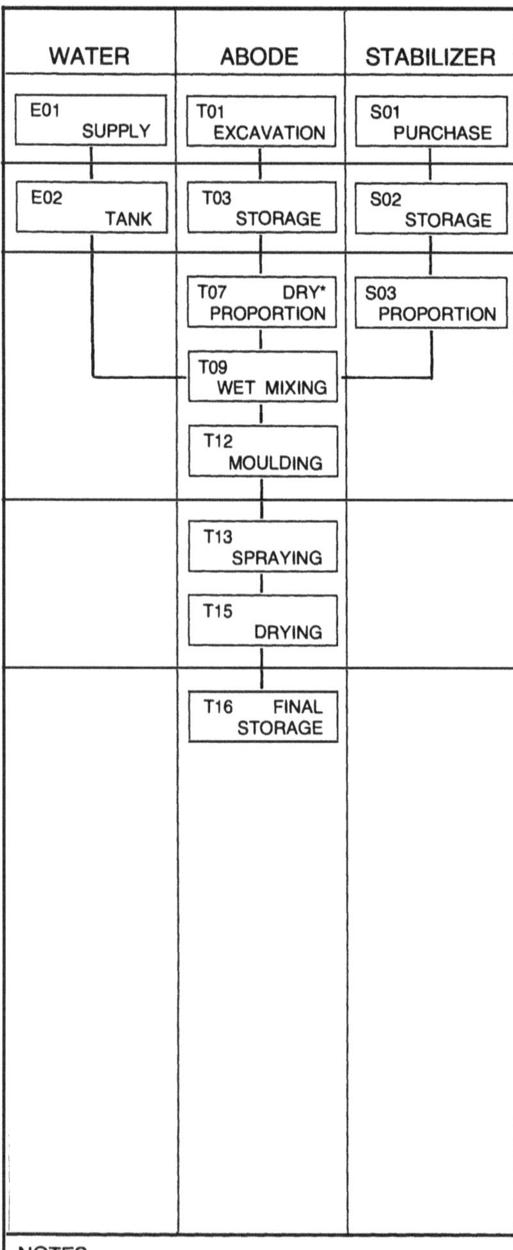

1. Production

The production of adobe bricks constitutes one of the simplest processes by which construction materials are made. The historical and geographical aspects of this process are extremely varied. As a result an almost infinite variety of production procedures could be described.

Adobes can be produced using liquid or plastic soil with or without any of a wide variety of moulds. Plastic earth can furthermore be used for production by extrusion.

In contrast to compressed blocks and rammed earth, adobe is malleable and fragile during the production process. Each brick must be dried individually. The total production area is thus very extensive.

If the process is mechanized, the first candidate is often excavation, followed by mixing and with moulding coming last.

2. Products

Adobes come in numerous shapes but the list is not nearly as long as that of the types of compressed blocks. The production technique is such that only massive forms are possible.

3. Production season

Each technique and every region throughout the world imposes its own characteristic limitations on production. The drying of adobes, as it is highly dependent on good weather, is the most important single limitation. Accordingly the production of adobes ceases in cold weather and sometimes in extreme heat as well. Many brickyards are set up on river embankments so as to be able to make use of flood deposits. Such a source of earth can impose limitations as the production area must be periodically cleared for the floods.

4. Crews and productivity

The size of crews and productivity are extremely varied. The outputs appearing in the table below take account of all the operations involved in the process, including excavation and storage.

	U/DAY	WORKERS	COST ($)
MECHANIZED	20,000	5 – 6	300,000
SEMI-MECHANIZED	10,000	5 – 6	50,000
MANUAL HIGH	2,500	4 – 5	0
MANUAL LOW	500	4 – 5	0

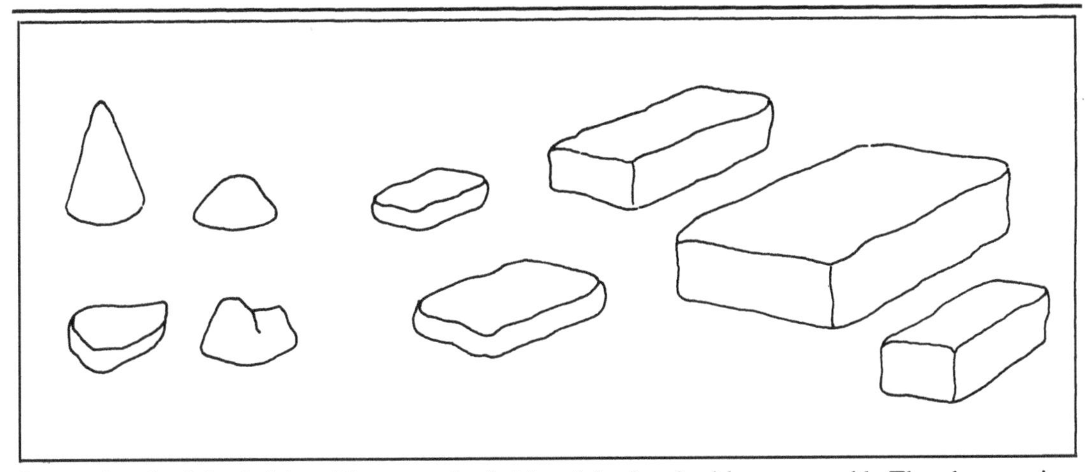

Conventional adobe bricks These can be fashioned by hand without a mould. They have various shapes, conical, cylindrical, pyriform or cubic. They can be moulded in wooden frames blocks or mechanically. In this case they are prisms, cubic or parallelepipeds and their dimensions highly variable, ranging in length from 25 to 60cm.

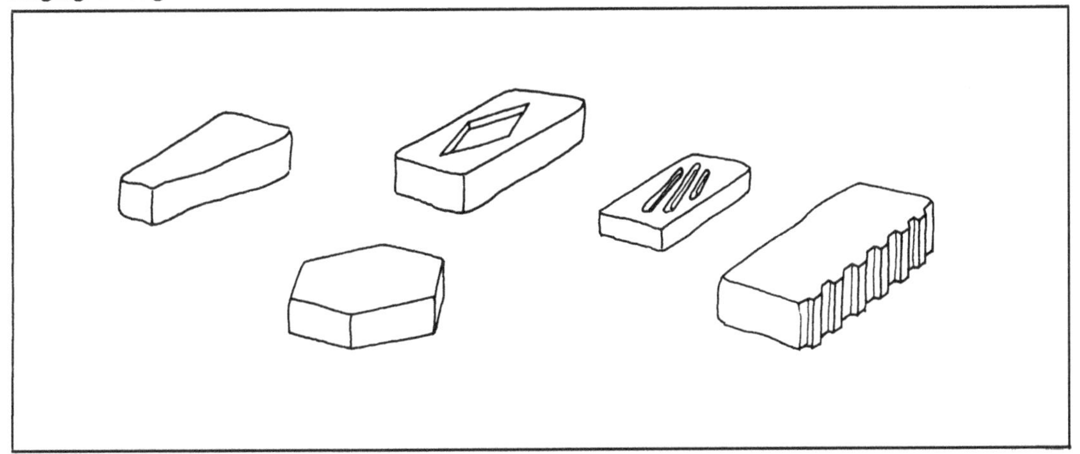

Special adobe bricks: These may be intended and designed for either conventional use or something out of the ordinary. Some adobes, for example, have grips so that they can be kept more securely in position (as in domes or arches), interlocked with other structures or have special decorations or markings.

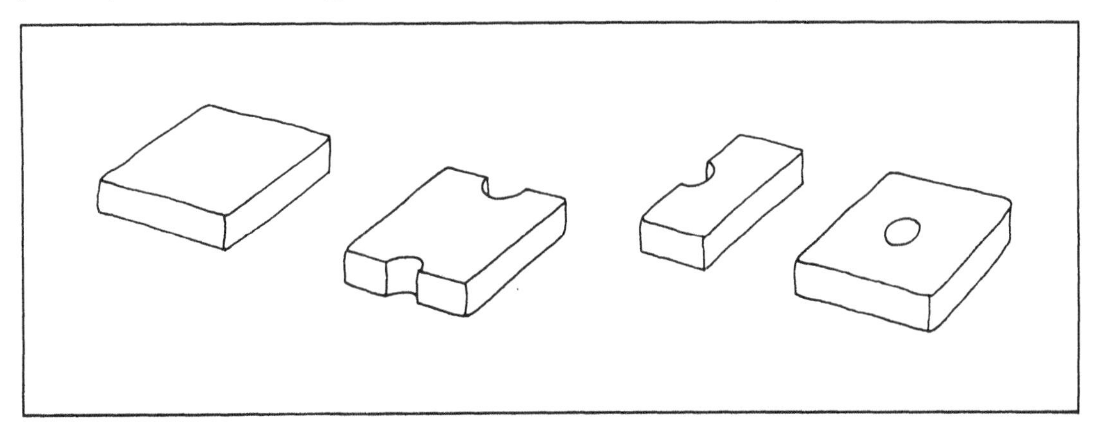

Anti-seismic adobe bricks Specially shaped, they are more resistant to earthquakes. Their special design makes them suitable for use with anti-seismic structural systems (e.g. reinforced walls).

Soil suited to making adobes has a rather clayey or very silty texture and is quite cohesive. This cohesiveness makes excavating the earth, whether dry or wet, a difficult task. The excavation sites are often waterlogged and muddy. The traditional method of preparing the earth is thus laborious and is done on foot. The soil must be prepared with great care in order to ensure a high standard of adobe. Today there are other methods of preparation, some of them mechanized. A third category, lying between the first two, utilizes draught animals. The last two preparation techniques are obviously more expensive than the first.

Chopping the straw

Plant fibres, generally straw, are frequently added to the soil. The stalks are cut using sharp-edged tools. However, manual and power straw choppers are commercially available which are capable of cutting large quantities of straw and other fibres into length of between 1 and 30cm. The normal price of such choppers starts at US$1000 for the manual type and $1500 for motor-driven ones. Fibre choppers can also be used to prepare plant cuttings destined for a methane digester.

Pugging or kneading

1. Animals

The preparation of the earth entails a long pugging operation. In many regions, animals that go round in a circle over a specified area perform this work by treading the earth with their hooves. Animals that can be used for this purpose include, among others, donkeys, mules, oxen, and horses.

2. Machines

The material can be pugged in a pit with mechanized plant such as shovel excavators and tractors that can combine the operations of excavation, mixing and transport. The pit should have a stable bottom and an incline so that the machine can get out. The manoeuvring space for the machine must be sufficiently large. The quantities mixed are enormous, being of the order of $10m^3/h$.

3. Pugging mills

Pugging can also be carried out in a pug mill. These can be set up in smallish drums and driven by a motor or else hauled by animals over a given area – two weighted truck wheels will serve the purpose. Wheel tracks should not be left in the soil and these can be avoided by devising a system that throws the soil back under the wheels so that it is constantly remilled. An improvised pugging setup costs only a few dollars while a mill in a containment vessel costs something of the order of US$2000. Such plant is very heavy and a typical output is in the region of $7m^3$ per day.

Mixing

1. Vertical mixers

The most common can be made using very basic materials: a few planks and timbers, ropes and steel wire, etc. They can be operated by animals. The lever should be at least 2.5m long, and the animal should not work more than five hours per day. Mechanical vertical mixers exist as well, and these begin at US$2000. They must be solidly built and the standard output is $10m^3$ per day.

2. Rendering mixers

As they are not very sturdy, these mixers should really be used with liquid soil rather than plastic soil. They are capable of a daily output of approximately $8m^3$. They cost US$1500 and upwards.

3. Linear mixers

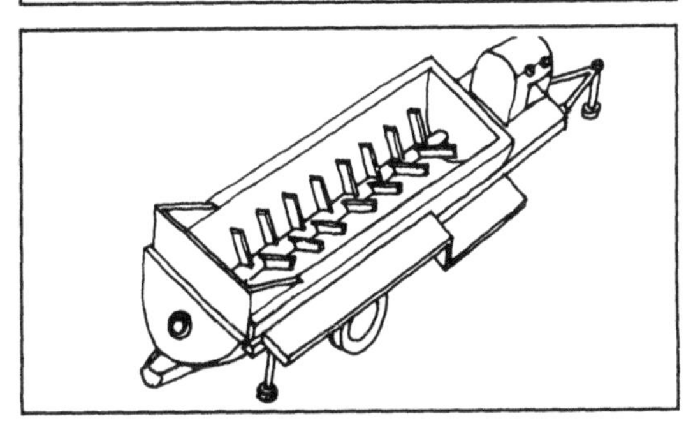

These are widely used in production units capable of medium and high output. There are a number of variations on them. For example, they can have a single or double shaft; they can be of the constant or discontinuous-flow type; they can be of heavy or light construction. Their output is very high and muddy earth can be dumped into a pit allotted for the purpose. The smaller linear mixers cost about US$1500 and are capable of outputs of 4 to $5m^3$ per day. The bigger mixers, which have been adopted from the ceramics industry, can cost upwards of $10,000. They have an output of $50m^3$ per day.

4. Concrete mixers

Although poorly rated, standard tilted-drum concrete mixers are capable of doing the job. Their output is low and resulting mixture often lacks homogeneity and suffers from lumpiness. Their main advantage is the wide range of models available, ranging from small to large, suitable for connection to a tractor PTO (power take-off), mixer trucks, and special wheeled plant.

5. Screw mixers

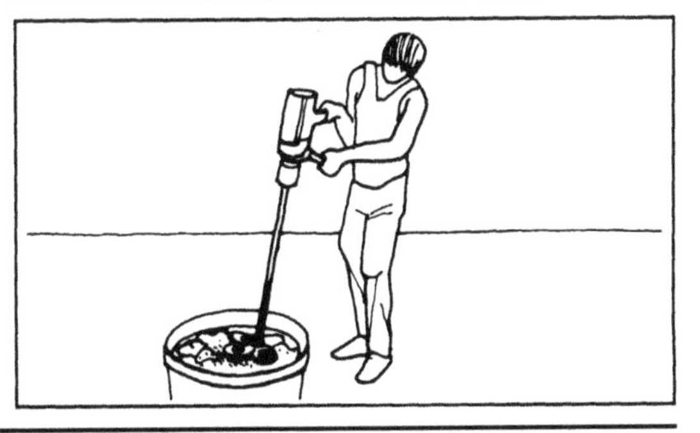

It is also possible to work with small quantities using drums provided with a screw of the sort used for paint and plaster. In this way it is possible to prepare 50 litres of mixture in ten minutes by making successive batches.

6. Planet wheel mixers

These are ideally suited to preparing mud even when it must be mixed with plant fibres. The smallest ones have a batch capacity of 100 litres, cost about US$3000 and have an output of $10m^3$ per day.

Small scale

Adobes can be produced with or without a mould. Very primitive production techniques are still practised today. Bricks made in this way do not have a very attractive appearance nor are the walls built using them particularly solid. Prism-shaped moulds are recommended. Semi-solid or semi-soft paste is required for manual shaping.

1. Semi-soft paste

The paste which has been put in the mould is lightly worked by hand and then immediately removed. In order for it to be removed easily, the mould must be cleaned and wetted beforehand. In this technique, called 'slop-moulding', the film of water that adheres to the mould facilitates release. The common type of mould has a single compartment and its dimensions are variable; it is up to 60cm long for the heaviest adobes. It can also have multiple compartments and be capable of moulding up to four adobes at a time. These moulds are made of wood or iron, some even being made of plastic. The bricks undergo considerable shrinkage and their quality must be carefully monitored.

2. Semi-solid paste

To produce better quality, denser and more resistant bricks, it is advisable to work with a semi-solid paste. The mould must be very clean, and then it is dipped into water and the inside sprinkled with sand. Using this technique, known as 'sand-moulding', a given quantity of earth is shaped roughly into the form of a ball, rolled in the sand and then thrown with force into a single-compartment mould. The ball is firmed up with the fists, care being taken not to neglect the corners. The excess is removed with a wooden guide strip. To facilitate release, only the earth coated with sand should adhere to the sides of the mould. There are many different types of moulds, some having bottoms and some not. The adobes are turned out of the mould onto the drying area. This technique means that the earth has to be stored near the moulding area and several moulds should be available. It is advisable to work standing at a table. There are even tables with built-in moulds and ejection levers. The adobe should be carried to the drying area on a small board (the bottom of a mould). The output using the moulding technique is of the order of 500 adobes per day.

Large scale

Large-scale production requires modifications to the technique.

1. Multiple moulds

These can be a ladder-like array in which moulds are juxtaposed or large parallelepiped moulds. In this way ten to twenty-five adobes can be produced at once. The earth should be able to fill the entire mould. It must therefore be more liquid, in the soft-paste state. Apart from this change in moisture content, the previous preparation remains the same. The earth is then poured into the moulds by means of a wheelbarrow, dumper, frontloader, or even straight from the mixer which, in that case, is self-propelled, towed, or mounted on a truck. The soil is then levelled with a kind of scraper so that it is evenly distributed within the mould, including the corners. Some time may be allowed to elapse before removal from the mould but usually this is done immediately after the previous step. The whole operation is then repeated without interruption. Large moulds must be cleaned properly either by allowing them to soak in water or spraying with a powerful jet of water. The cleanliness of the moulds and the moulding stage are crucial to ensuring the quality of the adobes. Owing to considerations of weight, the moulds should be made of wood or plastic as opposed to iron. They should be easy to manipulate by no more than two persons. The wood should be treated against rot and warping. The outputs possible with this moulding technique, with a crew of five or six workers, ranges from 8000 to 10,000 adobes per day.

2. Sawn adobes

It is possible to make a single very large adobe ($4m^2$ for example) with a mould consisting of four two-metre long planks. A soft paste is used. The resulting adobe slab is then cut into several small adobes with a taut wire saw on a wooden support or else using a plank with a studded edge. The output using this technique is similar to that of the above technique, and only a very modest investment is required, although the finish obtained is not so good. The moulding area must be absolutely flat.

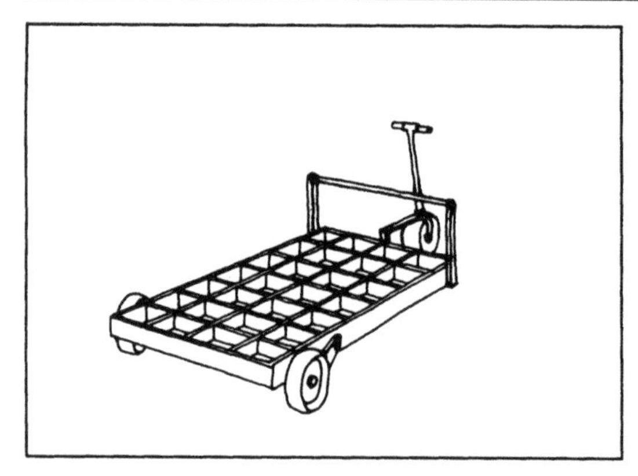

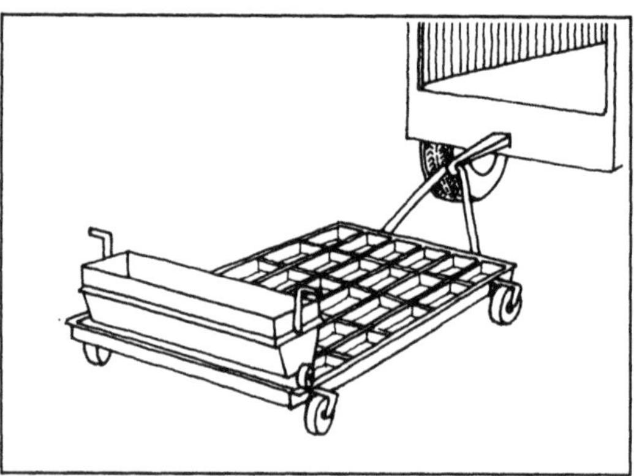

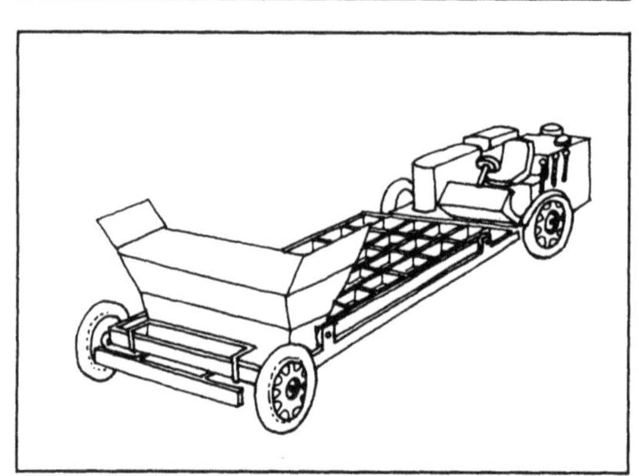

The difference between large-scale production using multiple moulds and a mechanized process is not all that great.

1. Moulding box

A metallic mould containing a large number of compartments is mounted on a frame on wheels. The mould is lifted by means of a lever arrangement after it is filled. The adobes are deposited on the ground. The wheeled mould is then pulled to the next moulding point. The moulding cart should be capable of being cleaned each time. A mobile hopper can be added and positioned above the mould for filling by a dumper. Soft paste is poured into the mould from the hopper which is drawn over the mould. The excess earth is removed with a scraper which can be fitted to the hopper. The standard output of such a system is in the order of 7000 to 10,000 adobes per day, and it has been refined with the development of the Hans Sumpf moulding box in the USA which was designed as an independent unit. This machine can achieve an output of 20,000 adobes a day. The adobes are deposited onto impermeable paper which is unrolled directly onto the ground on a gigantic production area. The adoption of a machine of this sort means that the entire upstream production plant must be modified to cope with the enormous increase in capacity. The total investment amounts to a few hundred thousand American dollars.

2. Cutter disc box

Cutting by means of a wire can be automated and the wire replaced with cutter discs. A hopper fitted with a barrel lays down a continuous sheet of soft paste which is cut into thin strips. The machine is stopped at a fixed distance from its starting point and the thin strips are cut transversally by another set of cutter discs. The output is very high, amounting to 15,000 blocks a day, and the investment is low. The production surface must be very flat and clean and the mixture highly homogenous and of an ideal consistency. The user of this system must therefore be absolutely sure of what he is doing. The machine in question costs about US$3000. To date only prototypes are known, but these appear to be most promising.

3. Extruders

The extrusion of adobes opens several very attractive possibilities. Applied to the manufacture of adobes, extrusion can serve as the basis of three principal processes.

Vertical extruder This consists of a vertical mixer provided with an extrusion nozzle. The system can be motorized or it can be driven a draught animal. The process, although giving good results, is used hardly at all nowadays. Fitted with an electric motor, the machine costs approximately US$2000. With a small mixer weighing about 500kg, output can reach 1500 bricks per day.

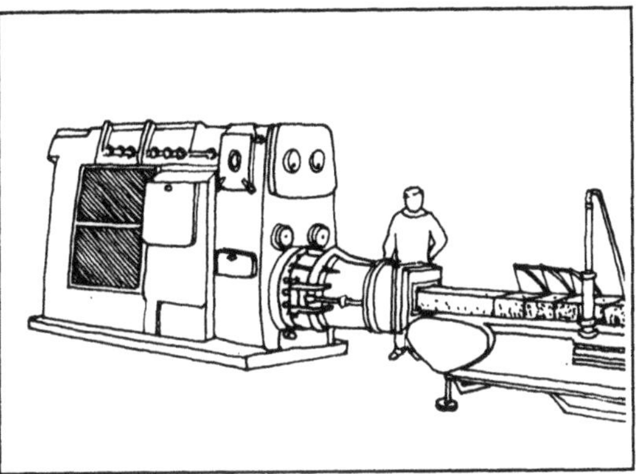

Horizontal extruder Adapted from the ceramics industry, this machine was widely used in the USA in the 1940s. It is still standard in India. Although it involves a heavy financial outlay, the system is efficient. It is capable of the same production rates as achieved in the brick industry for equivalent products. Nevertheless, the soil used for making adobes is sandier than that used for making burnt bricks. Consequently it is more abrasive, and a significant degree of wear resulting from friction must be allowed for.

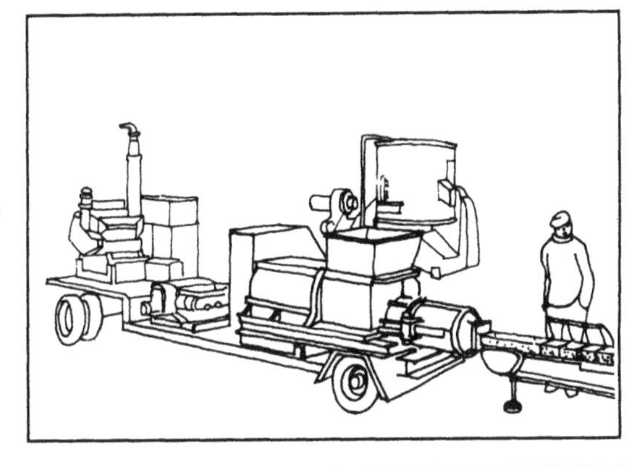

Mobile extruder Mobile extrusion units mounted on a frame on wheels are commercially available nowadays. These are heavy pieces of equipment weighing approximately thirty tonnes and combine a mixer, a generating unit and an extruder. Some units are already operational in various parts of the world for the production of burnt bricks. They could also be adapted to the production of sun-dried bricks. The system involves an investment of about US$250,000 for an output of between 2500 and 3000 bricks per hour.

4. Press

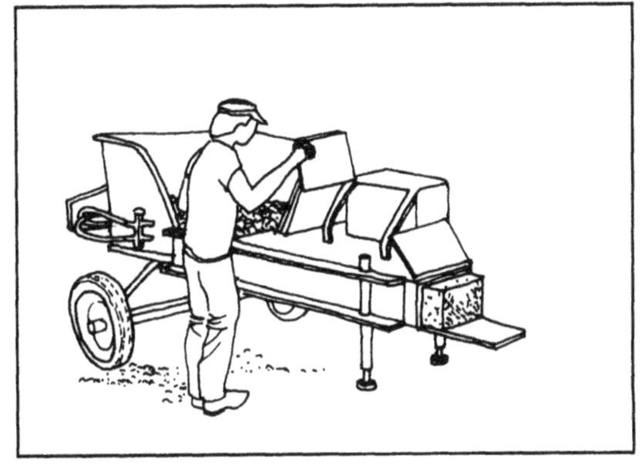

The traditional moulding table can be replaced by a press. The moisture content of the soil is not the same: the soil is either a semi-solid or solid paste. The required pressure does not exceed 2MPa. One or more holes 10mm in diameter are bored into the cover so as to facilitate the extrusion of any excess. Sometimes a small board is inserted into the mould and the earth ejected onto this for transportation. Output is much higher than it is for compressed blocks.

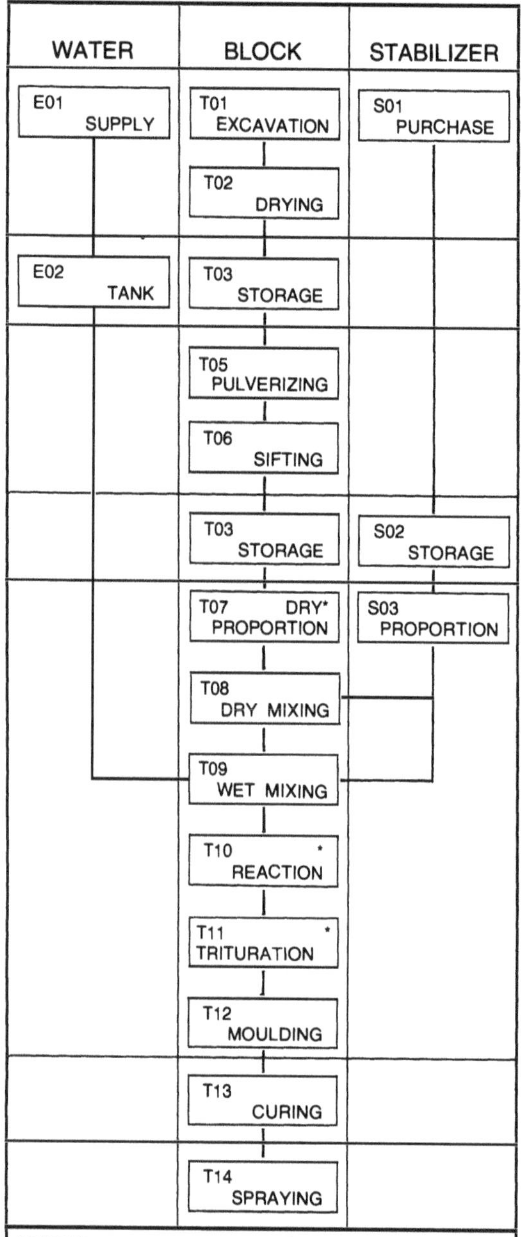

NOTES
– The above chart shows the phases involved in the production of cement in lime-stabilized compressed blocks.
– It is applicable to craft production of compressed earth blocks using presses of all description, but is not suitable for industrialized production.

* Stabilization makes these operations necessary.

1. Production

The production chart shown here applies to small artisanal brickyards with a daily capacity of 10 tonnes of blocks using two hand-presses and 25 workers. On the other hand, the production process used in large integrated production units has much in common with the lime-silica brick industry from which it is derived. The Latorex process (Denmark), for example, constitutes the essential operation of a 'turnkey' plant which, depending on its size, is variously capable of outputs of 2500, 4500, or 9000 blocks per hour. Between the manual press and the integrated plant there is a whole range of production units, with differing characteristics and capacities.

2. Products

These are as varied as the representative products of the ceramics and lime-silica brick industries as well as the concrete block sector.

3. Production season

The production area required is on the small side; the storage area for finished products can also be small as the finished blocks, as soon as they are removed from the mould, can be piled on top on one another to a maximum height of one metre. Compressed blocks can be made in any season on condition that certain precautions are taken with regard to storage during the first few days of curing in regions with harsh climates or with extremes of rainfall or heat.

4. Work crews and productivity

The size of work crews and their productivity is closely related to the degree of mechanization. Constructors' figures quite often correspond to the theoretical limit presses can achieve. However, in practice such theoretical outputs must often be reduced by 50% or even more. The values given below are for the standard outputs of brickyards fully equipped for all the preliminary operations (i.e. excavation and screening equipment) and subsequent pressing (i.e. storage facilities and cleaning equipment).

	U/DAY	WORKERS	COST (US$)
MANUAL	300	4 – 8	3,000
MOTORIZED	2,000	20	25,000
INTEGRAL	10,000	10	100,000
PLANT	60,000	15	2,000,000 and over

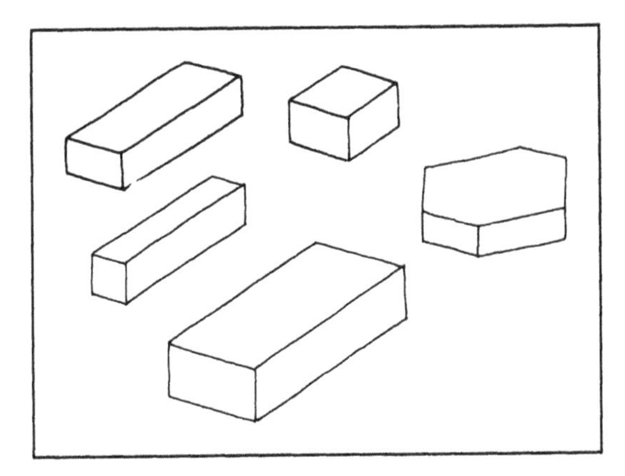

Solid blocks These are mainly regular solids and include cubes, parallelepipeds, multiple hexagons, etc. They are used in numerous different ways.

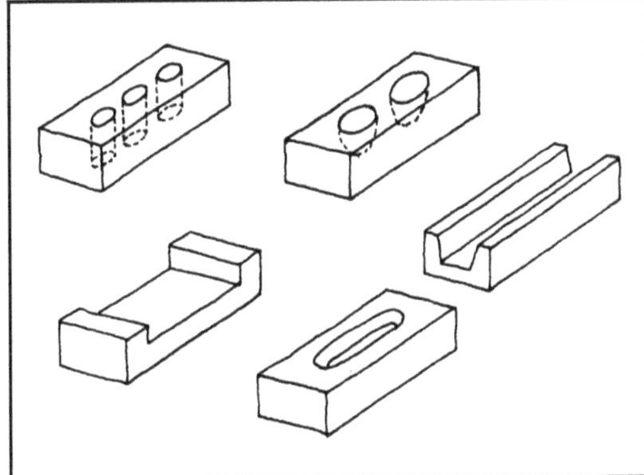

Hollow blocks Ordinarily voids represent 15% of the block but the hollow proportion of the block can reach 30% by using sophisticated processes. The frogs increase adherence to the mortar.

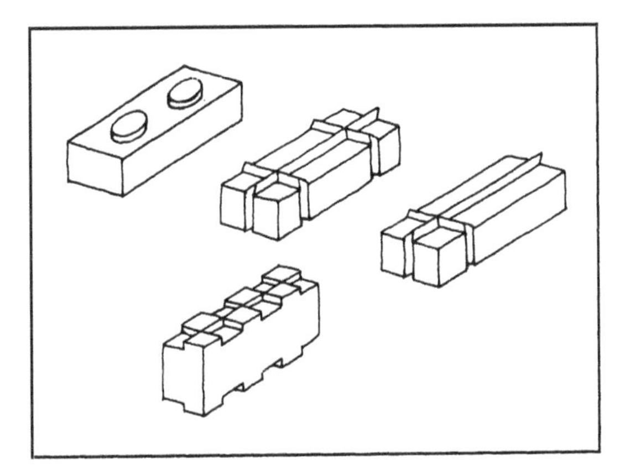

Interlocking blocks These make it possible to do away with mortar but require sophisticated moulds and relatively high pressures.

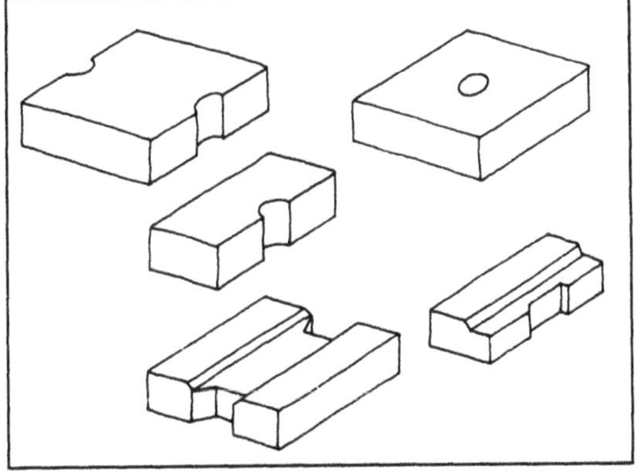

Anti-seismic blocks The shape of these enhances their resistance to earthquakes as well as their incorporation into anti-seismic structural systems such as tied blocks.

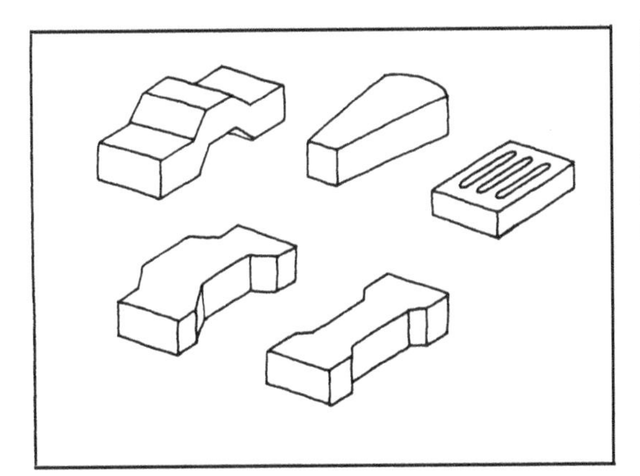

Special blocks These blocks are made on occasion for highly specific applications.

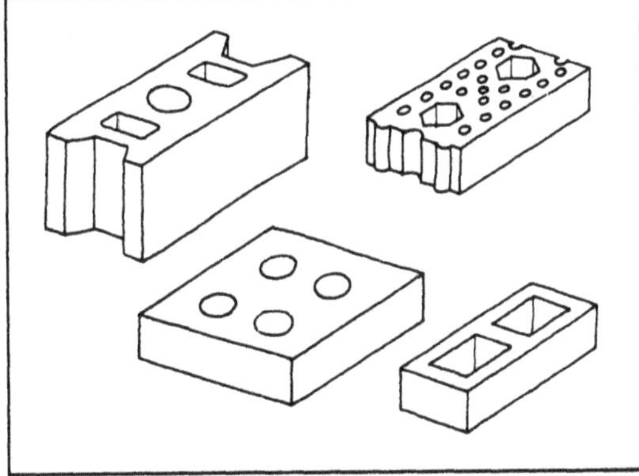

Perforated blocks These have the advantage of being light, but they require rather sophisticated moulds and higher than ordinary pressures.

In order to obtain uniform mixing of the mineral components, water and stabilizer, lumps have to be broken up which may have diameters of more than 200mm after excavation. Grains with a homogenous structure, such as gravel and stones, must be left intact and those having a composite structure (clay binder) broken up so that at least 50% of the grains are less than 5mm in diameter. The soil must be dry. Wet soil can only be handled by certain of the mechanized systems. Two basic approaches are used.

1. **Grinding** followed by screening. The material is pressed between two surfaces – a rather inefficient and tedious process in which useful stones are broken up. Only simple machinery is required.

2. **Pulverization** The material is hit with great force and disintegrates. The machinery required is complex but performs satisfactorily. At the delivery end, any large pieces left can be removed by means of a screen.

1. Pounding

Manual process; very slow; $1m^3$ per day per person; screening absolutely essential.

2. Jaws

Elementary mechanism – reciprocating motion; manual version. Output: 3 to $4m^3$/day. Weight: 150kg.

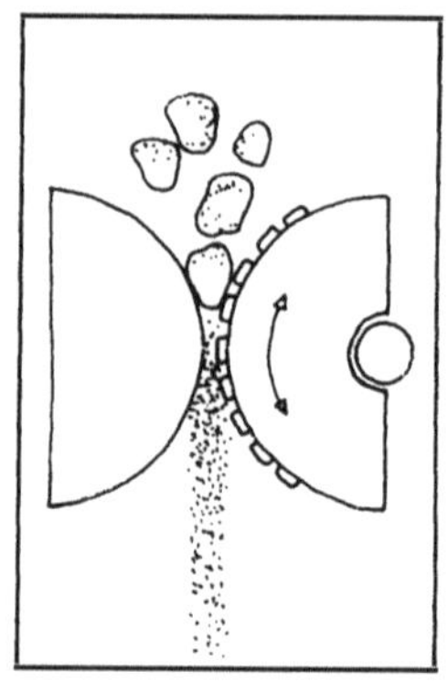

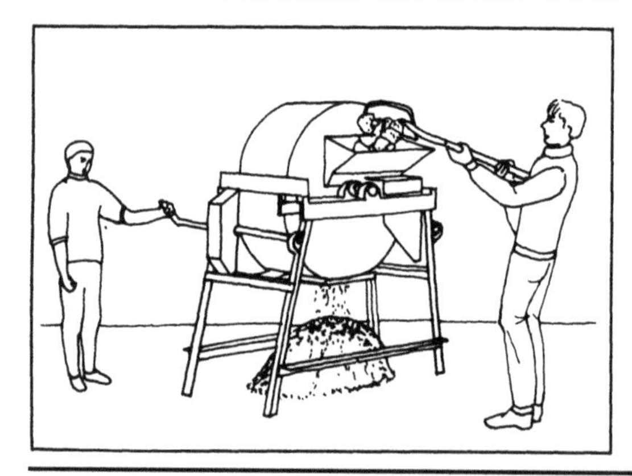

3. Carr

Four series of rods turning at 150rpm. Manual or motorized (electric) version, 1.5kW motor. Excellent mechanical efficiency, up to $10m^3$/day. Weight: 260kg.

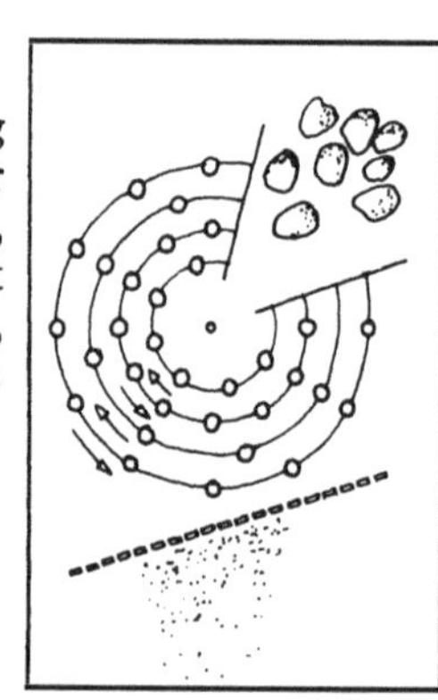

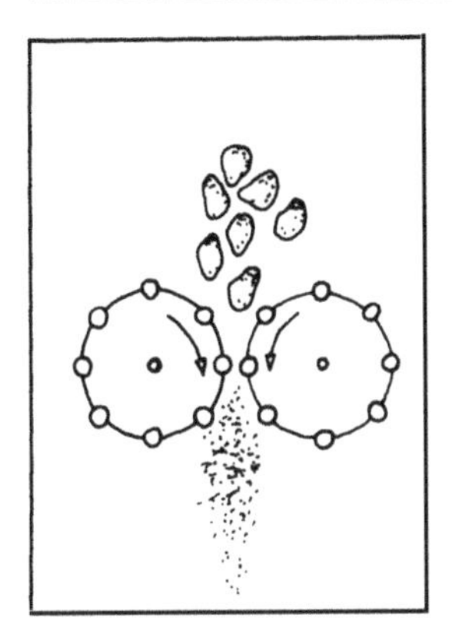

4. Squirrel cage

Very rapid rotation: 600rpm. 3hp electric motor (2.25kW). Output: 15 to 25m^3/day. Weight 150kg.

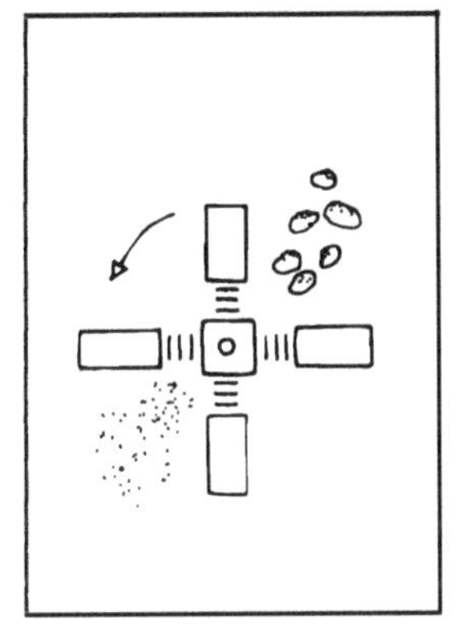

5. Hammers

Several spring-mounted hammers on a central axle beat the earth at a high frequency. 10hp electric motor (7.5kW). Output: 40m^3 per day. Weight: 200kg.

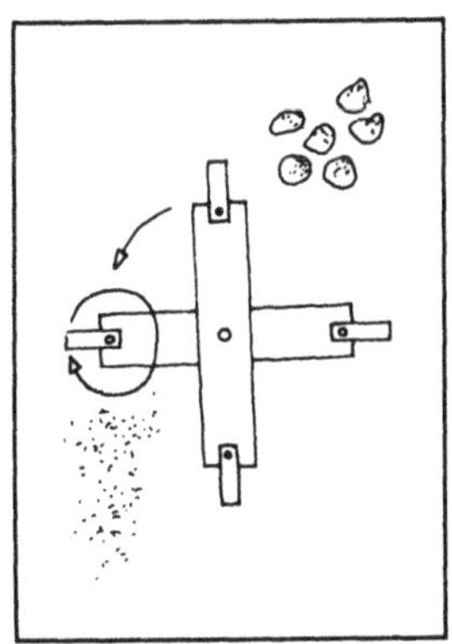

6. Screw

The same system as used in conventional composting machines. Indeed such machines can also be used if care is taken to avoid excess wear. Single screw or a set of screws. 5hp diesel motor (3.75kW). Output: 15m^3/day. Weight: 200kg.

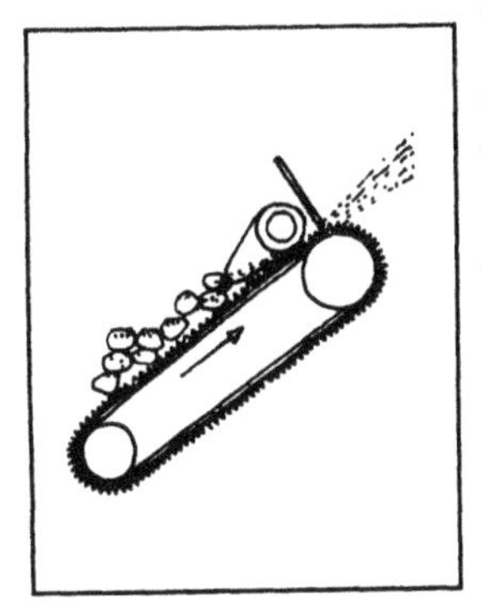

7. Toothed belt

Only machine with a hopper – highly efficient. 3hp motor: petrol (2.25kW). Output: 30m^3 per day. Weight: 100kg.

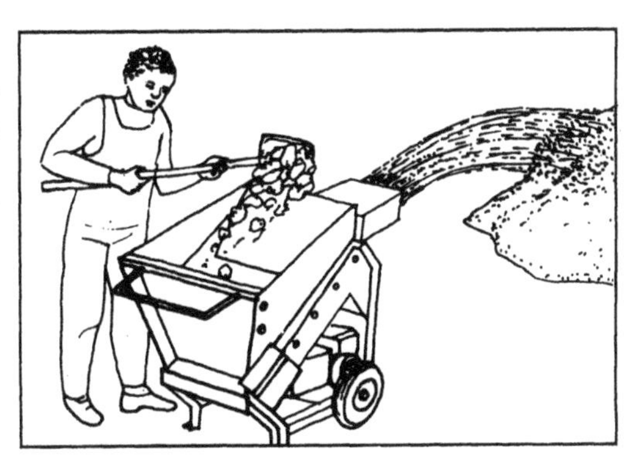

Screening

This operation is absolutely essential when: (i) removal of excessively large elements or organic matter is essential; (ii) after the structure of the soil has been corrected by an incomplete pulverization. In most cases grains with a diameter of between 10 and 20mm are passed – 10mm for presses sensitive to compression and between 20 and 25mm for those less sensitive to compression (hyper-compression).

1. Fixed screen

Set up either obliquely or suspended. The operation is manual and easy to carry out. Two basic operations: raw soil is thrown with a shovel against the sieve. The sieved soil is loaded into a wheelbarrow, unsieved material is rejected, or set aside for other use. Low yield: 1m^3 per hour per person.

2. Alternating screen

The simplest process consists of placing a frame sieve, on a pipe and a wheelbarrow. The sieve can also be suspended from a branch and set moving back and forth. A special manual tool has been designed along these lines by the Tallahassee school of Architecture, Florida, USA. It consists of a few planks, a cut-away barrel and chicken wire. The output obtained using the stationary sifting system is up to 2m^3/hour per worker.

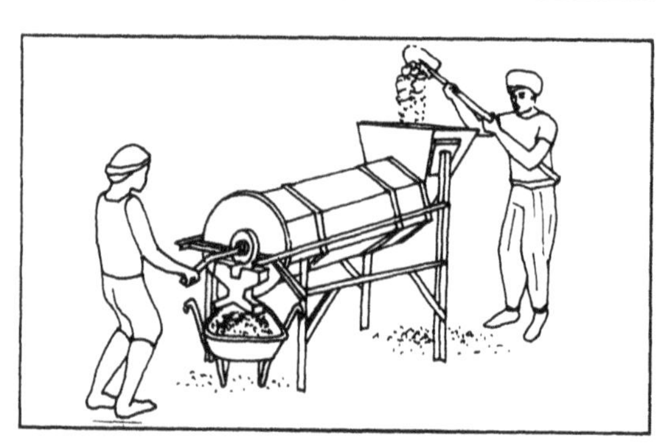

3. Rotary screen

A cylinder made of wire netting or metal is rotated either manually or mechanically. Its construction is very simple. It is possible to pass the soil through a number of stages in series and so separate into several fractions. Agricultural rotating sieves such as peanut sieves are suitable for the operation. Mechanical rotating sieves of all sizes ranging from 1 to 30hp are commercially available. Theoretically these sieves are capable of an output as high as 14m^3/hour.

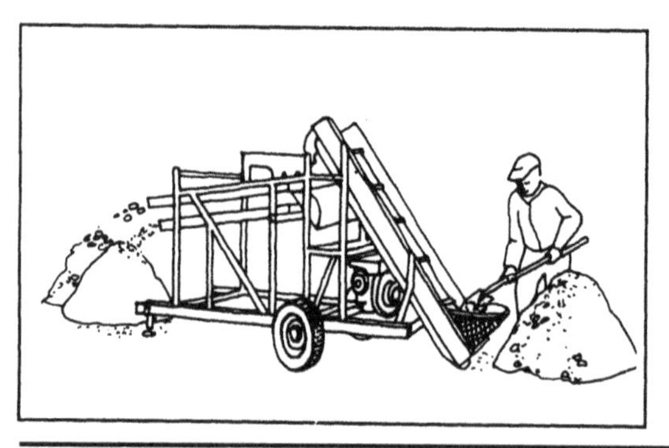

4. Vibrating screen

Either a single vibrating screen or a combination of several screens, usually superimposed, can be used in this process. This system offers the same advantage as the rotating sifter in that it makes it possible to separate the soil into several fractions, permitting its reconstitution. They are used in quarries. Vibrating screens of average size have outputs in the region of 5m^3 per hour.

Mixing

This is a particularly important stage. A uniform mixture is absolutely essential, regardless of whether stabilizer is used or not. Where manual labour is relied on, the heap of soil must be turned over at least four times. When a powerful mechanical mixer is available three or four minutes in the mixer is enough. It is important to mix the material dry first. Water should be added to the soil either with a sprinkler, or a mist sprayer, or by means of pressurized steam.

1. Manual mixing

Can be done by means of shovel, hoe, rake or any other simple tool. Output: 1–2m^3 per day per worker.

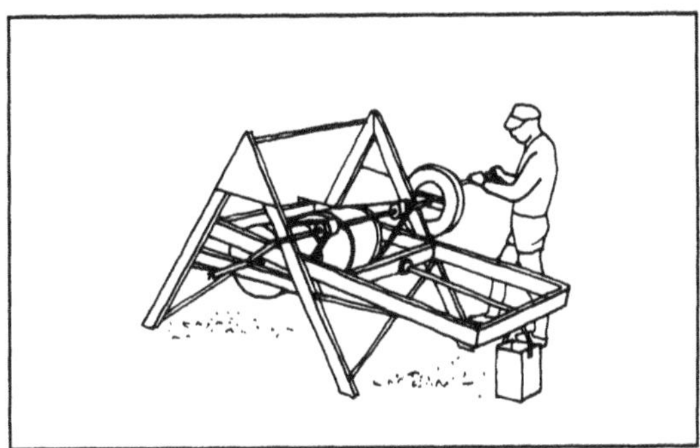

2. Manual mixer

Various systems have been devised which make use of a 200 litre oil barrel. The Tallahassee School of Architecture, Florida, USA is in the forefront of the development of such systems. Their output, at 1.5–2.5m^3 per day per worker, is slightly higher than that achieved using a shovel.

3. Motorized mixer

A motor can facilitate mixing, as this is a slow operation. Conventional concrete mixers, however, are not recommended because of the formation of lumps and crumbs in the soil.

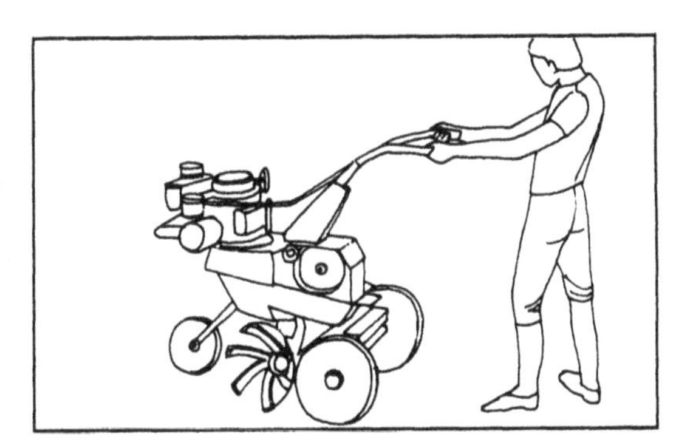

4. Motor cultivator

The motor cultivator is suitable for simultaneously crushing and mixing. This method requires plenty of space. The range of such machines available on the market is very broad with respect to size and power. Output: upwards of 4m^3/day.

5. Planet wheel mixer

This is the conventional mixer used for concrete extrusion. Small mixers are difficult to find (0.55kW). A 0.5hp (0.37kW) electric motor or a 0.75hp diesel motor is required to process ten litres of soil. A 180 litre planet wheel mixer has an output of 15m^3 per day.

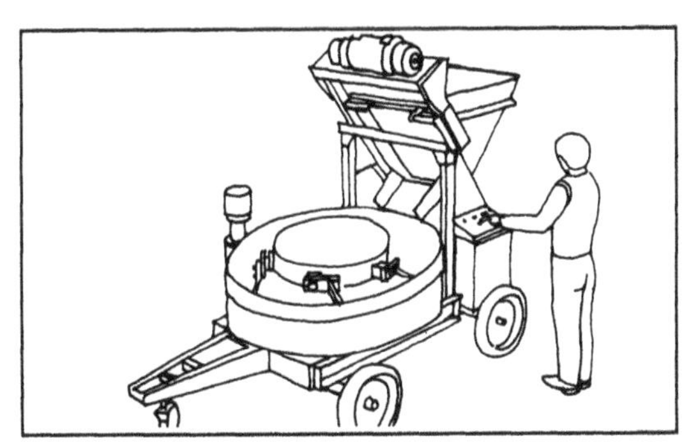

6. Paddle mixer

This is similar to a rendering mixer but sturdier. It works well with very dry soil but can break down if the soil is wet (12–15% moisture content). Required power: electric motors, 0.75hp (0.55kW) per 10 litres; diesel engines, 1hp per 10 litres (0.75kW). Output for a capacity of 150 litres installation: 8 to 10m^3/day.

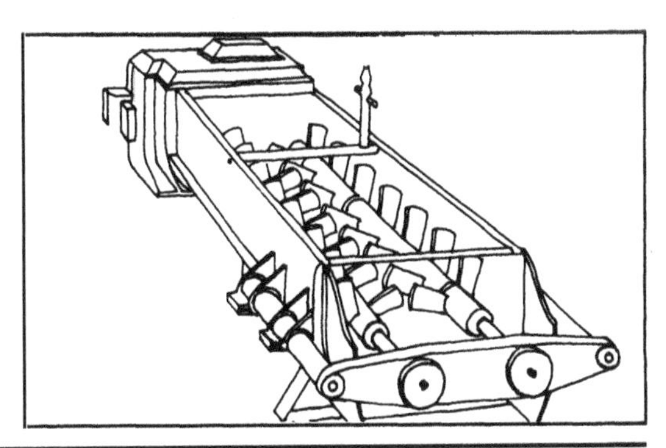

7. Linear mixer

A discontinuous helical screw shaft is fitted with either single or double blades. The shaft must be very sturdy. Extremely heavy and expensive, this type of plant is rarely used.

1. Basic problems

The mechanical strength of a soil is the result of friction between grains, which is in turn a function of its texture and structure. The friction attributable to texture, which depends on the grain size distribution, is a highly important factor, while structural contact, dependent on density, increases strength. In order to protect against water, a highly destructive agent, it is advisable to reduce the porosity of the soil by eliminating voids, which serve as channels for the water. Other reasons for compacting soil may be based in thermal considerations. This increase in density is achieved by compressing the soil with a press – a seemingly simple operation that involves several factors, which are discussed below.

2. Source of energy

Presses can be driven by human energy (the manual press) or animal energy (horse-powered equipment) or they can be powered by an electric motor or combustion engine or still yet by water (waterwheels, water turbines) or by wind. At present, only manual and motorized presses are available on the market.

3. Transmission of energy

Energy can be transmitted to the soil by means of levers, axles, connecting rods or swivels, pistons, and so on. There are three main categories of systems whereby energy is transmitted; mechanical, hydraulic and pneumatic.

4. Compression

Compression can act statically or dynamically. In the latter case it can act by impact or vibration. At present static pressure is still the more frequent. Dynamic impact compression is quite slow, and the mould is subjected to considerable stress. In addition, the thickness of the product is difficult to control and quite apart from the more serious problem of lamination. Dynamic vibration compression involves motorization and transmits a great deal of mechanical stress to the mould, not to mention its relatively high cost.

5. Effective force

This is the force available for compressing the earth which can be used at will, applied to a small area or a large one. The values put forward by press constructors thus do not convey an accurate idea of the performance of the press.

6. Moulding pressure

It is the pressure theoretically applied to the soil is expressed as the ratio between the effective force and the area over which this is applied.
Very low pressure: 1–2MPa
Low pressure: 2–4MPa
Average pressure: 4–6MPa
High pressure: 6–10MPa
Hyperpressure: 10–20MPa
Megapressure: 20–40MPa and higher.

7. Available pressure

This is the pressure that actually gets transmitted to the earth and it can be very different from the theoretical moulding pressure. The available pressure is dependent on the extent to which the plate has penetrated within the mould (or on the extent to which the space has been reduced). The available pressure differs according to the penetration position of the plate within the mould, where the space is diminished. The available pressure is different at each point of the compression stroke and depends on the mode of energy transmission. Each press gives rise to a set of characteristic curves, and in the case of manual presses, these curves vary with the weight of the operator; for hydraulic presses this curve is nearly parallel to the x axis.

8. Coefficient of dynamic effect

There is an inertial or dynamic effect which comes between the static measurement of the available pressure and the actual operation of a press, which profits from this momentum. A manual lever press thus has a coefficient of dynamic effect of 1.2 which raises the effective available pressure.

9. Effective pressure

This is the pressure that actually gets transmitted to the earth at the end of a compression cycle. It is the product of all the effects of the operation of the press, i.e. friction and inertia. Thus the effective pressure exerted by a small lever press (Cinva Ram) under ideal conditions is between 1.5 and 2MPa, whereas the literature speaks of pressures as high as 4.5MPa, giving a totally incorrect impression.

10. Absorbed pressure

Soil, while it is being compressed, absorbs the available pressure and becomes increasingly difficult to compress. Internal friction and the friction due to the surfaces of the mould increase. The absorbed pressure thus varies as a function of the reduction of space and the quality of the soil. The absorbed pressure curve for gravel soil is higher than that for fine soil. For soil to be compressed the available pressure must be higher than the absorbed pressure, otherwise the compression cycle will come to a stop.

11. Required pressure

The potential of a press is one thing, the quality of the soil another and the user's budget a third. The quality of the block increases linearly with increasing compressing but reaches a limit often between 4 and 10MPa where it becomes asymptotic or in some cases begins to fall off. Excessive moulding pressure can in fact have disastrous effects (layer forming, for example).

12. Compression ratio

Uncompacted soil has prior to moulding a density of between 1000 and 1400kg/m^3. Compressed it should have a minimum density of 1700kg/m^3 or in the region of the Proctor value. The compression ratio is the ratio of the height of the mould prior to compression to that after compression necessary to achieve the desired density. It has a value of 1.65 for manual lever presses but should ideally be equal to 2. This ratio is rarely achieved with mechanical systems, but manual or mechanical precompaction can remedy this drawback.

13. Compression gradient

Owing to the increase in the internal friction of the earth and the pressure on the surfaces of the mould, the material closer to the pressure plate is more thoroughly compressed than that near the other side. This variability in quality depends on the thickness of the final product. In order to obtain quality blocks, the height must be limited to 9 – 10cm at low pressures and 20 – 25cm under hyper-compression.

14. Method of compression

With simple static pressure the moulded thickness of good quality blocks should not be more than 10cm. A double compression exerted simultaneously on the two sides makes it possible to produce blocks 20 – 25cm thick. The least compacted earth is found in the middle of the block. This part is subjected to the least stress.

15. Speed of compression

Production requirements sometimes dictate high-speed operation. However, there is a limit of 1 to 2 seconds on the speed of compression, below which lamination may occur.

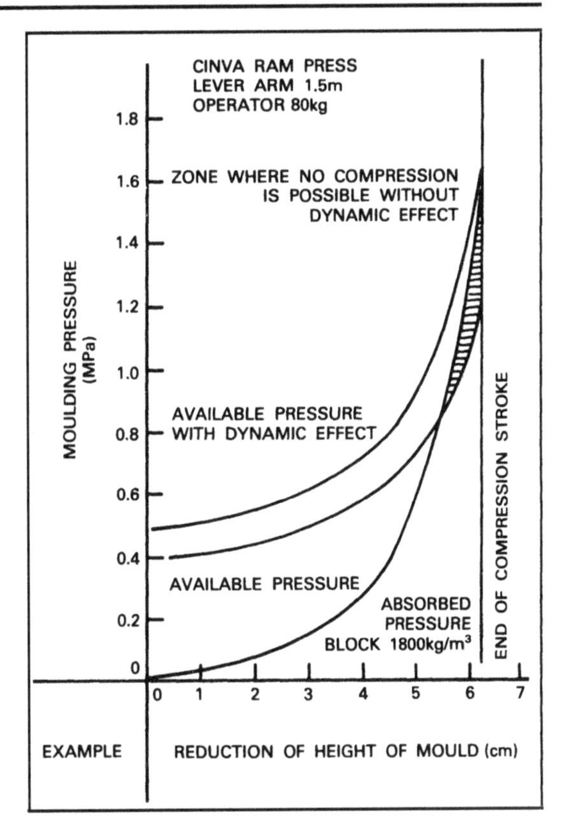

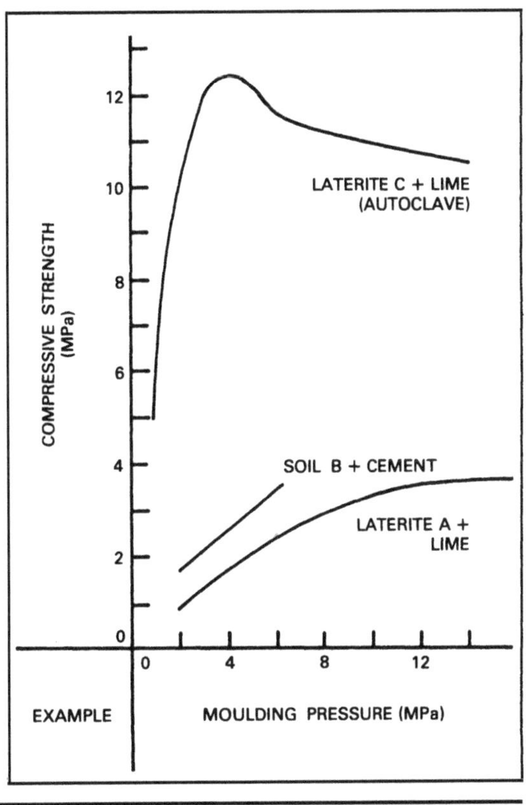

History and development

All things considered, the use of presses is a fairly recent phenomenon. Formerly, compressed earth blocks were made by means of manual tamping in a mould. This technique was derived from experience with rammed earth. In the eighteenth century, the Frenchman François Cointeraux designed the first press, called the 'Crécise' which was based on the wine press. At the time he prided himself on having invented a 'new pisé'. Manual ramming is still practised, sometimes even on a large scale. A recent example is the construction of the agricultural village of Maadher in Algeria.

It was only at the beginning of the century that the pressing of earth blocks was mechanized. This was first done with manual presses that had heavy covers (30kg). These were closed down with a great force due to the excess soil. These devices were even motorized. Then presses provided with a mechanical rammer appeared which were powered manually. This type of press is still to be found on the market, which is open to new ideas although none of these has met with any success as yet owing to the drawbacks inherent in the system. Since the beginning of the century press makers have been devising presses that make use of static force. The process was used to make burnt bricks and was more or less successfully adapted to the production of compressed blocks. The developers of the process, however, failed to make corresponding changes in the compression ratio and available pressure. Only in 1956 did the first press, specifically designed to produce compressed blocks, make its way onto the market. This was the Cinva-Ram press and it was invented by Raul Ramirez, an engineer at the Cinva Centre (Centro Interamericano de Vivienda y Planeamiento) in Bogota, Colombia. It has swept the international market. Its most important advantages are its mechanical simplicity, manual operation and lightness. Subsequently many other presses have emerged as competitors to Cinva-Ram. The 1960s saw the production of mechanical, automatic and hydraulic presses.

It was only in the seventies and the beginning of the eighties that a new wave of presses appeared, with new ideas now appearing almost every month. Some of the presses coming onto the scene artfully combine the principles of dynamic compaction by vibration and impact compaction, using in part road-building techniques. Vibrating presses were already used in the fifties in certain construction projects, as for example, in the Sudan and Burundi. They were then eclipsed only to re-emerge in France in the eighties.

Overview of the market

Nowadays, new press-makers favour hydraulic presses. Manufacturers with field and site experience favour the development of small sturdy mechanical presses that are simple, yet reliable and efficient. The difficulty of finding where to buy a press which existed only ten and even five years ago is a thing of the past. Nowadays the choice of pressing plant is very wide and reflects all the various technological trends at prices ranging from the very low to the very high. Unfortunately, these machines are not all without their short-comings, and the buyer is well advised to use a certain amount of circumspection in making a choice and to carry out preliminary test runs. Even so we may say that while the press is an essential tool, the selection of the soil is even more important. It is preferable to press a good earth with a mediocre press than to press a poor quality earth with a good press. This remark is of cardinal importance, as nowadays sophisticated hypercompression presses are available that add nothing in terms of the quality and performance of the equipment if the soil used is mediocre. The following trends should be taken into account in evaluating the modern market:

- to an ever increasing extent manual presses are nowadays mass-produced in small factories in the developing countries;
- motorized presses have just as great a market share in the developing countries as in the industrialized countries;
- in the industrialized countries the new owner-builder market and demand from small building contractors have stimulated the development of small on-site production units;
- there is no certainty that the promotion of large on-site production units would be economically feasible and such promotion would thus constitute an enormous risk.
- industrial production units have a certain market share in the industrialized countries and in some of the more advanced developing countries.
- 'turnkey' plants present enormous risks and do not appear to be any more feasible in the majority of industrial countries than they are in developing countries apart from the rare exception.

SYSTEMS				CLASSIFICATION OF PRESS TYPES		CHARACTERISTICS			
POWER SOURCE	SIZE	POWER TRANSMISSION	COMPRESSION ACTION			MOULDING PRESSURE	WEIGHT (kg)	DAILY OUTPUT (29.5/14/9)	PRICE RANGE (US$)
MANUAL	LIGHT	MECHANIC	STATIC	1	MANUAL PRESSES	VERY LOW	50 to 100	300 to 800	750 to 3 200
MANUAL	LIGHT	HYDRAULIC	STATIC	2	MANUAL PRESSES	HYPER	30 to 150	300 to 400	3 200 to 6 400
MANUAL	HEAVY	MECHANIC	STATIC	3	MANUAL PRESSES	LOW	200 to 500	400 to 1 000	1 800 to 3 600
MOTOR	LIGHT	MECHANIC	STATIC	4	POWERED PRESSES	LOW to MEDIUM	400 to 1 500	800 to 3 000	12 700 to 25 500
MOTOR	LIGHT	HYDRAULIC	STATIC	5	POWERED PRESSES	LOW to MEDIUM	400 to 1 500	800 to 2 000	12 700 to 82 000
MOTOR	LIGHT	MECHANIC	STATIC	6	MOBILE PRODUCTION UNITS	LOW to MEDIUM	1 500 to 2 000	800 to 3 000	12 700 to 25 500
MOTOR	LIGHT	HYDRAULIC	STATIC	7	MOBILE PRODUCTION UNITS	LOW to MEDIUM	2 000 to 4 000	800 to 3 000	32 000 to 109 000
MOTOR	HEAVY	MECHANIC	STATIC	8	MOBILE PRODUCTION UNITS	LOW	4 000 to 6 000	2 000 to 15 000	64 000 to 109 000
MOTOR	HEAVY	HYDRAULIC and MECHANIC	STATIC or DYNAMIC	9	MOBILE PRODUCTION UNITS	LOW to HYPER	4 000 to 6 000	1 500 to 7 500	82 000 to 182 000
MOTOR	HEAVY	HYDRAULIC	STATIC	10	INDUSTRIAL PRODUCTION	LOW to HYPER	2 000 to 30 000	3 000 to 50 000	109 000 to 2 700 000
MOTOR	HEAVY	HYDRAULIC and MECHANIC	DYNAMIC	11	INDUSTRIAL PRODUCTION	LOW	6 000 to 30 000	10 000 to 50 000	182 000 to 2 700 000

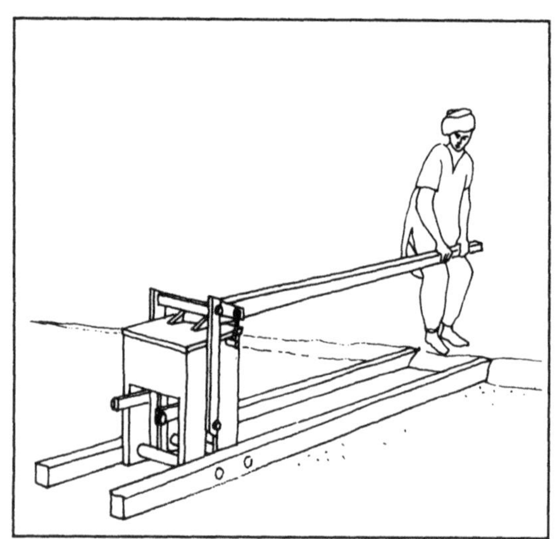

Light presses

Type 1: Mechanical presses

The advantages of Cinva-Ram type presses are obvious; they are light, have a certain sturdiness, low cost, and are simple to produce and repair. Their main disadvantages are as follows: they wear out prematurely (coupling rings), have only a single moulding module, can exert only low pressure and have a low output. Nevertheless, they are one of the best presses of their type on the market, and it is usually the copies which wear out prematurely. The skill that goes into the production of the Cinva-Ram is not always so well understood by its imitators. Nevertheless, this press could be improved. The following are some of the improvements designers have come up with: joining the cover to the lever (Tek-Block); better ejection (Stevin, Ceneema), greater moulding depth (Ait Ourir); better transmission of energy (Dart-Ram); fold-down cover (Meili); standard steel profile (Unata); dual compaction action (C + BI); compartmentalized mould (MRCI); production of perforated blocks (Cetaram). These technical improvements also aim at refining the production process which, in the last analysis, is relatively independent of the mechanical cycle of the press. Production is in fact determined to a greater extent by the mode of organization of the work, the mode of payment of the crews and the prevailing working traditions. So it is that the average output of a Cinva-Ram or similar press is 300 blocks/day although this could be increased to 1200 blocks/day. These presses are now produced in a number of countries including the USA, France, Switzerland, Belgium, Cameroon, Zambia, Tanzania, Columbia, New Zealand, Burkina Faso, Morocco, among others.

Type 2: Hydraulic presses

A small press, the Brepak, makes a major improvement to the Cinva-Ram. It was created by the BRE (England) and is marketed by Multi-Bloc. The swivel and rod system of the Cinva is replaced with a hydraulic piston which enables it to achieve pressures of 10MPa. The resulting blocks have identical dimensions to those made using the Cinva but are approximately 20% denser. The hyper-compression means that it is suitable for compacting highly expansive soils such as the black cotton soils.

Heavy presses

Type 3: Mechanical presses

These can produce pressures greater than the minimum threshold of 2MPa. These presses, being sturdy, do not wear out easily. The presses are easy to manipulate and take care of. They have interchangeable moulds. The fold-down cover of these machines allows precompaction. The back and forth motion from one side of the press to the other is eliminated. The design of the machine permits better organization of the work that is carried out around the press. On the other hand, these presses are heavy and are up to seven times as expensive as a Cinva-Ram or similar although the final price of the block is virtually the same. They are reliable and profitable. The Terstaram or Ceraman, which was adapted from the brick industry, was at first marketed under the tradename Stabibloc, S.M., etc. and Landcrete in Southern Africa. It is nowadays manufactured in Belgium and in Senegal, and before long will be manufactured in other African countries. Not all models are of the same quality. Speculation has affected sales, models often going for six to ten times their original price.

The Ellson Blockmaster is manufactured in India. There are other models of the same type exist, for example the CRATerre press, perfected and produced exclusively in Peru, and the Saturnia, designed by the ETH in Switzerland, while the Yuya, developed by Trueba in Mexico is suitable for the production of interlocking blocks.

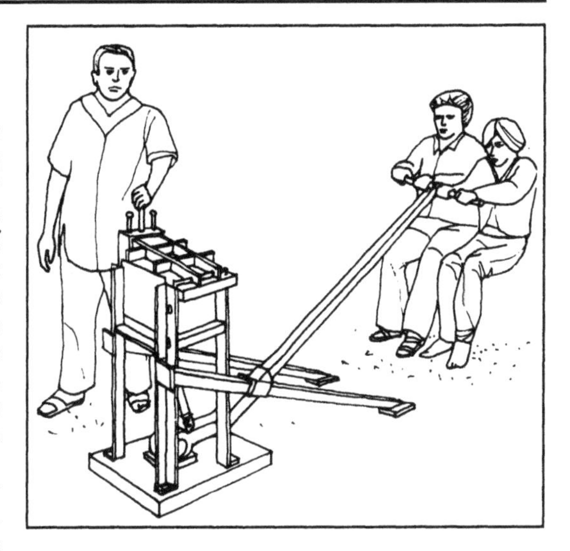

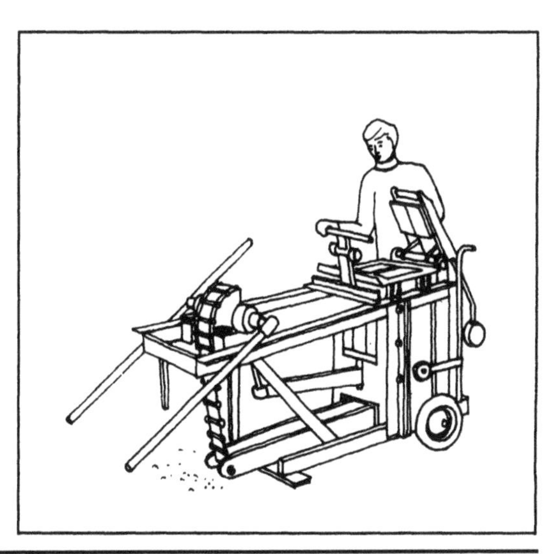

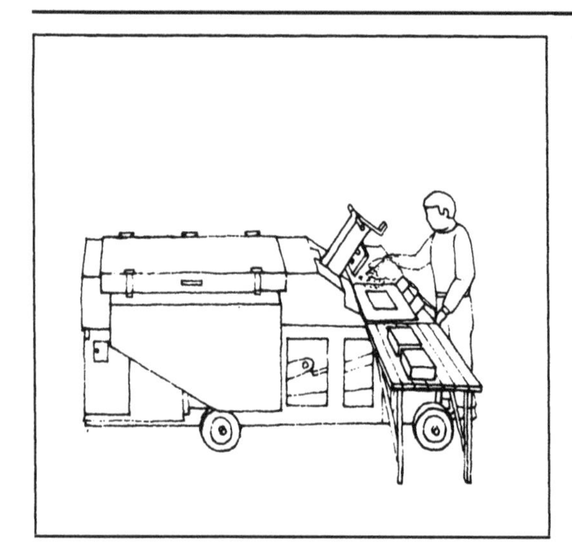

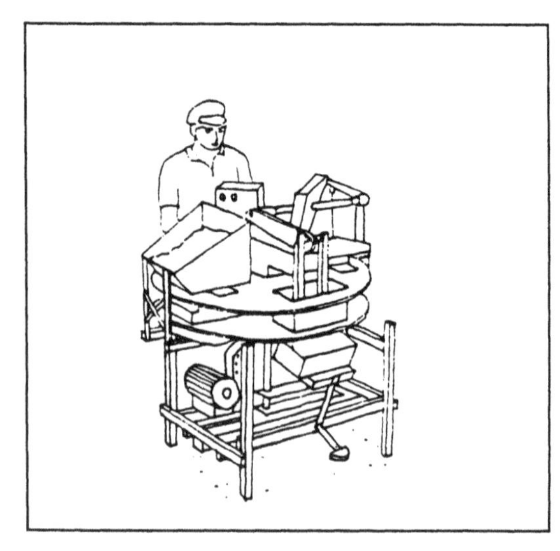

Type 4: Mechanical presses

These represent a new generation of presses which are currently available on the market and which appear destined for a bright future. Despite their cost, which is of the order of four to seven times that of heavy manual presses, their economic viability is excellent. Some of these presses, such as the Semi-Terstamatic, are direct descendants of the heavy manual presses and have benefited from the lessons learned on the older types. The Semi-Terstamatic was at one time on the market under the tradenames Majo and LP9 (Landcrete). Motorized mechanical presses belong to one of two groups: those having a fixed table and single mould, simple and sturdy, and those having a rotating plate and multiple mould (three or four), which under certain conditions raises the production rate. In the first case, the mould can be changed rapidly and cheaply, whereas with a rotating plate changing the mould takes more time and is costlier. The tables can be turned by hand (Pact 500) – a tiresome operation – or mechanically. The latter system requires a more sophisticated mechanism and more energy (Ceraman). Dynamic precompaction effected by lowering the cover becomes possible with the systems using a single mould and this confers significant advantages. By adjusting the tapered precompacting roller located between the feeding position and the compacting position, precompaction with rotating table presses becomes possible. The level of the earth should be slightly above the sides of the mould and this is only possible when the press has a feed hopper. The designers of this type of press have encountered major problems which were still not been resolved when the presses were brought to market: the soil may under no circumstances disturb the functioning of the machine by getting into sensitive areas; the safe operation of the machine must be assured, lest it be damaged; the press must not be allowed to operate in reverse, which would happen if the electric motor were installed backwards; where the available pressure is less than the required pressure (when there is too much earth in the mould, for example), the press will block; the removal of a half-compacted brick should slow production. Accordingly, the press ought to be provided with a compensating spring and a motor-release system.

Finally, these presses should be designed to give the user the choice of an electric motor, a combustion engine or another type of motor. These presses are very largely dependent on the up-stream production operations of screening, proportioning and mixing.

Type 5: Hydraulic presses

These are stand-alone presses capable of medium output and have a fairly high price. Hydraulic presses had a certain vogue in the 1950s but rapidly disappeared from the market (the Winget, for example, of which 125 were sold). New presses of the same type were launched in the seventies, but their reliability is disputed and they have brought as much disappointment as satisfaction. Nevertheless, hydraulic systems have the advantage, owing to the functioning of the piston and their compactness, of permitting a long stroke. It follows that compression ratios equal to, or greater than, 2 can be achieved. These systems can be easily adjusted to match the composition of the soil. They can also be provided with a hopper – the first step towards automation. Furthermore double compaction can easily be carried out with a hydraulic press as is the case with the Tob System press. It is also true, however, that the hydraulic press gives rise to several problems all of its own, such as a delicate hydraulic pump. Apart from this if the rotating plate is hydraulically driven as well, the oil reservoir should have a volume of at least 200 litres. Despite such a large quantity of oil, the temperature of the fluid can quickly rise above 70°C in tropical climates. This is the maximum permissible temperature, if all the hydraulic components are to function properly, apart from those that can tolerate a temperature of 120°C, but which are difficult to replace if they break down; the alternative is an oil-cooling system which makes the plant complex. The oil must be changed and is not always available. These presses may function well in the right circumstances, for example in a technologically advanced environment, but they often perform poorly in rural surroundings or even on the outskirts of cities in developing countries. Many models of this type of press have been built and the market sees the steady appearance and disappearance of models. Rarely have they been known to be reliable. They are manufactured in various countries including Belgium, France, the USA, and Brazil.

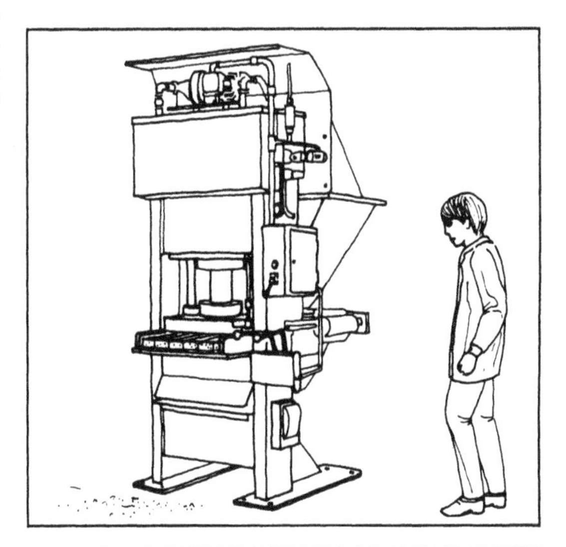

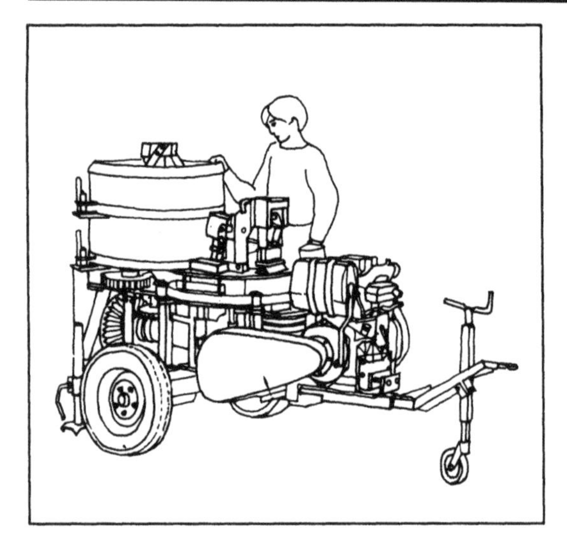

Power presses often necessitate a major mechanization of the upstream process. Accordingly, designers' research has proceeded along the lines of integrating all the plant equipment used in self-contained production units, which reflect fairly accurately current production trends. However, although the cost can be reasonable, the economic viability of self-contained production units continues to pose a problem. They do not all operate in the same way and all conditions must be optimal. Even in the industrialized countries, these machines operate in very tight economic conditions. In developing countries they are often uneconomic.

Light units

These offer the advantage of opening up a totally new market, in industrialized countries and in the urban areas of developing countries, namely leasing to do-it-yourself builders. Indeed these machines can be rented for the entire block production period at a relatively low price. Even so this type of machine still suffers from a few defects, mainly because of a lack of integration between the different types of equipment that have been brought together. An effort should be made to harmonize the outputs and cost of these machines to the different stages of production they integrate.

Type 6: Mechanical presses

At the present moment the Meili unit, which is manufactured in Switzerland, is the only example on the market of this type of press. There is not a very wide range of this class of unit and so far there is no totally integrated unit on the market. The pulverizer still suffers from defects.

Type 7: Hydraulic presses

The Earth Ram (USA), the Clu 2000 and Clu 3000 are only a few of the many examples of this type of unit. These machines are sometimes adapted from standing units. The principle of the design is attractive, but calculations of the cost show that, on large construction sites, it is more economic to purchase the production materials (pulverizer, mixer, press) separately. Unintegrated plant is not less efficient and it is not clear that integrated equipment is more convenient.

Heavy units

Some of the larger manufacturers have proposed entirely mobile units which can be taken everywhere, but which are very large and heavy. The corresponding annual production capacities are very high. The plant corresponds broadly to type 7, and at present there is a tendency towards the use of hypercompression. Only a few units of this type have been manufactured to date. The economic feasibility of these presses has yet to be demonstrated, and a thorough survey of the market should be undertaken before acquiring them.

Type 8: Mechanical presses

At present only a single unit of this type is known. Its design is based on the concept of a combination of existing units all mounted on a single chassis and goes by the name of 'Unipress'. This plant is ordinarily used for the production of burnt bricks, and attempts in Egypt to adapt it to compressed blocks have met with some major but not at all insurmountable obstacles. The plant is very sturdy.

Type 9: Hydraulic presses

These are represented as being all-purpose machines, but the models actually to be found on the market – the Al Niblack (USA) and Teroc (France) – have, all things considered, a fairly limited range of application. These units are not equipped with pulverizers or screens. The earth, which is deposited in a hopper, is premixed by gravity with a stabilizer by means of an integrated proportioning system and then moved by conveyor belt to the mixer where dry and wet mixing are carried out. A storage hopper distributes the earth within the mould where it is hypercompressed and then automatically ejected in the form of blocks. These units dispose of a system of slide moulds which cannot be used to produce hollow or cellular blocks. Being costly and having only a moderate output, their future appears to be confined to limited markets.

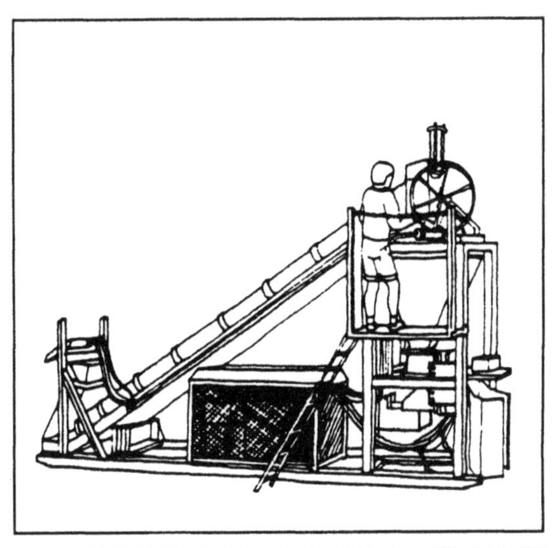

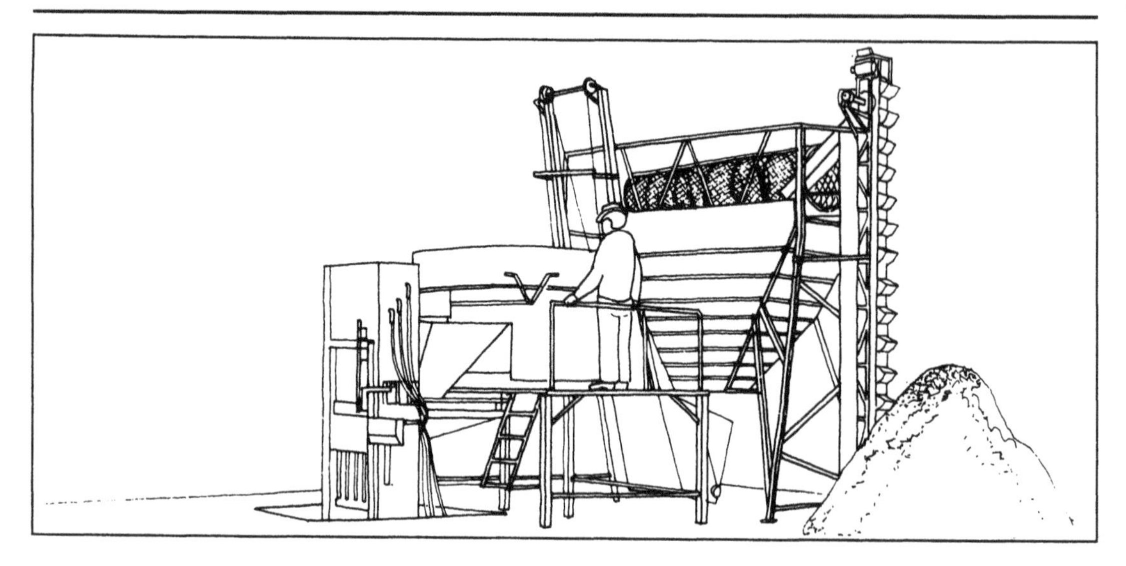

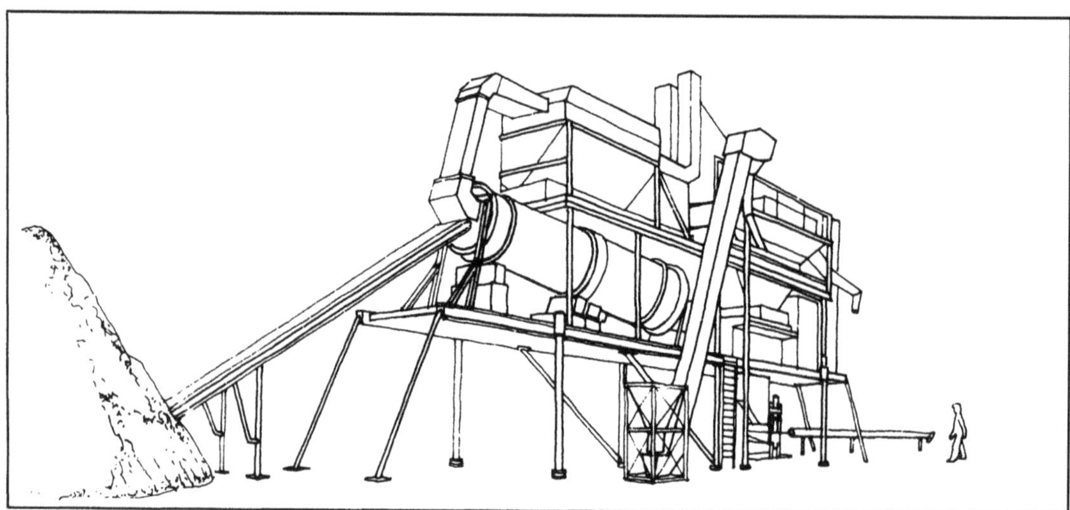

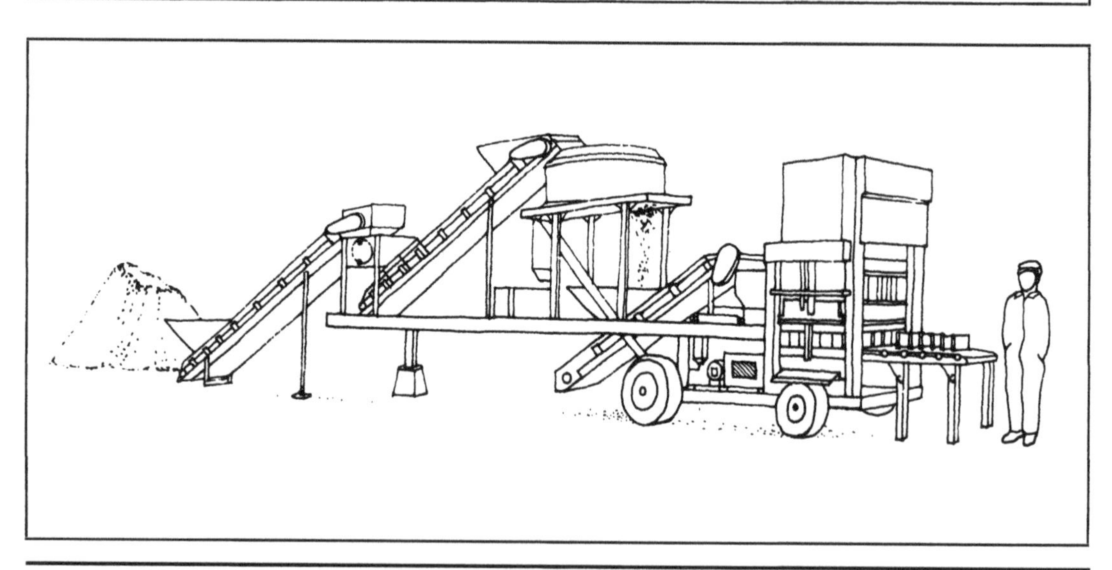

For several years now the market has seen the arrival of a whole range of fully equipped standing industrial production units of a limited size. These industrial units operate on the single or double static compression or dynamic compression principles. The list of products that can be manufactured by them is not limited to the small blocks which almost all the other presses can turn out. The list thus includes all the forms of concrete blocks and burned bricks which can also be made using stabilized earth, including hollow blocks and perforated bricks. This manufacturing equipment is as yet only intended for a limited market. Only massive construction programmes can ensure that the investments involved can be recovered and the cost of producing the blocks significantly reduced. This type of press is currently used in countries such as Brazil, Mexico, Algeria, Gabon, and Nigeria.

Type 10: Hydraulic presses

Wholly automatic hydraulic industrial units come in several sizes. At the small end of the range, units such as the Ceramaster (manufactured in Belgium) are fairly compact. This type is still in the prototype stage. Units of average size like the Luxor, formerly the Tecmor (Brazil) have been reconditioned and can handle production technology with ease. They use double compaction. Finally, the heaviest hydraulic industrial presses are veritable turnkey plants and are available in several sizes. Today it appears that only a few operational plants exist in all the world. Considerable secrecy surrounds the operation of these industrial units. The Latorex (Denmark) and Krupp (Germany) type units make use of a stabilization process based respectively on hydrated lime and quick lime. In both cases the technology has much in common with that of the lime-silica bricks industry from which it has been adapted. The pressures applied are in the hyperpressure and megapressure ranges. The blocks are dried in an autoclave and operation is entirely automatic. In Nigeria and the Philippines, major problems have been encountered with this type of heavy hydraulic industrial plant. The technology that has been developed is very sophisticated indeed, and requires faultless technical control and supervision of the organization of the work.

Type 11: Combined hydraulic and mechanical presses

These are equivalent to a true totally automated factory. A prototype of one is currently operational in France in particular, in the construction of the experimental 'Village Terre' on l'Isle d'Abeau near Lyon. These presses have been adapted from concrete block presses. The process combines mechanical vibration and hydraulic compaction technology. By this is meant vibration at a high frequency, low amplitude (1.5–2mm) and hydraulic compression at low pressure 0.2–1MPa. The frequency of vibration and the pressure of compaction can be regulated to match the soil. The production cycle consists of the following operations: filling the mould from a drawer, vibration, lowering of the plunger, raising the mould, removal of the material from the mould with withdrawal of the plunger, removal of the fresh product on a conveyor. The duration of the entire cycle is of the order of 40 seconds. This type of press can turn out solid and hollow blocks (20 × 20 × 40 or 50) at the rate of 1000–1500 per day or 2000–2500 solid blocks per day. This represents a decrease of approximately 50% with respect to concrete blocks. The smooth operation of these presses requires, moreover, an adequate technological environment and well-trained and experienced operators and maintenance personnel. A tendency towards a reduction in size is currently under way. There are also presses on the market which are not based on vibrated compaction, but on hammering. However, their output is problematic.

Construction in earth is at present becoming increasingly industrialized. This trend is reflected as much by the techniques that have been adopted as by the degree of mechanization developed. The same may be said of the very long list of commercial building products and components, which is just as long as that of the ceramic and concrete block industries. This tendency towards the industrialization of soil is fairly recent, having begun in the 1960s. Prefabricating plants, delivered as turnkey units, possess a degree of sophistication which is every bit as advanced as that of the most advanced plants, producing other building materials.

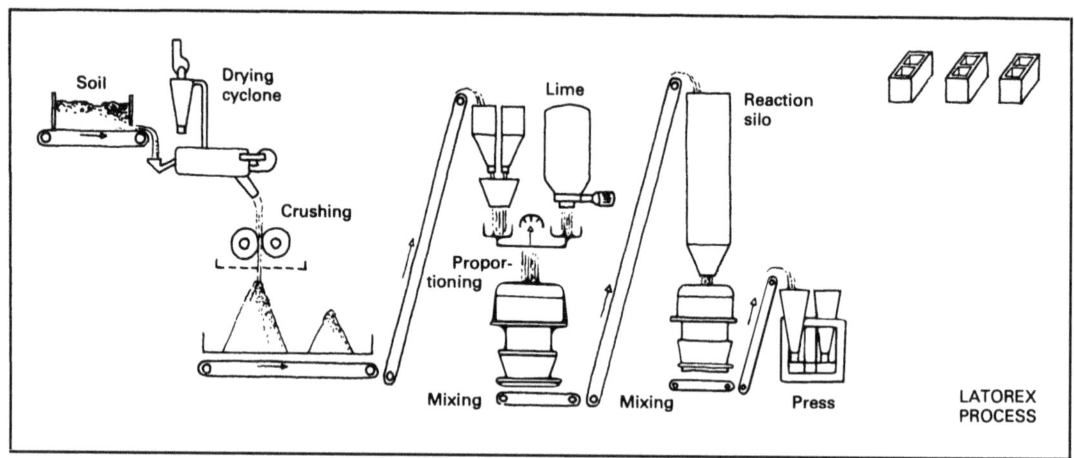

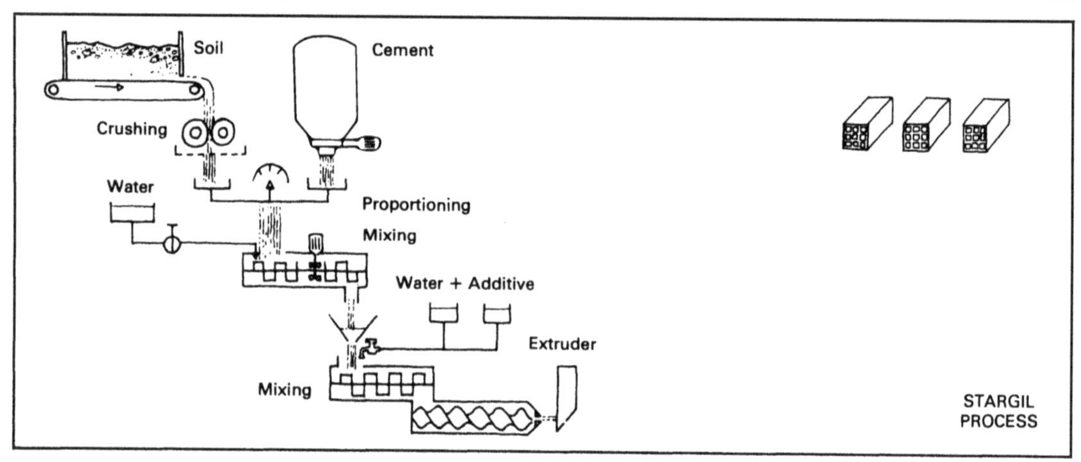

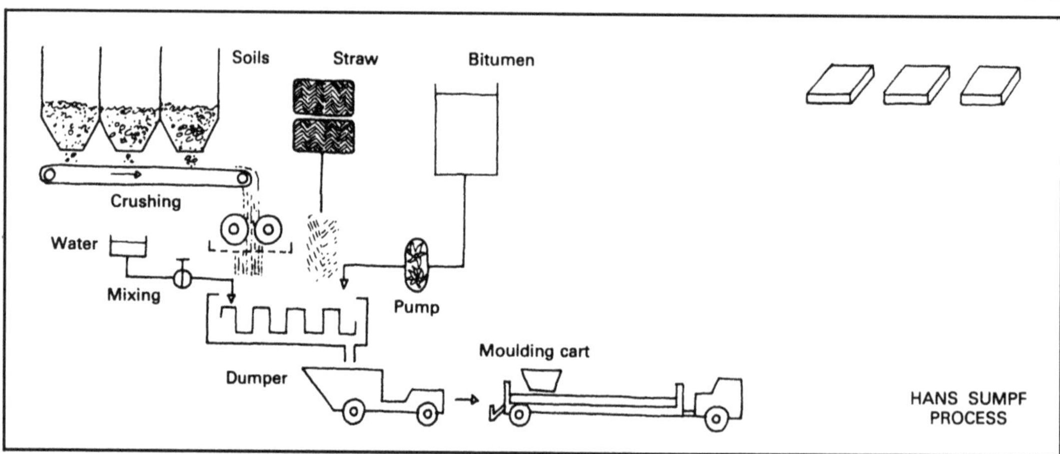

The main products made of industrial soil available on the current market include, on the one hand, compressed earth blocks, solid and hollow and in all shapes, and on the other hand, extruded bricks and 'asphadobe' or 'bitudobe'. These products are designed for use as the usual components of the building, and in the walls and floor, as well as for the finishing work such as the dressing of walls, decoration, floor coverings, and for surfaces. Extensive studies on manufacturing equipment and stabilization point to the further proliferation of applications as the creation of a true 'Soil' industry draws closer.

Compressed blocks

The most representative plants of this type of product are those of the Latorex and Krupp groups. The manufacturing process has been adopted from the lime-silica brick industry. The basic raw material used is laterite and the stabilizer is lime. These plants aim at the optimization of each stage of production; i.e. drying the raw material, proportioning, mixing, the reaction with the stabilizer, pulverization, compression, curing in an autoclave, and final drying and palletization.

ORIGIN	DENMARK
GROUP	LATOREX
INVESTMENT	US$800,000
PRODUCTION	5 tonne/h

Extruded bricks

The manufacture of hollow and perforated components in extruded earth has already become established on an operational basis. The processes were primarily developed in France at the INSA in Rennes (Stargil), and at the CTTB (Simarex). The extrusion plant disposes of all the facilities found in the a modern brickworks except for the kiln and artificial drying rooms. The earth is mixed with cement to form a base paste to which plasticizers (molasses) are added. This paste is drawn off and extruded and transformed into perforated bricks. By not burning, a 40–65% energy saving is achieved with respect to burned bricks.

ORIGIN	FRANCE
GROUP	CHAFFOTEAU & MAURY
INVESTMENT	US$2,500,000
PRODUCTION	7 tonne/h

Asphadobe

The manufacture of adobe bricks stabilized with bitumen has become a highly mechanized process, especially in the United States. The Hans Sumpf plant in Fresno, California is the best known of its type and has been operating for decades. The block laying machine can produce from 10,000 to 20,000 blocks per day. The upstream production is fully automated. This type of production unit is proliferating and has even been exported (Sudan).

ORIGIN	USA
GROUP	HANS SUMPF CORPORATION
INVESTMENT	US$750,000
PRODUCTION	20 tonne/h

One notes a general tendency to oversimplify matters in various discussions of the subject of earth construction. According to this simplistic view, it is a matter of excavating the earth, compressing or moulding it and that is all. Such an account, which might satisfy the uninitiated builder must not, however, be allowed to deceive him, and it must be rejected when it comes to realizing large-scale

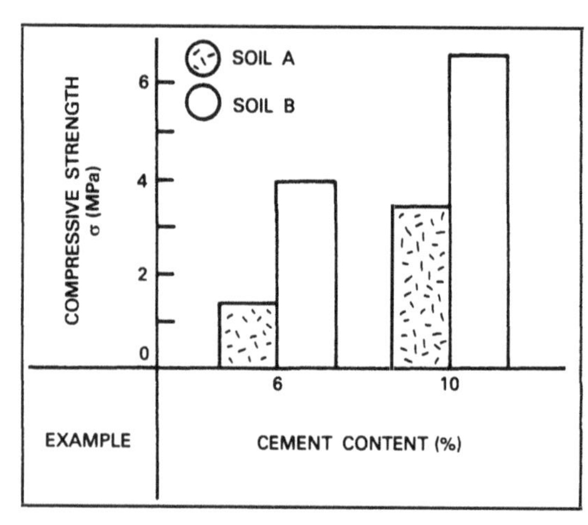

1. Choice of soil

The quality of product achieved with correctly selected or even improved soils is far superior to that possible with a mediocre soil. Similarly, stabilization is optimized when a well-graded soil is used, which implies that significantly lower percentages of stabilizer are required than is the case for a poorly graded soil. The stabilization of a mediocre soil should be regarded neither as an ideal solution nor a miracle. Getting a good soil does not always imply the need of using a distant borrow. It does, however, imply the ability to recognize the best soil locally available.

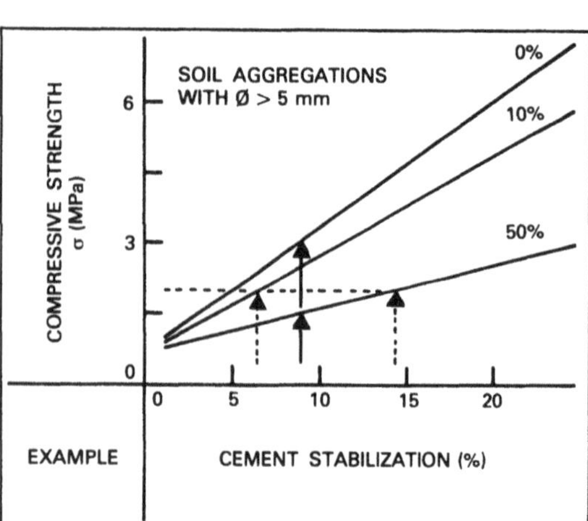

2. Pulverization

Given that it optimizes the proportioning of the stabilizer and thus facilitates production and renders the product economic, the pulverization operation is essential to the manufacture of stabilized compressed earth blocks.

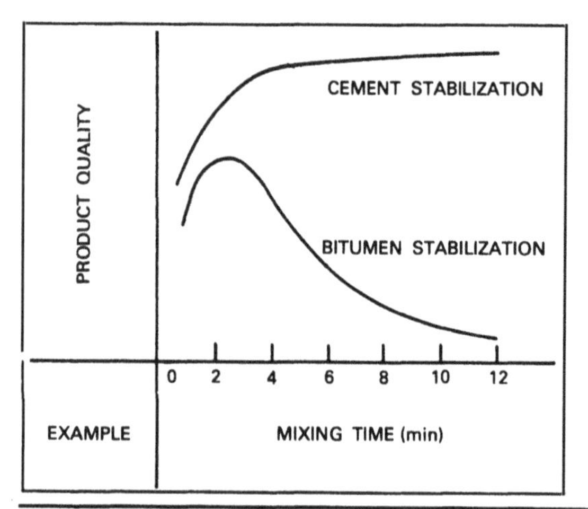

3. Mixing time

Depending on the equipment or the mixing technique used, the duration of mixing operations varies considerably. It has been demonstrated time and again that it is of cardinal importance to respect a minimum mixing time. For example, the minimum mixing time for cement-stabilized blocks is between three and four minutes with a risk of losing 20% of the effectiveness of the stabilization if it is less. With regard to asphalt-stabilized adobe, there is an optimal mixing time above or below which the stabilizer becomes considerably less effective.

projects. It should be pointed out that production parameters have a considerable effect on the quality of the product, output and economy of the operation. Optimizing production can be decisive for the economic viability and acceptability of the product. It does not necessarily follow that the answers lie with mechanized, sophisticated techniques or intense investment. Far more important is know-how, which is not always the result of experience, but rather of training.

4. Hold-back time

The elapsed time between mixing and moulding can be very important. For example, in the case of cement-stabilized concrete blocks, if the hold-back time is not reduced to a minimum, there is a danger that the cement may set prematurely, resulting in the formation of aggregates and an undesirable effect on the mechanical resistance of the blocks. A delay of one to two hours can result in the loss of half the quality of the product. In contrast, in the case of lime stabilization, this longer hold-back time, given the slow reaction of the lime with the air, enhances the quality of the blocks. The hold-back time plays a role in the organization of the production process.

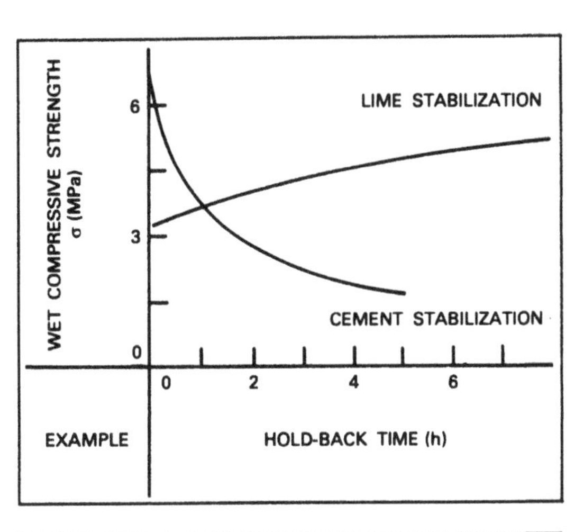

5. Moulding method

The ultimate deformation strength of the material (approaching 5%) depends in large measure on the method of moulding and compaction. In certain cases moulding in a press can result in a five-fold increase in the deformation strength with respect to manual moulding using a plunger. In other cases it has been found that moulding by kneading gives superior results. It is important, then, to choose a moulding technique with the right criteria in mind.

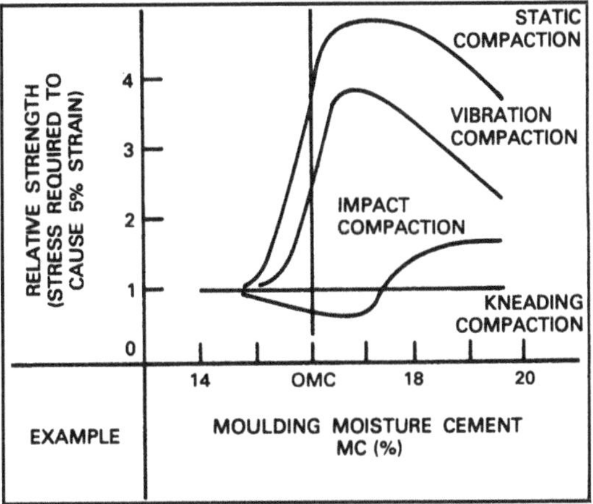

6. Drying method

If drying conditions are poor, it is well known that blocks with low cement stabilization suffer a dramatic loss of quality. If the blocks are intensely stabilized, almost two-thirds of the effective resistance can be conserved. On the other hand, about the same quality using half the cement can be expected under optimal drying conditions. These conditions play an even greater role when stabilization is carried out using lime. Too many working brickyards pay insufficient attention to the drying stage.

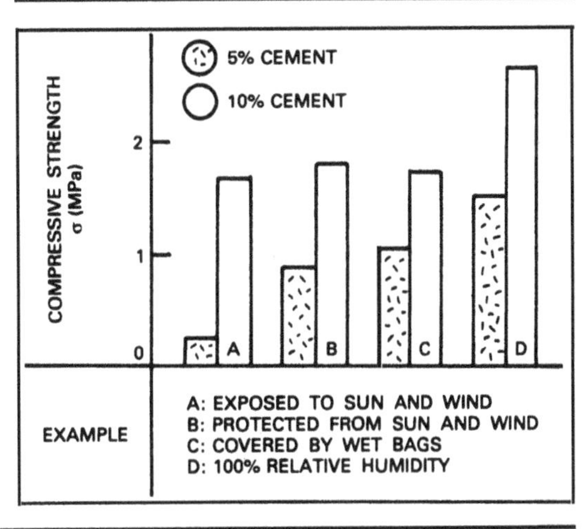

Adeten. *Etude et Expérimentation de la Construction en Terre à Vigneu*. Grenoble, UPAG, 1976.
AGRA. *Recherche Terre*. Grenoble, AGRA, 1983.
AGRA. *Recommandations pour la Conception des Bâtiments du Village Terre*. Grenoble, AGRA, 1982.
Altech. *Catalog*, 1984.
An. *Adobe Solar Project*. China Valley.
An. *Manual Prático de Construção com Solo-cimento*. Salvador, CEPED, 1978.
An. 'New portable adobe making machine now on the market.' In *Adobe News*, Albuquerque, Adobe News, 1980.
Anker, A. *Naturbauweisen*. Berlin, Deutsche Landbuchhandlung, 1919.
Atelier Nord (Platbrood), *Catalog*, 1984.
Barrière, P. *et al. Optimisation de la Mise en Œuvre du Pisé*. Clermont-Ferrand, Université Clermont II, 1979.
Bogler. 'Lehmbauten unter Verwendung von Trümmersplitt'. In *Naturbauweisen*, Berlin, 1948.
Ceratec. *Catalog*, 1984.
Cinva-Ram. *Catalog*, 1968.
CINVA. *Experiencas sobre Vivienda Rural en Brasil*. Bogota, CINVA, 1961.
Colzani, J.H.; 'Archéco'. *Tob System*. Priv. com. Toulouse, 1983.
Comet Opera. *Catalog*, 1974.
Consolid AG. *Catalog*, 1983.
Couderc, L. Priv. com. Pierrelatte, 1980.
CRATerre. 'Casas de tierra'. In *Minka*, Huankayo, Grupo Talpuy, 1982.
CTBI. *Catalog*, 1984.
Dansou, A. 'La terre stabilisée, matériau de construction'. In *Bulletin d'Information*, Lome, Centre de construction et du logement, 1975.
Doat, P. *et al. Construire en terre*. Paris, Editions Alternatives et Parallèles, 1979.
Dynapac. *Catalog*, 1982.
Ellson. *Catalog*, 1978.
Fauth, W. *Der praktische Lehmbau*. Singen-Hohentwiel, Weber, 1948.
Fernandez-Fanjul, A. *Rapport pour une meilleure Connaissance du Comportement et de l'Utilisation du Geo-béton*. Abidjan, LBTP, 1974.
Frey, R.P.; Simpson, C.T. *Rammed Earth Construction*. Saskatoon, University of Saskatchewan. 1944.
Gonález, J.M.V. *La Tierra Estabilizada su Utilización en la Producción de Componentes para la Construcción*. Panamá, CEFIDA, 1980.
Gravely. *Catalog*, 1983.
GTZ. Priv. com. Eschborn, 1983.
Guillaud, H. *Histoire et Actualité de la Construction en Terre*. Marseille, UPA Marseille-Luminy, 1980.
Gumbau, J. Priv. com. 1983.
Hays, A. *De la Terre pour Bâtir. Manuel pratique*. Grenoble, UPAG. 1979.
Hellwig, F. *Der Lehmbau*. Leipzig, Hachmeister und Thal, 1920.
International Institute of Housing Technology. *The Manufacture of Asphalt Emulsion Stabilized Soil Bricks and Brick Maker's Manual*. Fresno, IIHT, 1972.
ITDG. IT workshops product information sheet. ITDG, 1983.
Kahane, J. *Local Materials. A self builder's manual*. London, Publication Distribution Co-operative, 1978.
Kern, K. *The Owner Built Home*. New York, Charles Scribner's Sons, 1975.
Küntsel, C. *Lehmbauten*. Berlin, Reichsnährsthand, 1919.
La mécanique régionale. *Catalog*, 1983.
Latorex. *Catalog*, 1983.
Les Atelier de Villers-Perwin. *Catalog*, 1977.
Liner. *Catalog*, 1974.
Lunt, M.G. 'Stabilised soil blocks for building'. In *Overseas Building Notes*, Garston, BRE, 1980.
Luxor. *Catalog*, 1984.
Lyon, J.; Lumpkins, W. *Large Scale Manufacturing of Stabilized Adobe Brick*, Self Help Inc, Los Alamos, 1969.
Maggiolo, R. *Construcción con Tierra*. Lima, Comissión ejecutiva inter-ministerial de cooperacion popular, 1964.
Markus, T.A. *et al. Stabilised Soil*. Glasgow, University of Strathclyde, 1979.
Meili. *Catalog*, 1983.
Meynadier. *Catalog*, 1981.
Miller, L.A. & D.J. *Manual for Building a Rammed Earth Wall*. Greeley, REII, 1980.
Miller, T. *et al. Lehmbaufibel*. Weimar, Forschungsgemeinschaften Hochschule, 1947.
Minke, G. *Alternatives Bauen*. Kassel, Gesamthochschul-Bibliotek, 1980.
Morando. *Catalog*, 1977.
Moriarty, J.P. *et al.* 'Emploi du pisé dans l'habitat économique'. In *Bâtiment International*, Paris, CIB, 1975.
MTD. *Catalog*, 1982.

Muller. *Catalog*, 1979.
Musick, S.P. *The Caliche Report* Austin, Center for maximum potential, 1979.
Perin, A. Priv. com. Marseille, 1983.
Pliny Fisk. 'Earth block manufacturing and construction techniques'. *2nd Regional Conference on Earthen Building Materials*, Tucson, University of Arizona, 1982.
Pollack, E.; Richter, E. *Technik des Lehmbaues*. Berlin, Verlag Technik, 1952.
PPB SARET. *Catalog*, 1983.
Proctor, R.L. *Earth Systems*. Priv. com. Corrales, 1984.
Quixote. *Catalog*, 1982.
Reuter, K. *Lehmstakbau als Beispiel wirtschaftlichen Heimstättenbaues*. In Bauwelt, 1920.
Riedter. *Catalog*, 1983.
Ritgen, O. *Volkswohnungen und Lehmbau*. Berlin, Wilhelm Ernst und Sohn, 1920.
Rock. *Catalog*, 1978.
Serey, Ph.; Simmonet, J. *Etude de la Presse Cinva-Ram, Étude de l'Influence de la Compaction sur la Qualité du Géobéton*. Abidjan, LBTP, 1974.
Shaaban, A.C.; Al Jawadi, M. *Construction of Load Bearing Soil Cement Wall*. Baghdad, BRC. 1973.
Simonnet, J. *Définition d'un Cahier des Charges pour la Conception d'une Presse Manuelle à Géobéton Destinée à la Côte d'Ivoire*. Abidjan, LBTP, 1983.
Smith, E. *Adobe bricks in New Mexico* Socorro, New Mexico Bureau of Mines and Mineral Resources, 1982.
Sonke, J.J. *De Noppensteen*. Amsterdam, TOOL, 1977.
Souen. *Catalog*. 1983.
Sulzer, H.D.; Meier, T. *Economical housing for developing countries*. Basle, Prognos, 1978.
Tecmore. *Catalog*, 1982.
Tibbets, J.M. 'The pressed block controversy'. In *Adobe Today*, Albuquerque, Adobe News, 1982.
Torsa. *Catalog*, 1978.
Trueba, G. Priv. com. Mexico, 1983.
Venkatarama Reddy, B.V.; Jagadish, K.S. *Pressed Soil Blocks for Low-cost Buildings*. Bangalore, ASTRA, 1983.
Webb, D.J.T. 'Stabilized soil construction in Kenya'. In *Colloque L'habitat économique dans les PED*, Paris, Presses Ponts et Chaussées, 1983.
Webb, D.J.T.; BRE. Priv. com. Garston, 1983.
Williams-Ellis, C.; Eastwick-Field, J. & E. *Building in Earth, Pisé and Stabilized Earth*. London, Country Life, 1947.
Wolfe, A. Tallahassee. Priv. com. Florida A & M university.

The Chateau Escoffier in France, of traditional rammed earth construction (Hubert Guillaud, CRATerre-EAG)

10. DESIGN GUIDELINES

An earth structure may be exposed to bad weather, which represents one of the major hazards leading to the deterioration of these structures.

Careful examination of traditional earth structures, in all parts of the world, reveals the skills of their architects in solving the serious problems involved in preserving the durability of structures exposed to water risks.

These are tricks of the trade, ingenious and cunning structural systems, whose effectiveness is often remarkable. Nowadays, however, these tricks have often been forgotten or misunderstood and ignored by modern builders. Often these traditional devices have not been retained because they have not be noticed.

At the present time the quality of earth construction has deteriorated considerably, because traditional know-how has been forgotten or is lacking. On the other hand certain solutions which may have been suitable in the past are no longer feasible because of changing economic circumstances and technologies.

Techniques which require regular maintenance are often ignored nowadays, being regarded as unacceptable in a modern context.

Even so, there are many, many solutions to the problems arising from the nature of the soil itself, its inherent fragility, or its protection. The catalogue of possible solutions is so enormous that the problem is more one of making a good choice.

The chosen solution should above all make it possible to take a rapid decision, and provide a relevant and economical choice of solutions, which guarantee the safety and durability of earth structures.

Presence of water in buildings

Compared to other types of structures, those built with earth soil are particularly vulnerable to water action. When water stands close to a building or penetrates it, the building can become uncomfortable or even unhealthy, and runs the risk of rapid deterioration.

In order for water to take effect, the following three conditions must coexist:

1. Water must be present on the surface of the building;
2. There must be an opening in this surface, such as a crack or window, which allows the water to enter;
3. There must be a force – pressure, gravity or capillary action – facilitating the entry of water into the opening.

The elimination of conditions favouring the harmful action of water ensures that the structure remains healthy and lessens the risk of the deterioration of the building due to chronic damp. This, however, is not always so easy. The action of water on walls can be diminished by building good foundations and base courses, protecting the tops of walls and reducing susceptibility to condensation. Cracks and possible paths for water through the surface of walls can also be eliminated by regularly maintaining the outer skin of the building. On no account, however, may the surface of earth walls be made impermeable, as they must be allowed to breathe, and be permeable to movements of water vapour. It is also possible to act directly on the penetrating forces but such action is delicate and depends, for example, on the capillarity of the material.

The best strategy, and the most effective, consists of keeping water away from the vulnerable parts of the building, i.e. the earth walls. When this is borne in mind, the meaning of the old Devon saying, 'All cob wants is a good hat and a good pair of shoes', becomes clear. In fact this recommendation on how to dress a home is a good basic formula for quality earth construction, avoiding chronic damp. Water which strikes wall surfaces (e.g. rain) is not particularly serious if it subsequently evaporates. When, however, water penetrates inside the walls and accumulates there, it is very serious indeed.

Precautions

Typical chronic damp problems can be avoided by adopting a good approach to the design and execution of the earth structure. It is in fact a matter of 'knowing how to build earthen structures properly'. There is, however, an unfortunate tendency to clad buildings with a view to increasing the resistance of 'the earth' to water (excessive protection of the material) as opposed to the proper approach which consists of making the 'building' resistant to water. The most typical effects of water can be summarized with reference to the drop of water: impact, run-off, standing, absorption, seepage, splash. Dampness, which is a second stage in the action of water, acts in more harmful ways, namely by permeability, and capillary action. These effects become more pronounced as the material increases in porosity.

The most fragile points in earth structures and most vulnerable to water action and moisture are:

- the bottom of the wall;
- the top of the wall.

There are other localized weak points as well, such as the reveals of openings, parapets of terraces, gargoyles, and bonds between different materials, such as between earth and wood. These are the points which require special care and regular maintenance.

Mechanisms and effects

1. Foundations Capillary rise at the base of walls, beginning in the foundations, has several origins, such as seasonal changes in the groundwater table, retention of water by the roots of shrubs, defective sewers, a lack of drains for the building, standing water at the foot of walls. Persistent dampness can bring about a weakening of the base of the walls. The material passes from the solid to the plastic state and the walls can no longer bear loads, increasing the likelihood of collapse. Dampness favours efflorescence by salts such as NaCl, $CaSO_4$ and $NaSO_4$, attacking the material and causing hollows to form. Insects and rodents attracted by the damp conditions can cause further deterioration to the wall.

2. Base course Above ground level, the base of the walls may be eaten away for any of the following reasons: water splash from gargoyles, water thrown up by passing vehicles, washing of floors indoors, surface condensation (morning dew), runoff at the foot of the wall (gutters too close to the wall), surface rendered impermeable (water-tight walkway or rendering) preventing evaporation or encouraging condensation between the earth wall and the waterproof rendering, the growth of parasitical flora (moss) and efflorescence.

3. Walls Water infiltrates through structural cracks (settlement, and shearing) and shrinkage cracks caused by repeated dry-wet cycles, unfilled holes left by formwork clamps, and by defective mortar joints: capillary action, and hollowing of walls.

4. Water runs off at the junction of reveals and earth walls (support, lintel) and infiltrates between the masonry of the reveal or the wooden frame and the soil; localized deterioration.

5. Rain and variations in temperature can bring about the decomposition of the material: clays are washed out, reducing the cohesion of the soil.

6. When an earth wall is protected with a rendering which prevents the movement of water vapour, condensation on the cold surface of the wall (indoor walls in summer; outside wall in winter) or condensation between the wall and the rendering cause the wall to deteriorate.

7. Water may penetrate at the point where floor or roof beams pass through earth walls.

8. Water runs and gets in where poorly designed gargoyles pass through walls and are unprotected at their entry and exit. Accumulations of earth can stop up gargoyles, resulting in standing water, absorption and capillary action.

9. Parapets unprotected by a projecting cap or which are cracked or covered with a defective rendering encourage water run-off and infiltration. Objects placed against parapets such as plants requiring watering and poorly drained terraces can cause water and dampness to be retained.

10. Cracked terraces and damaged surfacing facilitate infiltration.

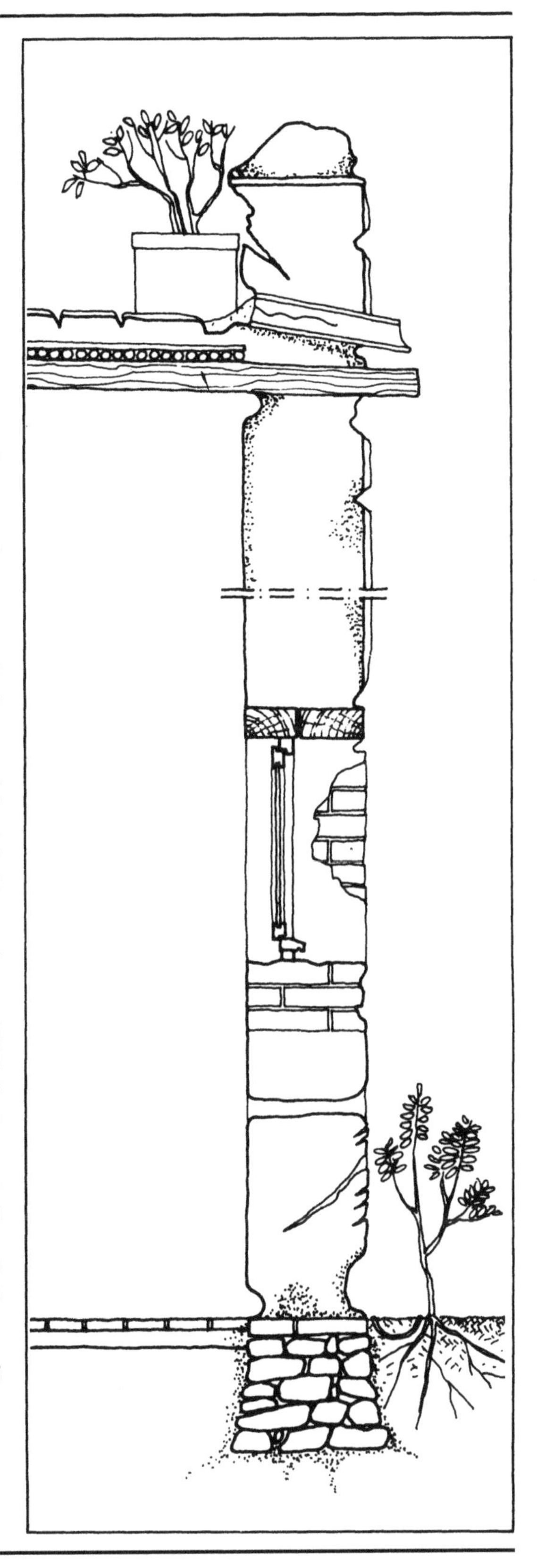

Like all structures, those built with earth may be subject to the effects of structural defects which occasionally cause irreversible damage. The use of soil as a construction material demands scrupulous respect of codes of good practice for the material and building systems. Structural defects may, however, be the result of causes which have nothing to do with the earth itself. These may include problems related to the site – e.g. settlement and earthslips – natural disasters – earthquakes and hurricanes – which can have very serious effects on structures, particularly when they are poorly designed, badly built and carelessly maintained.

Typical structural defects

Usually, these first become apparent as a crack in the building. Nevertheless there are also faults of a physico-chemical nature (decomposition of the material) and defects due to external agents, such as the action of living organisms.

Structural cracking These involve the structure of the building and usually arise from construction defects, subsequent modifications to the building or accidents. The capacity of the material to resist mechanical stress is exceeded. Such stresses include great compression, penetrating forces, tension, bending, and shearing. They may be localized (e.g. bonds between soil and various 'hard' materials, downward load due to floors, openings, etc.) or exerted in the body of the walls: e.g. ground subsidence, poor foundations.

Shrinkage cracks These are usually the result of the neglect of quality control of the earth used (e.g. excessively clayey earth) or during construction (earth too moist, drying too quickly). Shrinkage cracks can be easily recognized, being vertical and regularly spaced (e.g. every 0.5m to 1m in rammed earth). Shrinkage may also result from marked variations in relative humidity: repeated cycles of wetting and drying.

Bulging High mechanical stresses – resulting perhaps from a sudden forward movement of an untied wall or excessive localized loads – may cause a wall to become distorted (e.g. an outward bulge). Such distortions are often accompanied by cracking, although not always as soil has significant creep characteristics.

Collapse can be provoked by a build-up of stresses which weaken the structure, or by a loss of strength of the material (e.g. caused by chronic moisture). Occasional or accidental stresses may also play a role, such as the subsidence or caving-in of the ground, soil heave, tremors due to vehicles or earthquake.

Decomposition of the material Water, damp, great heat or frost may cause the chemical and mineralogical structure of the soil to undergo changes, lose its coherence and disaggregate. The inclusion of organic parasites and salts can transform the structure of the material.

Principal causes of structural defects

- Stresses on the material to which it is unsuited, such as tensile and bending stresses. Earth only functions well in compression. Other stresses require other materials: wood, concrete, and steel (used as ties, lintels, etc.)
- Chronic damp, decreasing the strength of the material, even in compression.
- Construction on a poor site which cannot stand up to the loads transmitted to it, or on moving ground (slip, uneven settlement, heave and swell).
- Poor design of the building: under-designed or off-centre foundations, inadequately braced walls, untied walls, walls which are too high, walls with too many openings, or made of composite materials; excessive loads in the form of floors, roofs, occupancy and point loads; construction systems unsuited to the use of earth as a building material.
- Poor construction: poor quality material (e.g. unsuitable soil, poor bricks); poorly implemented construction techniques (e.g. mistakes in the bond, vertical cracking along joints); incorrectly mixed mortar; poorly designed openings, no ties, no protection at the top and bottom of walls (damp).
- Related causes: climatic influences (e.g. wind action on damp wall, loss of material). Action of living organisms: plant parasites (mosses, lichens), rodents, insects (termites).

Structural defects: examples

1. Foundations There is a danger of chronic structural defectiveness when structures are erected on unstable or weak sites (heterogeneous soil, landfill, soils subject to swell or subsidence, for example). The risk increases when the foundations are under-designed; lack sufficient strength or are not properly under the load, if they are poorly constructed (joint cracking in masonry foundations or poorly constructed rubble foundations), or are badly drained (water and erosion). Tree roots and and gardening activities at the foot of the walls, insects and rodents (especially when foundations are made of soil and already suffer from chronic damp) can cause damage to foundations and debilitate walls to the extent of eventually causing their collapse.

2. Base course Above ground level, the base of the walls in particular is exposed to water attack. On chronically damp sites where walls may be weakened, and the cohesion of the material diminished, wind, plant parasites, deposits of soluble salts and rodents and insects can aggravate the undermining of the base of the walls once this has started. The base course is also exposed to accidental damage due to vehicles, livestock, or periodic work (road works, agriculture, etc.).

3. Bad construction of soil walls can greatly weaken the structure of the building; e.g. in the case of rammed earth, if care is not taken to see that ramming sections overlap (joints above one another or not adequately staggered) vertical cracks will develop through the joints. The holes made by the formwork clamps if left unfilled will weaken the walls and favour infestation by pests which attack the material. Variations in the soils used or uneven compaction can cause differences in the density of the compacted layers, causing piercing or layer formation in the material. If the moisture content is much less than optimal, the cohesion of the material may be diminished; if it is too high, there will be more shrink cracking.

4. Perpendicular to openings, descending loads can shear window breasts (cracks at right-angles to the plumb of the jambs). Poorly designed lintels can bend and give rise to cracks that weaken the wall.

5. Poorly bonded brick walls or brick walls laid with bricks of different qualities, size and and strength, or with a poor mortar (e.g. weak brick to mortar bond) are more fragile and may crack.

6. Poorly anchored joists (e.g. insufficient penetration) or the absence of a system for distributing localized loads (such as wooden boards or stabilization) can cause shear, cracking, or failure of the material.

7. If roof loads are not taken up by ring beams, the walls may crack.

8. On damp sites, if gargoyles are maltreated and the parapet is inadequately protected, structural defects will soon set in.

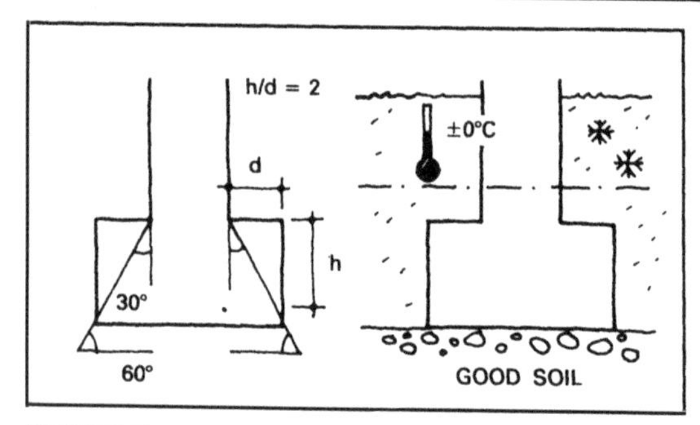

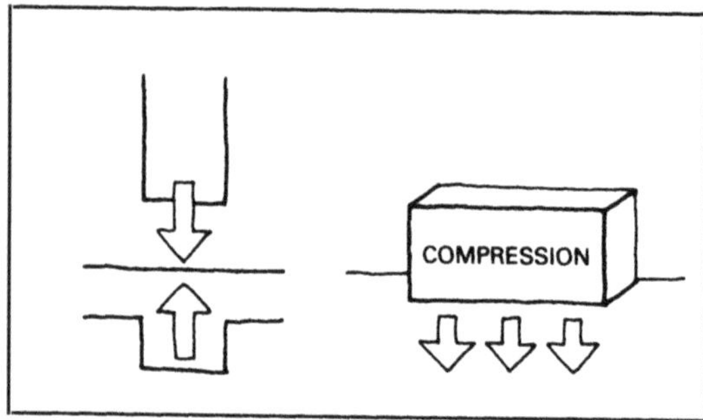

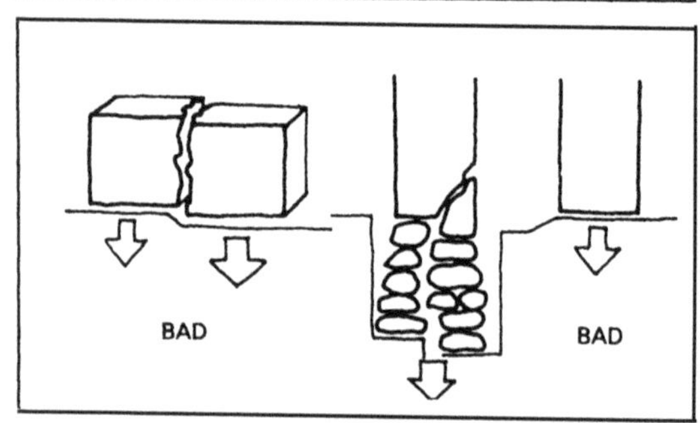

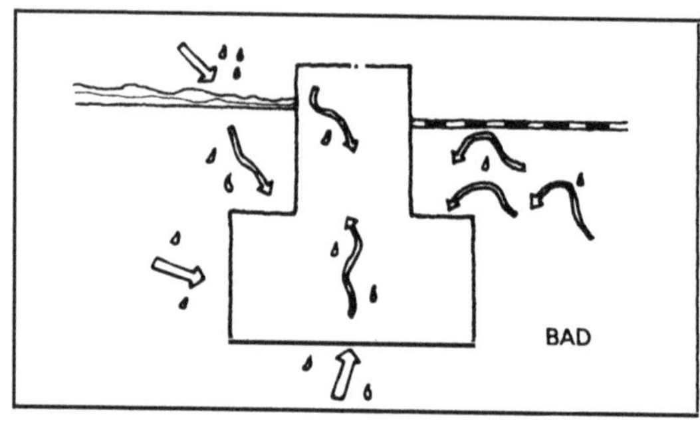

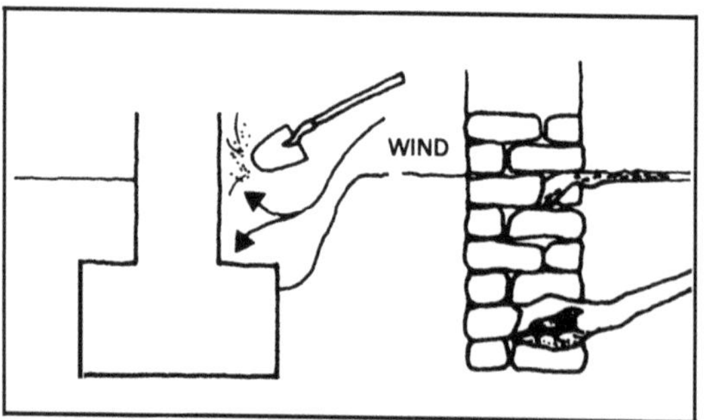

General principles of design

Unburnt soil structures with solid walls made up of smaller units such as bricks, adobes or compressed blocks, or monolithic walls made of cob or rammed earth can be approached along the same lines as conventional masonry. These are heavy structures built on shallow foundations (footings) or moderately deep foundations (ground beams, pads) designed according to well-established rules. Conventional foundation systems and materials are perfectly satisfactory.

Foundations should be deep enough to be:

- constructed on good soil. Be particularly wary for expanding soils or soils liable to severe subsidence (e.g. black cotton soils);
- protected against the action of surface water and damp;
- protected against frost;
- protected from wind erosion which can wear away foundations (in severe storms);
- protected against the effects of works in the vicinity (roads, gardening, agriculture);
- protected from rodents and insects (e.g. termites).

Specific restrictions and problems

- Soil structures with solid walls are heavy. For a one-storey rammed-earth house with earth roofing and terrace, the downwards thrust is of the order of 0.1MPa. Many soils have strengths close to this figure, or lower, or very slightly greater (the range is from 0.05 to 0.15MPa).
- Soil is only really effective as a building material when it is compressed and offers little resistance to tension, bending or shearing. The risks of differential settling must therefore be reduced to a minimum and loads properly transmitted to the foundations. Subsidence should be uniform and a situation where posts and walls have separate foundations should be avoided.
- Soil is very sensitive to water. Foundations of soil structures should be protected from water by:

- draining surface water;
- draining the surrounds of foundations;
- preventing infiltration;
- not hindering drying.

Solidity of foundations

Foundation blocks must be solid and be capable of providing effective transfer of loads to the soil without themselves being affected. To do this they must be made of strong materials and not be sensitive to water.

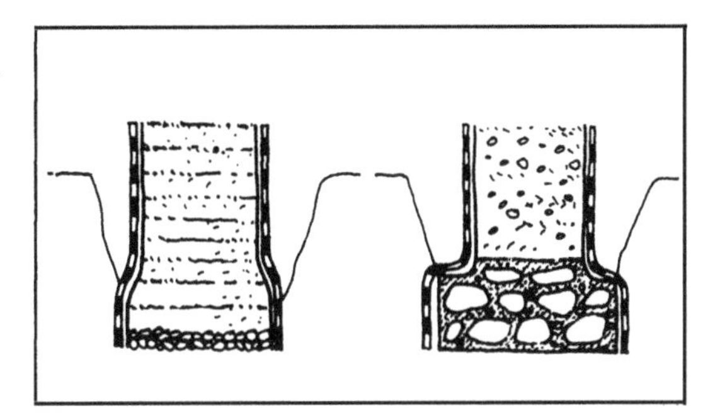

Materials

1. Stabilized earth

- Stabilized earth is not recommended and should only be used in exceptional circumstances on dry and well-drained sites. If it is the only possible solution, the foundations are built in stabilized rammed earth in an open trench, or in compressed blocks.
- Stabilized earth foundations on blinding concrete or on stoned pitching or a layer of sand. A coarse concrete or reinforced concrete slab in the bottom of the trench represents a considerable improvement.
- In wet regions earth foundations, even if stabilized, are out of the question. If there is no other alternative, steps must be taken to protect the surface or make it waterproof (coating with hard materials, waterproof skin, etc.)

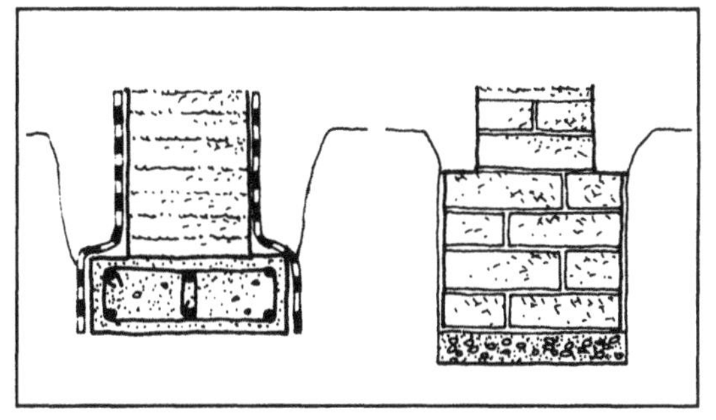

2. Other materials

All other materials are suitable.

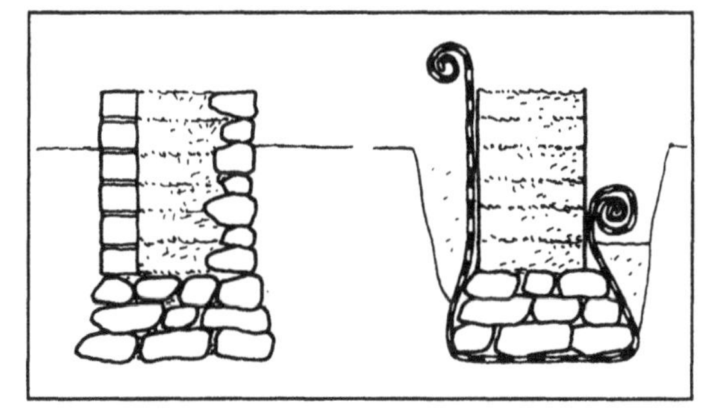

- Foundation slabs can be built in stone. In this case rubble can be used, laid as blockwork, over which a mortar is poured. They can also be coated with mortar and tightly packed against one another. Care must be taken to lay the rubble in a good bond to avoid cracking along the joints, by staggering vertical joints.
- The foundations can also be built in coarse concrete. The rubble is in this case embedded in the successive layers of concrete enveloping each layer of stones, covering these to a depth of at least 3cm. The stones do not touch.
- Burnt bricks are also suitable for making good-quality foundation slabs. Good quality non-porous burnt bricks should be used, and care should be taken to construct the bond properly.
- Finally, foundation slabs can be built using reinforced concrete and modern techniques.

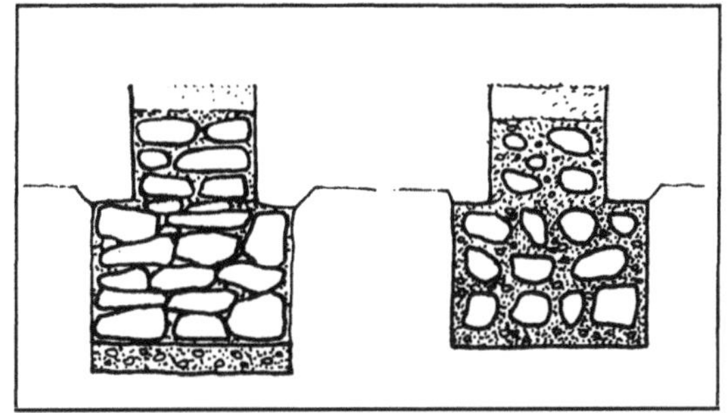

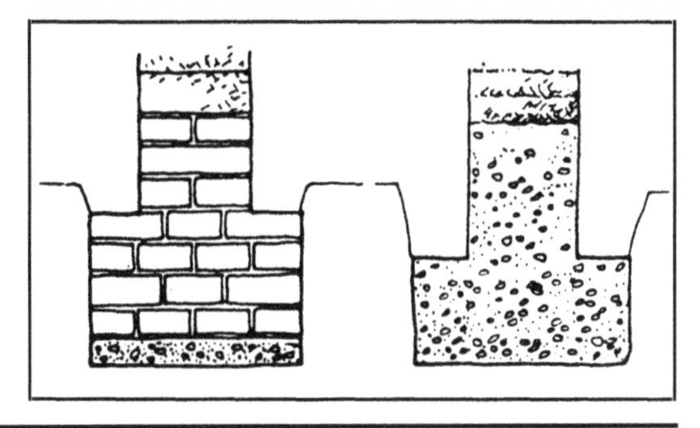

Protection from water

Earth, even when stabilized, is still highly susceptible to water, which diminishes its properties. It is thus advisable to the greatest possible extent to remove surface and underground water in the vicinity of buildings made of earth so as to avoid capillary rise through the foundations.

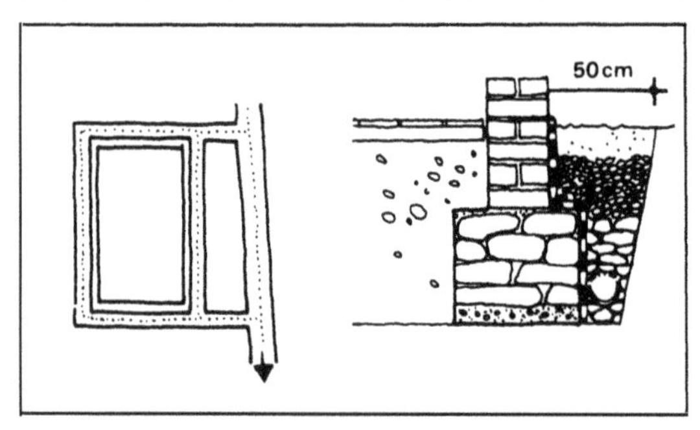

1. Drainage

Good peripheral drainage is essential if water is to be kept away from the building. It must be constructed with the greatest of care to ensure its effectiveness. Drains should be built during excavations at the bottom of the trenches, close to the foundations or at a short distance (1.5m) from the foundations. A channel (in burnt clay or some other suitable material) is laid at the bottom of the trench, which collects water and removes it by means of regular gradient. The drain is then filled with stones and gravel to create a filter system.

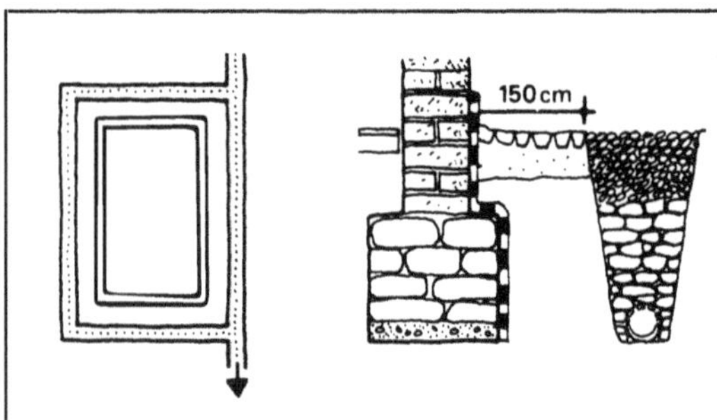

2. Gradient and gutters

The soil outside the building is specially arranged. A gradient of 2cm per m or more allows surface water to run off into a properly designed gutter some distance from the wall. Waterproofing the soil should be avoided (impervious pavement, etc.) in order not to hinder the evaporation of moisture in the soil. It is better to spread gravel over a narrow strip. Trenches should be backfilled in compacted layers sloping towards the outside.

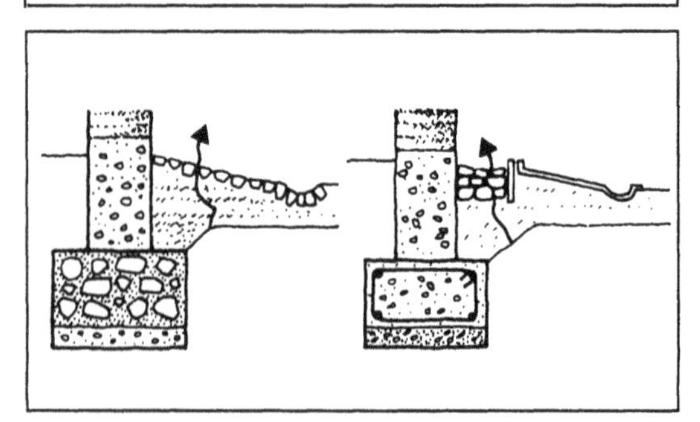

3. Moisture barriers

These can be, among others, vertical screens on the outer surface of foundations or horizontal screens serving as an anti-capillary course between the foundations and the base course. Such moisture barriers must be perfectly continuous and may not be cracked or defective. A damp course can be made either of water-repellent cement (500kg/m^3) or a bituminous product.

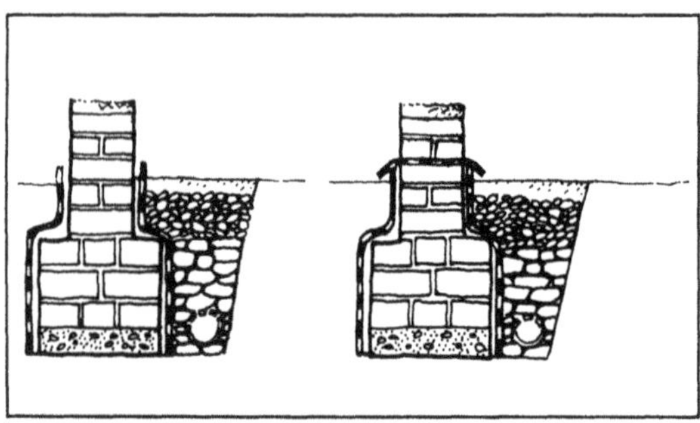

4. Waterproofing

Where daub or clay-straw is used it is wise to take steps to treat the wood, particularly posts fixed in soil. Wooden posts should be embedded in the stone or concrete foundation slab. Care must be taken to drain around the structure.

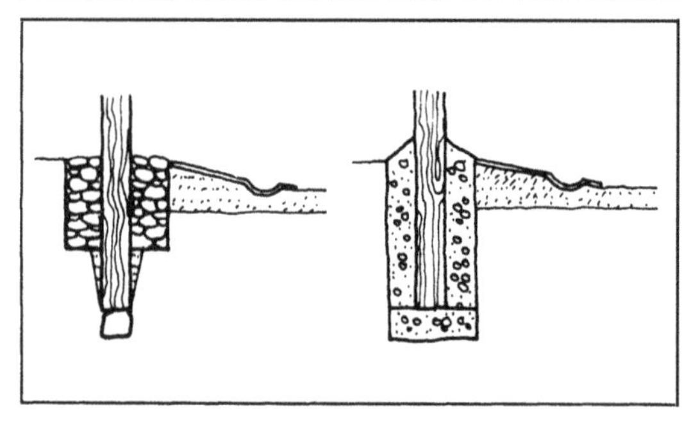

Foundations on unstable soils

Soils in arid regions are often very unstable. Alluvial and dark tropical soils in particular are very expansive. The instability of these soils is mainly due to water action reducing their cohesion. A special treatment of the soil in the foundation or special foundations may be required.

1. Expansive soils

- Keep water at a distance: a peripheral drainage is essential as well as a gradient sloping outwards (at 5cm per m) at the foot of the wall.
- Dig trenches down to good soil, compact the bottom of trenches and backfills under slabs and close to the building.
- Build rigid foundations: stones, reinforced concrete, piles, on fills of coarse gravel and stones.
- Stabilize the soil so that it will be less sensitive to water.
- Erect sufficiently flexible structures: wooden or metal frames.
- Construct extremely heavy walls so as to counteract heave.

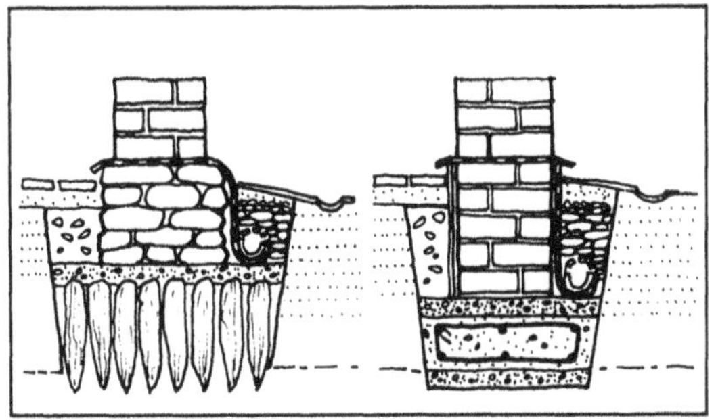

2. Soils lacking in cohesion (e.g. loess)

- Provide for the escape of water and drain the edges. Do not hinder water evaporation.
- Ram the soil and/or stabilize. One method is to flood the soil first so that it is already packed once it has dried.
- Erect floating structures: floating slabs, sole plates. Stabilize the soil around the sides and under these floating structures.

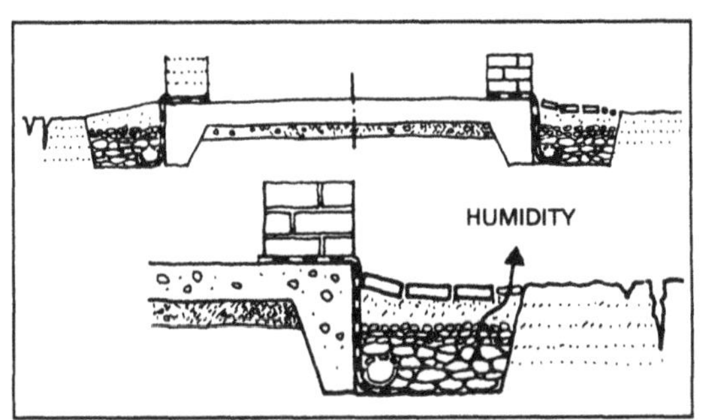

Termite protection

In the humid tropics, buildings are often devastated by termites digging tunnels through them. Timber is very susceptible to such attack, but masonry materials, such as mortar and cement, are also not excluded. Dampness and heat are favourable conditions for termite infestation. The following precautions should be taken:

- Combat dampness: drainage
- Keep borders of the structure clean at all times.
- Isolate the structural timber from the soil: build on studs or piles built in masonry. Treat wood: harden with fire, impregnate with old sump oil, or creosote.
- Plug cracks in masonry
- Paint the base courses white in order to make boreholes visible.
- Create a chemical barrier by treating the soil with anti-termite insecticides.
- Stabilize foundations or base courses by incorporating crushed glass into the soil.

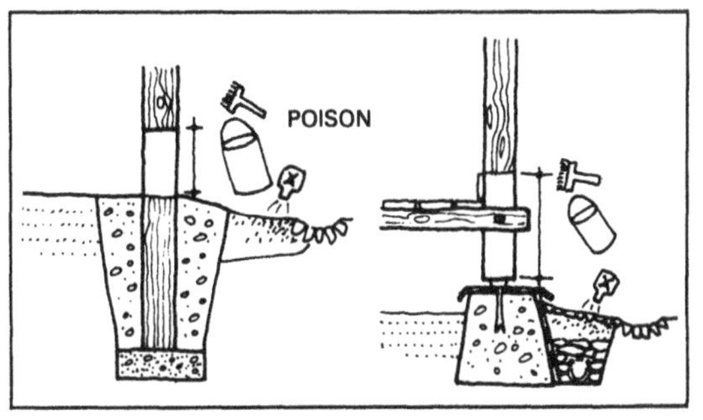

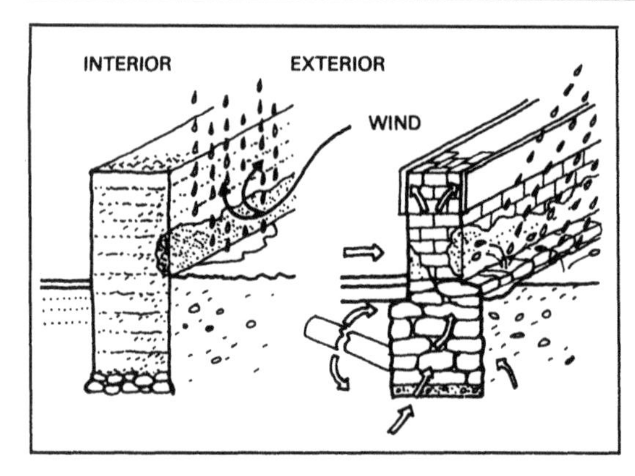

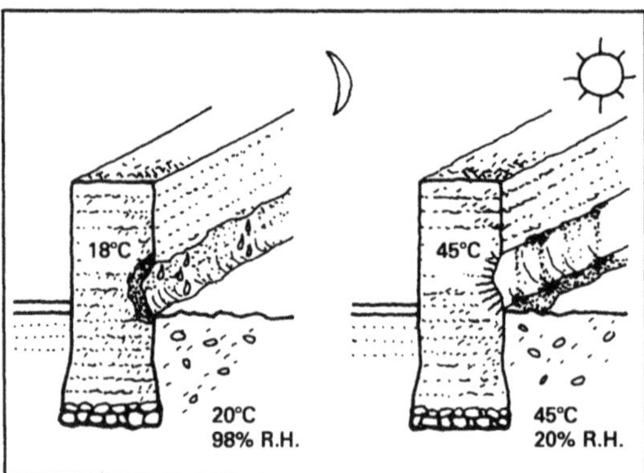

In regions with a rainy climate or exposed to natural water disasters (tropical cyclones, flooding, etc.), the base course is an indispensable element that ensures that water is kept out of the structure as a whole.

Constraints

1. Water attack

The base of an earth wall is highly vulnerable to water erosion by: water splash, standing water, pore water. Similarly, condensation (dew, even in desert climates) and repeated wetting and drying can cause significant erosion of the base of the wall. Water action can cause the following:

- A change in the state of the material, which loses its cohesion and strength by constantly passing from the dry to the wet state.
- Salt deposits which alter the mineral structure of the soil: break-down of the clay matrix.
- Splitting due to frost in the cold season.
- Colonization by parasitic flora (moss, lichen) which hold damp and facilitate infestation by insects and rodents.

2. Rise in ground level

Ground level near structures may rise as a result of farming activity, road works (raising of the roads in urban areas), the creation of waterproof footways, the laying out of flower beds along the base course, and by placing objects against the wall. Such a rise in ground level involves the more direct exposure of the earth wall which is gradually deprived of the protection afforded by the base course.

3. Erosion due to accident or inappropriate design

- Human activity such as shocks from motor vehicles, carts, miscellaneous objects.
- Activity of livestock: livestock brushing against the base of a wall, boring of galleries by rodents or termites, birds' nests.
- Plant growth on the surface as the result of damp or intentional planting bringing about chronic dampness and salt deposits, and efflorescence.

Principles of protection

1. Wearing layer

This basic technique is particularly suitable for soil base courses exposed to water erosion – e.g. splashing, condensation, etc. A solid layer of soil is applied to the base of the wall, and will be affected before the base course is. Regular maintenance of this protective layer is essential.

Wearing layers of this sort should preferably be accompanied by a system for draining run-off water away from the base of the wall collected: e.g. a ditch or a gutter.

This procedure is just as good for buildings with overhanging roofs as for those with flat roofs topped by parapet walls.

Traditional architecture has made extensive use of this principle of protection: earth architecture in Mali, Morocco, Saudi Arabia, and elsewhere.

This protection technique offers the advantages that it is economical and easy to use.

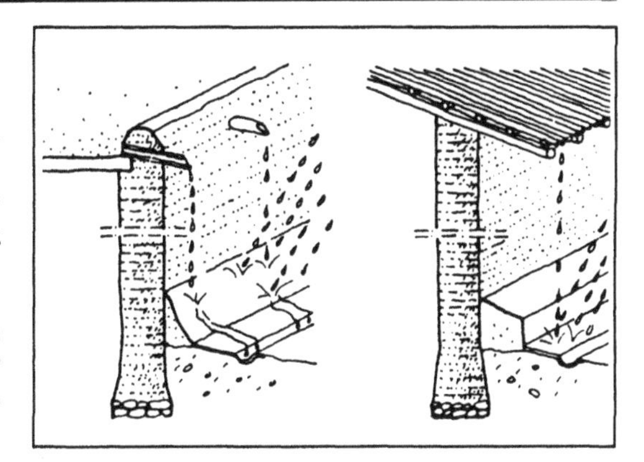

2. Material deposited at the base

Stones, bricks, wood, matting, tiles, etc.

These protective materials are either deposited against the bottom of the wall or incorporated into its base.

The thickness and height of the added protective material should be enough to allow good evaporation of pore water.

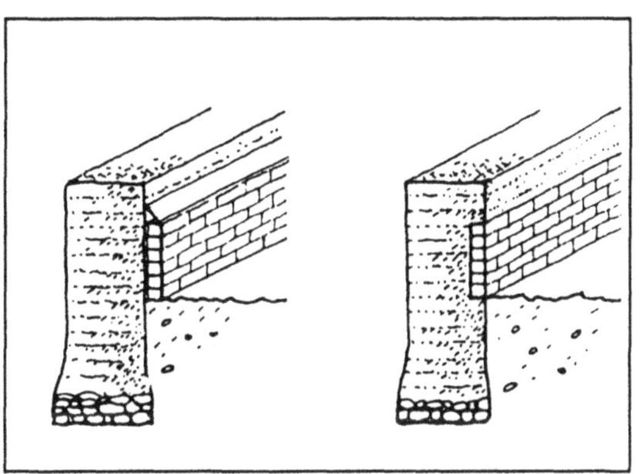

3. Base course

All known solid materials can be used: stone, brick, concrete.

The base course should be sufficiently high to cope with the constraints imposed by local eroding agents (water, wind), and follows the lie of the land and the layout of the surroundings: terraces, steps, etc. Care must be taken to provide a waterproof layer between the base course and the earth wall.

4. Height of the base course

This depends on local rainfall patterns, the likelihood of flooding, the overhang of the roof, and the evaporation of the water accumulated in the base of the wall.

- Dry regions: 0.25m
- Average rainfall: 0.40m
- Heavy rainfall, small roof overhang: 0.60m or greater.
- Area susceptible to flooding (e.g. at the side of a water course): 0.80–1.00m, to allow a good evaporation of pore water absorbed by the base.

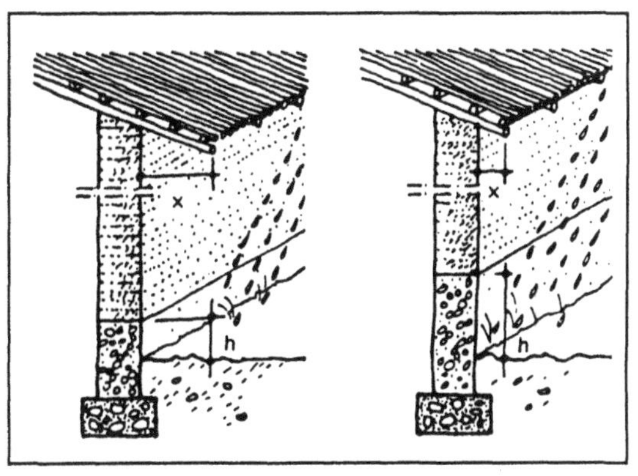

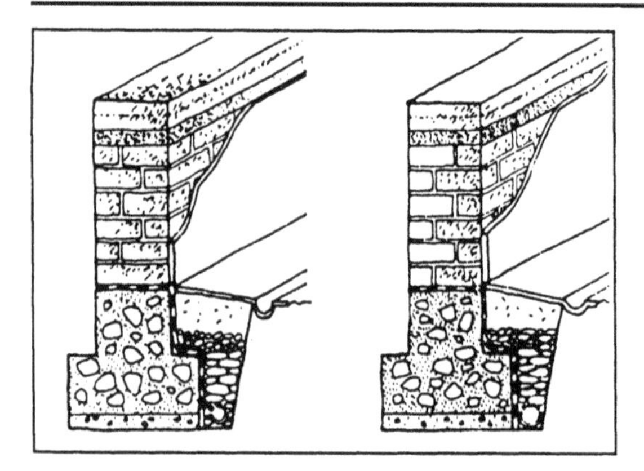

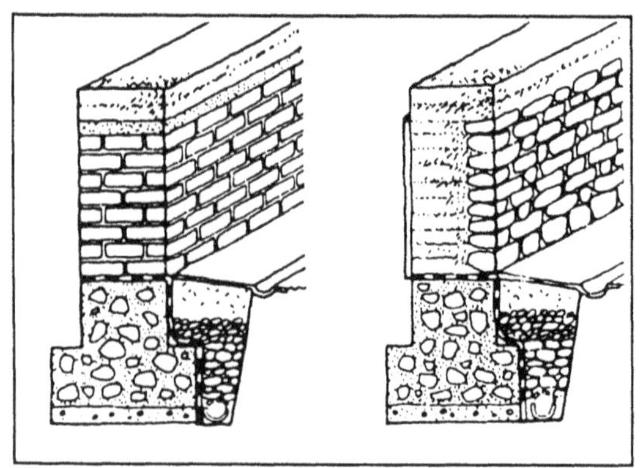

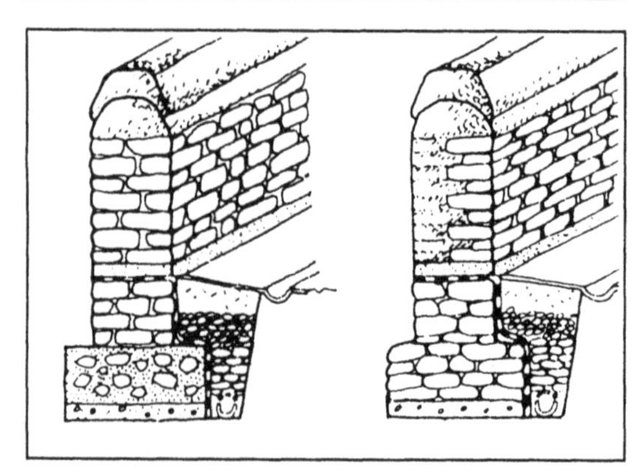

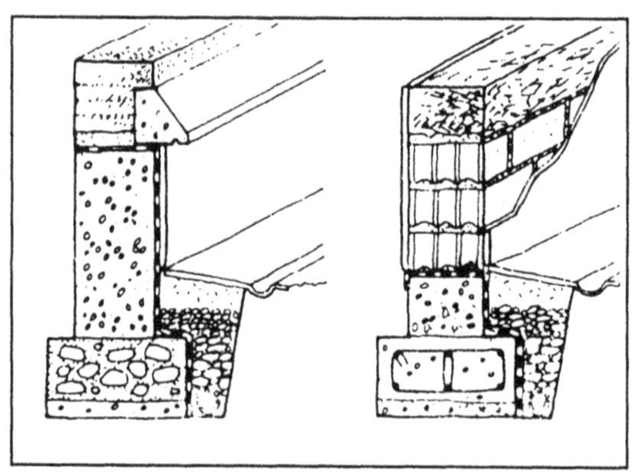

1. Stabilized brick

This material can only be used very infrequently, when it is certain that the site of the structure dry, well drained and protected against infiltrating water.

The entire base course can be built in stabilized brick, observing the rules of sound construction (bond in particular) or only the facing. In the latter case the difference in strength between the stabilized and non-stabilized brick must not be too great.

It is recommended that this type of base course is finished with a rendering which allows the passage of water vapour. The rendering must end above the level of the ground if capillary absorption is to be avoided.

The edges of the structure must be laid out with care: slope away from the building, peripheral drainage belt, drainage ditch for surface water.

2. Burnt brick

Burnt brick used for this purpose should not be porous. If this is so, the same precautions as for stabilized brick are adopted. In addition, burnt brick can also be used as a facing only or to make a protective coating of bricks, and stones. This type of facing is an attractive solution for the repair of the lower part of a building being restored.

3. Stone

Depending on its permeability, stone is regarded as a more or less waterproof material. Good quality stones can be left visible, but care must be taken to rake out the joints so as to prevent water from infiltrating. Raking-out facilitates the bonding of any rendering.

4. Concrete or concrete elements

Concrete, when correctly mixed (250kg/m^3) can be considered to be waterproof but it is advisable to protect it with a damp-proof membrane; particularly where it is buried or where the concrete is on the lean side.

Concrete blocks, whether solid or hollow, can be used for base courses but are best for light earth walls made of straw-clay, for example. A damp-proof membrane screen is essential.

5. Stabilized rammed earth

Rammed earth can be stabilized throughout the entire thickness of the base course or the surface alone can be stabilized. The second procedure has the advantage of being more economical but requires very careful execution. The non-stabilized and stabilized rammed earth are rammed simultaneously, layer after layer, the stabilized earth being rammed against the form on the outer facing of the wall.

A surface coating can also be applied by laying down a vein of lime mortar in each compacted layer. The mortar is thrown with a trowel and its thickness controlled by a mark on the formwork to ensure good finishing of the wall.

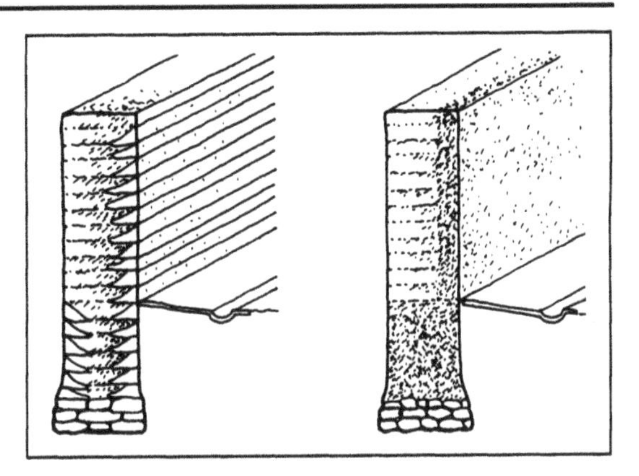

6. Cladding

Cladding (slabs of wood or shingles, nailed onto laths) can be fitted against the base course. This type of work often requires a wooden lattice to carry the cladding. An air space increases the effectiveness of the system by permitting ventilation and evaporation of moisture which might be retained behind the cladding.

Woven reeds and straw matting make cheap cladding but must be regularly maintained and periodically replaced. Cladding has the added advantage of offering thermal protection in warm climates, particularly when ventilated.

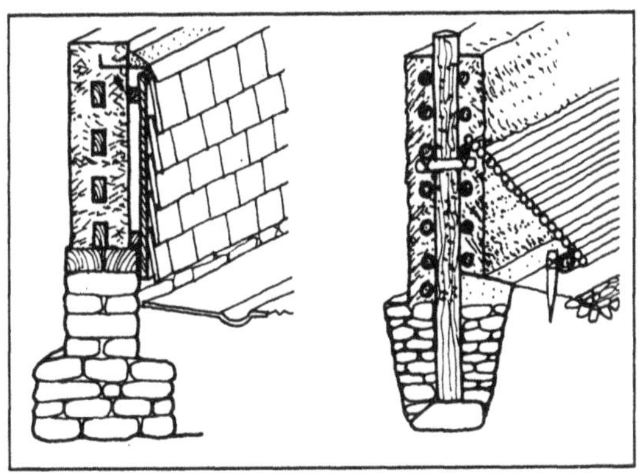

7. Surround or footway

This can be a simple timber positioned lengthwise in front of the base course, immediately below the roof gutter to catch any drips. Bottles, tin cans or stones driven into the earth are just as satisfactory.

Upended posts, held by cross-pieces and retaining an earth pavement form a more elaborate system. The pavement should have slight gradient and be provided with a drainage ditch.

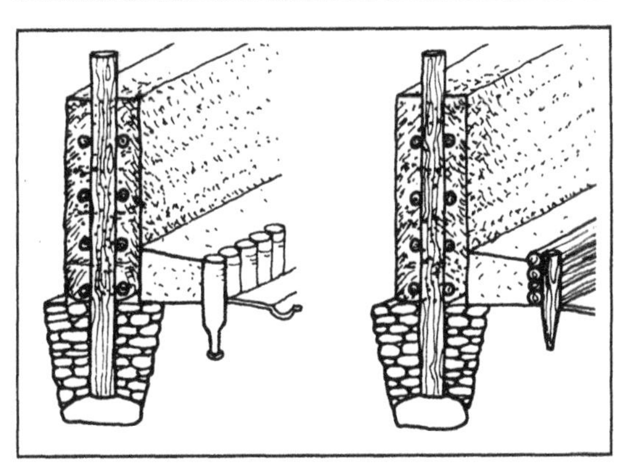

8. Encrusted renderings

A rendering covering the base course and encrusted with gravel or stone chippings constitutes a good wearing layer.

9. Temporary protection

During the rainy season simple tiles or flat stones positioned against the base of the wall form satisfactory and extremely cheap protection.

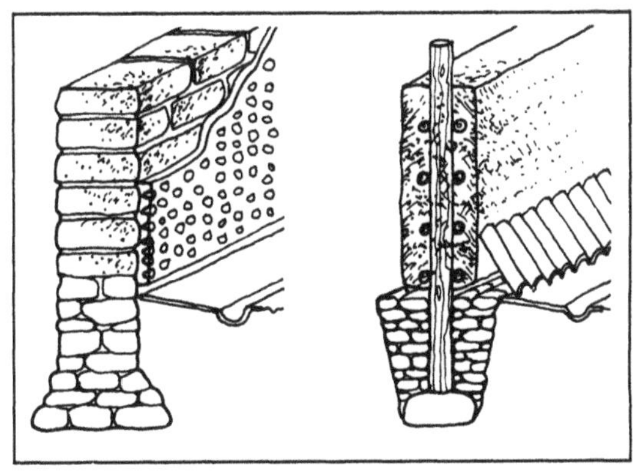

Wall systems

Wall systems in unburnt earth are highly diverse, but when the variables are ignored, the rules of building in earth are universal. Of primary concern to these rules is the compatibility between the mechanical stresses to which the material and the system of the structure are subjected and the performance and characteristics of the material. All designs must thus bear the 'logic of earth' in mind. The rules attach importance to the bond of masonry (brick) and the coarser bond of rammed earth, and the susceptibility of the material to water, making protective measures necessary. Apart from these fundamental rules, however, it is worthwhile to think in terms of the building technique chosen and the limitations of the site. As there are several building methods for each technique, certain details of execution and suitable construction methods will have to be thought out from the very beginning of the project.

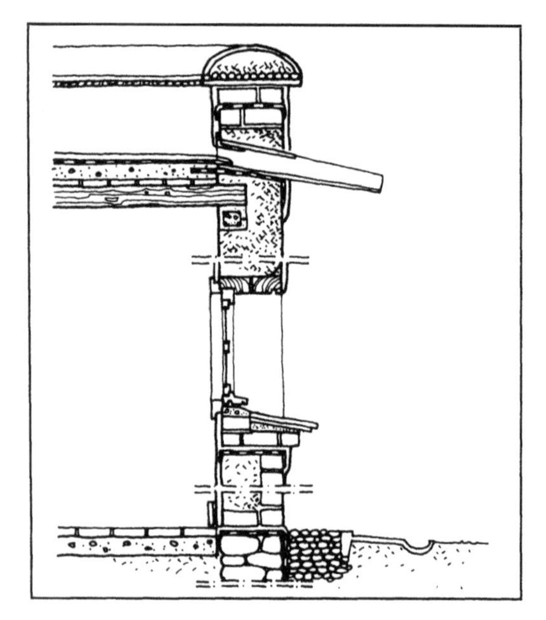

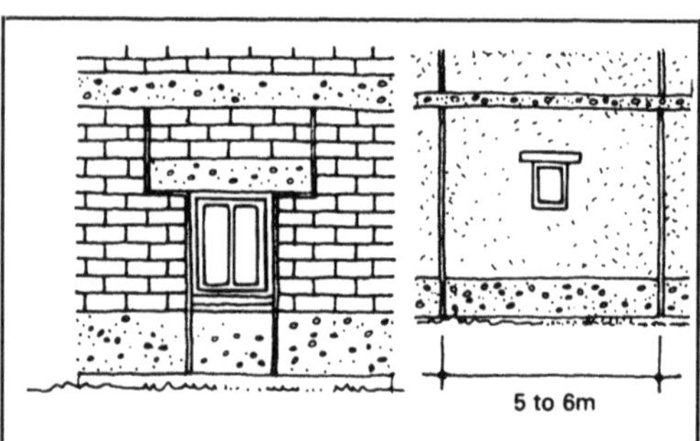

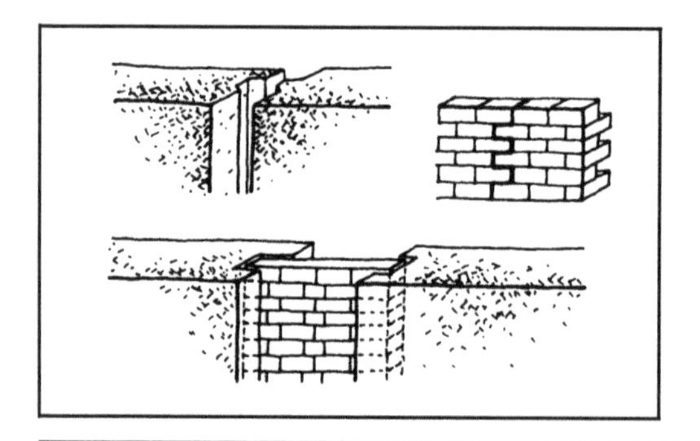

1. Mechanical behaviour

Earth functions well in compression but has low tensile, bending and shear strength. The following must thus be avoided:

- off-centre loads
- bending and the possible bulging;
- point loads.

Attention must also be paid to:

- the size and stability of the main walls and dividing walls, pillars, and buttresses, the support of arches and vaulting
- the bond used in masonry and
- the frames and closures.

2. Dimensional design

Building experience has established an empirical relationship for earth brick walls, where the thickness of walls should be at least one tenth of their height. The minimum thickness of rammed earth walls in single-storey structures should be taken as 30cm, while with a second storey it should be 45cm. Similarly the distance between partition walls or buttresses or expansion joints (dry joints planned in advance or kept strictly away from openings) should not exceed 5 to 6m.

Bond Earth walls made of adobes or compressed blocks must satisfy the same requirements with regard to bond as brick or stone walls.

Sensitivity to water Walls must be protected with particular attention to the prevention of capillary rise at the base, condensation of vapour on cold walls, and humid rooms, as well as rain, frost and snow.

Susceptibility to impact Earth walls or parts of walls exposed to erosion due to wear – bottoms of walls, corners, tops of walls, parapets, reveals in openings, etc. – must be protected by a rendering or 'hard' masonry work.

Fire resistance Surface hardening occurs with a risk of smouldering where wood is incorporated in the structure (ties, for example).

Mortar

The quality of mortar and the care taken in laying can considerably enhance the strength of the walls. The use of a cement-stabilized soil mortar increases the compression strength of adobe walls by 25% and doubles their shear strength. Stabilization can double friction and quadruple the adhesion of mortar to the brick. A failure to fill vertical joints decreases the compressive strength by 20 to 50% and eliminates all bending and shear strength. The use of a mortar which is too liquid must be avoided because of the excessive shrinkage and lack of adhesion of such material which decrease the stability and strength of walls. These considerations are of particular importance in earthquake zones.

1. Performance

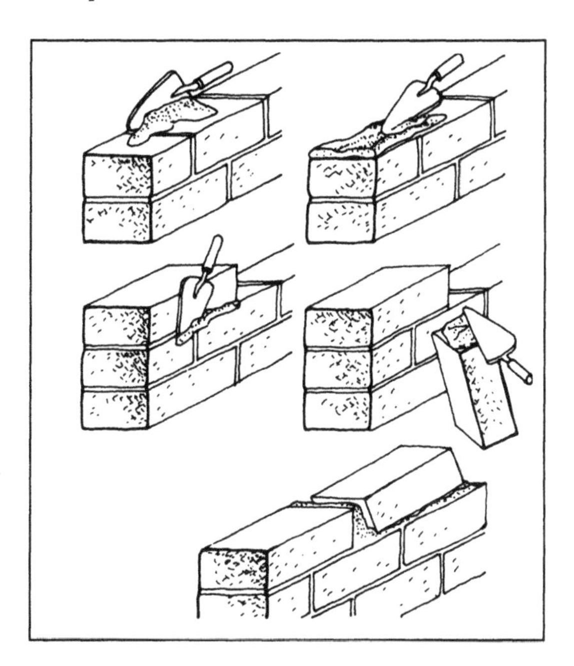

The mortar used for joints should have the same compressive strength and erosion resistance as the bricks. If the strength of the mortar is less, erosion, and infiltration of water will occur and bricks will deteriorate. If the strength of the mortar is greater than that of the bricks, the bricks will erode, water will stand on the exposed surface of the mortar, causing further erosion of the bricks.

The mortar should be checked in prior tests for shrink, adhesion, erosion and compressive strength.

Stabilized mortar must be used for stabilized bricks. The texture and water content of the mortar are different from those of the brick: the proportion of cement or lime in the mortar must be increased from 1.5 to 2 times in order to achieve the same resistance as the stabilized brick.

The shrinkage of the joints causes a horizontal shrinkage of the wall of 1 to 2mm per 5m.

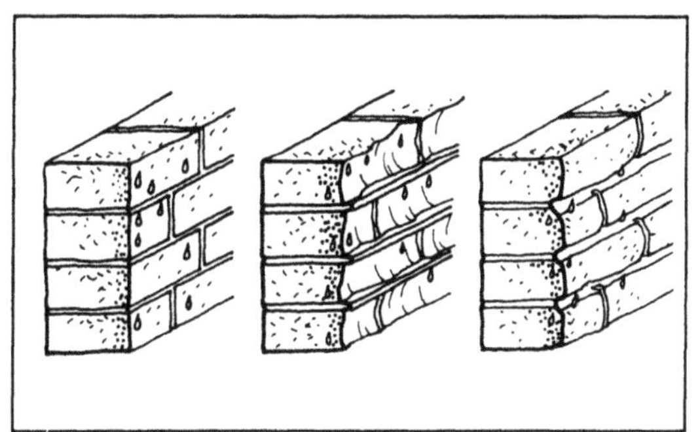

The settling of the joints under the load causes a vertical shrink of the wall of 1 to 2cm per 3m.

2. Good practice

Texture of the mortar Sandier than that of brick with a maximum grain size of 5mm, but preferably from 2 to 3mm.

Thickness of joints Horizontal and cross-joints should have a maximum thickness of 1 to 1.5cm, with a maximum tolerance of 2cm for adobes.

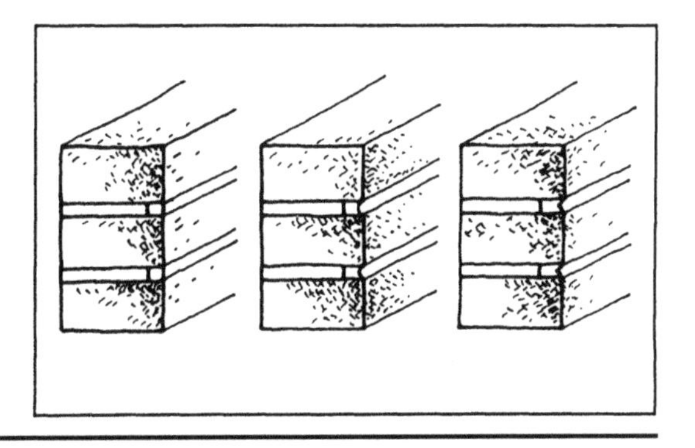

Execution Stabilized bricks must be pre-soaked and the bed thoroughly wetted. Mortar must be spread over the joint faces of the brick and the right quantity must be used. The brick is laid down rather than tapped. Protection against the sun and wind is recommended.

Jointing Immediately after laying, rake out to a depth of 2 to 3cm, joint up flush with mortar, firm with a jointing tool.

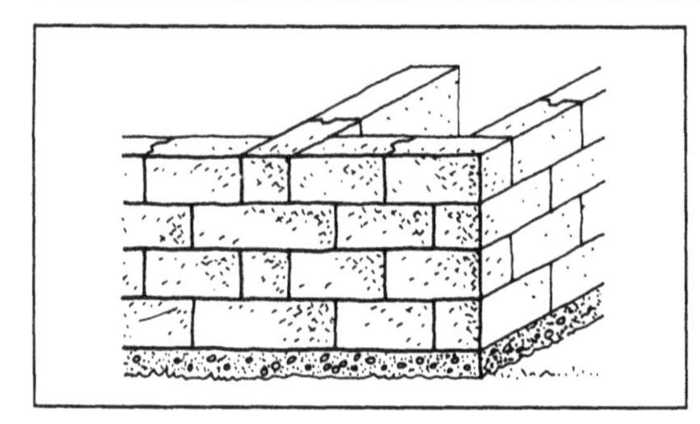

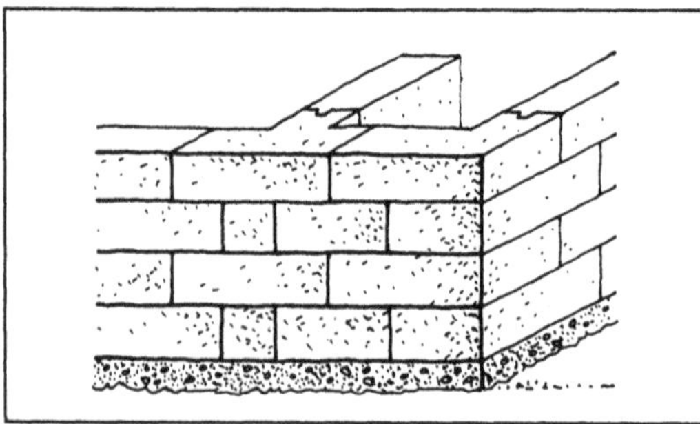

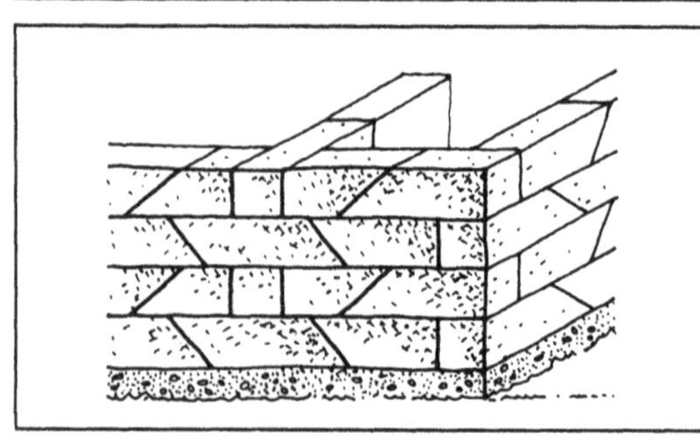

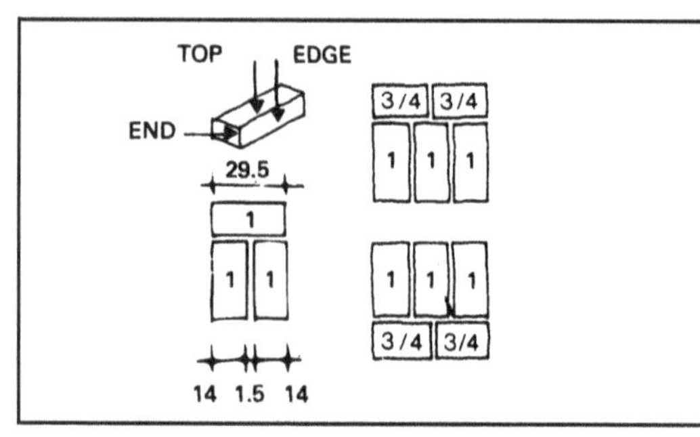

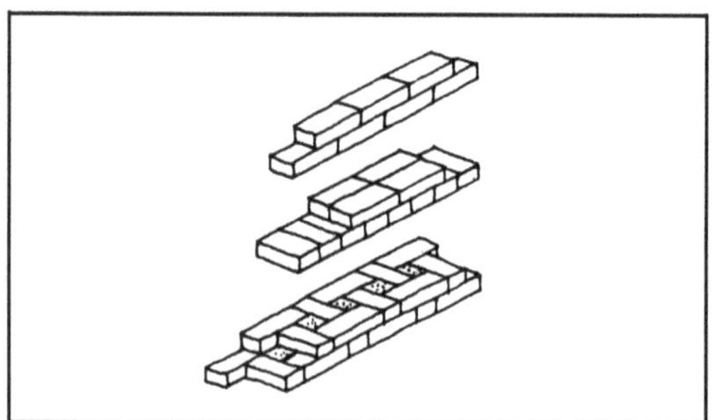

Bond: bricks and rammed earth

The bond of adobe and compressed block walls must be carefully worked out (position of the joints). A poor bond can cause structural faults: e.g. vertical cracking. The rules which apply to plain brickwork are the same as those for burnt brick. Rammed earth moulded in adjustable forms should be treated as masonry with a large bond, where the cross-joint must be staggered at least a quarter of the length of the form and provided with perfect toothing at corners and wall joints. The dimensions of the forms must accordingly be adapted to the bond planned in the design of the project in relation to the extreme dimensions of the design. The joints can be straight (Morocco, Peru, for example) or inclined at an angle (France, for example) and, if possible, grooved for better bonding.

Bricks: terminology and sizes

Bricks or earth blocks have six surfaces: a top and a bottom, two ends and two edges. The bonds in which the bricks are laid are named after the places where they are widespread and the pattern they make on the surface of the brickwork:

Header bond The brick is set in such a way that its greatest dimension is contained in the thickness of the wall, one of its ends appears on the surface.

Stretcher bond The greatest dimension and one edge are visible.

Tile bond The greatest dimension and one face are visible: to be avoided since the brick is set in the cleavage plane.

Heading bond The two ends of the brick are visible on both surfaces of the wall.

Modulated bond Brick walls are usually laid using complete bricks, but three-quarter bats and half-bats are also used. The bonds in pillars often use three-quarter bats.

Bond

The distance between two cross-joints on the surface of the wall, from one course to another, should be not less than a quarter of a stretcher. Superimposed joints result in vertical cracking – poor workmanship which must be avoided. The following is the rule for overlapping of joints:

In the thickness of the wall, bonded rising joints running in one direction should overlap by no more than three-quarters of the length of a brick. The sum of the length of overlap of vertically bonded through joints should not exceed the length of a brick.

Conventional bonds

1. Square bricks

Square elements are often used in adobe construction in South and Central America: typical dimensions are 40 × 40 × 9cm. Heading bond is the bond most commonly used for these adobes. The corners require the use of a half-bat for good toothing; a bigger, rectangular brick helps to prevent weakening of the corner. The same applies to the bonding of outer and partition walls.

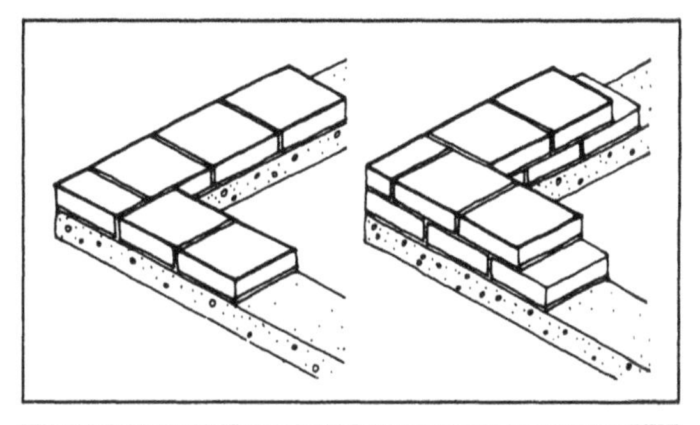

2. Rectangular bricks

The most commonly used rectangular element, produced by the great majority of presses on the market, has the following nominal dimensions: 29.5 × 14 × 9cm. Variations in the thickness of the block do not affect the bond in the chosen bedding plane. This type of brick allows the construction of thin walls only 14cm thick in stretcher bond, and walls 29.5cm thick using header bond or stretcher and header bond. A three-quarter bat is required for the construction of right angles. Such bricks allow the construction of walls 45cm thick, which offer the benefit of greater thermal inertia and of being able to assume the thrust due to arches, vaults or domes. The bonds in walls 45cm thick are as varied as in walls which are not so thick. Walls may be header and stretcher bond where three-quarter and half-bats are used for the corners or in headers or stretchers, using three-quarter and quarter-bats for the corners.

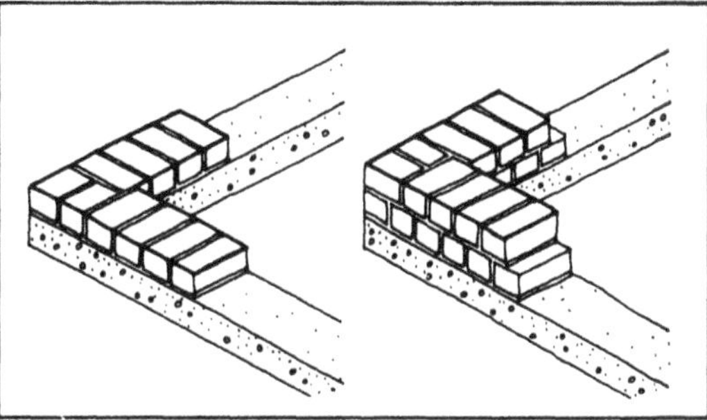

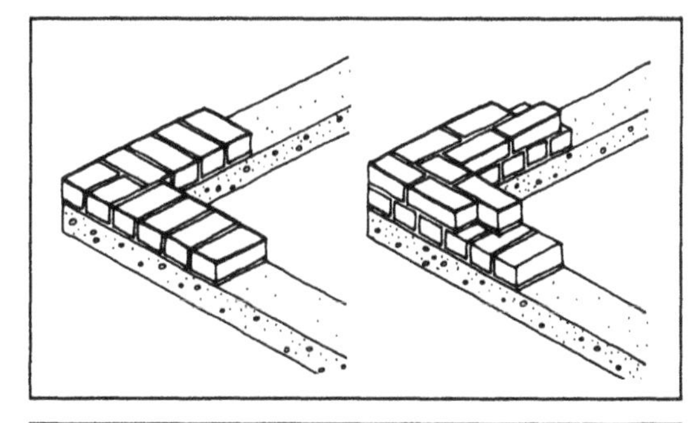

Thicker walls – e.g. 60cm – require the use of three-quarter, half- and quarter-bricks for corners and joints between walls. Very thick walls, in view of their greater complexity, are more difficult to execute, and are less economical because of the greater quantity of materials used and lower speed of laying the brick. Other construction methods should be sought for thicker walls in plain earth, such as rammed earth or cob.

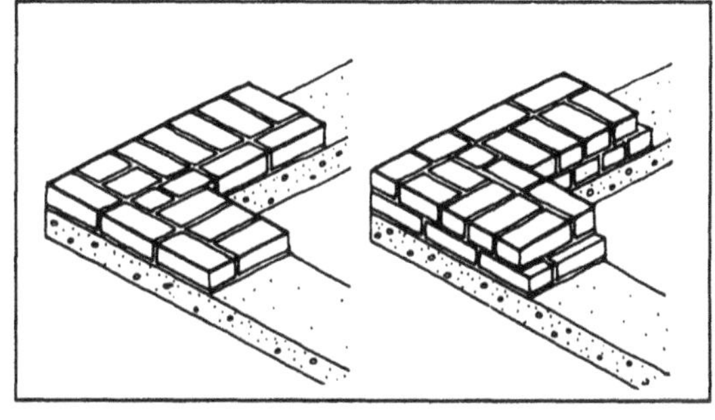

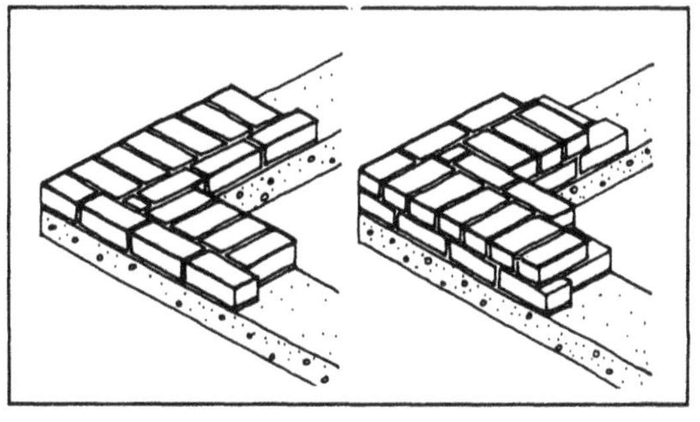

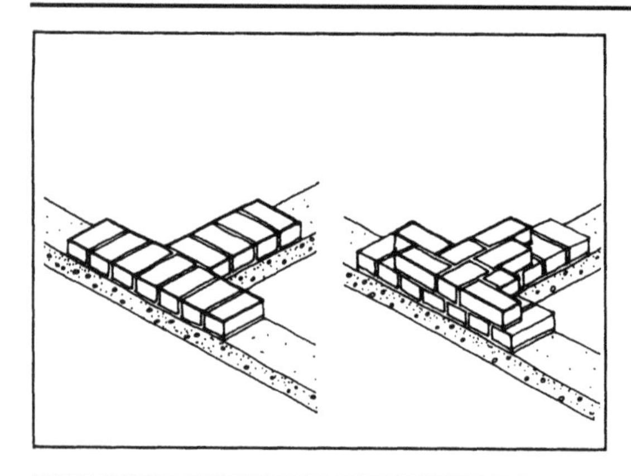

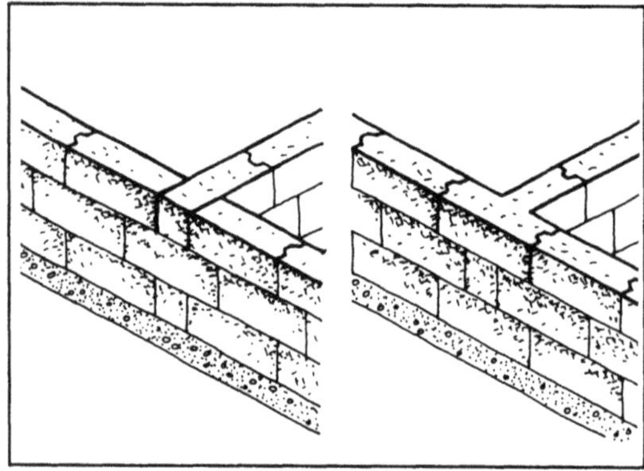

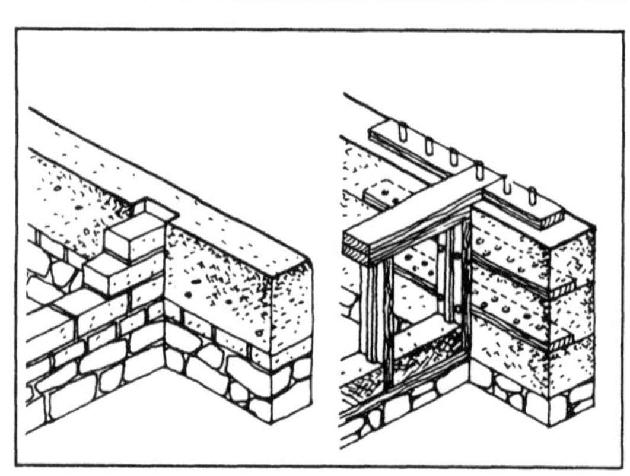

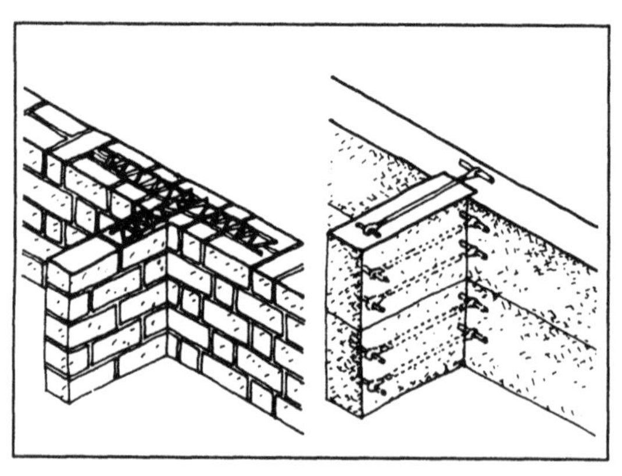

Bonding between walls

Good structural bonding between the walls – e.g. outer walls and partition walls – is essential if the structure is to be strong and stable.

The best way to bond walls together depends on whether the walls to be bonded are made of identical or different materials.

1. Identical materials

In brickwork and blockwork, the bond between walls should be perfect, in order to ensure good toothing. Wall bonds must be designed in accordance with the rules of the masonry bond and the overlap must be adequate to prevent vertical cracking through the perpends. The complexity of the bond between walls depends on their thickness, and the use of three-quarter bats and half-bats is fairly common. Where a thick wall and a thin wall are to be bonded a vertical groove in the thick wall can be provided into which the thin wall can be fitted. Horizontal strengthing must, however, also be provided (bars, wood, netting). These strengthen the 'T' bond between the walls, which may be installed every five or six courses of bricks. Ties should also be constructed to link walls continuously at floor level. In rammed earth construction the wall bond can be rammed in one piece by means of a special 'T' form, or the sections can be toothed in two directions: one section in two of the partition wall extending into the outside wall. A slot in the outer wall can also be provided, but in that case the wall bond must be strengthened horizontally in all directions and for all sections.

2. Different materials

In this case toothed wall bonds are not recommended since the difference in function and strength of the materials could cause cracking. The best procedure is thus to make a slot in the thicker wall. For bonding light wooden frame dividing walls to rammed earth walls, the best course is to embed boards in the rammed earth and to screw the dividing wall to the boards. The last wooden post for the partition can also be sunk into the wall.

Corners

The stability of the corners is largely responsible for the stability of the structure. It is usually in the corners of earth houses that structural cracks which jeopardize the structure can be observed. These cracks may be caused by the different rates of settlement of the ground and the structure (in the case of poor foundations) and it is thus the corners which suffer most as well as the other wall bonds. Such fissures, however, may also be the result of poor bonding between the walls.

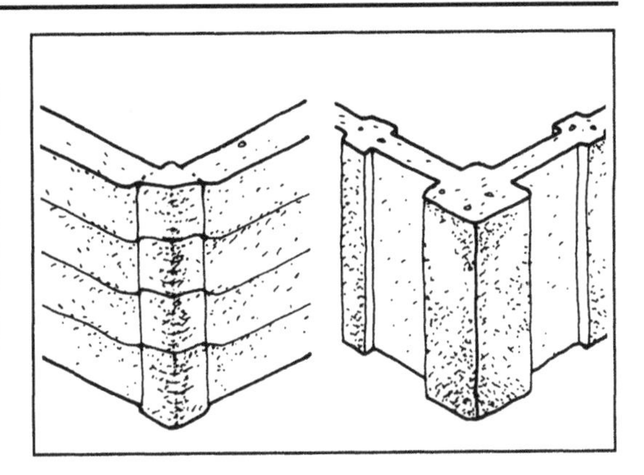

1. Corners in earth

In adobe or compressed block masonry the construction of corners requires the strict application of the rules of the bond. Furthermore, breaking bricks into smaller parts must be refrained from so as not to weaken the corner. If at all possible, a three-quarter bat is the smallest element which should be used. A half-bat is acceptable if necessary, but a quarter-bat is too small.

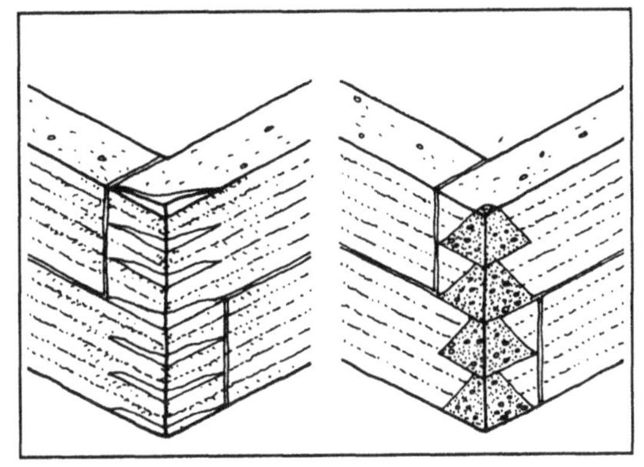

In rammed earth, the corner should be toothed in both directions of the wall being bonded, or, alternatively, every other section can be toothed. The corner can be formed in a single piece with a special L-shaped form, and in this case care must be taken to ensure that the displacement of the cross joints is great enough between sections. Do not neglect to bevel corners.

In cob there are numerous examples of thickened corners. These are very acceptable and constitute a form of wearing layer.

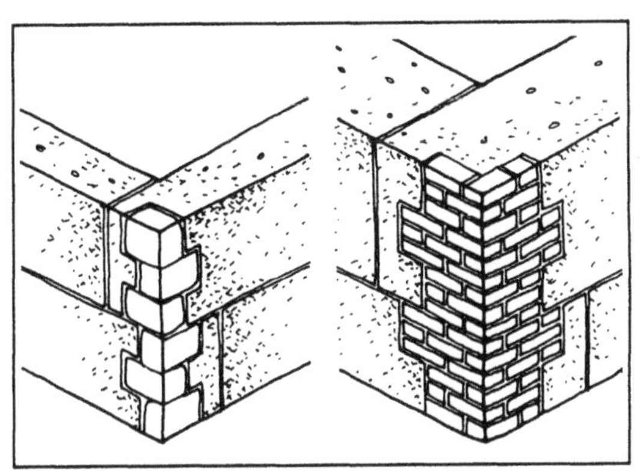

2. Corners in hard materials

There are numerous methods of strengthening corners. These involve the use of stone, burnt brick, or lime or cement mortar. These materials are included in the outer corners which are exposed to erosion. They are placed in the form (in the case of rammed earth) as the earth is being rammed. The corners take the form of rectangular or triangular toothing to ensure good bonding between the 'hard' material and the soil. A layer of mortar between the two materials enhances their bonding. The strengthening of corners with layers of mortar must be carried out carefully and uniformly.

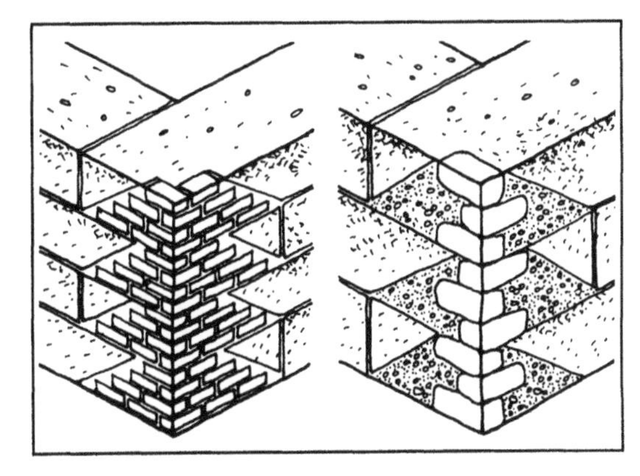

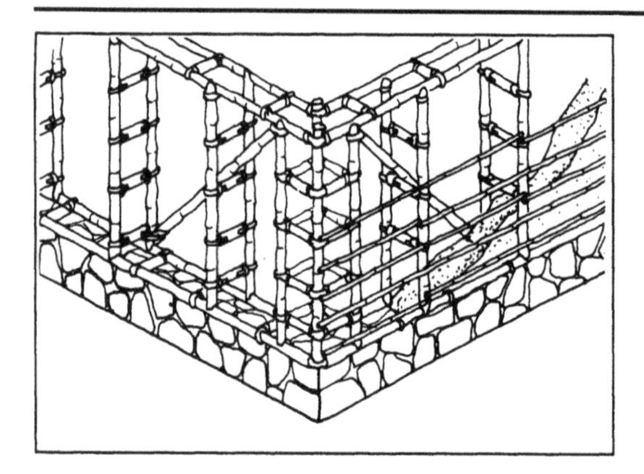

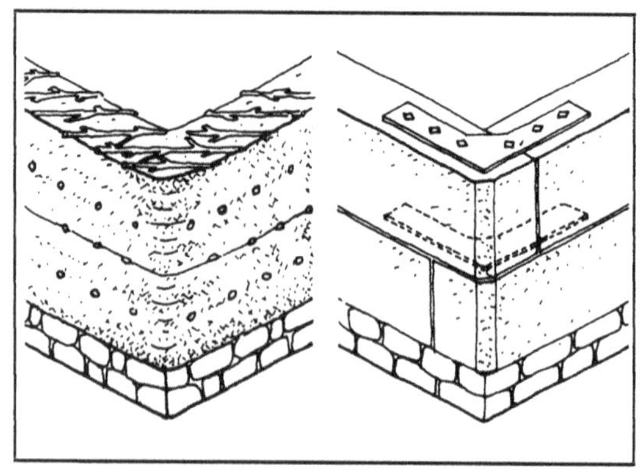

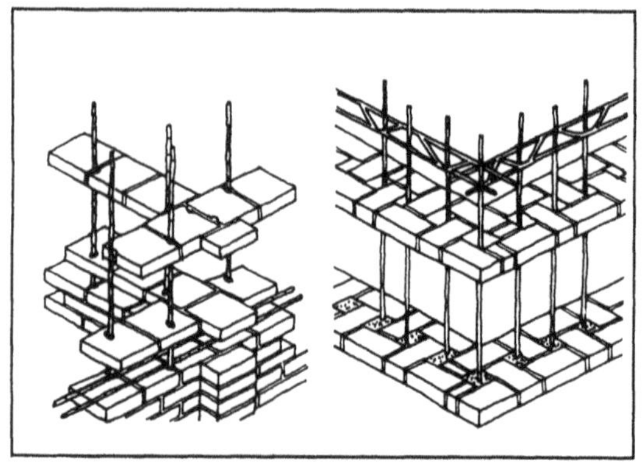

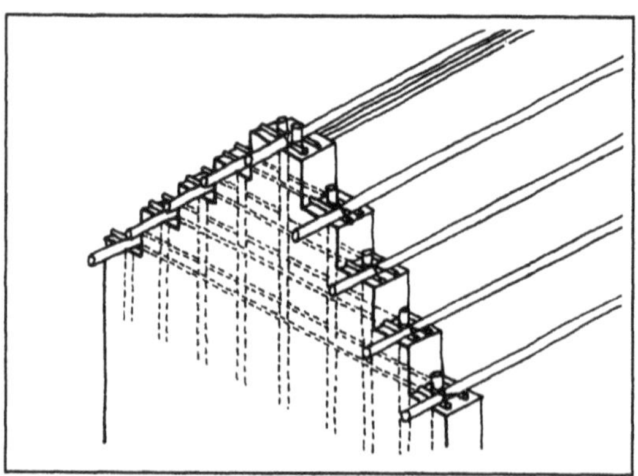

Strengthened walls

1. Frames and structures

To cope with stresses other than compressive stress, builders have developed methods of reinforcing walls. Particularly well-known are the techniques which use studding, daubed wattle, clay-straw infills together with wooden frames or structures which form an integral part of the walls.

2. Ties and reinforcements

Horizontal and vertical ties are the most frequently used reinforcement systems. These are sometimes localized and positioned in the weakest parts of the walls, such as corners, or reveals of openings. They are made of wood or iron (e.g. sills), metal lattice work or netting (in corners).

A rendering on wire-netting serves as a reinforced 'skin' which must not be used to conceal structural defects, such as vertical cracking. If it is used for such a purpose, it could prove ineffective. Earth rammed in gabions is also used to reinforce walls.

For thin walls (29.5 × 14 × 9cm bricks in stretcher bond) the use of buttresses integrated into façades around door and window openings can be considered (simple abutment outwards). The walls are also tied horizontally at floor or roof level.

For gable ends a pillar should be integrated in the axis of wall, worked into the bond and properly toothed into the wall. This pillar stiffens the wall, provides better resistance to wind loading and takes up the load of the ridge-pole. Ties at the base of the gable transmit the thrust from the roof.

Reinforcement systems for earth walls have been developed primarily with a view to improving the earthquake resistance of soil structures. The majority of earthquake regions have imposed standards for vertical and horizontal strengthening (e.g. Peru, Turkey, USA). The solutions used make use of wood or steel ties embedded in the walls. Corners and reveals of openings are reinforced in particular.

Ring-beams

Ring-beams are particularly important in ensuring the stability of earth structures. In fact, cracks and breaches in walls are due in particular to the following:

- Different rates of settlement;
- Shrink, swell, thermal expansion;
- Rotational or shear stress (e.g. openings and and bonds in walls);
- Strains due to floors;
- Lateral wind pressure, from pitched roofs and arches, vaults and domes.

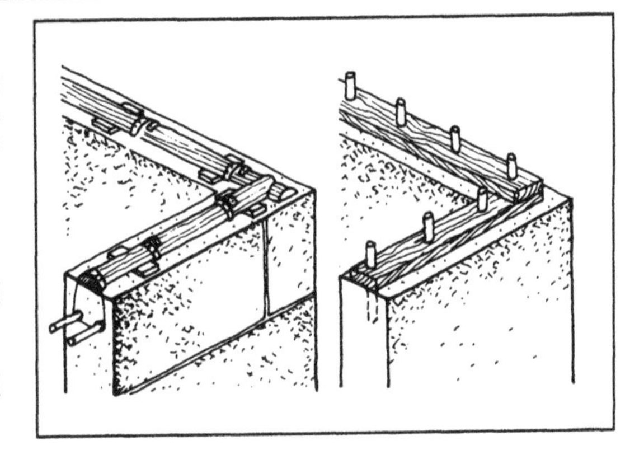

Tying affords a means of controlling these harmful limiting factors since it provides a continuous girdle for the walls in every direction.

To be effective, ring-beams must be rigid and unyielding (tensile strength).

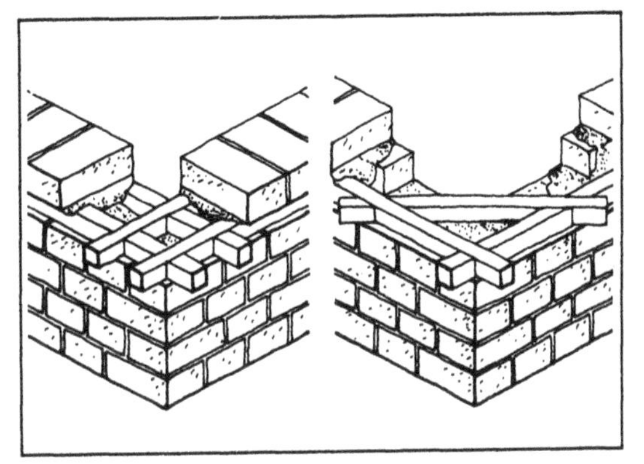

Ring-beams can be used for other purposes such as: even distribution of loads, wind-bracing, continuous lintel, support and anchorage of floors and the roof. Intermediate tie systems also exist. These are used in the sills and lintels of openings and are, in particular, encountered in regions prone to earthquakes. In the main, however, ties are confined to the floor and the edges of roofs, to transfer loads and thrust.

Materials for ring-beams

The main materials used are wood, steel and concrete. These materials must have a high degree of adhesion with the soil in order to ensure the effectiveness of the tie.

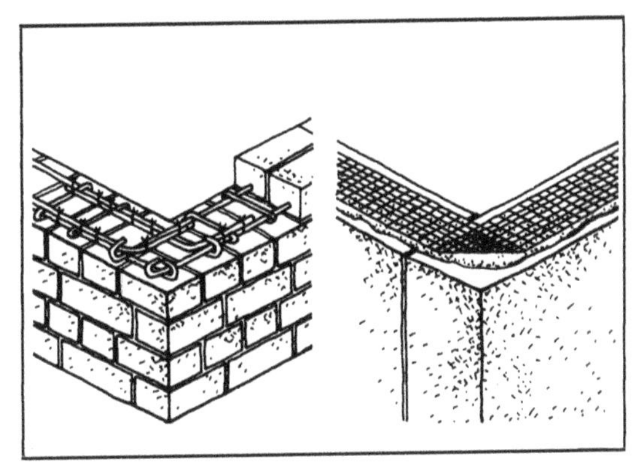

Wooden ring-beams are often set in the thickness of walls, after immersion in mortar, or anchored by means of steel or metal collars. Economic and quite effective solutions consist of using local wood such as bamboo or eucalyptus. Wood is prone to water and fire damage and damage by termites if it is not specially treated. It is desirable to use treated and dry wood from which the bark has been stripped.

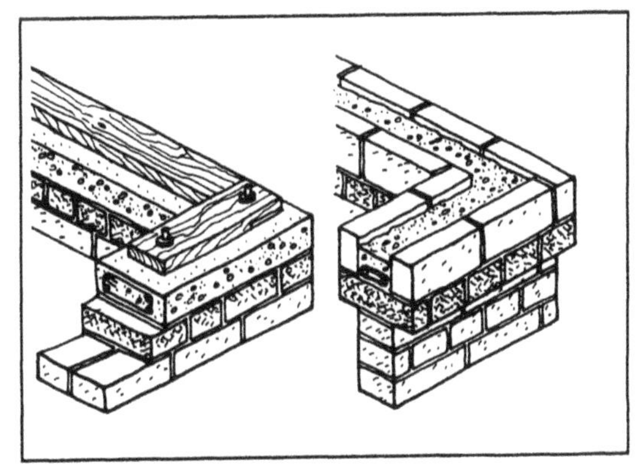

Steel ties must be fastened properly, especially at the angles of walls, and adequately coated with mortar and concrete. Metal lattice ties also exist. It is advisable to cast the tie in reinforced concrete upon a layer of stabilized earth in order to ensure that the concrete adheres well to the earth and that there is no deterioration due to contact with wet material.

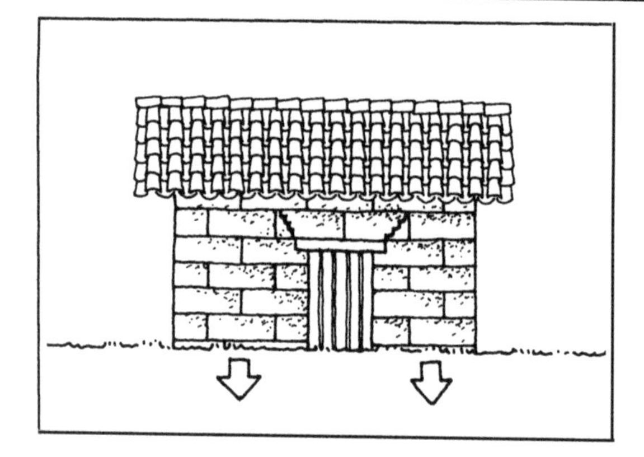

The structural bond between the frames of openings and earth walls must be given particular attention lest cracking occur, which would give rise to rapid erosion, especially if there is the added problem of chronic damp.

Structural faults

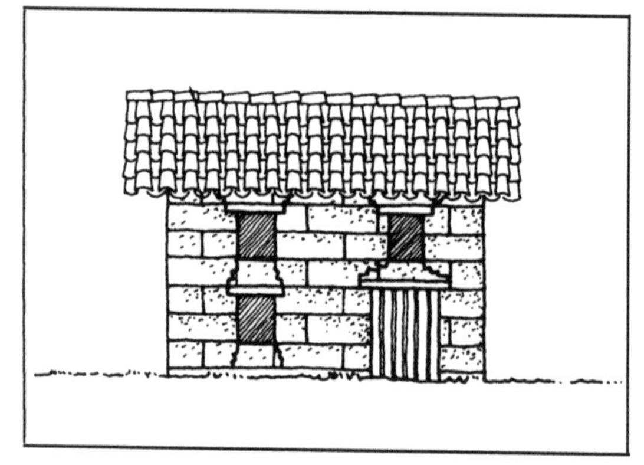

Failure of sills to adequately take up shear stress, undersized lintels or oversized openings cause cracking in reveals. In addition, the weakening of reveals and earth walls can be observed when a wall shrinks on drying (e.g. rammed earth, cob). Certain classic errors must therefore be avoided, such as:

- Oversized openings: overloading of lintels and different rates of settlement;
- Accumulation of openings in the same wall section and an excessive variety of sizes, weakening the wall.
- Openings too close to a corner of the building: bowing of the corner.
- Openings too close together, with a pier of insufficient strength: bowing of the pier.
- Weak jambs.
- Insufficient anchoring of lintel and sill: shear.
- Structural error in a wall near openings: e.g. vertical cracking.

Water damage

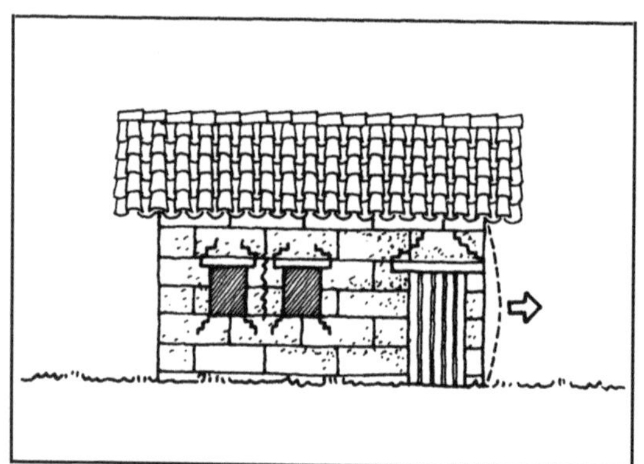

Damage due to cracking gives access to water erosion by run-off, splash, infiltration, and standing water. The weakest points are the bonds between the following: lintels/soil walls (anchoring), jambs/walls (toothing and bond) and sills/walls (anchoring). Similarly the rebates and reveals must be strengthened as well as all sealing around frames, and hinges of doors and shutters.

It is advisable to provide the following:

- Drips under lintels and sills, on outside walls, or a system incorporating flashing which water cannot penetrate. Avoid unsuitable projections on lintels and jambs.
- Solve all condensation problems (e.g. thermal bridges).
- Stabilize earth walls in the vicinity of reveals (particularly under sills).
- If possible, cover or render the reveal in the outside wall.
- Provide waterproofing under sills.

Mechanical aspects

The reveals of openings should be generously designed: heavy lintels and sills and stable jambs. Point loads must be taken up. The openings can be dressed in wood or masonry (taking care not to increase differential stresses between the frame and the wall). The opening can be cut after the wall is dry, but the lintel must be fitted beforehand.

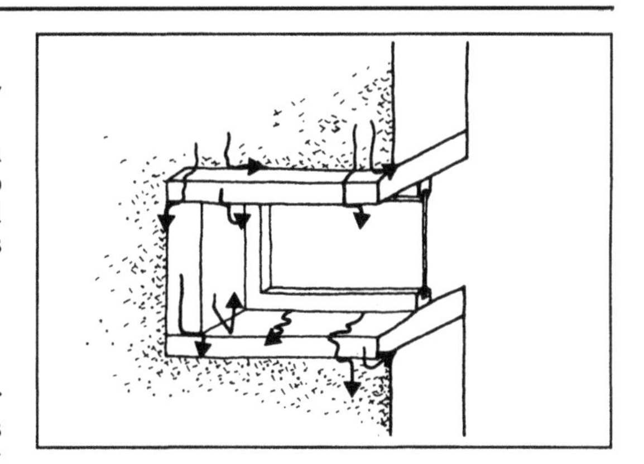

1. Lintels

These can be made of wood or concrete, stone, or brick. It can be cast *in situ* or prefabricated. Lintels are subject to considerable stress. The lintel must be supported in the wall over at least 25cm, and more for large openings. It is moreover advisable to increase the compressive strength of supporting lintels: stabilize jambs or dress them in a masonry.

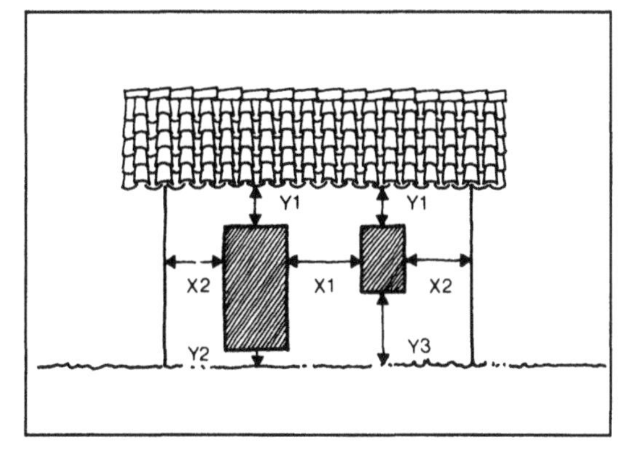

2. Sills

Loads transmitted by the jambs must be properly taken up: lengthen the sill and add reinforcement under the sill. To prevent shearing of the window breast, it is best to use dry joints between the breast-wall and the wall proper or else use independent infilling after the construction of the wall: e.g. rammed earth wall, sun-dried brick breast-wall. It should not be forgotten to plug the dry joints of the breast-wall once the wall has completely dried and is stable.

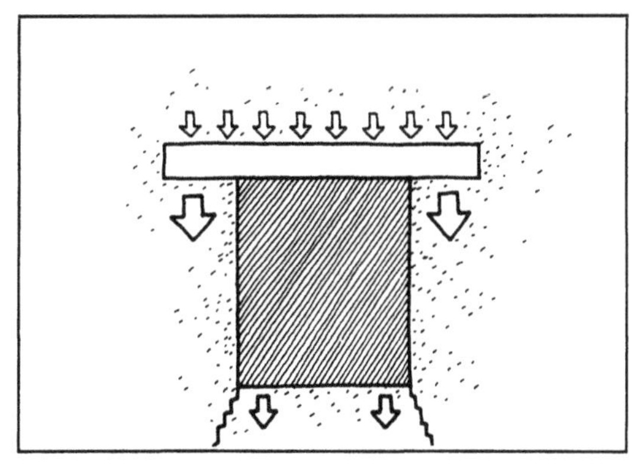

3. Dimensions

The following are guidelines, and do not exclude variations in the design of openings.

- The ratio of apertures to solid sections in any one wall should not be greater than 1:3 and should be as evenly distributed as possible. Avoid concentrations of apertures or excessively large openings.
- The total length of the openings should not be more than 35% of the length of the wall.
- Conventional spans do not exceed 1.20m.
- The minimum distance between an opening and a corner is 1m.
- The width of a pier should not be less than the thickness of the wall with a minimum of 65cm. Piers less than 1m wide cannot bear loads.
- The ratio of the height of the breast-walling under sills and above lintels to the width of the bay should be adequate. The reveals can be strengthened: e.g. reinforcement under the sill.

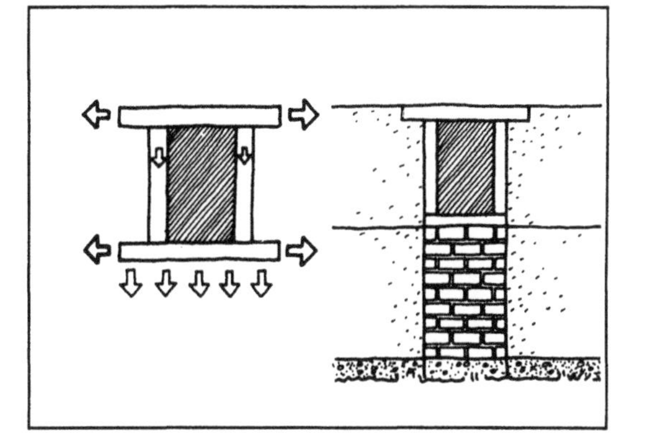

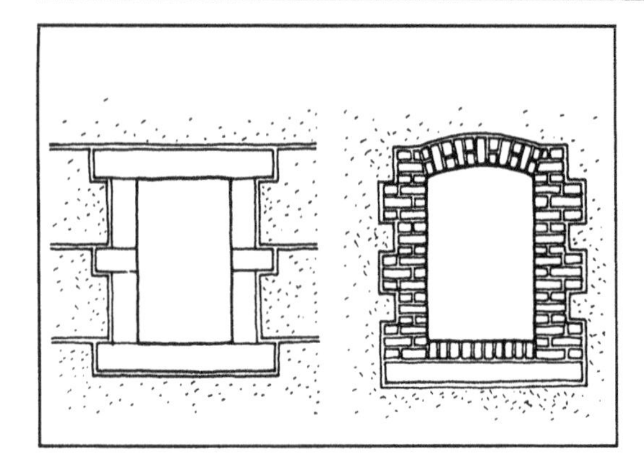

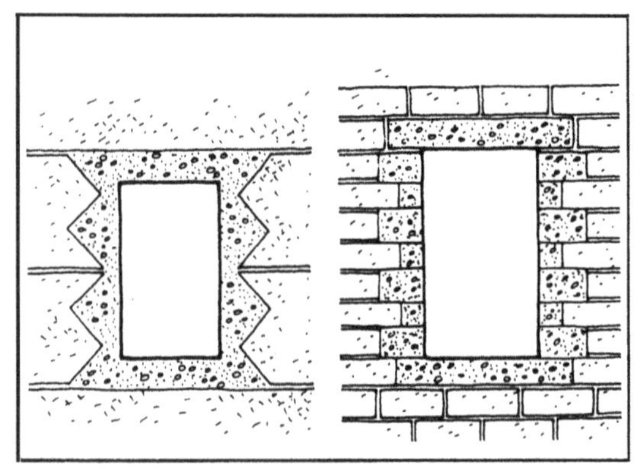

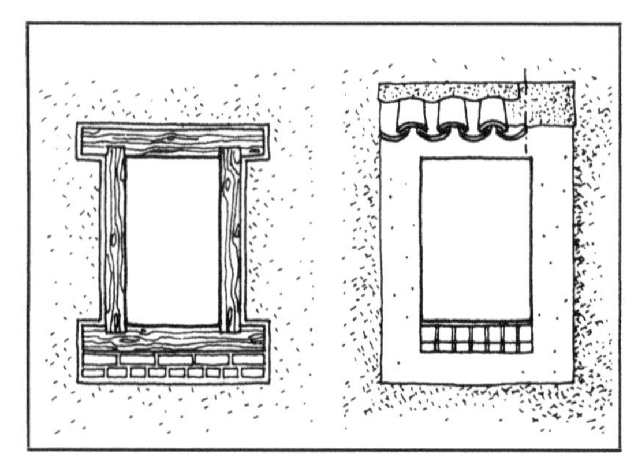

The reveals of openings can be dressed with 'hard' materials so as to ensure that loads and stresses are properly taken up. This procedure facilitates the anchoring or fixing of joinery, but the provision of such hardening is difficult: problem of the bond between 'hard' materials and earth walls.

Materials

– Stone or brick lintels can be in the form of straight lintels or a lintel course, or variously shaped arches: dropped, segmental, basket-handle, semi-circular, etc. The jambs either consist of solid elements or are built up (e.g. brick); the jambs can be secured in place in the walls with toothed anchors. The toothing can be either rectangular or triangular and the bond with earth walls is secured by a seam of mortar applied between the masonry reveal and the earth wall. Sills are either solid (e.g. stone or concrete) or built up (e.g. bricks laid on edge).

– Concrete makes it possible to build monolithic reveals, cast while the earth walls are being erected or prefabricated. The bond between the concrete and the earth (toothing) must be good and projections of concrete with respect to the bare earth wall should be avoided.

– Wooden reveals are very widespread. The lintels, jambs, and sills constitute a rigid frame. This can either be set on the outside of the wall or form a solid mass extending through the entire thickness of the wall. Adequate penetration of the lintel or sill through the earth wall ensures good structural bonding of the materials. Nevertheless it is advisable to place boards on a mortar bed or even on brickwork. In regions with a damp climate it is better to render these wooden frames (roughing or studding of wood) and provide flashing above the lintel.

– Reveals can be made of stabilized earth or be strengthened. In walls made of stabilized compressed blocks the lintel can be replaced by a brick arch. The jambs must be perfectly bonded while the breast wall can be independent, constructed after the walls and arches. Small openings can be of corbelled construction, if carefully bonded. In rammed earth walls the lintel can be made of wood or be a brick arch and the jambs reinforced by lime mortar beds or triangular toothing in mortar. Care must be taken to bevel projecting angles so as to reduce erosion.

Protection of reveals

Reveals should be protected against water and wind erosion. This can be very pronounced on a wall subject to cracking. Such protection can be provided by giving careful attention to the construction of reveals. It can moreover be enhanced by surface stabilization or rendering round the reveals.

For dwellings with storeys above ground level, openings on walls facing the prevailing winds are more exposed than on the ground floor, particularly at sill level. The breaking wind creates eddies which are particularly marked near window breasts and lintels of ground floor openings. It is thus advisable to stabilize exposed parts. The sill of the upper opening should not protrude too much from the wall (wind erosion). The seal between the sill and the earth wall must not be forgotten, nor should the flashing above the lintel and the throating under the lintels and sills be neglected.

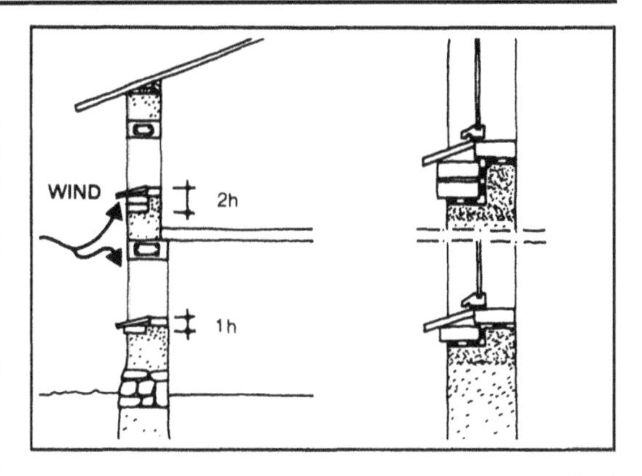

Fixing subframes

When it is planned to fix subframes for windows and doors directly in earth walls care must be taken to provide solid anchoring as the shocks and vibration resulting from frequent use can cause cracking and loosening. Pieces of wood, to which the joinery can be attached, can be embedded into the earth walls. The fastening of these boards (planks in the form of jambs or special blocks) is done by means of metal holdfasts, by nailing and setting after coating in mortar, or by means of barbed wire ties which are progressively sunk into the wall as the jambs are built.

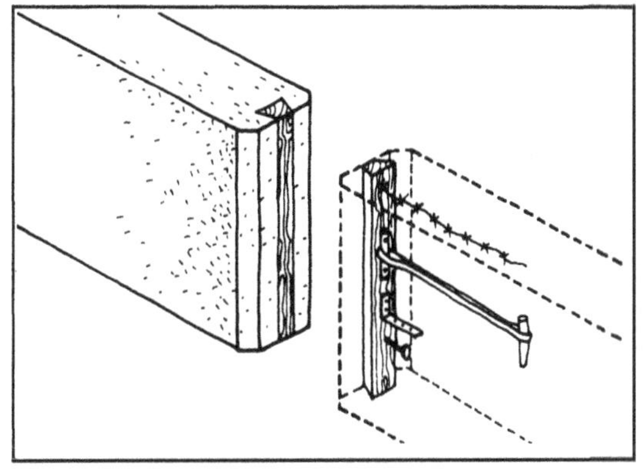

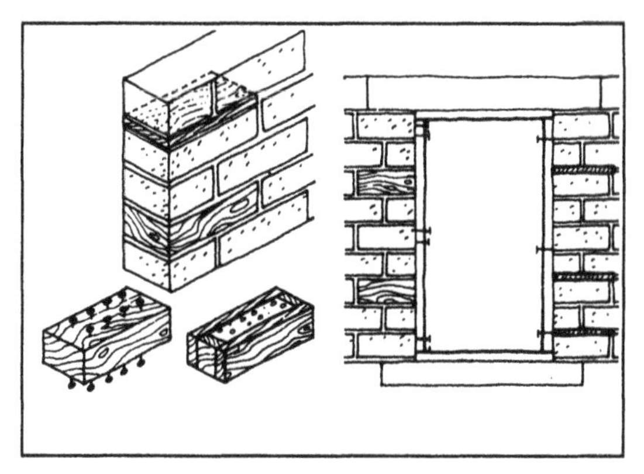

Joinery

The joinery should be made with great care and should be provided with throating on the underside if it projects beyond the wall. If the joinery is recessed, the sill must be provided with a weathered surface and be well inclined so that the water can be drained away. Care must be taken to fix the shutter hinges securely in the wall. The damp-proofing under the sill must not be forgotten.

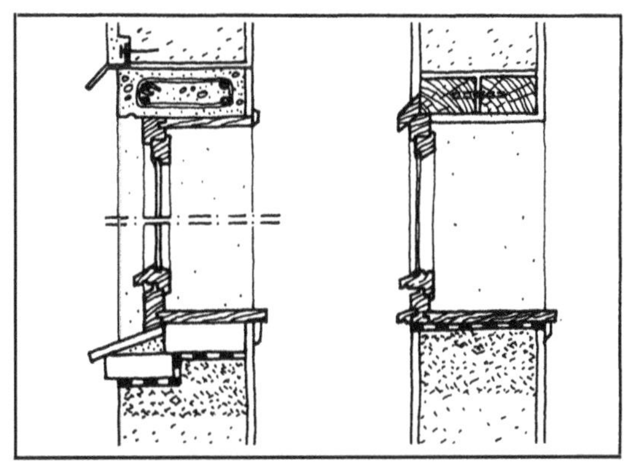

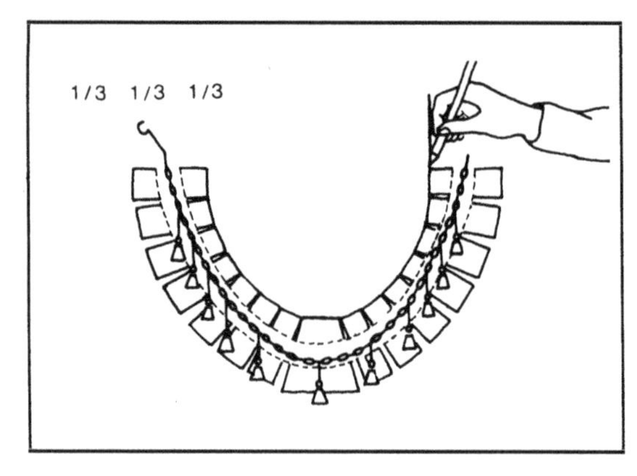

Arches are good for bridging openings in walls, as the soil is under compression.

Varieties of arch

The shapes best adapted to building in earth are:

- semi-circular and drop semi-circular
- basket handle, both full and drop
- ogee arch and drop ogee
- tudor, elliptical and flat arch
- catenary and parabolic

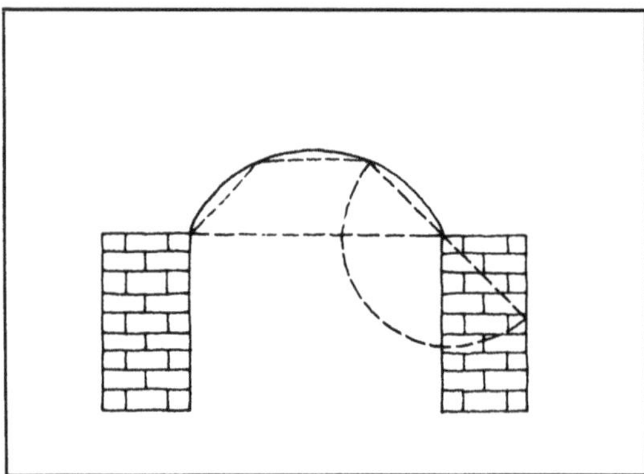

Arch design

The lines of thrust within the arch can be evaluated by various methods: calculation, graphically, or by simulation. The plot of a catenary (suspended wire) simulates an arch bearing its own weight, but does not simulate the situation of a load-bearing arch. The resultant curve of the pressure line should stay within the middle third of the arch. If the curve leaves the middle third, there is reason to fear cracking.

There are two main solutions to the problem of arch design:

– The shape of the arch is determined in advance. The plot of the thrust lines should pass within the middle third of the arc and the arch is given the required thickness.
– The thrust lines are determined and the shape and thickness of the arch are adapted to them.

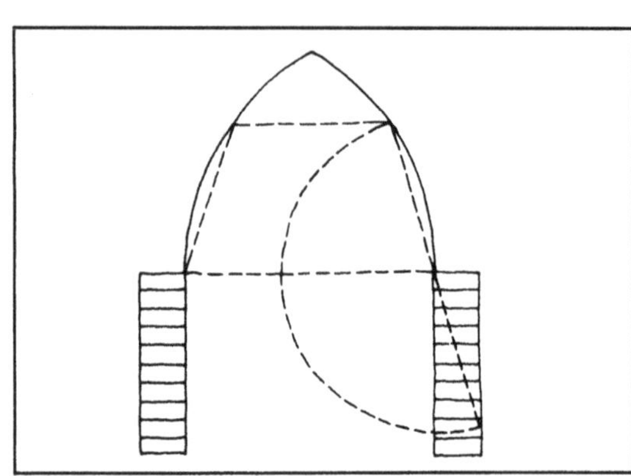

Even so, the calculations do not exactly reflect reality as the masonry loading the arch is not passive. It has its own internal cohesion and forms a corbelled arch (this can be observed in walls which have collapsed, and the span may be as much as 7m). Any plot of thrust lines is therefore hypothetical. It is thus essential, especially for large spans, to build the arches with great care to avoid cracking and loosening.

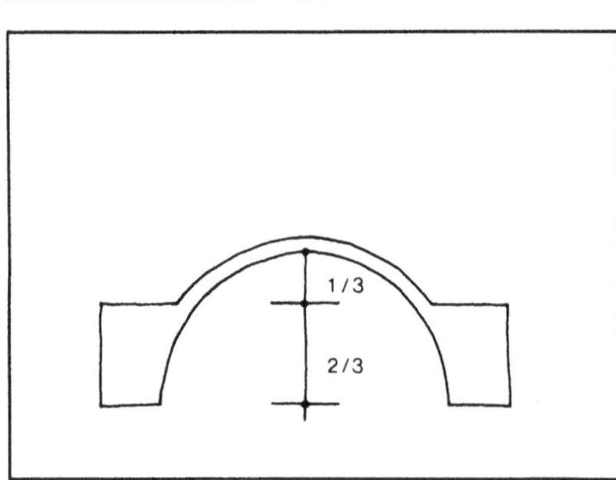

Pier design

Arches transmit great thrust onto the springing points and onto the piers supporting these. Consequently, the latter must be solid and stable. These thrusts can be calculated or estimated graphically (e.g. the Maxwell-Cremona polygon of forces). It is also possible to estimate the size of the piers by an empirical method: the prolongation of the first third of the arch should always fall within the pier. When two arches having the same plot meet on the same pier, the thrusts cancel one another out. In this case, only the descending vertical load is considered.

Building arches

1. Arch rings

Depending on the span and loads involved, arches may be built using a single, double or triple ring. A light formwork permits the construction of the first ring, which serves in turn as a base for the second, and so on. However, the load distribution of these consecutive rings must be correct, and they must be bonded to one another so that no cracking occurs. The actual construction of the bond is a more delicate task.

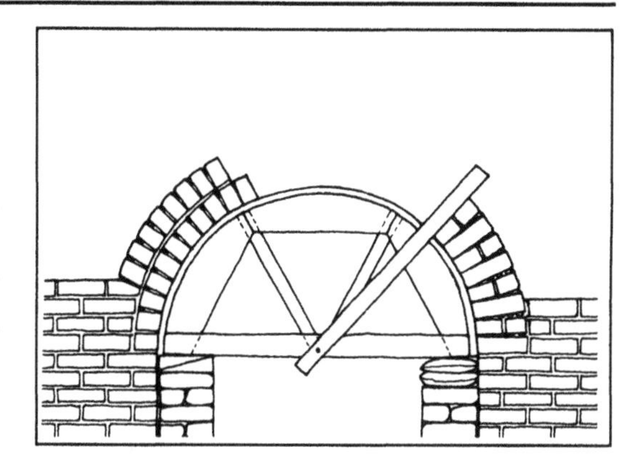

2. Formwork

A corbelled arch requires no formwork (with a shape approaching that of an ogee arch) but other sorts require the support of formwork. The formwork can be wood or metal, temporary masonry or something even lighter (e.g. formwork made from palm trunks with a coat of rendering). The use of wedged wooden blocks or small sandbags allows the formwork to be removed without stressing the arch (risk of cracking). The formwork should only be removed when the structure is stable and safe.

3. Masonry

Arches made of adobes or stabilized compressed blocks must be well bonded: the heads of the joints between the masonry and the arch must correspond. Construction proceeds symmetrically from the two imposts of the arch and ending with the keystone. The imposts are recessed in the wall and are cut so as to determine the direction of the arch and to take up the thrust as well as possible. In principle the mortar is not taken into account when constructing the arch: the bricks are regarded as if they were laid dry. The bricks touch at the intrados, on the formwork and are wedged at the extrados by a pebble. The packing of the joints must be carried out with great care. Special voussoirs can be used (trapezoidal or interlocking) which prevent slipping. These bricks are better suited to large spans, and the arch and the masonry which it bears are constructed simultaneously. In large arches the keystone is ideally made of a material which is cut to size on the spot and will be quite wide (40cm). When pebbles are used for wedging the formwork can be removed immediately. If only mortar is used, it is necessary to allow a drying period before daring to remove the forms.

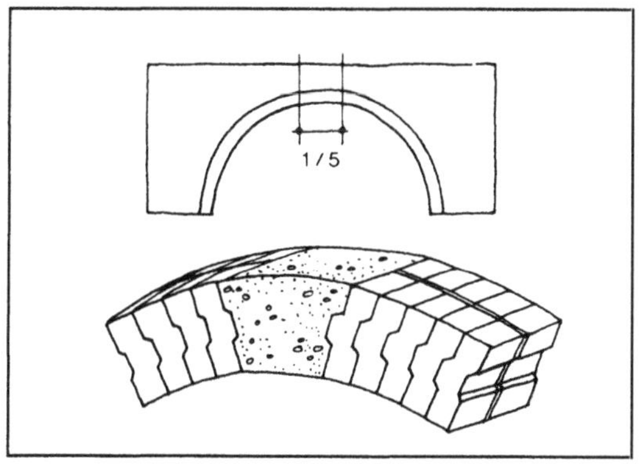

4. Doors and windows

They either have to adapt to the shape of the arch, or a beam has to be placed under the spandrel, and the space in between filled in with light masonry.

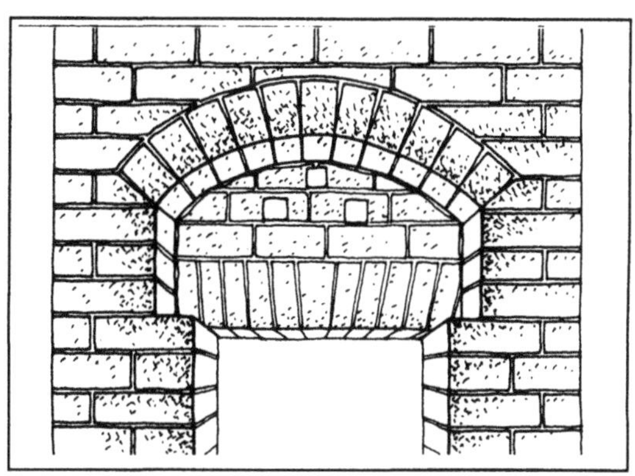

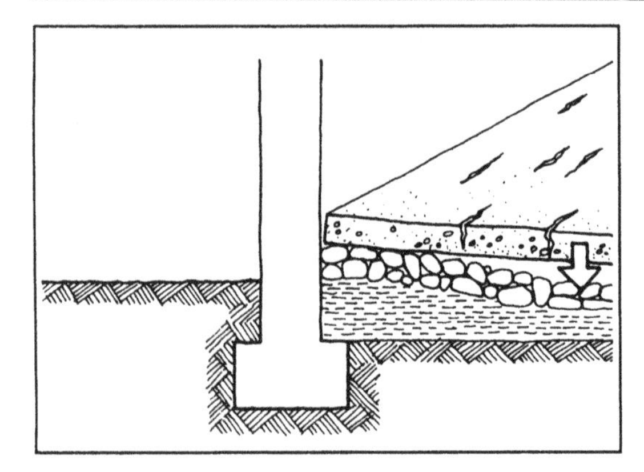

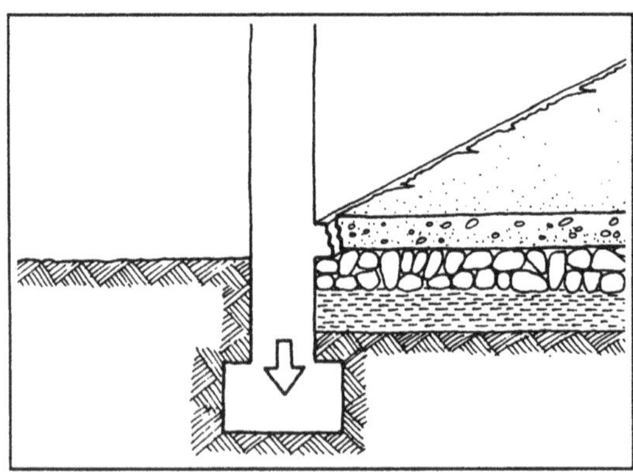

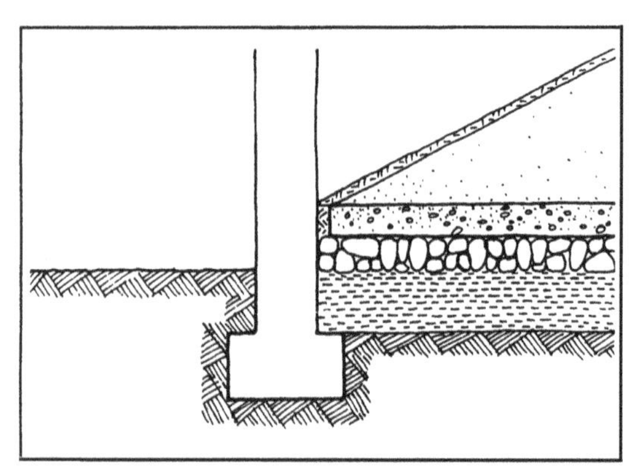

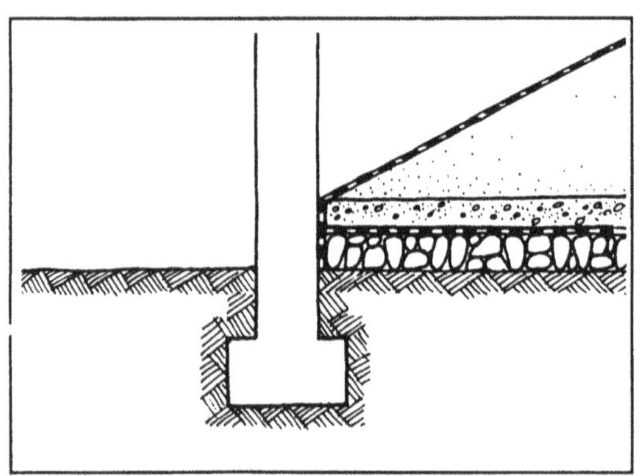

By tradition, earth floors are widely used in earth construction. Such structures, when well executed, are strong, attractive, sound and economic. This type of floor is hardly used any more in modern construction but deserves to be revived. When badly executed, earth floors constitute a health hazard.

To construct an earth floor, certain rules must be followed and a few precautions taken since the floor must be able to resist perforation, wear, attack by water, standing and live loads. It should distribute these loads evenly and transmit them to the ground; and it may furthermore be necessary to enhance the load-bearing capacity of the ground. The floor should moreover have insulating properties (thermal and acoustic), be easy to maintain, be attractive (texture, colour) and be able to accommodate services (e.g. electricity supply), remain dry and be pest-free.

Earth floors are found chiefly in ancillary buildings and spaces: sheds and outhouses, for example. In cellars their practicability depends on the permeability of the underlying ground and the level of the groundwater (at least 3m) and how well the cellar is ventilated. In general, earth floors can be used for dry and well-ventilated spaces, on well-drained and dry soil. They are also used for living areas (e.g. living-room, bedrooms), but this demands very careful finishing and a more elaborate design (insulation, for example). Their strength, going up to 35MPa for point loads, is high.

Different rates of settlement

Floors and walls function differently; walls can support heavier loads than those applied to floor slabs. Consequently the floor should be made structurally independent of the walls so as to avoid cracking due to different rates of settlement. The material of the damp-proof membrane against the foundation slab must be flexible so as not to be altered by the different stresses applied by the two structures. The preparation of the underlying soil upon which the earth floor is constructed must be painstaking so as to avoid settlement: levelling, back filling if necessary, and compaction.

Dimensions

The thickness varies depending on the load-bearing capacity of the underlying ground, the stresses due to the applied loads and the chosen finish. A floor comprising a foundation in pitching, a sand profile and covered by rammed earth can be 45cm thick.

Protection against water

Earth floors must be protected from water in general and from capillary rise in particular. Depending on the contour of the ground, it may be either the floor is permeable and allows the water to evaporate or it is water-tight like the foundation. The water-tight membrane must be laid with the greatest of care and this earth must not subsequently cause the barrier to deteriorate. The latter must connect up with the water-tight membrane of the wall. It is thus preferable to provide a layer of sand on which to lay the water-tight barrier. The finished level of the floor will be above the finished level of the base course which will also have been damp-proofed (specially between the base course and the earth wall). It is furthermore desirable to provide finishing for the joint between the floor and the wall by means of a skirting board. The fastening of this skirting board must not damage the vertical damp-proof membrane. The other forms of water attack can result from the daily use of water: washing floors, condensation (thermal bridges), or defective piping. Care must be taken to find a solution to all these problems.

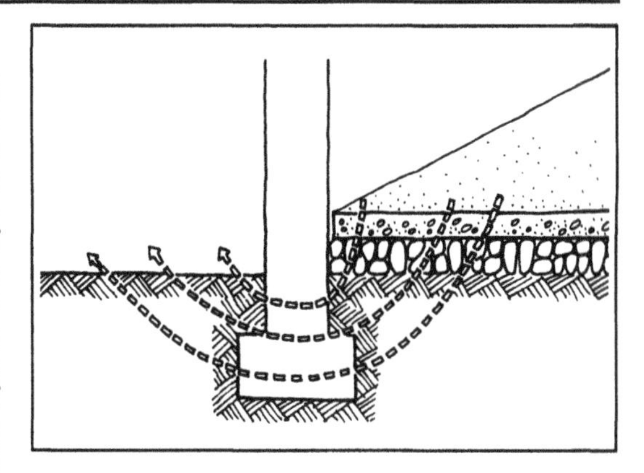

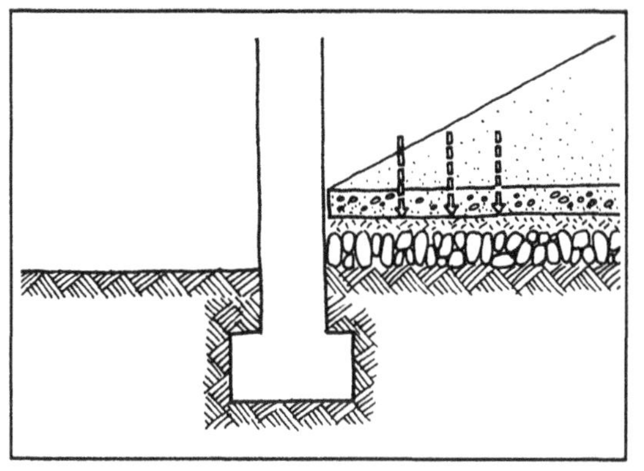

Thermal insulation

Earth floors do not provide much insulation, particularly around the edges. It is thus appropriate to envisage putting down a peripheral thermal insulation. Care must be taken to use an insulating material which will not easily perforate.

Durability

1. **Stabilization** Either the entire floor mass or just the surface layers (more economic) can be stabilized. The same additives are used as for the production of stabilized earth and procedures are similar. Vegetable glues (e.g. white glue) can also be added. Stabilization considerably enhances the resistance to water and wear of earth floors.

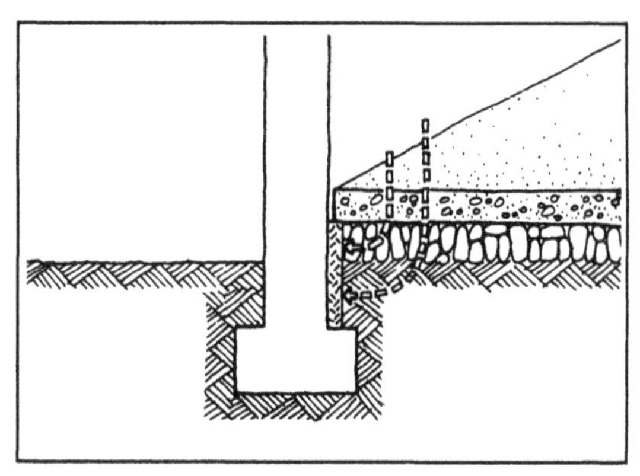

2. **Surface hardeners** Dry earth floors are traditionally hardened with animal (horse) urine and ox blood. Ox blood is sprinkled with cinders and clinkers and then beaten. Silicates of soda (floor not completely dry) or fluosilicates (dry floor) can also be used. Sump oils or a mixture of turpentine and linseed oil will also serve the purpose. A clean finish can obtained using polishing wax.

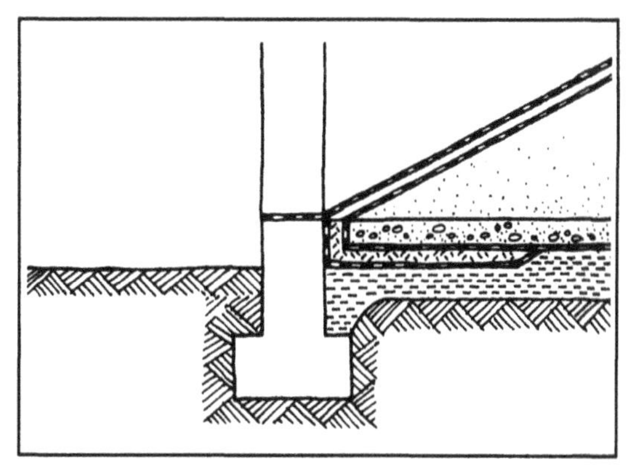

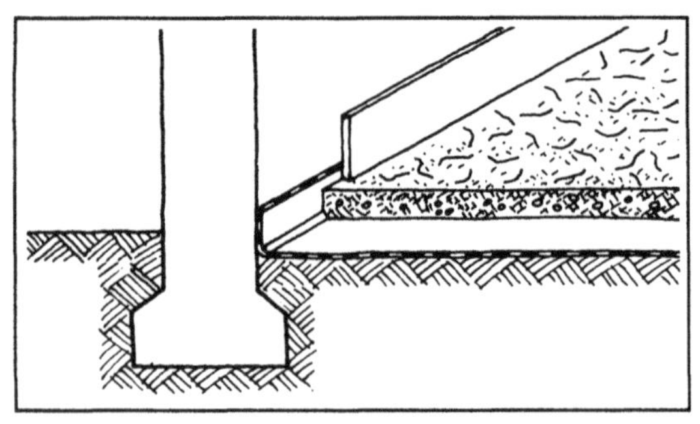

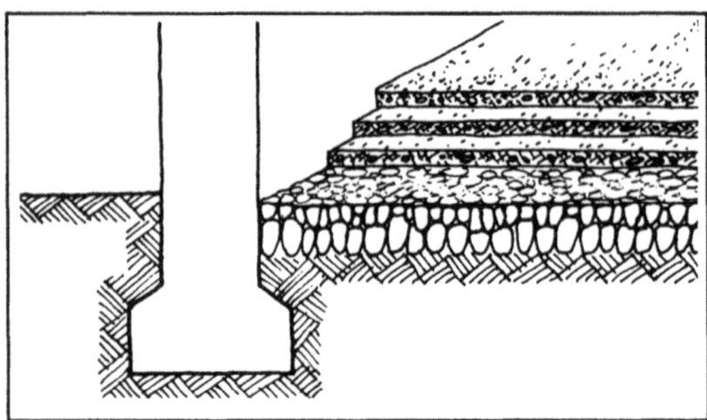

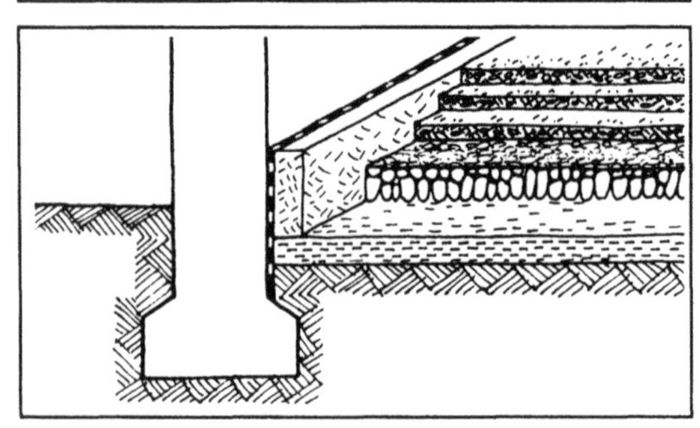

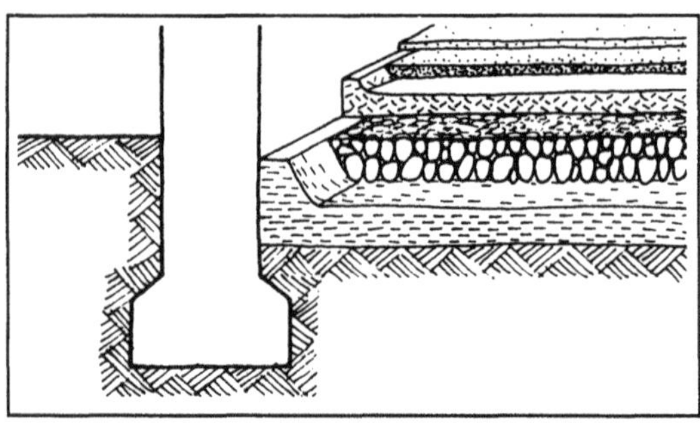

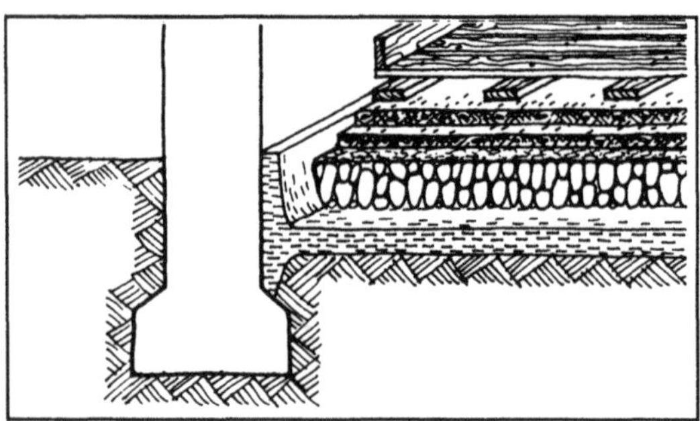

Execution

1. Preparation Before constructing an earth floor the underlying ground must be prepared. The layer of earth containing vegetable material and humus is carefully cleared and all organic residues eliminated from it. The upper layer of the earth is rammed except when the soil is subject to swell or when its load-bearing capacity is adequate. The ground must be dry before the floor is begun.

2. Waterproofing An anti-capillary barrier can be made with a 10cm layer of clayey earth applied in the moist state in carefully rammed layers. Each layer must be dry and fine cracks filled before being covered. The top layer is finished with a flattening hammer after ramming. A bitumen or lime-stabilized sandy soil can replace the clayey soil. Stabilized earth or a plastic sheet can also be used. Whatever solution is adopted, the damp-proof membrane must rise up along the base of the wall.

3. Stone filling Dry-stone pitching, 20 to 25cm deep, makes a good foundation. The largest stones are put down first, followed by tipped gravel. Dry stone can be replaced by gravel and coarse sand or by ballast.

4. Insulation The pitching is covered by an insulating layer, which can be a 10cm layer of straw-clay. If this layer is too pliant, it can be covered with a bearing course which distributes the loads.

5. Load-bearing This layer can be made of clayey earth mixed with cut straw or clay-straw containing hard and finely chopped straw (stalks 4 to 6cm long) to which cement mortar is added, at a rate of one volume of cement for every six volumes of washed sand. The load-bearing layer should be 4cm thick and always superstabilized so that it will harden after it has cured. Such curing sometimes requires moistening (damp sacks).

6. Finish Finish can be made with a thin layer of cement grout to which fine sand is added or has been given an oil-based treatment. A light filler in the form of sawdust can also be added to the grout. Proportions are as follows: one volume of sand, one volume of cement, one volume of sawdust. To mineralize the sawdust it is soaked beforehand in a lime or cement slurry and dried. The final stage of the finishing can be the colouring or waxing of the floor.

Monolithic floors

Floors can be constructed using standard adobe mortar stabilized with bitumen. First the soil is properly prepared, then the mortar is poured and spread out with a screed after the floor guides have been positioned. Shrink cracks must be filled with a finer adobe mortar and the final touches are made with a trowel. After drying, the surface is treated with a mixture of turpentine and linseed oil and allowed to dry for a week before putting down each layer. The floor can then be waxed and polished.

The earth-stabilized concrete mortar (one part cement to six to eight parts sandy soil) is also suitable. It is put down in a 5cm thick layer marked out with pegs. Contraction joints must be left every 1.5m. These must be deep and well defined (3cm deep and 0.5cm wide).

Other types of monolithic floors are constructed using rammed earth, sensitive to damp, stabilized rammed earth or rammed stabilized clay-straw. Insulation around the edge and a finishing screed (more costly) can also be provided. These earth floors are not attached to the walls. A flexible joint can be made in clay-straw.

Floors made from elements

Earth-stabilized bricks or tiles can be laid on a 2cm thick layer of mortar: one part cement to six parts of sandy soil. Care must be taken that the tiles adhere properly to the mortar. The joints can be filled with cement grout. The blocks used should be superstabilized or coated with a wearing layer. These blocks can also be laid on a layer of sand and the joints filled with sand.

The shape and even the colour (addition of oxides to the earth) of the paving blocks can be worked out in advance. The shape, however, should always be kept simple so as not to complicate the moulding operation, especially when a press is used. To obtain good wearing resistance, all the corners of the blocks should be equal to or greater than 90°. They should be from 6 to 9cm thick, and can be a given a two-layer treatment with a superstabilized wearing surface.

Floors made of clay-straw paving blocks (40 × 40 × 16cm) coated with a wearing cover and hardened (e.g. with fluates) are insulating.

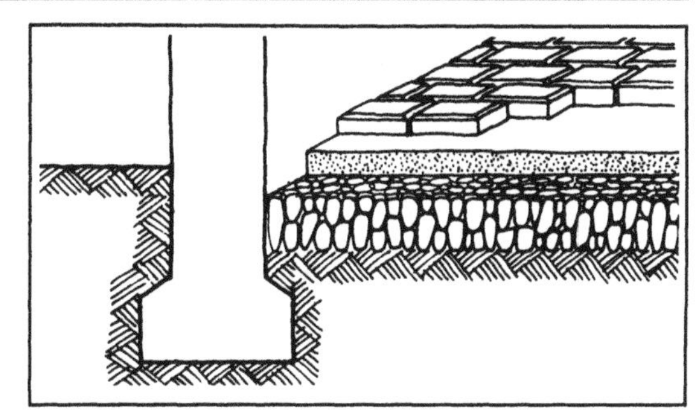

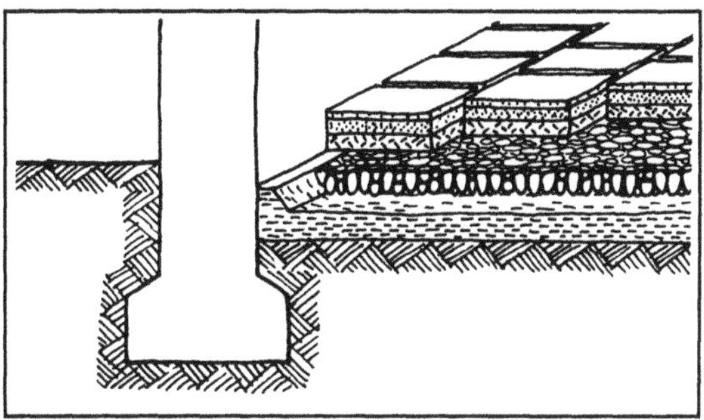

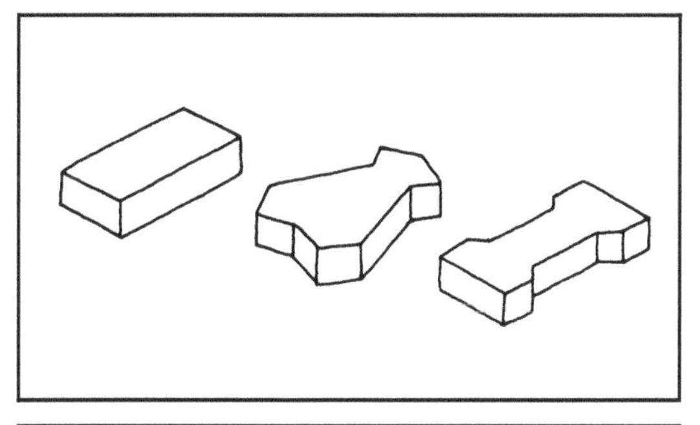

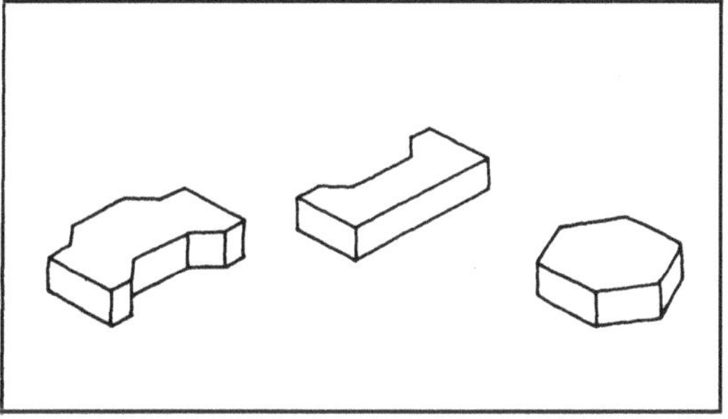

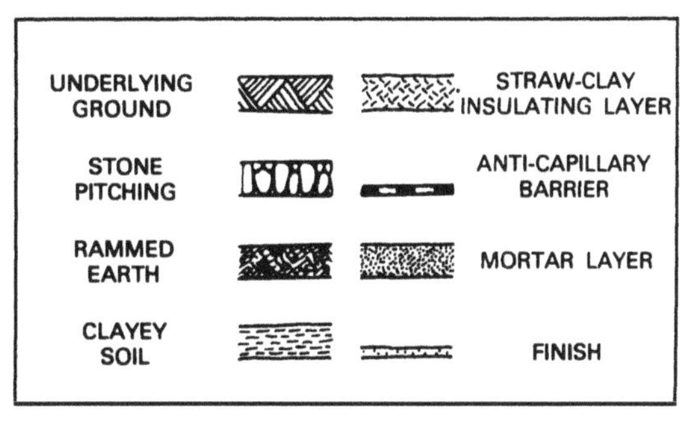

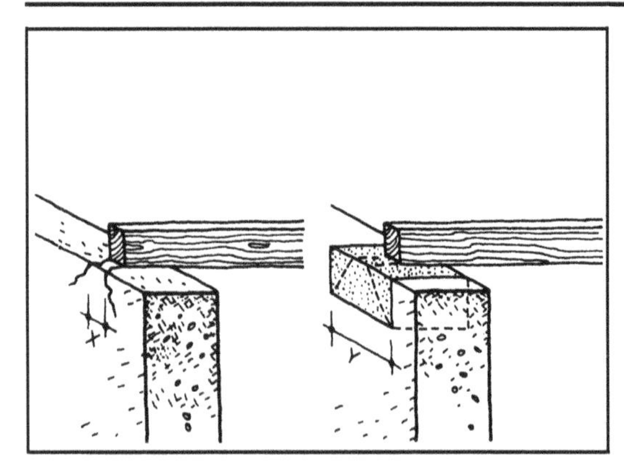

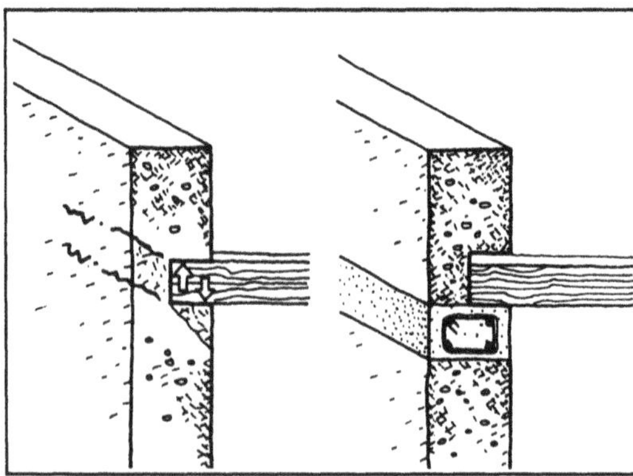

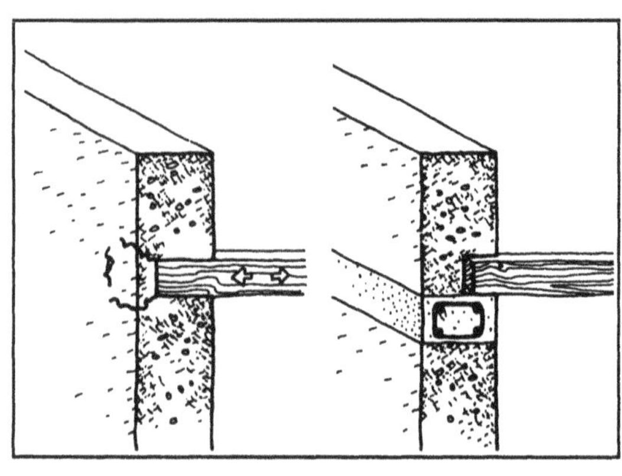

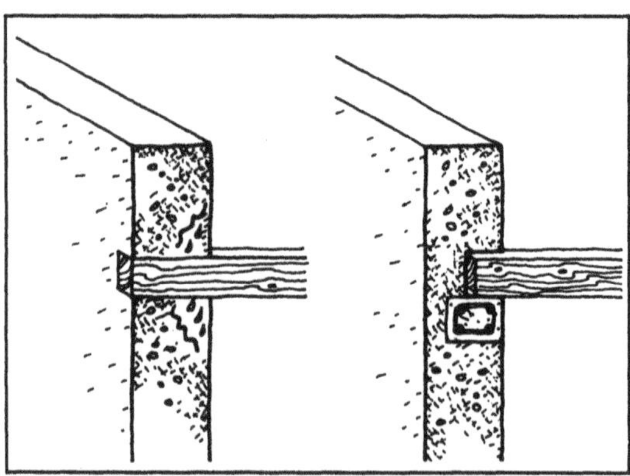

Raised floors can be at various levels:

- above a basement or sanitary space;
- at the top of a house, under the roof;
- on the flat roof.

A raised floor is made up of different parts:

- the upper surface (floor)
- the structure (woodwork or slab)
- the underside (ceiling)

Each of these parts imposes conditions on the soil used for its construction and also gives rise to specific problems related, in the main, to support. In principle, however, all the traditional kinds of floors can be used.

Conditions and limitations

The upper surface or covering must be able to resist wear and be durable, flat, pleasant to look at, easy to maintain and in the case of flat roofs be water-tight.

The structure, woodwork or slab must be able to support live and static loads, resist point loads, be very rigid and transmit loads to the supports properly.

The underside, or ceiling must be pleasant to look at.

Floor-wall bond

This the support with provides the bonding between the floor and the walls or pillars and the transfer of loads. Four sources of defects can be identified:

1. Indentation The support is under-sized and does not distribute the loads uniformly: differential tension and cracks. The surface area of the support must be increased and transfer the loads towards the centre of gravity of the wall. Supports which weaken the longitudinal section of the wall are to be avoided.

2. Rotation The floor sags and undergoes rotation upon its support: rising at the edges, eccentric loading, compression of the inner edge, and cracking. Care must be taken to establish a satisfactory load/bearing ratio and to install ties.

3. Dimensional discrepancies These may be thermal in origin or due to differential stress between the floor and its support. Direct contact between the wall and floor is avoided by not fastening them to the wall, and horizontal ties.

4. Discontinuity Discontinuous thermal insulation causes condensation.

Execution

Tolerances must be allowed when inserting floors in walls. Precautions must be taken against the rain or floors must be laid after the building has been roofed.

Support

There are various solutions for the problem of the bond between floors and earth walls with support being either independent of the wall or on the wall.

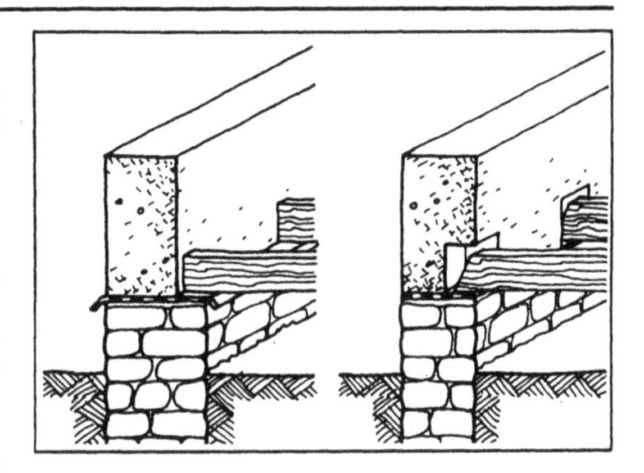

1. Support independent of the wall

On the foundations Studs supporting the beams are fixed on notches on the foundation footings. This is satisfactory for floors above cellars or sanitary spaces even though fixing the studs may be difficult.

On the base course This is made wider than the wall to take the beams. The base course must be levelled and and the beams must not touch the wall. The base course can be widened more cheaply by locating beams in niches in the bottom of the wall, at base course level. Here again the beams do not touch the wall. In both systems a horizontal anti-capillary membrane must be laid, and a skirting board fitted as finish. For reinforced-concrete slab floors it is best to use a widened base course since penetration of the wall by the slab will weaken the base of the wall.

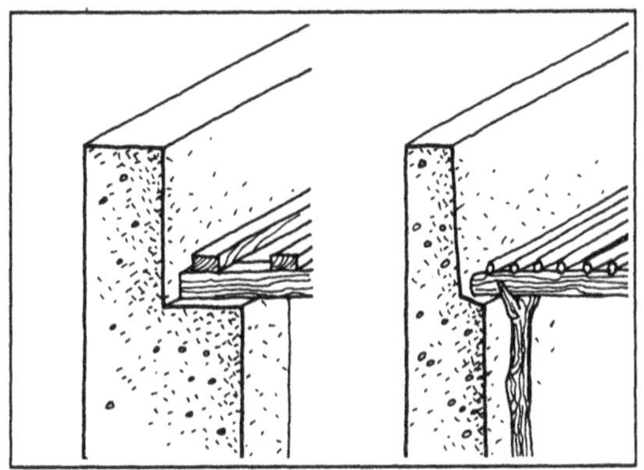

Beside the wall This method makes use of wooden corner pieces or masonry projections from the wall. This is the preferred solution for heavy floors and non-load-bearing walls: e.g. wattle filled in with clay-straw or cob.

2. Support in the wall

In this system there are the following dangers: point loading (beams) or continuous weakening (slabs) of the wall, rotation of the support (excessive freedom of movement of the beams or sagging of floors), different degrees of thrust on the wall.

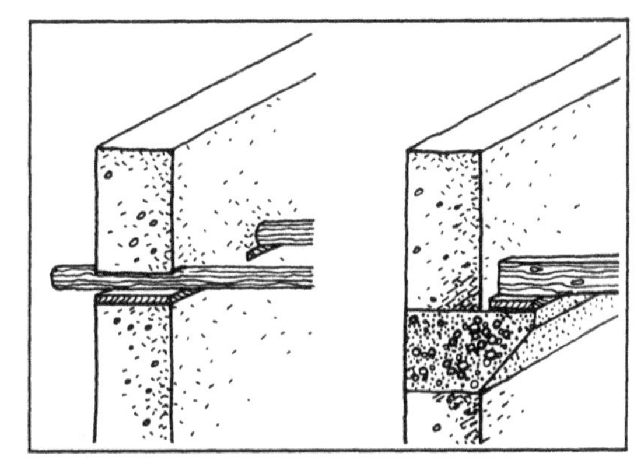

Point loads should be distributed by the ring-beam or a carefully designed (weight of loads transferred at an angle of 45°) wooden, stone or concrete sole plate or through ties. Space for the beams can be hollowed out or provided when building the walls. Treated wood should be used for wooden floors, and contact between wood and earth is to be avoided using bitumen felt or water repellent mortar, flexible material.

3. Support on the side of the wall

The thickness of the wall is reduced where the floor passes through it. Here too care must be taken with point beam supports or a ring-beam provided. Avoid placing the floor on a beam along the wall as there is a danger of it being torn out.

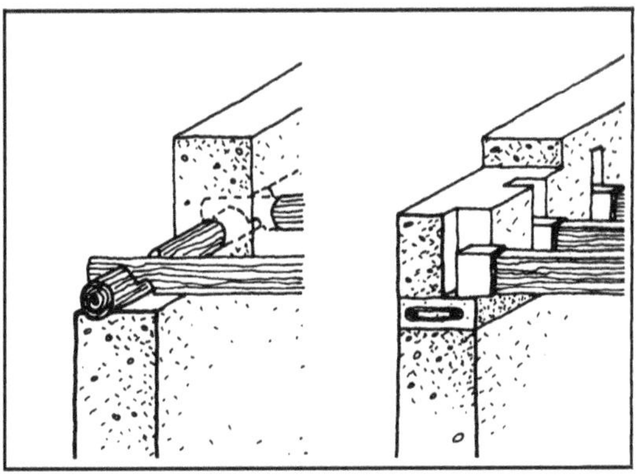

Despite its rather low tensile and bending strength, earth has been used to build raised floors. The results are quite remarkable, especially when the material is stabilized and/or reinforced. The main drawback is the weight of earth floors which makes them particularly dangerous in earthquake regions. In addition, these systems generate heavy loads in load-bearing walls. It is thus desirable to reduce the distance between the joists of earth floors (distance between 60 and 90cm) in order to obtain better distribution of the loads over the walls. Among the materials used for floors, clay-straw has the advantage of being light and having respectable tensile, shear and bending strengths as well as being a good insulator.

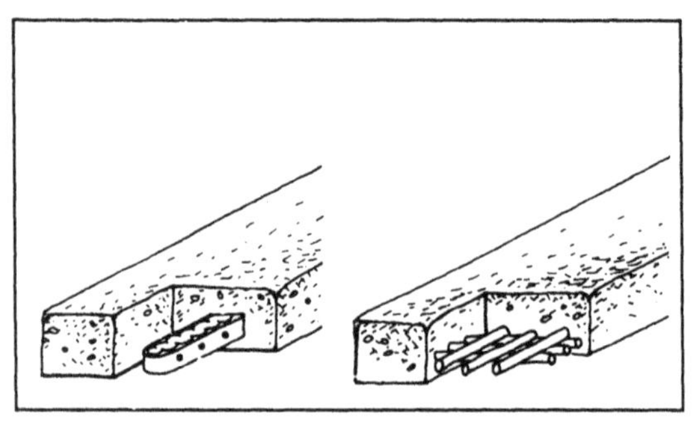

1. Earth used for the structure

Reinforced soil concrete slabs The traditional systems in which soil floors reinforced with wood are used have been the object of research aimed at modernization. This research has examined bamboo (USA) and galvanized steel (France, Senegal). The results are far from satisfactory as the systems are still very heavy ($500kg/m^2$).

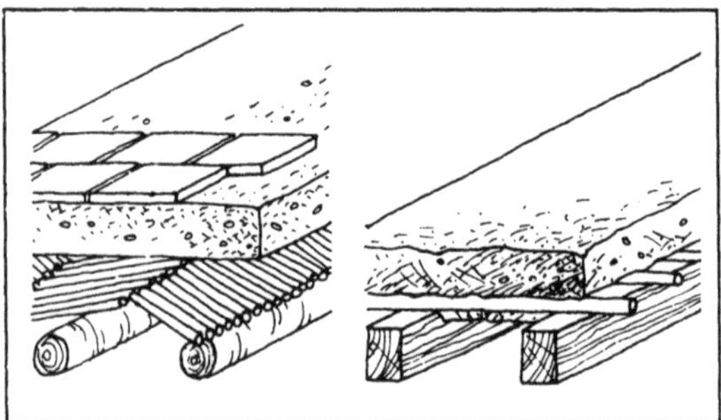

2. Earth used for the surface

Structure Wooden structures, consisting of beams overlaid with boards, wooden logs or sometimes even flat stones. To reduce dust, woven matting, strong paper, or straw is laid between the support and the earth, which is generally rammed (10–15cm). These floors weigh as much as $250kg/m^2$. Those laid with packed clay-straw on laths spaced 10 to 15cm apart are lighter ($150kg/m^2$). The finish is either a stabilized clay-straw wearing layer, or a rendering or even a surface coating.

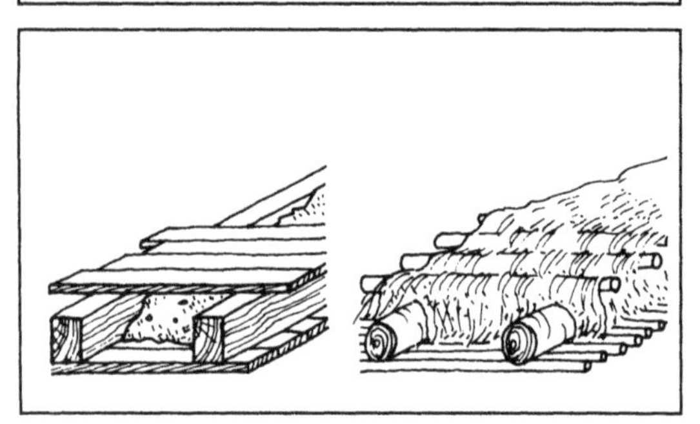

3. Floors with earth infill

These systems are frequently used to enhance acoustic insulation.

Loose earth Many wood joist floors and wooden board floors, with an underside of boards or reed screen, are packed or plugged with loose soil. The earth used should be absolutely dry.

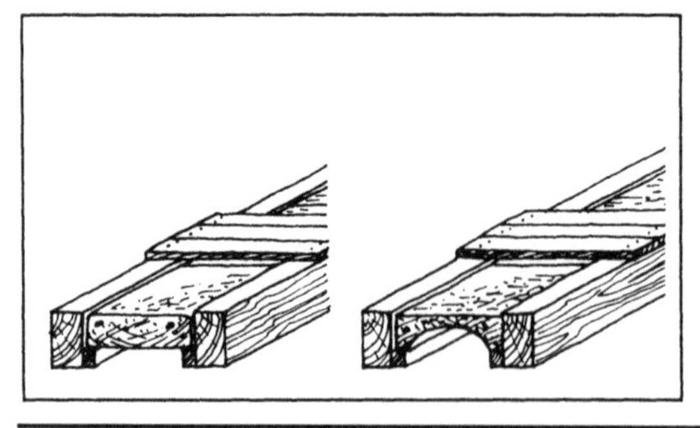

Prefabricated panels Usually these are made of daub or clay-straw and are used to fill the underside of the floor without adding to its bearing capacity. As a result, the distance between joist beams can be quite large – 80 to 90cm – and the prefabricated elements can be either short or long: 0.4 to 1.2m for a thickness of 15cm and a weight of 35kg to 120kg per element. The shape of the floor cavity makes it possible to lighten the floor.

4. Vaulting and load-bearing infills

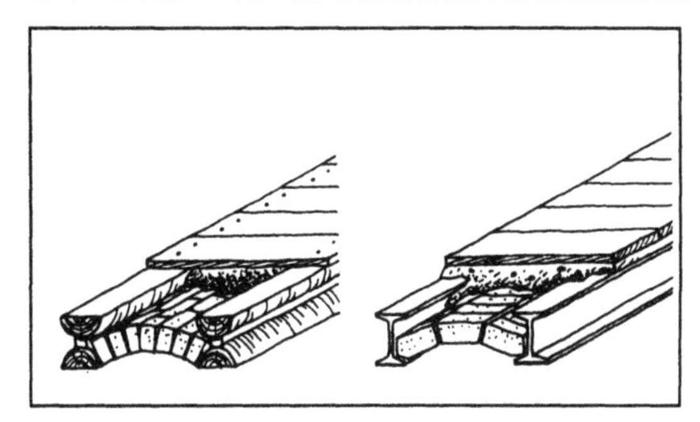

Earth-brick vaulting Floors constructed using brick vaulting put the soil in compression with bending stresses being taken up by beams in timber, steel or even reinforced concrete. The space between the beams ranges from 0.5m for the smaller systems to 2m for the largest ones, which sometimes require the use of metal ties. The brick vault rests on the lower flanges or the flanks of small beams. A slight deflection (one-tenth of the span) allows the small beams to take up the stresses well. These systems are fairly heavy (400kg/m^2). They can be constructed with formwork, usually slip forms, with the bricks being laid on edge or flat in two layers and with staggered joints; or without formwork: Nubian style slanting bond. Lighter, perforated bricks and tiles, shaped to fit on beams, or procedures which rely on wedging and require no formwork can be used.

Cased vaulting The vaulting is made of clay-straw, and is built using sacrifice formwork in extremely dense reed matting giving a light floor (150kg/m^2). Reed is an excellent key for ceiling renderings. Flat vaulting in clay-straw, internally strengthened with round timbers, is heavier (220kg/m^2). These systems provide good thermal and acoustic insulation.

Prefabricated infills The load-bearing element is still a long or short beam, but they are not spaced as far apart from one another (0.5m). Clay-straw rough masonry is reinforced by two wooden rods, solving problems of bending. A concrete or even stabilized earth slab (6 to 10cm) for distributing loads, and which can be reinforced with netting, joins the beams to the infills. Instead of infills earth reels in straw-clay can be used. These systems are light (150 to 200kg/m^2).

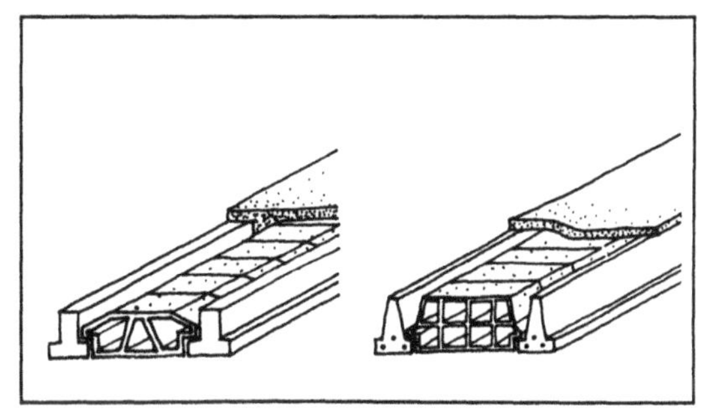

Hollow infills Made of compressed stabilized earth or extruded earth, this product is currently under test.

5. Other systems

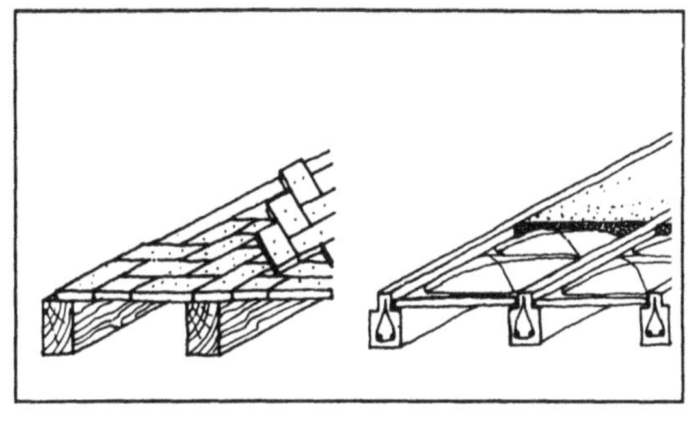

Hyperbolic infills Stabilized earth, on short reinforced concrete beams, has been tested in Pakistan. Loads can be as high as 1220kg/m^2. Research has been conducted in the United States (Max-Pot).

Thin tiles Thin superstabilized bricks with a rough-cast plaster finish (2 layers).

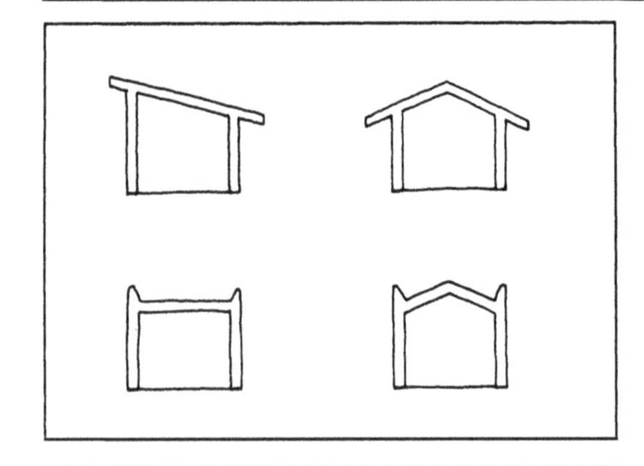

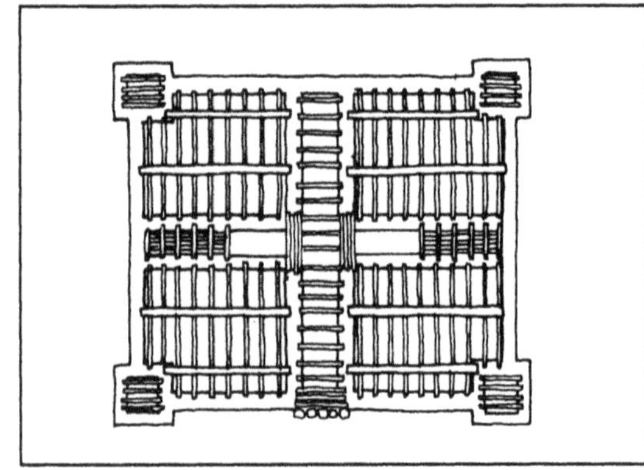

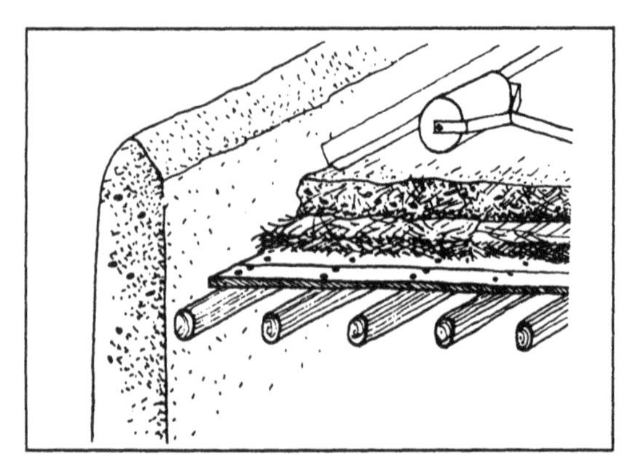

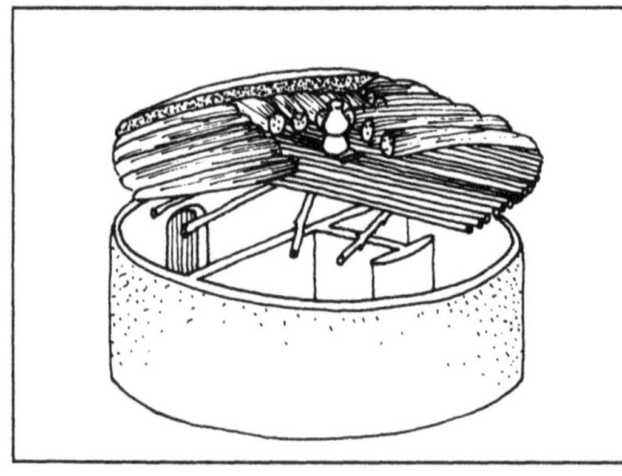

Roofing is the most costly part of a structure and can absorb as much as 70% of the total outlay. In addition, first-rate workmanship is required for the system to be durable. By definition, a flat roof has a maximum gradient of 10. A gradient of 1 to 2 is sufficient to drain off water. Flat roofs are common in hot regions but are not suitable for regions prone to tropical cyclones – there is danger of the roof being torn away by pressure differences. Generally, normal floor systems are best for flat roofs, but these should be improved with waterproofing surface protection. Systems which use steel troughs or fibre-cement covered with soil are not recommended because of the low bearing capacity of these supports and the infiltration problems connected with overlapping supports. Flat roofs made of earth encompass a great number of systems in which the earth has several different functions: protective coating, wearing layer, damp-proofing, provision of gradient, load distribution slab, thermal insulation and buffering, etc. The design of these earth roofs depends on the use to which the system is to be put, i.e. whether the roof is accessible or inaccessible.

Weight

Flat earth roofs are very heavy, ranging in weight from 300kg/m^2 to as much as 500kg/m^2. They are therefore not suitable for use in earthquake regions. This weight results in the transmission of large loads to the walls and involves large quantities of material subject to bending stresses, such as wooden, steel and concrete beams. Spans must be kept small which makes careful planning essential. Yemenite and Moroccan architecture abound with examples where this weight limitation on the design of spaces is very apparent.

Protection against water

Flat earth roofs are particularly sensitive to water. Torrential rains can destroy a roof. Care must thus be taken to ensure adequate waterproofing: special soils (kaolins, saline soils), stabilization, other water-tight materials and, above all, appropriate design and constant maintenance.

Thermal functions

Because of their immense thermal inertia, these roofs are suitable for hot, dry climates. If the roof requires insulation, stretched matting, loose straw, clay-straw or seaweed can be used. Limewashing increases solar reflection: 90% compared to 20% for roofs left bare.

Draining off water

Waterproofing This can be accomplished using common materials: e.g. bitumen felt and plastic sheeting: or layers of rammed clayey earth, or of stabilized earth: e.g. a well-compacted lime screed. Bitumen felt or plastic waterproofing must be covered with soil to prevent ultraviolet damage. Drain roofs and drain water against the direction of the prevailing wind to prevent splashing. Waterproofing must be assiduously maintained.

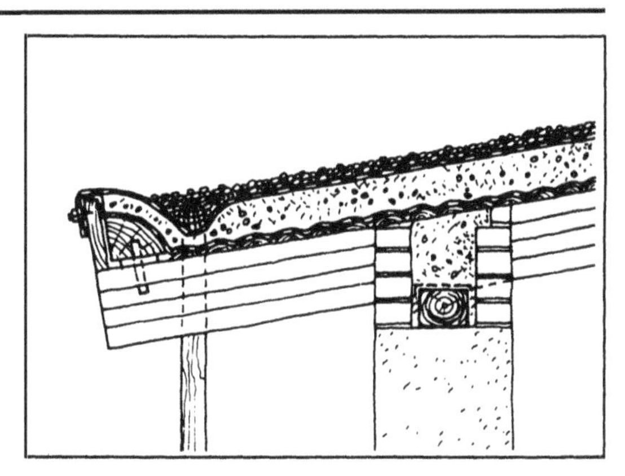

Gutters and downpipes Care must be taken not to fix the roof gutters against the walls but instead along the eaves of the roof. Sections should be generous and have a steep enough gradient. Drain water from each section of the roof individually and allow it to escape at the end of a straight gutter. Never use a bend: as there is a danger of overflowing and blocking. Connect roof gutters directly into downpipes.

Use broad sections for downpipes and avoid fixing them against the wall. Do not use fragile or perishable materials (e.g. plastic).

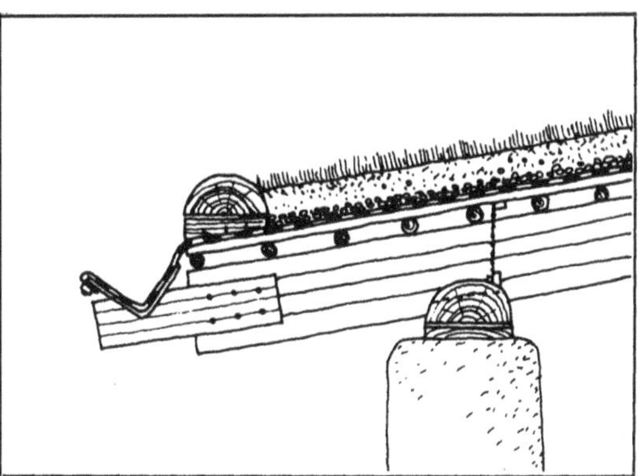

Water should not be discharged at the foot of the wall: provide a rainwater shoe which faces towards a drain or gutter. Do not place a rainwater barrel at the foot of the wall. Maintain the downpipes: attend to repairs without delay.

Downpipes built into or recessed into the wall should be avoided. Avoid decorative downpipes and descents: e.g. hanging chains (splashing on windy days).

Gargoyles Known on the American continents as 'canales' these require a great deal of care and must be provided with a wide throat on the roof to lessen the danger of clogging or obstruction. The mouth should be well away from the wall: 0.5 to 1m.

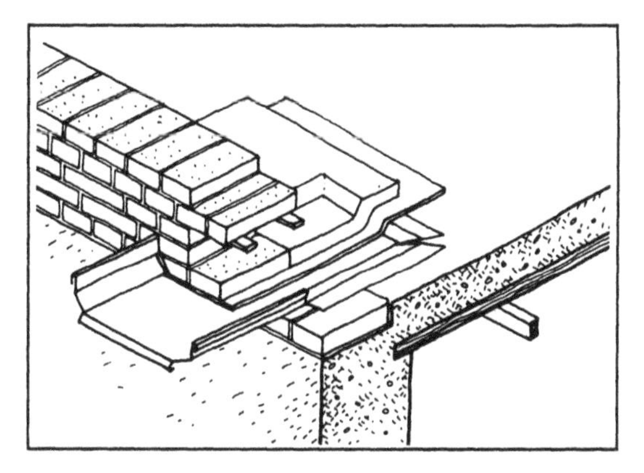

The overlap of the waterproofing in the throat must be carefully constructed and the gargoyle should be provided with a water-tight sleeve where it passes through the parapet. A hole in the parapet is even better.

The gargoyle must be solidly installed and impossible to dislodge. Gargoyles should drain away from the prevailing wind.

Quality materials are used : zinc, galvanized iron, glazed earthenware. When used, wood should be carved out of a single piece rather than constructed – the wood should be treated for rot. No gargoyles should be provided above buttressing, windows or projections.

The bottom of a wall must be protected against splashing and the water drained into a gutter.

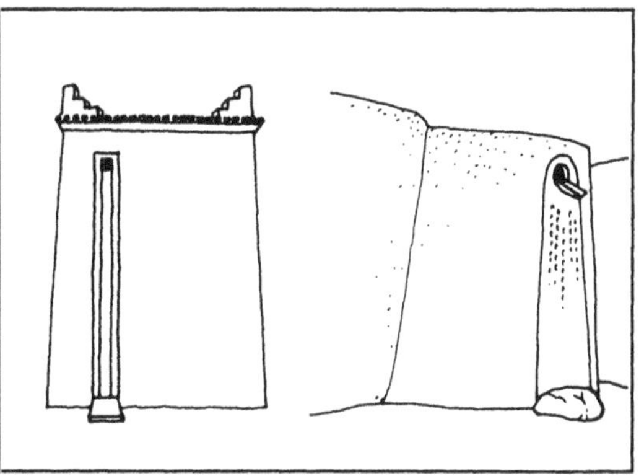

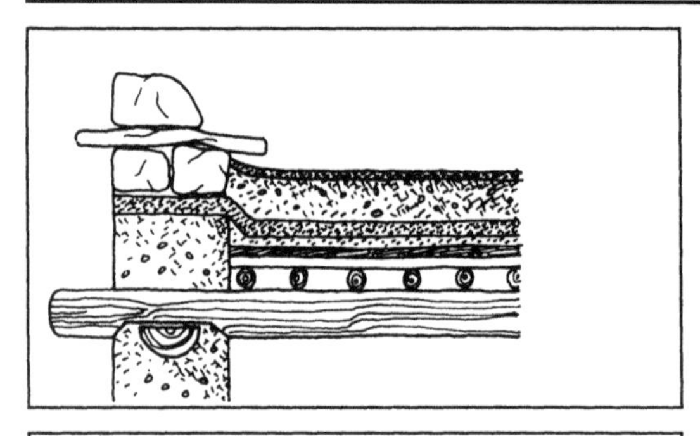

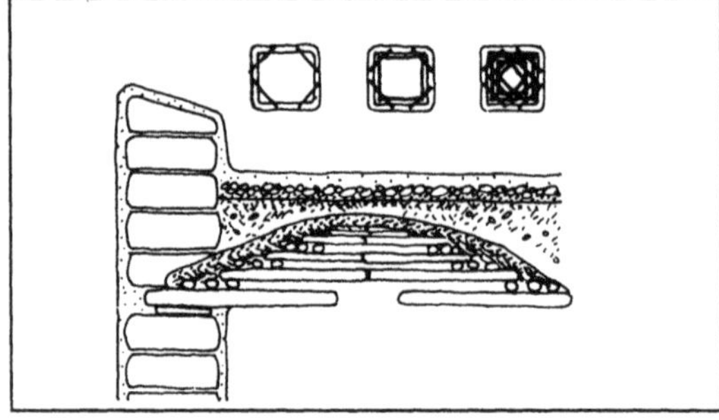

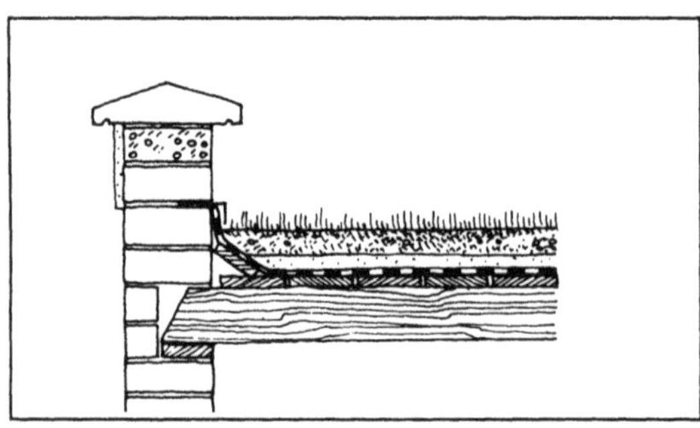

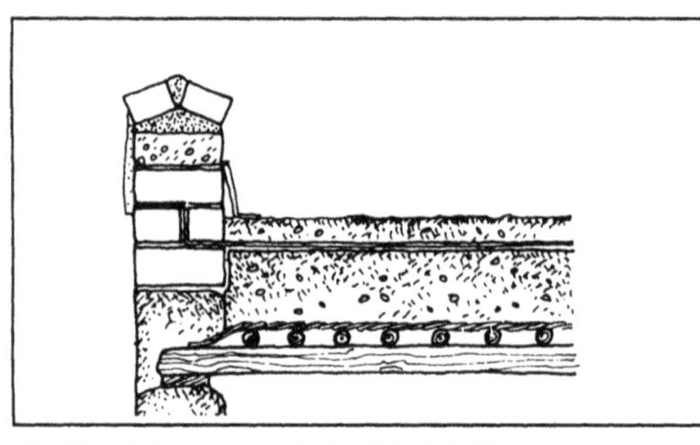

The main drawback of flat earth roofs is that the waterproofing is often defective. In traditional construction, special soils are used: powdered soil from termite hills (Africa), salt clay mud (the 'corak' of Turkey) or soils to which natural stabilizers have been added: e.g. animal dung, shea butter. There are also systems where layers of sand and clay are alternated, to make a thick layer. These roofs are heavy and their water-tightness can only be guaranteed if the upper layers become saturated. The quality of an earth roof depends above all on good execution: ram the earth in several layers, seal all shrinkage cracks and compact by beating. Pebbles can be tamped into the surface layer. This makes a good wearing layer that is resistant to rain-impact erosion. It is still preferable to decrease the thickness of the earth (less weight) and to treat or stabilize the soil. Regular maintenance is absolutely essential.

1. Soil layers

If these soils are treated or stabilized, effective stabilizers must be used. The disastrous example of sump oil is well known. Stabilization with bitumen is more effective for this type of job than lime or cement.

2. Renderings

A screed of lime and coarse sand, carefully sealed and in which the micro-fissures have been sealed by a regularly refreshed limewash, is effective. Rubber-based paints can also be used, but one must guard against condensation in damp or poorly ventilated rooms.

3. Bitumen felt

This can be placed under rammed earth which is then covered with pebbles and planted moss. It can also be fitted above the rammed earth but must be protected against heat with gravel. The connections with parapets and gulleys must be carefully maintained.

4. Plastic sheet

A polyethylene sheet under the soil is just as effective as bitumen felt. The soil can be sown with grass with non-penetrating roots. The joints between the roof and the walls must be made with care.

5. Paving, tiles, flags

Accessible roofs can be paved with all sorts of surfaces. The gradient must be steep enough and the water effectively drained towards gargoyles and downpipes. Use a water-repellent mortar.

Effects of wind and rain

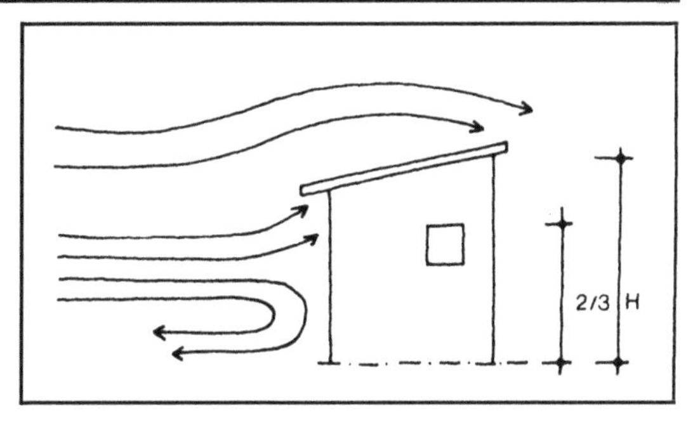

Observations of airflow on walls exposed to prevailing winds show that at about two-thirds of the height of the building above the ground the airflow splits into two, with a rising and a falling current, which cause eddy effects on the bare surface of the wall. These effects are all the more pronounced if obstacles such as ledges, projections, porches and eaves oppose them. Research conducted by the BRE in England has shown that the size of the projections has a direct effect on the extent of the erosion due to the combined action of wind and rain. The effect is concentrated in the area the projections are meant to protect, over a distance of at least 20cm. The effect is all the more marked if the projection is narrow (5–20cm) and higher on the wall. The most extreme case are the eaves. To limit erosion, the projection should be made wider (at least 30cm). Furthermore, it is absolutely essential to render the area above and below the projections for a minimum distance of 20cm.

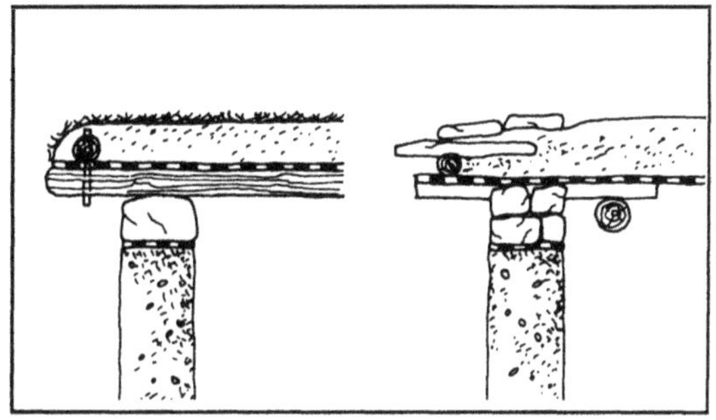

Roof overhang

The function of a roof overhang is manifold: the retention of the soil of which the roof is made, to decrease erosion at the top of the wall, cause water to be discharged away from the base of the wall, protect the surface of the wall against vertical rain and provide shade.

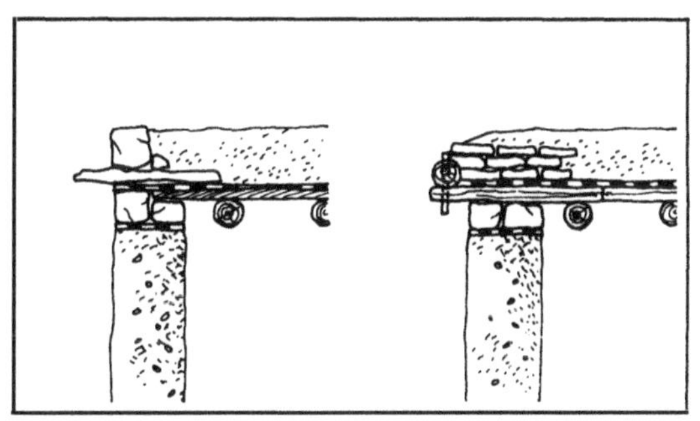

Wide eaves (at least 30cm) are often encountered in traditional earth architecture. They provide greater stability and resistance to pressure differences caused by the wind. Such eaves can be fitted with gutters of generous proportions, which should be firmly fastened.

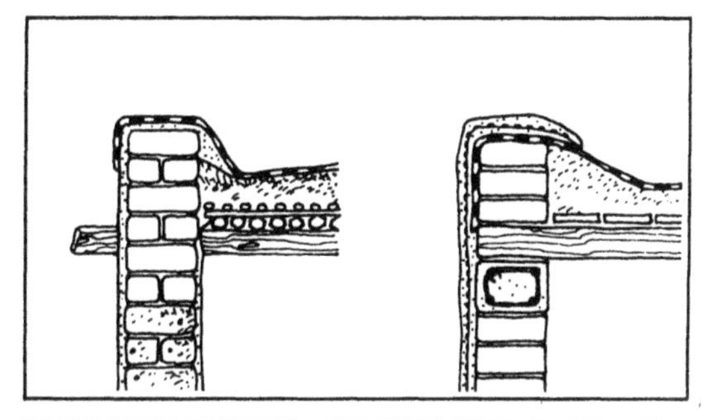

Parapets

A parapet allows the roofing soil to be retained and gives greater control over the drainage of water from the roof by guiding the water to the gargoyles and downpipes. A parapet also serves as a railing for accessible terraces and affords effective protection against pressure differences due to wind. In addition, the parapet plays an aesthetic role. It should be heavy and constructed of durable materials and/or protected from erosion either by a coping (30cm minimum) made of a conventional material (stone, brick, tile, reeds and earth) or by a, preferably waterproof, rendering. The upstands of damp-proofing against the parapet and around gargoyles should be very carefully constructed.

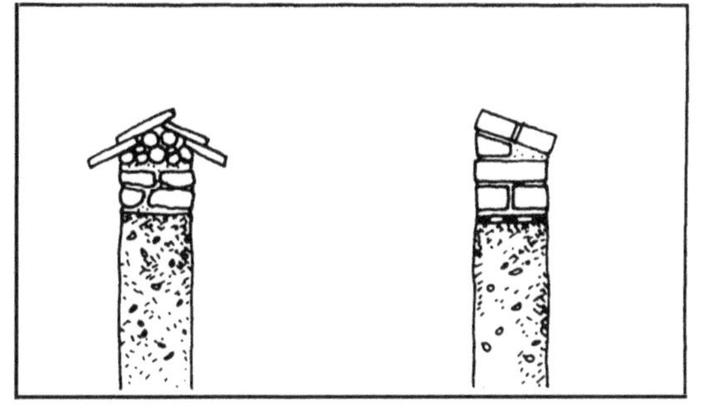

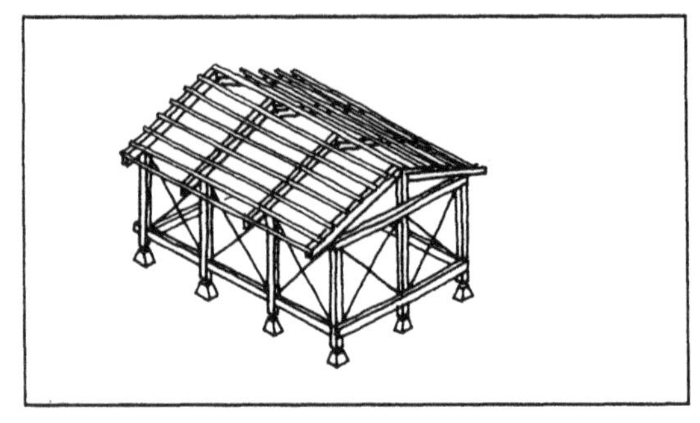

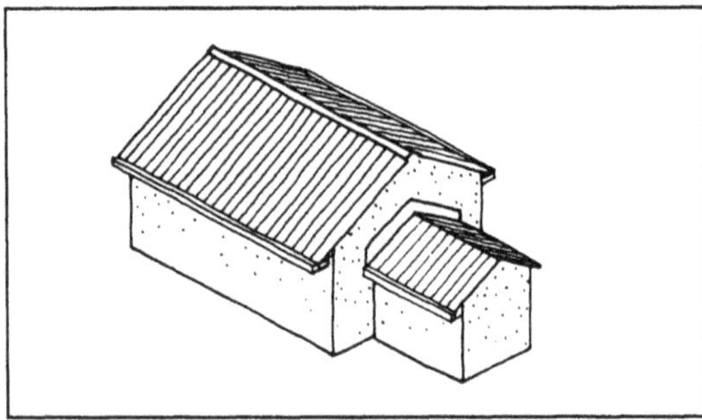

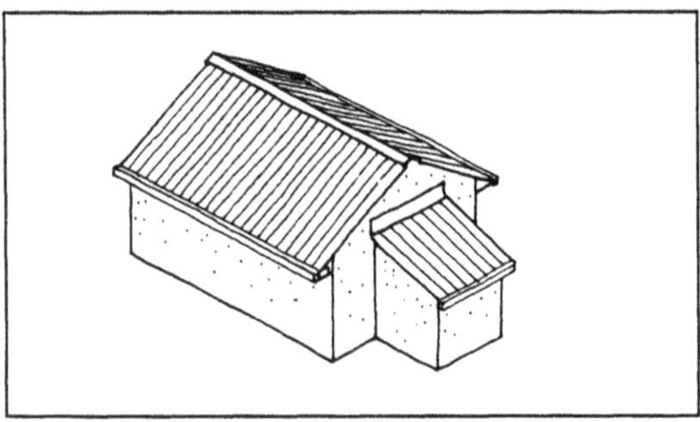

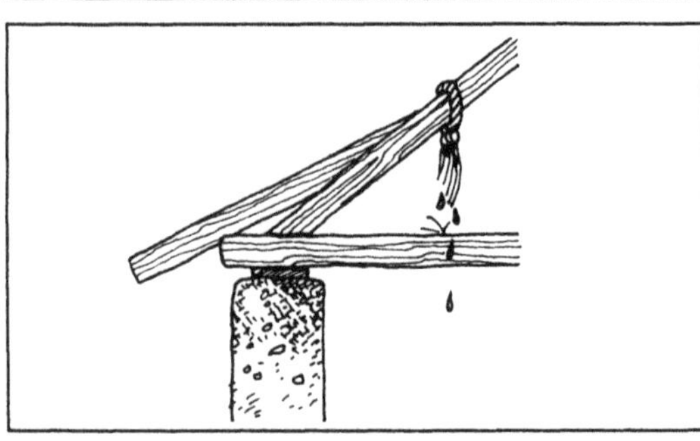

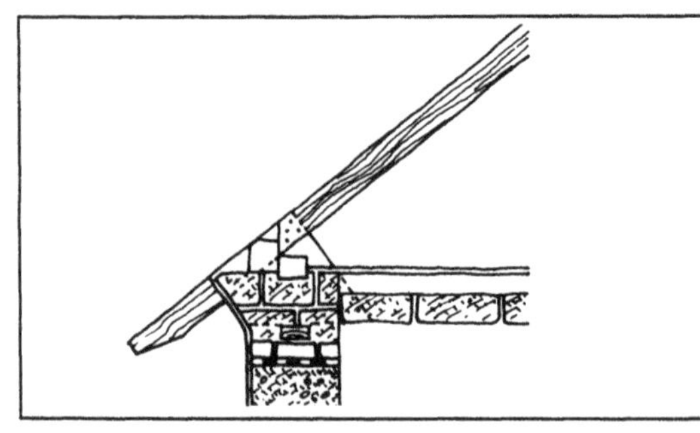

Pitched roofs with broad eaves (minimum 30cm) drain rainwater very well and are particularly well suited to earth structures. These roofs are also suitable for use in tropical cyclone regions (minimum gradient 30°). This applies in particular to hipped roofs (i.e. four slopes) which are to be preferred in such regions to gable roofs (two slopes). Hipped roofs offer better protection against wind and rain and economize on wall material but make the structure more complicated. These elaborate roofs are costly in materials and manpower and it is not unusual to see that the pitch is reduced. This reduces the surface area and economizes on the roof frame and decking. Hipped and gable roofs make it possible to build 'umbrella' and 'parasol' structures erected before raising the walls and enabling the work to be carried out in a sheltered area. This system does away with the limitations of load and thrust on earth walls since the roof structure bears directly on the foundations and slabs. There is, however, a redundancy of structure and materials which are only acceptable in richer economies or in large construction projects.

Protection against water

The roof must be built rapidly and not too long after the rest of the construction work. If it rains during the construction of the roof, the tops of the exposed walls must be well protected. One must always allow for the possible failure of the roof and make the tops of the walls water-tight. There is a suitable pitch for every roof covering. Reducing this pitch may lead to leaks as the result of inadequate drainage, standing water or infiltration by water. Saw-tooth roofs are to be avoided as are roofs with two adjacent slopes having low edges, unless the use of wide sloping gutters is envisaged. The joints of damp-proofing must be properly executed. Walls of which the upper portion forms a gable are to be avoided as they make flashing and mortar infilling necessary and this is not always reliable. The flashing for roofs supported by gable walls should be executed by means of a strip of rendering or wearing surface consisting of stones or burnt bricks which affords protection against water splash.

Loads and thrust

Horizontal thrust must be either eliminated or properly taken up: there is a danger of walls bowing or cracking around openings. Loads should be uniformly distributed on ring-beams. Reinforce gables of structures with thin walls: reinforcement, ties, posts.

Anchoring pitched roofs

Anchoring roofs to walls is absolutely essential if the dangers of causing deformation of the roofing and roof-loss in strong winds are to be avoided. This precaution is of cardinal importance in regions prone to tropical cyclones. As a general rule, the anchoring system must be very solid and properly designed. No economies on materials are justified. The roof must be fastened to outside walls, gutter-bearing walls and gable walls, but can also be anchored to dividing walls. If the structure of the building is reinforced with concrete or wooden posts, anchoring is fairly easy. It is, however, better to anchor roofs on continuous ties than on isolated supports.

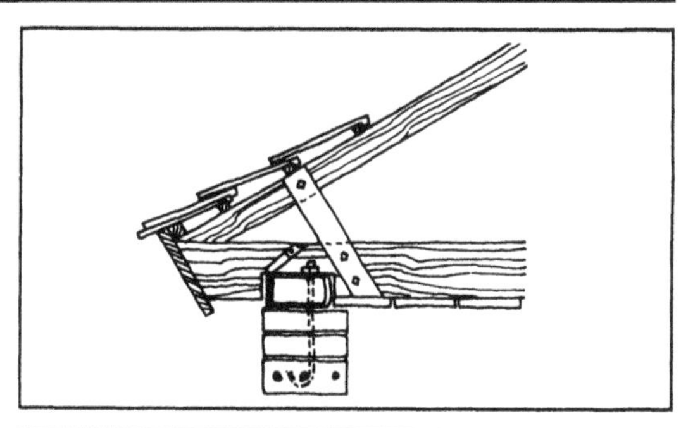

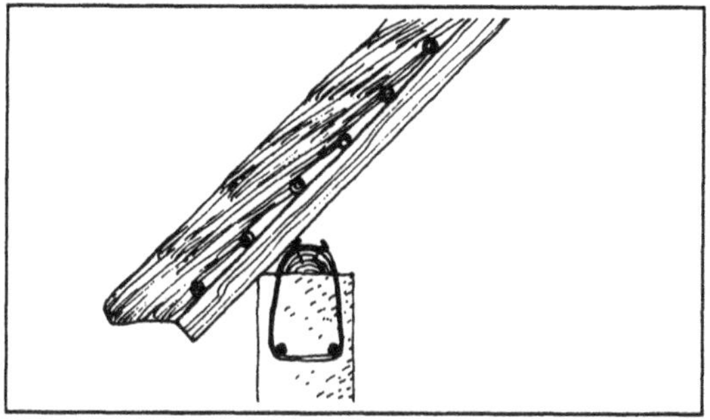

1. Anchorage on gutter-bearing walls

Wood A wall plate on a gutter-bearing wall can act as a ring-beam as well as an anchor plate. This continuous wooden support must be extremely well secured: e.g. collars, anchor-bars bolted and fixed to the wall. When the ring-beam is lower than the finished top of the levelling course below the roof on the gutter-bearing wall, the lower purlin can be fastened to the ring-beam by clamps or metal or wooden beam ties. This same ring-beam can be used to anchor corbels or metal shoes to receive a facing beam for a porch or verandah, even a future extension of the building (India).

Concrete A concrete ring-beam makes it possible to provided iron ties for a lower purlin or roof beam.

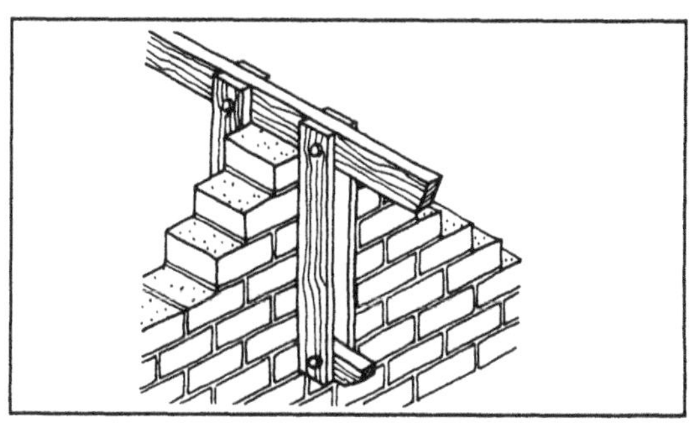

2. Anchorage on gable walls

Hipped roofs are anchored to side walls in accordance with the principles already set forth. For gable roofs with purlins extending into the gables, the purlins can be anchored to the ring-beam, which is an extension of the one in the gutter-bearing walls or an alternative ring-beam at the height of the support of the purlins. Metal or wooden tie-bars fixed to wooden corbels at the level of the ring-beam can also be used, either above it (the weight bears on the ties) or below it (tensile stresses due to wind are exerted on the ring-beam). This solution to the problem of anchoring to the ring-beam avoids overloading the gable itself, which is the weakest wall in the structure. Systems incorporating clamps or wooden or metal ties increase the stiffness of the gable.

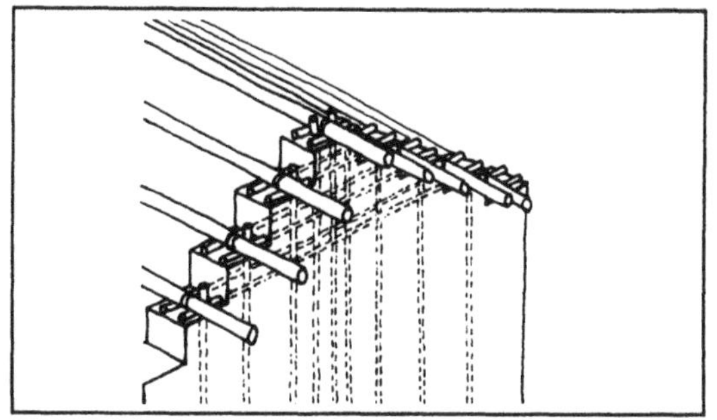

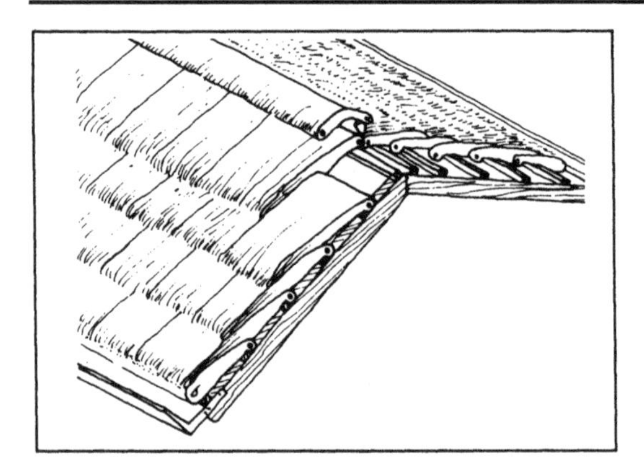

1. Straw-clay shingles

These shingles consist of a layer of straw – visible on the roof – held in two layers of clay of which the second is visible on the underside. The clay is protected from the rain by the straw which is protected from fire by the clay. The shingles are easy and economic to produce. Their sizes varies from 90cm to 1.20m and their width from 45cm to 1.5m, although a width of 60cm is easier to lay.

The average weight of a 20cm thick roof in this material is 50kg per m^2.

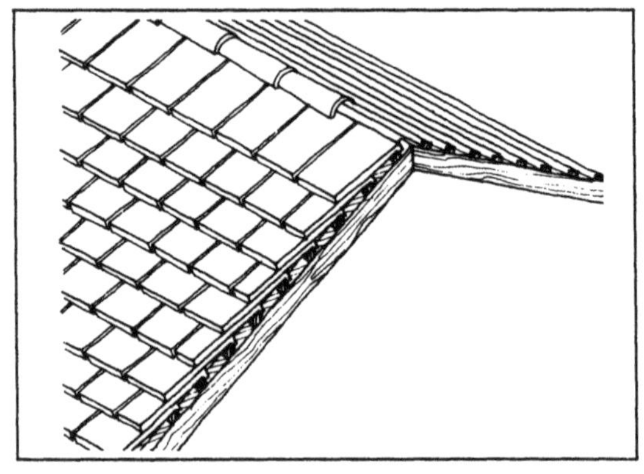

2. Tiles

Two types of tiles are used: flat single-lap tiles (e.g. Brazil) and curved tiles (e.g. France) in strongly stabilized clayey earth. These materials freeze easily except when they are overstabilized in which case they are no longer really economical.

Clay tiles made with phenolic resins are still in the experimental stage.

3. Roof mortar

In certain countries (e.g. Tibet) highly clayey soils are used instead of mortar to secure slates or even roofing flags. This waterproof mortar (saturated clay) can also be used to secure curved tiles or Roman tiles (e.g. Burundi).

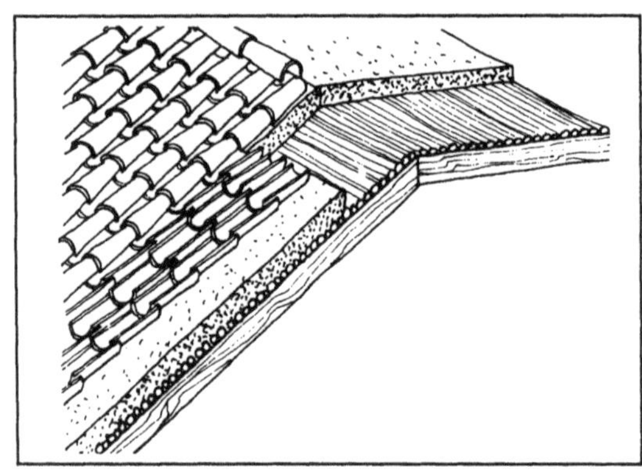

4. Earth layers

The systems are similar to those used for flat roofs: rammed earth or clay-straw. Damp-proofing in bitumen felt is laid over the soil or over thin wooden planking over the soil. There are also roofs made in bitumen-stabilized clayey earth reinforced with a netting (screed or shingles). These roof claddings are heavy.

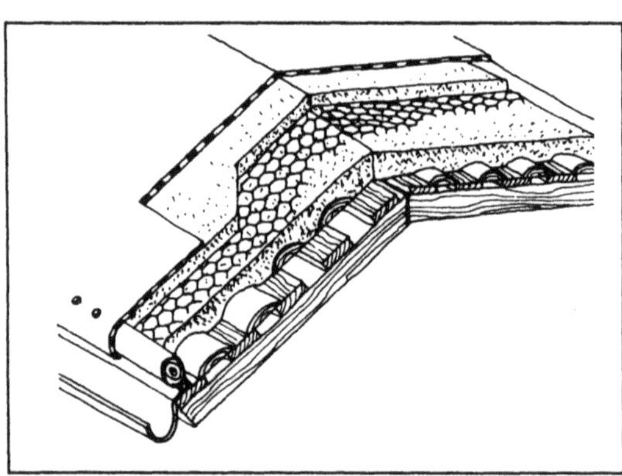

5. Grass roofs

Such roofs have several advantages: purification of the air indoors, thermal insulation and buffering, control of indoor comfort by condensation and evaporation, sound-proofing. The thickness of the soil in which the grass grows is at least 20cm resulting in heavy loads (strong absorption of water), making it necessary to design a special roof frame. The damp-proofing must receive top priority. In traditional systems, treated wood or flat stones covered with bitumen were used. Nowadays plastic damp-proofing is available which is put down in a single layer without joints. This type of damp-proofing must be non-flammable and able to stand up to solar radiation. The pitch of this type of roof lies between 5 to 45, an angle of 20 being preferred as this pitch greatly facilitates drainage.

6. Earth reels

These consist of a clayey soil mixed with long fibres and rolled around spindles in the form of reels. The resulting reels are laid between the purlins when still moist and pressed against one another. The spaces between the rows of reels are packed with daub. After drying, the cracks are filled with clay applied with a mop. A 2cm thick layer of soil mortar stabilized with finely chopped fibres and lime is then applied and this layer is then covered with a roofing felt and covered with sand. This type of roof weighs about 200kg/m^2 (including the frame which is on the light side). Its use is not recommended in areas infested by termites.

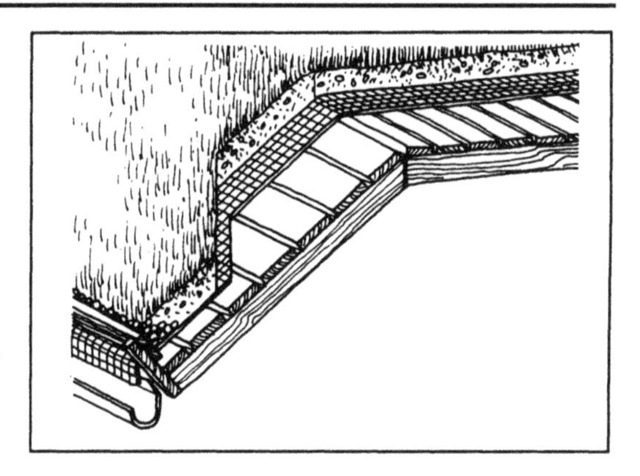

7. Panels

These relatively thin elements are in clay-straw or daub and reinforced with wooden sticks. They are laid directly on simple rafters. They have a low bearing capacity. The panels are held in place on the bottom edge of the roof by a board which takes up the thickness of the finished roof. The ridge pole is covered with a batten and a layer of clay-straw. The roof is given a screed of soil mortar mixed with chopped fibres and stabilized with lime before being covered with bitumen felt.

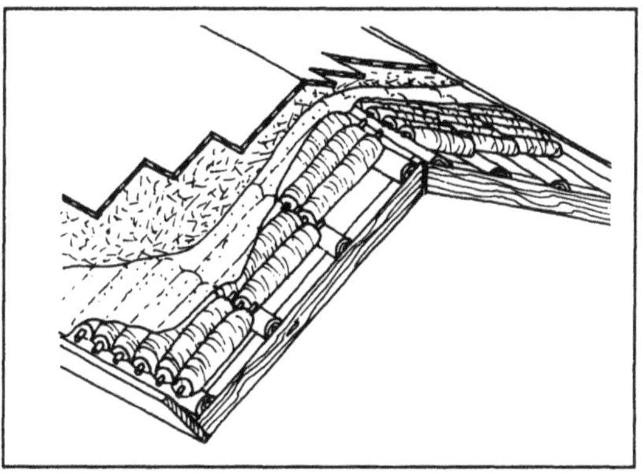

8. Brick roofs

This uses the principle of the tile floor. It is built in small flat bricks given a rough-cast of plaster. Both the bricks and the workmanship must be of a high standard. The system consists of two layers of bricks, the first of which is coated by a layer of plaster and the second by a screed in sand and cement mortar. Tiles set in mortar finish the roof.

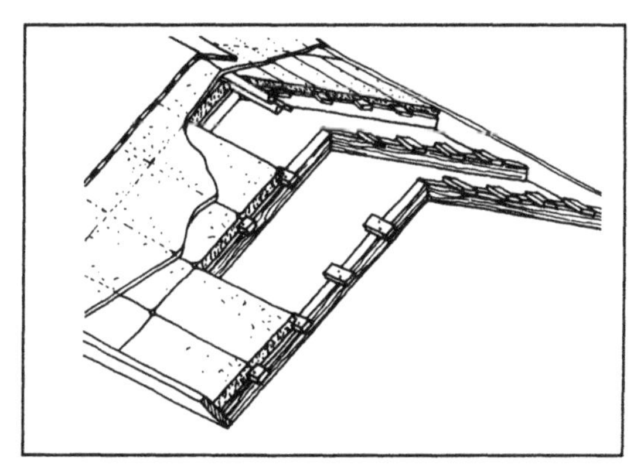

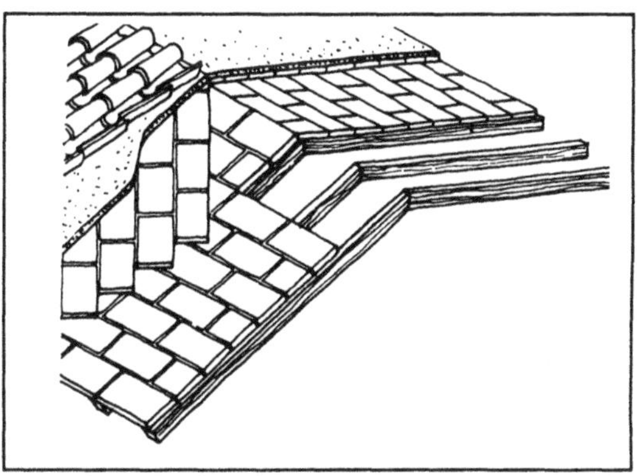

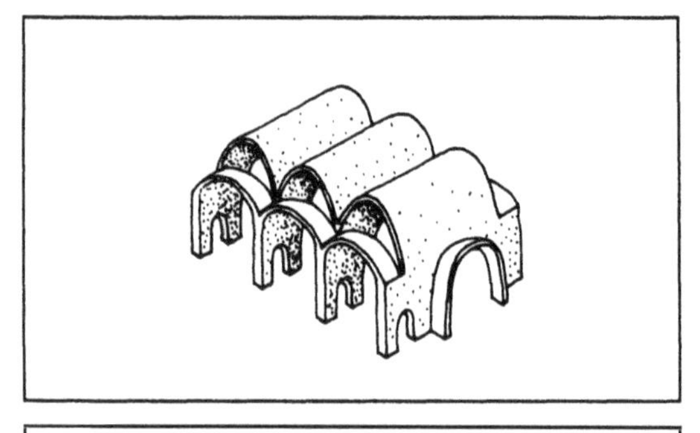

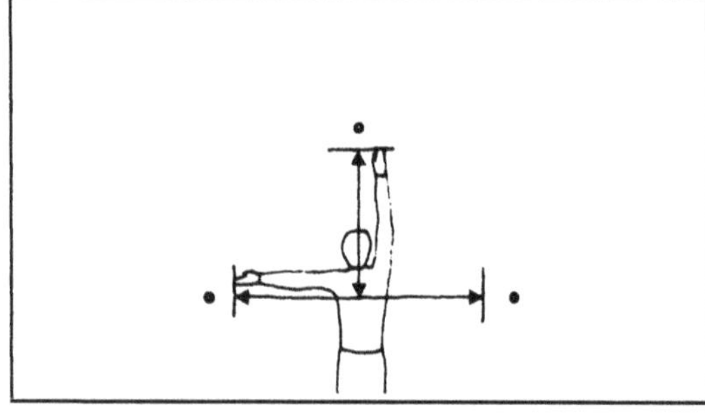

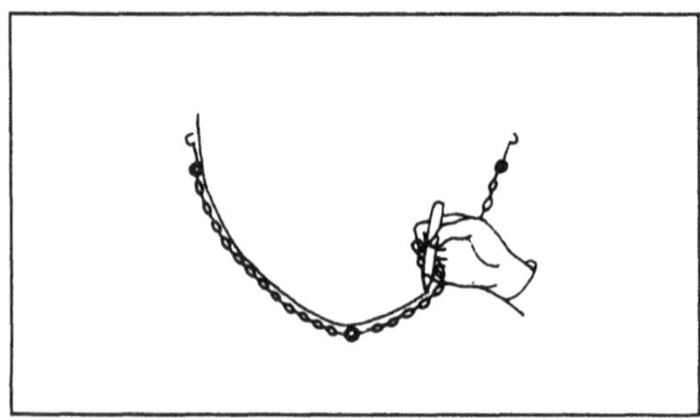

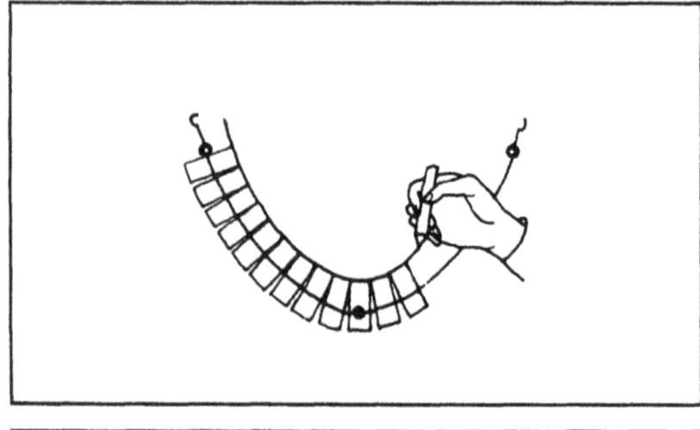

In many regions where good structural timber could not be obtained, builders came up with the idea of the vaulted roof to cover spaces. Developed very early on in Egypt and made of sun-baked earth brick (e.g. the granaries of Ramesseum in 1290 BC XIXth Dynasty), the arched roof went on to dominate construction in the majority of countries in the Near and Middle East over the centuries. Earth roofs of this sort are still a feature of the architectural scenery of many towns. Made of earth, the vault has the advantage of exploiting the high compression strength of the material and the good transfer of roof loads to vertical walls.

Areas of application

A large space can be covered by a single continuous vault or by a series of small transverse vaults. These two systems, which are very common, are associated with a variety of shapes of vaults. The use of the vault sometimes poses problems of cultural acceptance, lack of adaptation to regions in which there is heavy rainfall or structural behaviour during earthquakes. There are, however, appropriate solutions to these problems.

Shapes

There are numerous shapes: vaulted, co-ovoid, groined vaults, cloister vaults, ogee or ribbed vaults. The most common, however, is the barrel vault of which there are numerous shapes: round, continuous, basket handled, transversal, ringed, ogee or skewed, spiral semi-circular (e.g. stairs), and winding. The catenary arch is another shape in widespread use in Nubia (whence the term the Nubian arch) and made famous by the buildings of Hassan Fathy. All these vaults may be plain, dropped, or raised.

Marking out

Mathematical design methods can be used. However, e.g. for the catenary arch, it is more practical to construct a plot by using a light chain. The span and the rise are predetermined by three points in a vertical plane through which the chain must pass. The curve which is plotted is inverted and it must be confined to the middle third of the thickness of the vault. The shape of the formwork must be corrected to the required thickness of the arch: a half thickness on each side of the curve.

Dimensions

Span Vaults with spans of up to 6m and 15cm thick have been built in stabilized compressed blocks; whereas spans of vaults in rammed earth rarely exceed 2.5m. In Iran, the most commonly found span is 4m. This is the maximum for earthquake regions, and for such regions it has also been established that the ratio of span to length should be equal to 1.5 × width. If the value of span to length is greater than this, there is a danger that the vault will resonate and shatter (according to research conducted by the University of Baja California, Mexico).

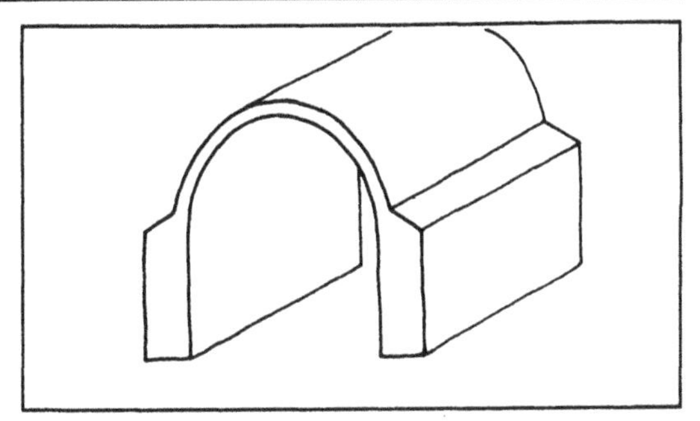

Rise The less the rise, the greater the lateral thrust, the greater the rise, the lower the thrust, approaching a limit value of zero. For catenary vaults the conventional ratio of span to the rising indicates that the rise should be equal to 50cm plus half of the span. In earthquake zones it is advisable to limit the rise to between 20 and 30% of the value of the span.

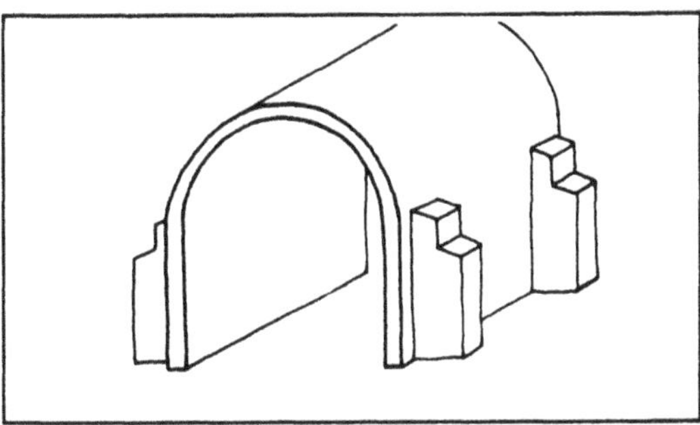

Thrust If the rise is very large the thrust on the walls can be enough to cause them to collapse. The methods used to determine the dimensions of arches and piers can be used to determine the dimensions of walls. Above all wall bases and wall foundations must be carefully designed. There are several ways of ensuring the stability of vaults.

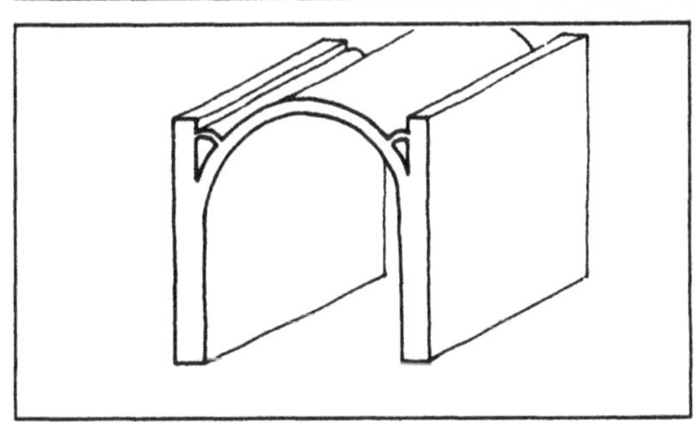

- Thick walls require a great deal of material and should not be weakened by large openings.
- Buttresses which must be supported by good foundations linked to the wall foundations.
- Prestressing load (parapet or pinnacle), which transform lateral thrust into vertical thrust.
- Tie bars which eliminate thrust; these must be evenly distributed and securely anchored.
- Ties taking up thrust and which should be able to resist the bending forces induced by thrust.

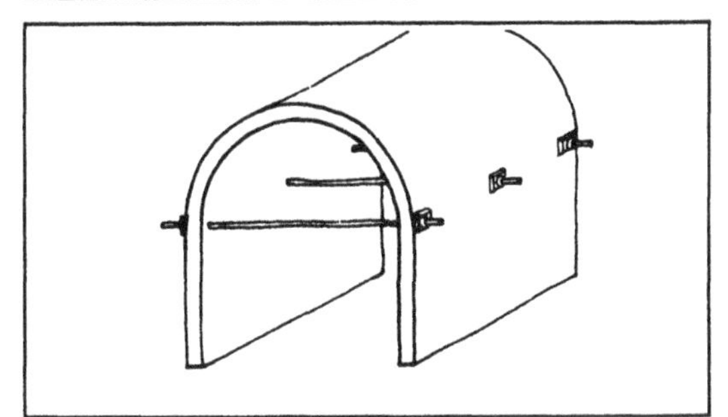

Protection against water

This is the major problem with vaults in regions where rainfall is heavy: infiltration through fine surface cracks (result from expansion in repeated hot and cold cycles), inadequate drainage at the crown and haunch (water accumulates and permeates the material: the slope must be made steeper). Vaults must be carefully protected from rain by means of damp-proof renderings, such as washes, bitumen felts, elastic paints, and by regular maintenance.

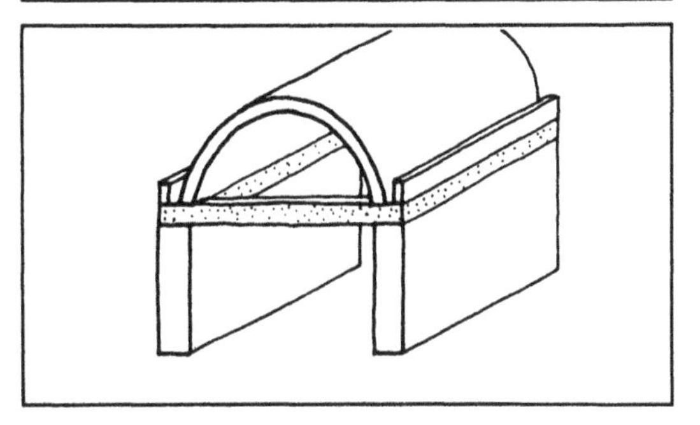

Construction systems

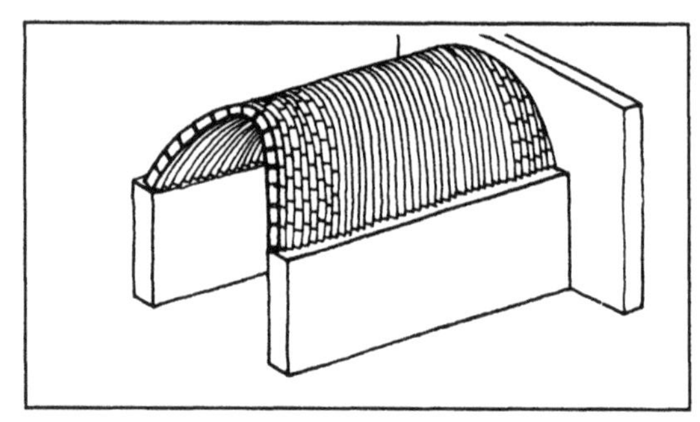

1. Bricks – without formwork

The 'Nubian-style' method exploits the mutual friction of the bricks and the adhesion of earth mortar, the courses being inclined at an angle of between 10° to 15°. The bricks must be light (e.g. adobe rich in straw) and not thick (5–6cm). Thinnish compressed blocks with grooves to enhance their adhesion can also be used. The main difficulty is in respect of the curvature (confinement to the middle third). A template for the construction is a must for inexperienced builders.

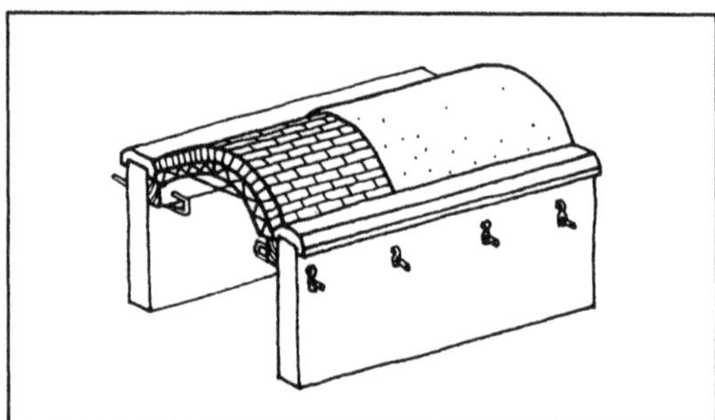

2. Bricks – with formwork

The formwork makes it possible to work parallel to the walls. It can be heavy and fixed or light and sliding. In the latter case, the builder proceeds in successive layers of bonded bricks or even in rings. The setting of the keystone of the arch must be carried out with care.

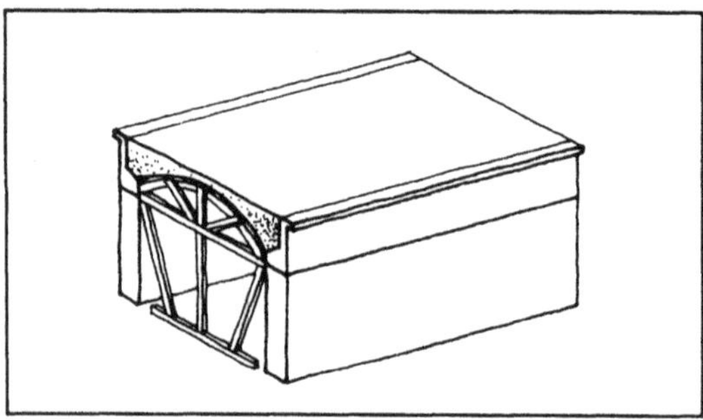

3. Rammed earth on formwork

Although rarely used, this construction system does exist. Monolithic arches made of rammed earth and on formwork are generally dropped and held in position by ties. The layer of earth above the extrados is given a gradient. The formwork is very heavy and fixed (ramming stresses) and cannot be used for several days.

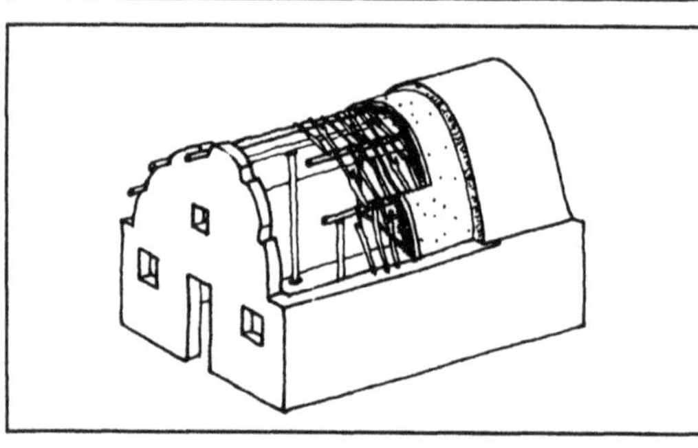

4. Mud over frame

An arched structure consisting of interlaced branches is covered with mud from earth to which cow dung has been added. The rigidity and water-resisting qualities of the system are precarious.

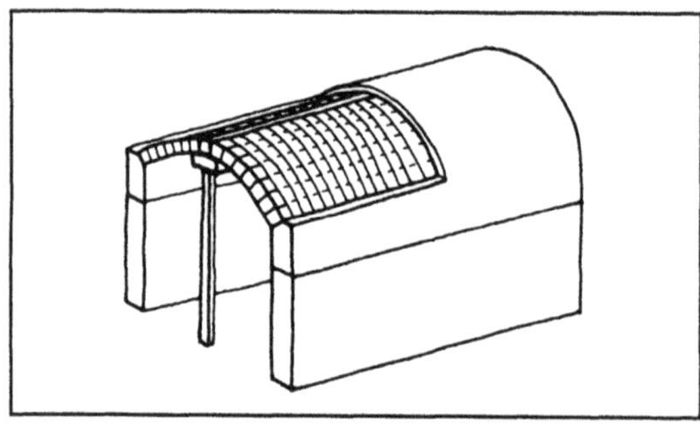

5. Prefabricated elements

Hollow elements are prefabricated by double-acting hydraulic presses. The system is highly sophisticated and requires the use of concrete bars to provide the prefabricated vault with a high degree of stability. The vault is assembled in small half vaults assembled on a ring.

Construction of brick vaults without formwork

1. Starting the vault

The springing of the vault can be supported in two ways.

- Against a vertical support: a suitable gable wall on which the entire curve of the vault is inscribed. This supporting wall must be heavy and perfectly stable and not have any defects in bond which could cause cracking under the thrust of the arch.
- Against a horizontal support: the vault is supported by the last courses of the gable walls and starts like a dome. The courses are progressively altered until an angle of inclination of 15° is reached. The bond between the gable walls and gutter-bearing walls can describe a rectangle, a triangle or half-hexagon or even a semi-circle. The latter is recommended for earthquake zones since the vault behaves as a monocoque. Vaults supported by a vertical gable wall risk breaking because difference in the amplitude of vibration between the wall and the vault during an earthquake.

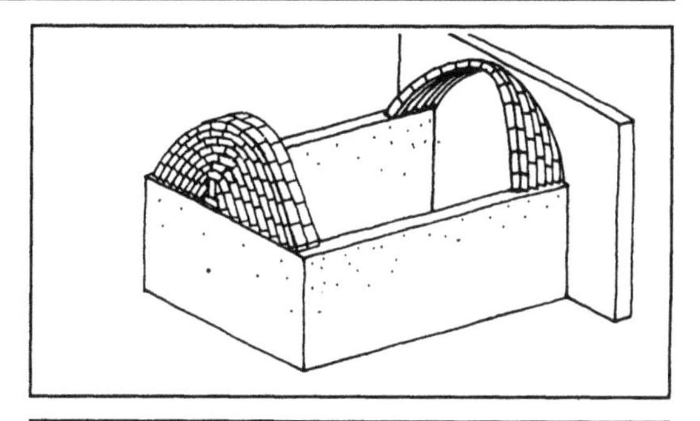

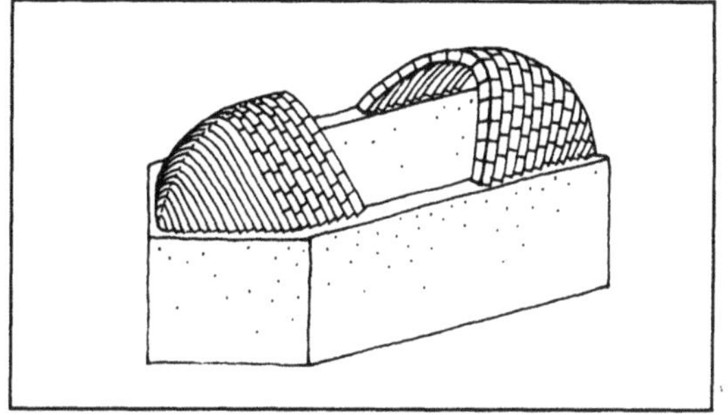

2. Ending the vault

At the end of the vault the angle of inclination of the courses be retained, but it is preferable to gradually true it up until the vertical is reached. The vault can then be supported on another vertical gable wall or the tympanum can be closed with masonry. The two gables can also be started simultaneously and the courses joined in the middle of the arch.

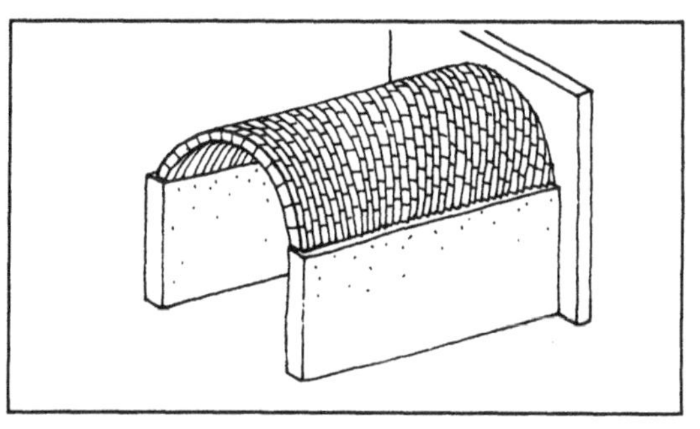

3. Profile of the vault

This is uniform in appearance if the procedure of progressively canting the courses, the bricks being laid flat is employed. Highly attractive profiles can be obtained, however, by using a wheat-ear or herring-bone bond. Complex bonds must be executed with great care and the bricks snugly wedged into the joints at the level of the extrados with small pebbles.

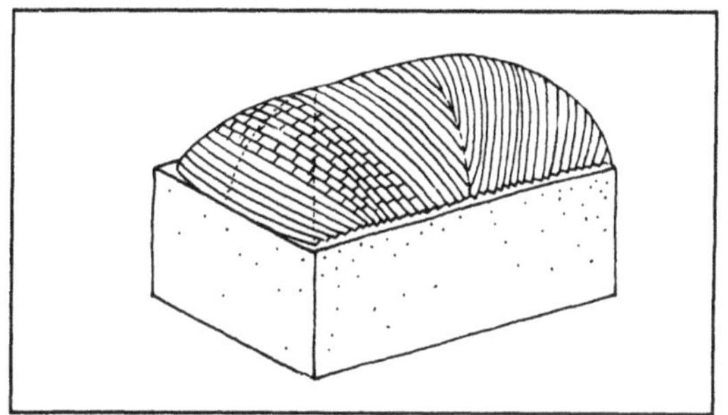

Workcrews and productivity

In Nubia, one master mason assisted by four mates can cover from 12 to 15m^2 a day. If the crew is inexperienced, output is much lower. In the latter case it is advisable to use a light template (e.g. made of steel wire), which facilitates execution and gives an even curve.

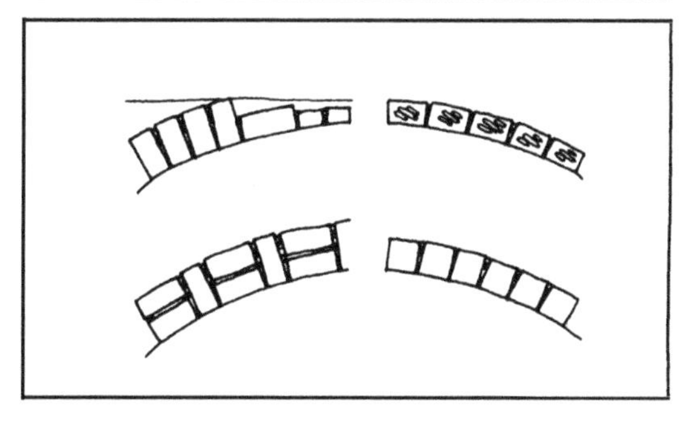

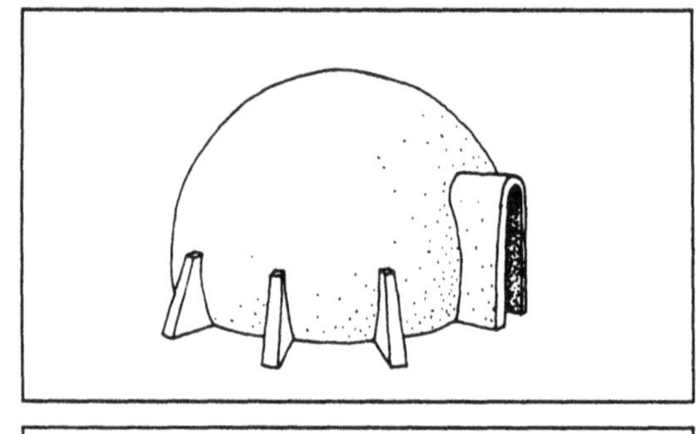

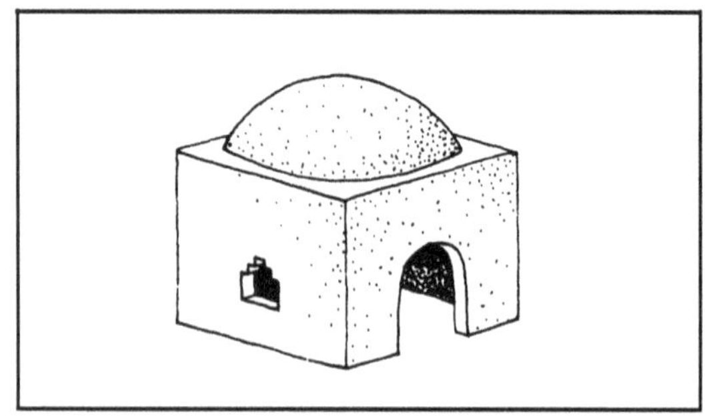

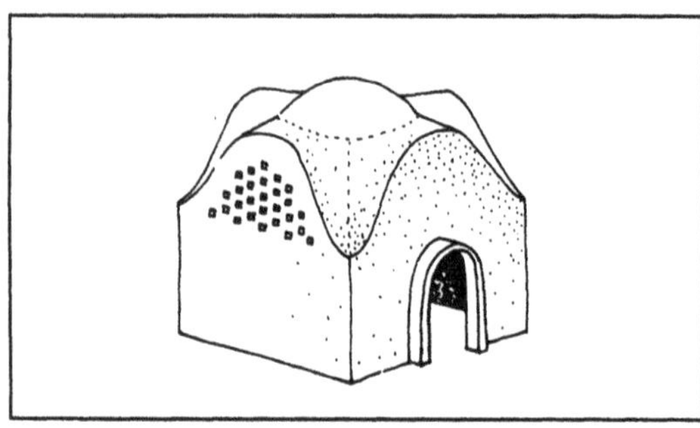

The cupola is a vault of circular cross-section the intrados of which is generated by the rotation of an arch about its vertical axis. Like the arch and the vault, the advantage of the cupola is that it uses the material in compression.

Areas of application

The most usual function of the cupola is to totally cover enclosures. Like the vault, the cupola is not suited to regions with heavy rainfall. Cupolas perform well in earthquakes; it is the walls that are rather fragile, and which must be well-designed or reinforced: buttresses, tying. From an architectural point of view, the cupola poses certain problems with respect to the design of the enclosed spaces: considerable heights are achieved with unflattened cupolas. The normal diameter of adobe cupolas is about 4m (Iran), but diameters up to 12m have been achieved using stabilized compressed blocks. Another problem connected with cupolas is their acoustic resonance which renders them unsuitable for certain purposes unless special measures are taken such as elaborate bonding, stretched materials, or sophisticated design. Simply shaped cupolas built in isolation are easy to construct, but when a project involves a series of cupolas, arches and vaults, the site must be well-organized and the building crew experienced. It is also possible to construct two-layer cupolas where, for example, the extrados is raised and the intrados dropped – thus providing an insulating cushion of air between the two layers and a better gradient for draining off rain water. Openings can also be made in cupolas without affecting their stability, except in earthquake zones. Traditionally, cupolas are waterproofed with a screed (e.g. bitumen, lime) or by surfacing with burnt brick and ceramic tiles (e.g. Iran).

Shapes

These are defined in terms of elevation and plan. The elevation encompasses virtually all the shapes of arches and vaults. Plans can be circular, elliptical, square, rectangular (oblong cupola), etc. There are also half-cupolas or half-rounded vaults.

Marking out

Simple hemispherical cupolas are easy to set out but more elaborate plans necessitate the use of geometric formulas of considerable complexity (e.g. mosques in the Iranian tradition).

Decoration

Whether internal or external, done by means of highly elaborate patterns of bricks, rendering, painting, or tiling, there is no limit, except taste, to the ways in which cupolas can be decorated.

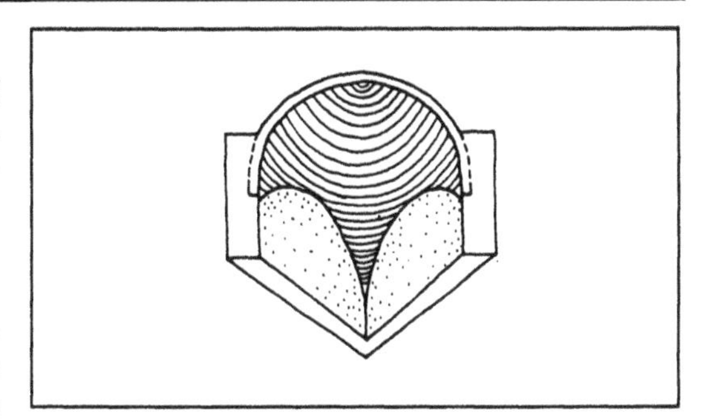

Supporting cupolas

The supporting structure can vary greatly depending on the plan, and the height of the walls, the diameter of the cupola and its centering must be adapted to this. A spherical cupola on a circular plan poses no problems, but square, rectangular or pentagonal plans give rise to complex setting out and necessitate the construction of intermediary supports such as squinches and pendentives.

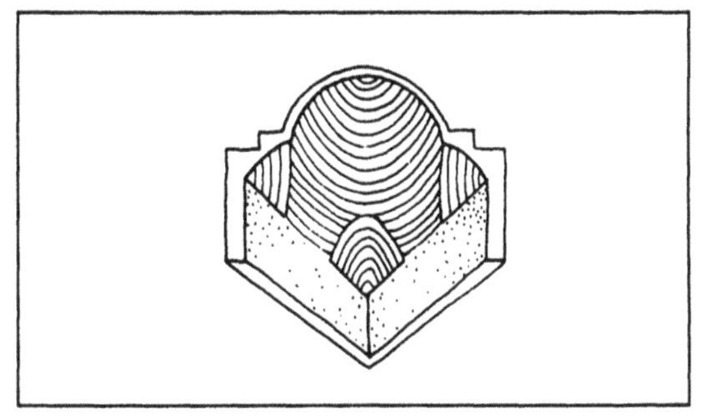

1. Squinches

These are small vaults that provide support. Usually they are conical, half-domed or semi-circular with a face arch. They are built at the inside corners of polygonal compartments and serve to multiply the number of sides to the corner with the result that the form approaches a circle and the start of the cupola is facilitated. The use of squinches tends to raise the height of the cupola.

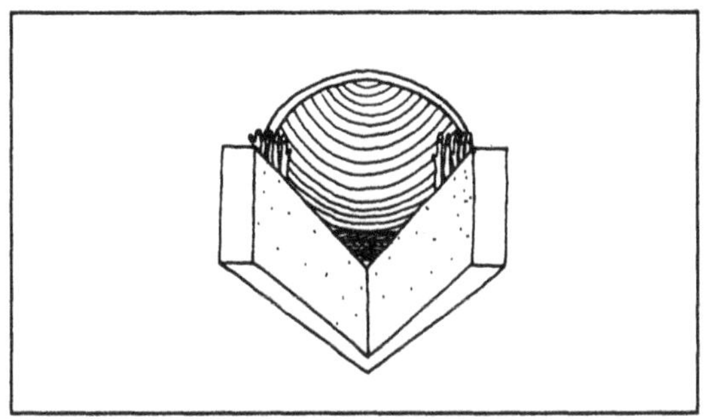

2. Pendentives

A pendentive is a squinch the intrados of which is a concave spherical triangle. The use of pendentives lowers the height of cupolas. Cupolas using pendentives on oblong compartments are well known from Byzantine and Renaissance architecture.

3. Lintels

These are positioned at the inside corners and made of wood, steel or concrete. A smaller amount of material is used but the system needs materials with a high tensile strength.

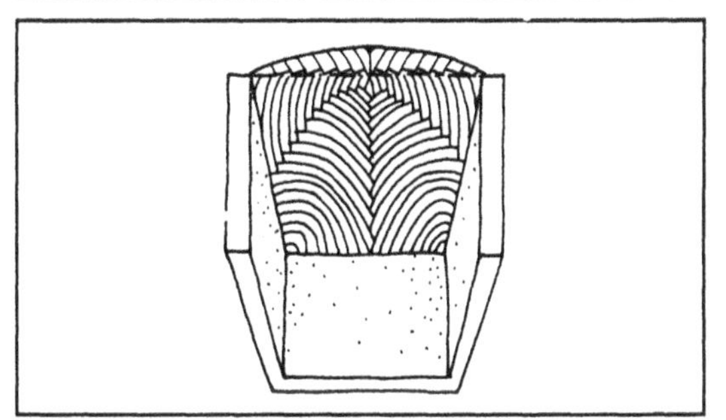

4. Starting at the corners

The cupola starts at the four corners simultaneously as if it were a vault. Where the courses meet the angle between them may be 45°.

5. Starting from the walls

The work is started as if it were a vault, but it is difficult to join at exactly the same height and with the same curvature.

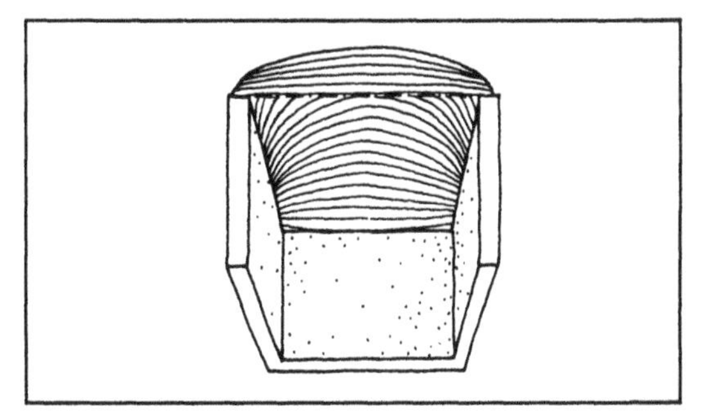

Construction systems

1. Monolithic cupolas

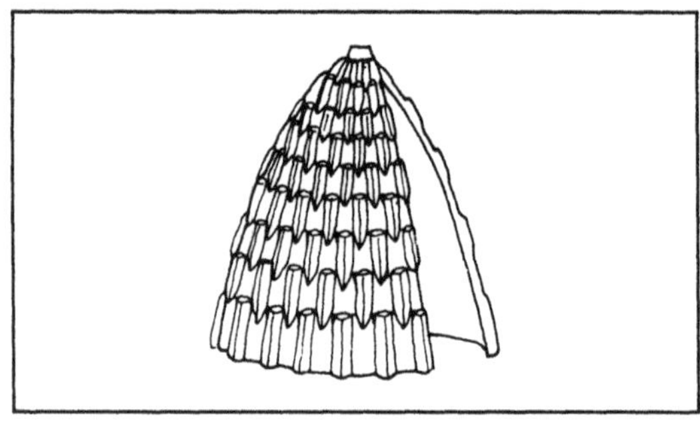

These are domes built by hand, using pottery techniques adapted to architecture. A case in point are the famous shell-shaped dwellings of the Mousgoum tribe of Cameroon, but which are becoming increasingly rare. This consists of a highly cohesive specially prepared earth shell, 15 to 20cm thick at its base, 5cm thick at the top and 7 to 8m high. It is constructed without scaffolding, but a surface relief gives footholds while at the same time dividing run-off, effectively reducing erosion. Modern versions using rolls of extruded earth have been tested at the T.H.K. (Germany), where the construction of a rammed earth dome has also been tried with success.

2. Corbelled domes

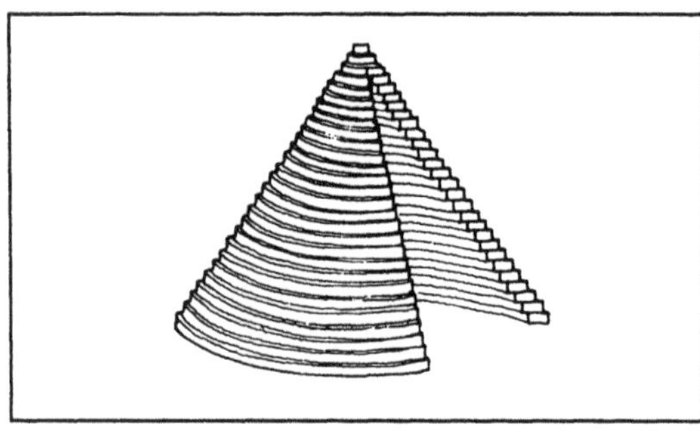

These are erected by building inwards on successive horizontal courses, which hold thanks to the adhesion and cohesion of the mortar. This principle has been adapted to conical and pyramidal shapes (projects in Honduras in the 1980s), but the execution is very slow since it is necessary to stop for drying every two or three courses of bricks.

3. Domes with inclined courses

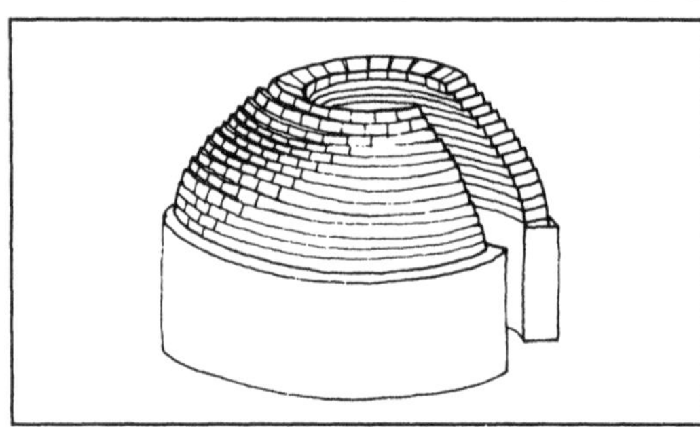

This involves the construction principle developed by traditional Nubian architecture (Upper Egypt) and adopted in designs by Hassan Fathy (village of New Gourna project). The construction system is very cunning and rapid.

4. Cupolas on sacrifice formwork

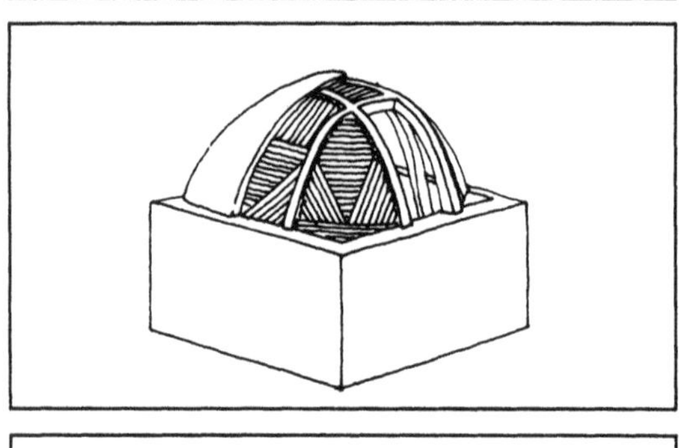

This method is typical for the architecture of Niger and Nigeria, among others. The dome is built using brushwood and small branches and then covered with earth, which makes the structure rigid and waterproof. Inside the wood ribs are also covered with earth.

5. Cupolas on formwork

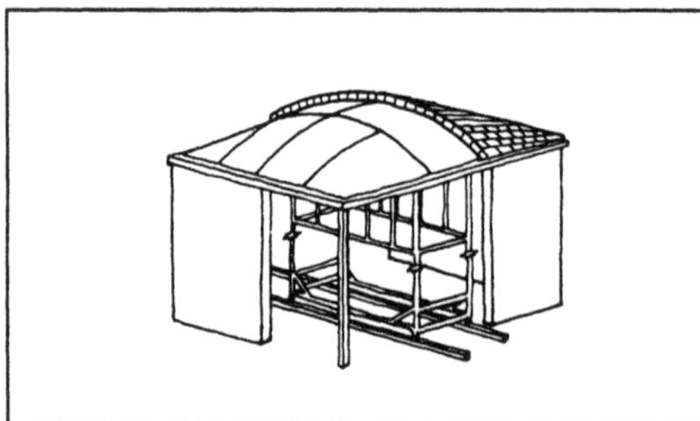

Formwork is indispensable for extremely flattened cupolas. In India, funicular shells (which transmit uniform vertical forces) with a span of 5.2m and a rise of 60cm have been constructed. The formwork must be light, mobile, and easily dismantled to be effective.

Construction of cupolas

1. Courses and bonds

These are erected using independent closed rings. They can also be built as a continuous spiral starting from two or three different points: the courses overlap without connecting with one another. Using this method two or three experienced bricklayers can construct a dome 3m in diameter in one day. The opening of the joints in the extrados is plugged with small pebbles. The bricks can also be laid obliquely to increase friction: trapezoidal elements correct the oblique orientation and serve as keys. As in the case of vaults, the courses are inclined at an angle ranging from 10 to 15°.

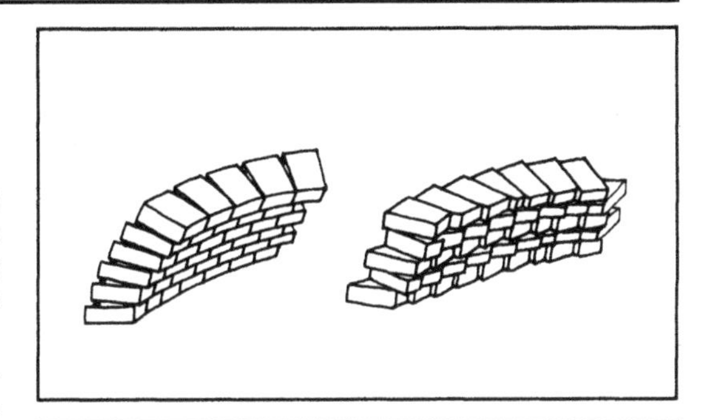

2. Bricks

Small, square or rectangular, light (high straw content) adobes or stabilized compressed blocks, between 5 and 6cm thick, are the types of bricks normally used. Keying or raking enhance adherence. The standard thickness of the cupolas ranges from 10 to 30cm, 15cm being a frequently found value for 3 to 4m spans.

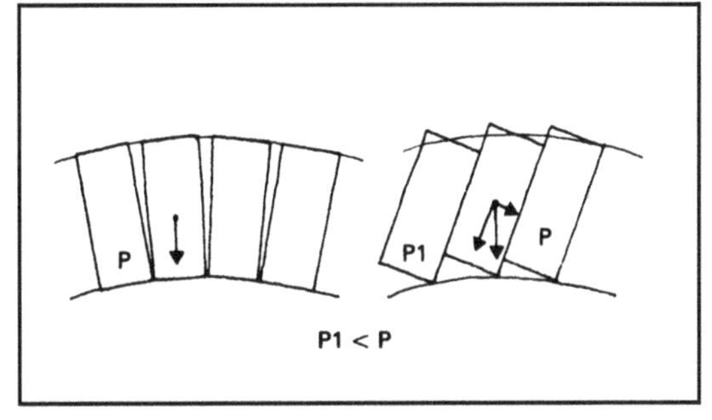

3. Finish

When carefully laid and the joints well raked out, a simple limewash is enough for the internal finish. For the exterior, a fairly thick rendering will assure an attractive finish and good waterproofing.

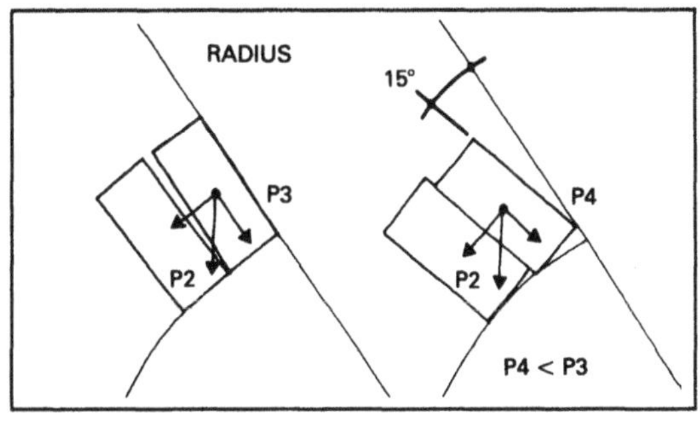

4. Tying

If this is planned, it is positioned at the base of the cupola, above the pendentives or the squinches and cast in-situ in the thickness of the cupola or even carried through onto the extrados.

5. Tools

Traditionally, this type of cupola is constructed without tools. The mason's eye, aided by a marking-line attached to the focus of the cupola and the mason's wrist, is sufficient. Several tools have been developed which free the hand: the 'cupola tracer'. They usually consist of an adjustable central device that determines the shape of the cupola, indicating the position of each brick as well as showing the correct inclination to each brick.

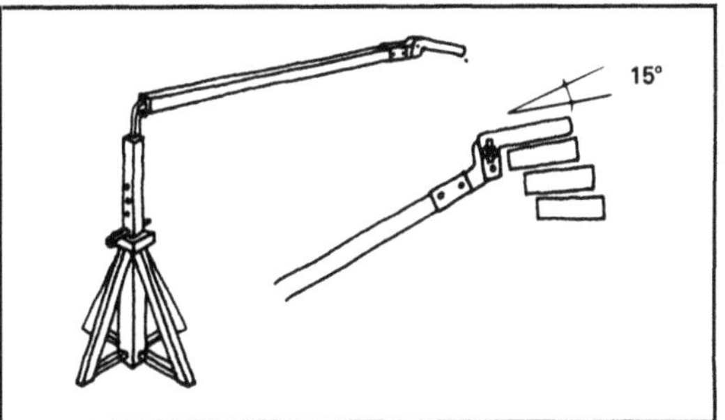

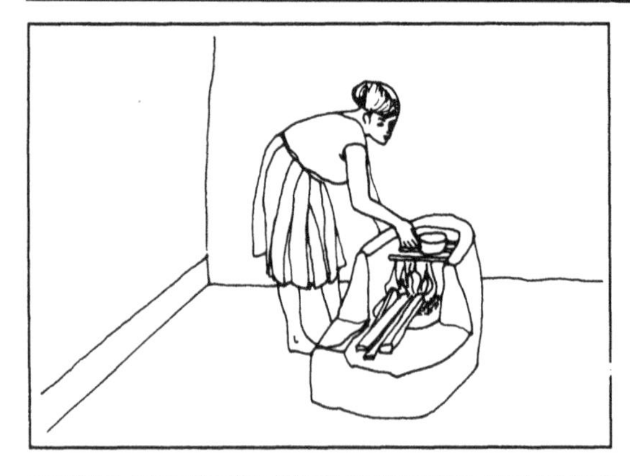

Area of application

Earth can be used in for an immense variety of applications, including the construction of kilns and furnaces, fireplaces and flues. Such equipment functions well but requires regular upkeep; without it, it is liable to deteriorate rather rapidly. One is inexorably led to the conclusion that burnt earth materials are better suited to the building of such items of equipment and that they are clearly more durable. The use of earth as a monolithic material, e.g. rammed earth or cob is acceptable, but is not recommended for making fireplaces or flues, because of cracking and shrinkage problems. On the other hand, constructors will find stabilized compressed bricks or very compact adobes serviceable.

Hearths

There are two possibilities: either the hearth and fireplace surrounding it are constructed using sun-dried earth or the fireplace is constructed using burnt materials and the hearth built in raw earth. If the earth is very clayey, the firing of the fireplace flue will cause shrinkage cracking which must be plugged with the greatest of care. If the earth is less clayey, the heat will cause drying and the material will lose its cohesion and become friable. These are the problems associated with burning at low temperature; a temperature ranging from 600° to 900°C is necessary to produce a quality burnt material.

Fireplaces and flues

While it is true that the inner facing of kilns can be made in raw earth as opposed to the outer facing for which burnt material is necessary, it is also true that firebacks must be made of refractory material. Internal cracking in a kiln is acceptable; it is less so in the case of an indoor chimney. The mortar used can be made from unstabilized earth to which ground burnt brick has been added, the latter serving as sand. All of the cracks in the hearth must be plugged with care. Flues in unburnt earth are not recommended. Earthenware incorporated into the thickness of the walls or built outside, i.e. external flues, is preferable. This precaution guarantees the stability of the flue and helps to prevent noxious gases from escaping through the cracks. Inside, ducts can be rendered in raw earth but chimney pots on the roof should be in burnt brick.

Firing up

No fire can be kindled until the masonry of the oven or fireplace has completely dried. The first fire should be moderate and slow-burning; an intense fire is liable to damage the structure. Cooling should also proceed slowly. With the second firing, the temperature should be raised slightly so as to bring about controlled smoking (e.g. wet leaves) at the end of burning when the fireplace is hot and has expanded. In this way any possible escape of smoke can be detected through cracks which will then be located and plugged. Where earth and metal meets (e.g. the cover of the oven) a little salt can be applied, as when heated salt hermetically seals any cracks.

Applications

1. **Ranges** Numerous agencies and institutes specialized in the relevant technology have published details of improved models of earth kitchen ranges. Some of these, such as the *herl chula* (India) or the *pouna*, *lorena* and *singer* are well known. These earth furnaces consume 10% to 20% less wood than conventional open-hearth ovens. The common feature of the majority of these furnaces is a heating tunnel which passes through a mass of earth.

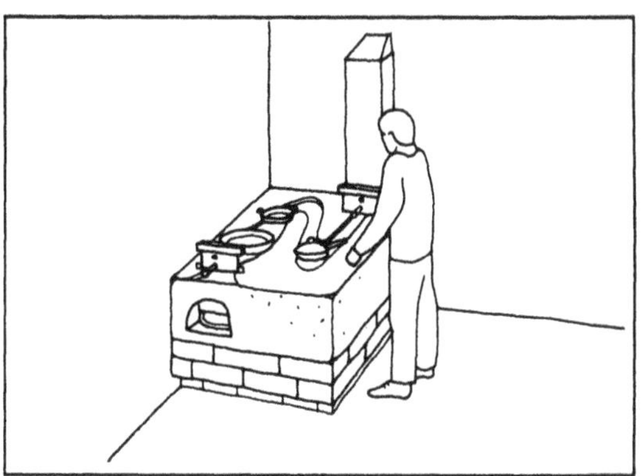

2. **Ovens** These are very often a vault, or a conical dome, (e.g. the ovens of the Pueblo Indians). The walls must be quite thick and the height no greater than 60cm for the heat to be retained. The base of the oven can be made of sun-dried or burnt brick, on a layer of sand and ground glass. Such a layer retains heat very effectively.

3. **Open hearths, fireplaces** These are often built in adobe and their construction poses no major problems. Open-hearth fireplaces are often installed in every room of the adobe homes of the American South-West where they are hallmark of the so-called 'Adobe Style'. It should not be forgotten that an open hearth made of earth is heavy and requires good foundations, and to take the normal fire-hazard precautions for building open hearths: such as not fitting a wooden beam or lintel across the flue. It is better to build hearths in refractory brick.

4. **Boilers, stoves** The Chinese and Korean K'ang are well known examples of these: a mass of earth is heated by the flue gases, and slowly releases the warmth.

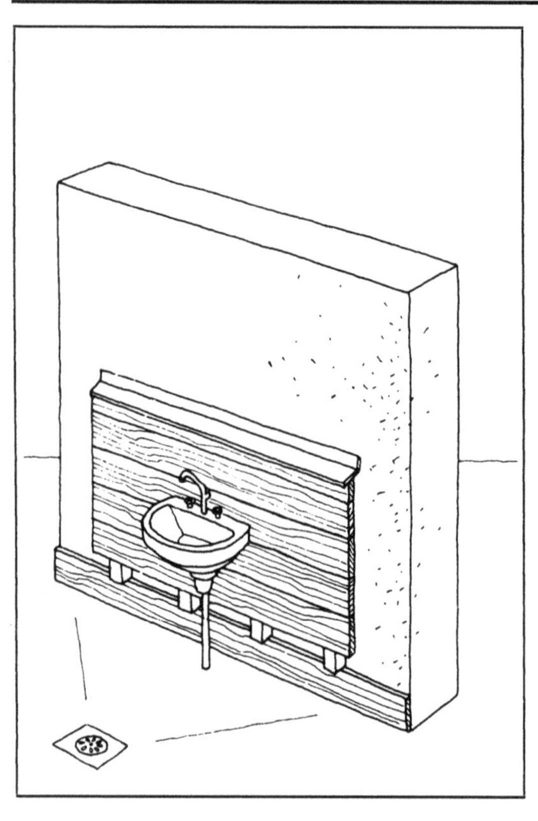

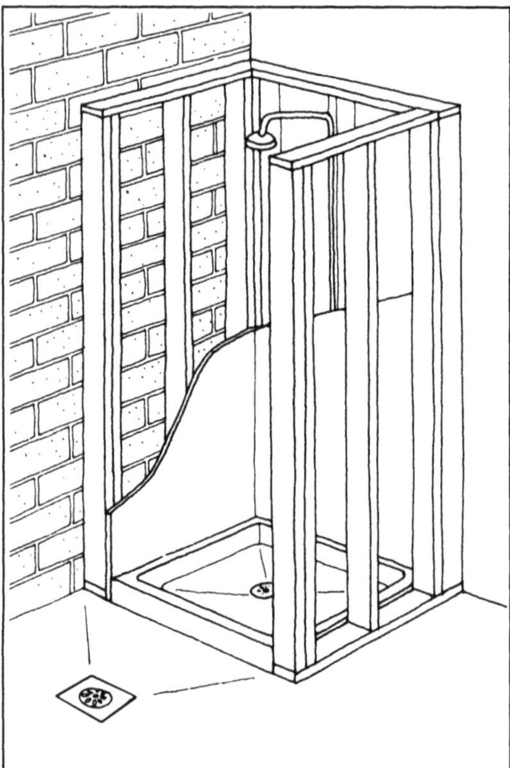

Plumbing

The installation of supply and drain pipes in earth homes must be carried out with special care, as constant moisture can have very serious consequences. In principle, every attempt should be made to centralize the plumbing to the greatest possible extent so that it can be easily examined and maintained, while avoiding adding to the risk of defects which would be difficult to locate. All waste water must be carried outside the house well away from the foundations. Gutters and manholes must be carefully maintained. Inside the house attention must be paid, in particular, to the fittings in rooms containing plumbing installations (kitchen, water closets, and bathrooms) as there is a considerable likelihood of damp. They must be ventilated and provided with easily accessible floor drains. Floors must be adequately sloped for drainage. The showerbaths must stand away from earth walls and these must be protected by a ventilated waterproof coating (danger of condensation). Attention must be paid to ducts that may cause condensation: e.g. air conditioning and heating. It must be possible to drain any such condensate away. The slope of ducts and drip channels must be adequate or else the ducts must be insulated.

1. Siting pipes and ducts

The incorporation of pipe and ducts into earth walls is not advised. The best course is to make use of some part of the structure which is not made of earth (stone or concrete base courses, vertical and horizontal tying, wooden frames) for securing them. Such fastenings must carefully constructed to prevent leaks.

2. Fastenings

The anchoring points of pipes, ducts and appliances must be planned in advance. Wooden blocks of a good size firmly anchored in the walls will allow the fastening of brackets, hooks and rings. Wash basins, sinks, water-heaters, expansion tanks, etc. can be fixed on, for example, ventilated wooden structures. Walls must be protected against any possibility of splashing: protective facing and plinths.

Electricity

An earth house, like any other house, is supplied with electric power by means of a connection with the existing mains. If the mains supply is underground, connection poses no problem and can be effected at terminals and in cabinets (external meters). In the case of large housing complexes the connection of the electrical networks to the houses must be worked out in advance. On the other hand, when the mains supply is from overhead wires, the problem of the connection of the mains supply to the house requires careful attention. The wires are very heavy, and once they are taut, induce tensile stress and vibrations. Consequently, the anchoring of mains supply brackets directly to the earth walls themselves is not recommended. They should rather be anchored to parts of the house that can resist tensile stress, e.g. vertical and horizontal ties.

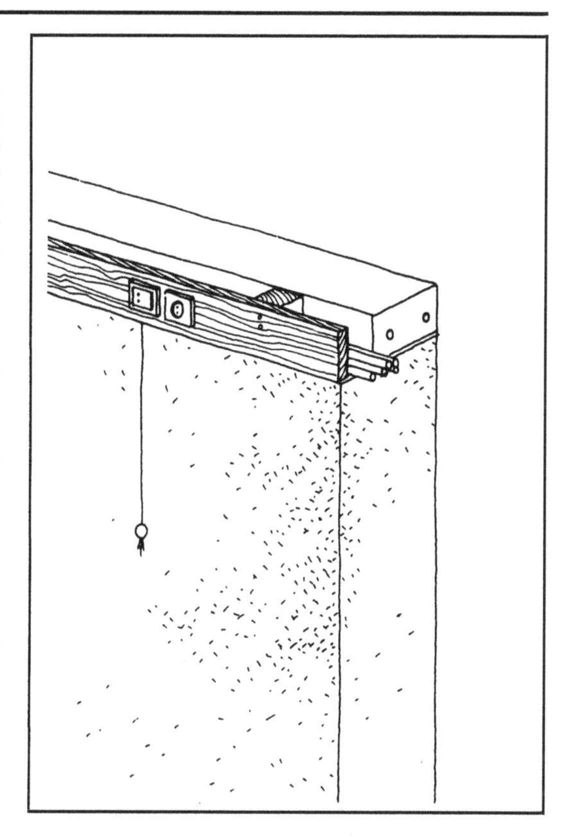

1. Wiring

It is advisable to incorporate sheathed wiring as well as all shielded conductors into the thickness of the floor slabs or floor screeds (widespread practice) or into 'hard' base courses where recesses or anchoring points will have been provided. Maximum use must be made of materials other than earth – wood, concrete – for fastenings: e.g. the wooden framework in clay-straw structures. Generally speaking, grooves should not be made in earth walls for wiring but this is a viable solution if a rendering is provided. In that case the grooves are plugged with earth or mortar and finished with rendering on netting. It is wise to make use of plinths and door frames or to provide a cable duct on the ring-beam or in the thickness of floors.

2. Fastening

Switches and sockets can be flush-fitted or surface mounted. If they are flush-fitted they must be very deeply embedded; properly designed fitting must be used for this purpose. The bedding can be in plaster. Surface-mounted fittings should preferably be installed on the base course, on plinths, on the subframes of openings, and vertical wooden frames. When mounting fittings on earth walls, small wooden blocks integrated in advance in the walls and embedded in deep plaster should be provided for mounting purposes.

Renovation

The renovation of an earth structure, like that of all structures in other materials, consists, by definition, of giving a new look to a structure marked by years of use. Sometimes major operations are required to wipe out these ravages of time, particularly if it is planned to thoroughly revamp the building. Less far-reaching works will be required when what can be described as improvements are carried out. The renovator is, however, immediately confronted with a major decision: should the causes of deterioration be tackled, thus providing a lasting repair, or should only the symptoms be dealt with, thus allowing the basic defects which are destroying the core of the structure to remain?

Restoration and rehabilitation

Restoring a structure entails dealing with existing damage and directly curing defects in the framework of the building: structural defects – cracking due to different rates of settlement (bad foundations), cracks due to poorly underpinned loads (e.g. lack of ties) – and water-related damage: e.g. hollowing at the base of the wall owing to capillary action. Such restoration can involve major work on the underground work of a building: e.g. redoing foundations, construction of a base course, and rebuilding partition walls.

Apart from the restoration of the framework of the building, restoration of minor structural work may be required, e.g. openings; and installations: plumbing and electricity, and other equipment.

If the project involves the restoration of a historical structure 'to its original condition', the restoration process must respect the historical background of the structure. If, on the other hand, the restored structure is to be used for anything more than as a witness to history, it is better to speak of rehabilitation. In this case the restoration must give greater thought to the future occupants of the structure, who may have very different requirements than the original occupants. The original layout of the building may be restored if the objective is a historically accurate restoration. If the rehabilitation of the building is envisaged, the potential occupancy scenario(s) for the structure will affect all stages in the restoration and impose spatial, formal and aesthetic limitations, which themselves depend upon the structural limitations and the original layout of the structure: position of walls, extent of structural damage, etc.

Rehabilitation more than restoration is primarily concerned with the practical responses which are made to the needs of using the structure, rather than responding to the historical image of the building. Apart from the foregoing clarifications of the problem, restoration very quickly becomes subject to a number of budgetary limitations as the cost will be very different for the two different courses. In fact, the rigorous pricing of the work involved assures the real feasibility of the operation and, above all, its completion within an acceptable time limit. Such estimates must be made by an expert since it is undoubtedly more difficult to restore or rehabilitate an earth structure properly than to erect a new building. Determining the extent of deterioration, the ways to cure it, and the restoration techniques used has a direct effect on the cost of the operation. This estimate must not be made lightly but with a full knowledge of the problems and solutions. This is why the programme of operations must be formulated step by step: historical research, accurate surveying, assessment of damage (particularly for damp, which may necessitate the drying out of the walls, restoration of level, and the drainage of the natural ground), scenarios for future use, details of construction systems, and renderings.

The budget will determine possibilities and available means, especially, among others, for the restoration of a building's underside – foundations, base course, wall drainage, floors, frames, openings – which can be very expensive. Priority must in any case be given to the major problems: e.g. absorption of the water causing the damage, and the reinforcement of the structure.

Conservation

The preservation of earth architecture of a historical nature has been undertaken by many countries, who seek in this a way of conserving and restoring pride in their architectural heritage, which is of considerable cultural value. Apart from the restoration of isolated buildings, conservation is now also concerned with archaeological sites, some of which may be very extensive. The example of the Middle East comes particularly to mind. There, structures which constitute a living testimony to the important civilizations that grew up in Mesopotamia, Elam, Sumeria and Babylon, are threatened with destruction because of either a lack of maintenance or restoration techniques incompatible with the soil used in their construction, and which could well hasten the destruction of these sites. This problem is in fact fairly widespread and nowadays the very survival is at stake of the remains of all the great historical civilizations which have bequeathed a legacy of raw earth architecture to nearly all the continents of the worlds.

The conservation of archaeological sites or historic monuments made of raw earth poses extremely delicate problems since earth as a structural material requires appropriate restoration treatments and these are sometimes incompatible with those suited to other materials: problem of impermeable renderings, which may cause chronic damp – as a result of the barrier to movements of vapour in walls; as well as problems of know-how for the restoration of buildings using the original construction techniques. Apart from these problems which are specific to the use of earth as a structural material, there are the problems which are common to all operations aiming at the preservation of historic buildings.

Among these are the following:

– Should one refrain from transforming the appearance of the monument or site and attempt to conserve it in its current state no matter how advanced the deterioration? If it is decided to do so, intervention will consist essentially of protection techniques, providing shelter or stabilizing the material.

– Should one modify the environmental conditions which appear to be causing the deterioration of the site or building with a view to arresting the process which underlies such deterioration and is already underway? If so, drainage techniques which include levelling the ground and the reduction of gulleying (e.g. by planting) are indicated.

– Should one reconstruct the buildings either partially or completely and attempt to restore their original appearance? If so, demolition and reconstruction operations require great technical competence, if the decay or disfigurement of the buildings is to be avoided.

Nowadays there are international agencies which have brought about a proliferation of conferences and seminars on the theme of the preservation of sites and historical buildings. A case in point is the ICCROM or International Centre for the Study of the Preservation and the Restoration of Cultural Property, which was founded by UNESCO in 1959 and which acts as an independent, intergovernmental scientific organization. It has various functions, including the collection and study of technical and scientific documentation on the subject of the preservation of cultural property, the co-ordination and stimulation of research institutes working in this area, the provision of advice and the making of practical technical recommendations and contributing to the development of competence in the field. There is also the ICOMOS or International Council of Monuments and Sites, which organizes numerous international seminars: Yazd (Iran), 1972; Santa Fe (USA) 1977; Ankara (Turkey), 1980; Cuzco (Peru), 1983.

Some of the actions recommended by these international agencies, particularly with respect to archaeological monuments and sites, can be summarized as follows:

– a well defined conservation policy for the excavation of sites which are believed to contain earth structures;
– a detailed assessment and definition of the reasons for the deterioration of these structures;
– covering up, after the excavation, of archaeological structures which have to remain below the natural ground level;
– temporary protection taking into account the specific exposure condition;
– adequate drainage systems for those monuments which, after excavation, have to remain above ground level;
– roofs or other means of protection;
– regular maintenance.

The current research aims to establish principles of good practice which will lead in the long term to effective application of the proposed methods.

Iccrom, CRATerre and the School of Architecture of Grenoble established in 1990 a global framework for action: the GAIA project.

Adam, J.A. *Wohnlund Siedlungsformen im Süden Marokkos*. München, Georg D.W. Callwey, 1981.
ADAUA. *Chantier d'Essais*. Genéve, ADAUA, 1978.
AGRA. *Recommandations pour la Conception des Bâtiments du Village Terre*. Grenoble, AGRA, 1982.
Alva, A.; Chiari, G. 'Protection and conservation of excavated structures of mud-brick'. In *Conservation on Archaeological Excavations*, Rome, ICCROM, 1984.
An. 'Ausführungswarten der Decken'. In *Neue Bauwelt*, Berlin, 1947.
Auzelle. R.; Dufournet, P. 'Le beton de terre stabilisé'. In *Techniques et Architecture*, Paris, 1946.
Bardou, P.; Arzoumanian, Y. *Archi de Terre*. Marseille, Editions Parenthèses, 1978.
Beitdatsch, A. *Wohnhaüser aus Lehm*. Berlin, Hermann Hübener, 1946.
BRE. 'The thermal performance of concrete roofs and reed shading panels under arid summer conditions'. In *Overseas Building Notes*, Garston, BRE, 1975.
BRU. 'Fireplace in houses'. In *BRU data sheet*, Dar-Es-Salaam, 1974.
Carola, F. *Recherche de Systèmes Economiques de Construction*. Rome, CONSASS, 1977.
CRATerre. *Projet de 8 Logements de Fonctionnaires*. Grenoble, AGRA, 1982.
CRATerre; GAITerre. *Marrakech 83 Habitat en Terre*. Grenoble, REXCOOP, 1883.
Dalokay, Y. *Lehmflachdachbauten in Anatolien*. Technischen Universität Carolo-Wilhelmina zu Braunschweig, 1969.
Dellicour, O. *et al. Vers une Meilleure Utilisation des Ressources Locales de Construction*. Dakar, UNESCO-BREDA, 1978.
Denyer, S. *African Traditional Architecture*. New York, Africana, 1978.
Dethier, J. *Des architectures de terre*. Paris, CCI, 1981.
DIN. *DIN Lehmbau 18951–18957*. Berlin, DIN, 1956.
Doat, P. *et al. Construire en Terre*. Paris, éditions Alternatives et Parallèles, 1979.
Evans, I.; Boutette, M. *Lorena Stoves*. Stanford, Appropriate Technology Project of Volunteers in Asia, 1981.
Fauth, W. *Der Praktische Lehmbau*. Singen-Hohentwiel, Weber, 1948.
Foadey, S.M. *L'habitat en Afrique, contribution à l'étude des possibilités d'utilisation des matériaux locaux*. Liège. Faculté des sciences appliquées, 1978.
Fox, J. *Building with Zed Tiles*.
Galván Duque, H. Peña Tomé, E. *Cartilla de autoconstrucción para escuelas rurales*. Mexico, Conescal, 1978.
GATE. *Lehmarchitektur, Rückblick-Ausblick*. Frankfurt am Main, GATE, 1981.
GATE. *Low-cost Self Help Housing*. Eschborn, GATE, 1980.
Gérard, V. *De l'architecture traditionnelle à la construction scolaire*. Paris, UNESCO, 1976.
Hammond, A.A. 'Prolongation de la durée de vie des constructions en terre sous les tropiques'. In *Bâtiment Build International*, Paris, CSTB, 1973.
Harris, P; 'Earth roofs'. In *Adobe News*. Albuquerque, Adobe News, 1977.
Hays, A. *De la terre pour bâtir*. Manuel pratique. Grenoble, UPAG, 1979.
Herbert, M.R.M. 'Some observations on the behaviour of weather protective features on external walls'. BRE, Garston, 1974.
Herrera Delgado, J.A. *et al. La Tierra en el Arquitectura una Revalorizacion*. Mexicali, Universidad autónoma de Baja Califórnia, 1978.
Hölscher Wambsganz Dittus. *Lehmbauordnung*. Berlin, Von Wilhelm Ernst und Sohn, 1948.
Hughes, R. 'Material and structural behaviour of soil constructed walls'. In *Techniques and materials*, 1983.
ICOMOS. *International Council of Monuments and Sites*. Yazd, ICOMOS, 1972.
Innocent, C.F. *The Development of English Building Construction*. 1916.
Iyad Ruwaih; Orhan Erol. 'Building damages caused by foundation failures in arid regions'. In *International Journal IAHS*, Pergamon press, 1984.
Kuba, G.K.; Madibbo A.M. *Polyethylene Waterproofing for Traditional Mud Roofs*. Khartoum, BRD, 1970.
Leroy, L.; Idabouk. *Etude d'une Voute Surbaissée en BTS*. Rabat, CERF, 1968.
Liétar, V.; Rollet, P. *Mayotte Habitat Social*. Grenoble, AGRA, 1983.
Markus, T.A. *et al. Stabilized Soil*. Glasgow, University of Strathclyde, 1979.
Matuk, S. *Architecture Traditionnelle en Terre au Pérou*. Paris, UPA.6, 1978.
McHenry, P.G. *Adobe and Rammed Earth Buildings*. New York, John Wiley and Sons, 1984.
McHenry, P.G. *Adobe Build it Yourself*. Tucson, The University of Arizona press, 1973.
Meunier, A. *Technologie professionnelle de chantier. Les matériaux de construction*. Paris, Foucher, 1958.
Middleton, G.I. *Build your House of Earth*. Victoria, Compendium Pty, 1979.
Miller, T. *et al. Lehmbaufibel*. Weimar, Forschungsgemeinschaften Hochschule, 1947.
Miller, T. 'Adobe or sun-dried brick for farm buildings'. In *Farmers Bulletin*, Washington, US Department of Agriculture, 1949.
Ministerio de vivienda y construcción. *Mejores Viviendas con Adobe*. Lima, Ministerio de vivienda y construcción, 1975.

Morales Morales, R. *et al. Proyecto de bloque estabilizado. Estructuras.* Lima, Universidad Nacional de Ingenieria, 1976.
Moreno Garcia, F. *Arcos y bóvedas.* Barcelona, CEAC, 1978.
Mukerji, K. *et al. Dachkonstruktionen für den Wohnungsbau in Entwicklungsländern.* Eschborn, GATE, 1982.
Musick, S.P. *The Caliche Report.* Austin, Center for maximum potential building systems, 1979.
Niemeyer, R. *Der Lehmbau und seine Praktische Anwendung.* Grebenstein, Oeko, 1982.
Pering, C. *Autoconstruction Organisée.* Lund, Ecole d'architecture de l'université de Lund, 1981.
Perrin, H; 'Université Officielle de Bujumbura, Burundi'. In *Schweizer Baublatt*, 1974.
Pollack, E.; Richter, E. *Technik des Lehmbaues.* Berlin, Verlag Technik, 1952.
Scarato, P. *Les Conditions Actuelles de la Réhabilitation des Constructions en Pisé.* Région du Dauphiné. Grenoble, UPAG, 1982.
Schild, E. *L'Etanchéité dans l'Habitation.* Paris, Eyrolles, 1978.
Schölter, W. *Dünner Lehmbauverfahren.* In Natur Bauweisen, Berlin, 1948.
Schultz, K. *Adobe Craft Illustrated Manual.* Castro Valley, Adobe Craft, 1972.
Smith, S. *La obra de fábrica de ladrillos.* Barcelona, Editorial Blume, 1976.
Sperling, R. 'Roofs for Warm Climates'. In BRE, Garston, 1974.
Structural Engineering Research Centre. *Houses for Economically Weaker Sections.* Madras, SERC.
Stedman, M. and W. *Adobe architecture.* Santa Fe, The Sunstone press, 1975.
Sulzer, H.D.; Meier, T. *Economical Housing for Developing Countries.* Basel, Prognos, 1978.
Torraca, G. *An international project for the study of mud-brick preservation.* 1970.
Torraca, G. *Porous Building Materials, Materials Science for Architectural Conservation.* Rome, ICCROM, 1982.
Trueba, G.Y. Coronel. 'Systema constructivo "YUYA"'. Priv. com. México, 1983.
US/ICOMOS. 'Recommendations of the US/ICOMOS-ICROM adobe preservation working session, Santa Fe', US/ICOMOS-ICROM, 1977.
Verwilghen, A. *Details of an Improved Method of Traditional Wattle and Daub Construction.* Panzi, Verwilghen, 1976.
Verwilghen, A. Priv. com. Antwerp, 1984.
Vita/ITDG. *Fourneaux à Bois Économiques pour Faire la Cuisine.* Mt Rainier, VITA, 1980.
Volhard, F. *Leichtlehmbau.* Karlsruhe, CF Muller GmbH, 1983.
Warren, J. *The Form Life and Conservation of Mud-brick Building.* 3rd International symposium on mudbrick (adobe) preservation, Ankara, ICOM-ICOMOS, 1980.
Wienands, R. *Die Lehmarchitektur der Pueblos.* Köln, Studio Dumont, 1983.
Williams-Ellis, C.; Eastwick-Field, J. & E. *Building in Earth, Pisé and Stabilized Earth.* London, Country Life, 1947.
Wolfskill, L.A. *et al. Bâtir en terre.* Paris, CRET.
Yurchenko, P.G. 'Methods of construction and of heat insulation in the Ukraine'. In *RIBA journal*, London, 1945.

The Kooralbyn Hotel Resort in Queensland, Australia, made of stabilized rammed earth (CEAC). (architect: Greenway Architects; Walling: Engineered Aggregates Australia) (David Oliver, Engineered Aggregates, CRATerre EAG)

11. DISASTER-RESISTANT CONSTRUCTION

The world is regularly hit by natural disasters: earthquakes, storms and floods. They occur in many parts of the world and their consequences are often disastrous, particularly in developing countries. In 1976 alone, seismic activity in the Philippines, Indonesia, Turkey, Italy and China caused the loss of more than 500,000 lives.

Although casualties due to the collapse of buildings frequently occur in urban areas constructed using materials and techniques which are considered strong and safe (such as reinforced concrete), they are much more common in areas of precarious housing, where the majority of structures are in earth. Great progress has been made in recent years in the field of disaster resistant construction. Unfortunately, all too often unsuitable methods of construction are used in disaster areas. This may be explained by the lack of time and means, but above all by the lack of proper information. The same mistakes are repeated everywhere, and always have been.

Nevertheless it has been clearly established that it is not the material on its own, whether earth, concrete or other, which is at fault. The main cause of damage is often the way in which the material has been produced and used in construction. It has been clearly demonstrated that a well-built and well-maintained house can withstand the majority of earthquakes, whatever the material of its construction. It is therefore possible to greatly reduce the risk of disaster with only a relatively small increase of the cost of construction.

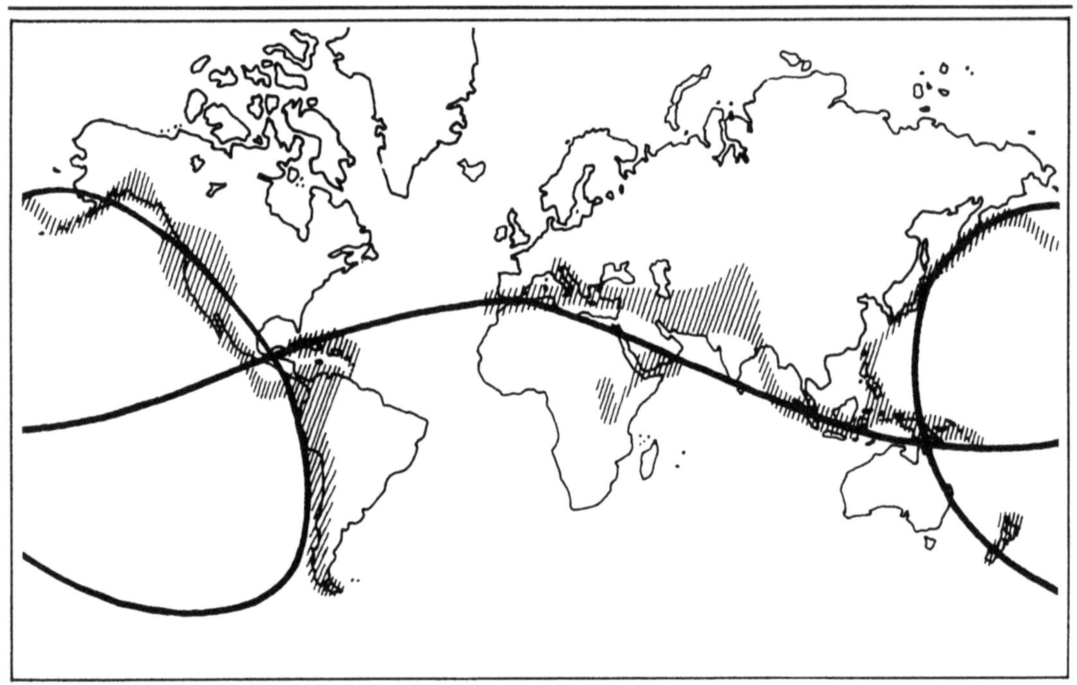

Earthquakes occur on land and under water along narrow belts situated on the edge or in the middle of oceans. These belts also demarcate vast areas of land and ocean where earthquakes are very rare, although not entirely absent (Australia, Siberia, Scandinavia, West Africa, Northern Canada, Brazil).

The regions of heavy seismicity are situated:

- along the continental margins, for instance along the Pacific Coast of North and South America;
- along island arcs like the one running through the Caribbean and in the Pacific Ocean the ones associated with the Aleutian and Kuril Islands, Japan, the Marianas and the Philippines, as well as Indonesia, the New Hebrides, Tonga, etc.;
- along the mid-ocean ridges: the mid-Atlantic ridge, the Pacific and Indian ridges, etc.;
- outside these narrow belts, within continents, in regions with a more widespread seismicity: from China and Burma to Turkey, in the Alps and the Mediterranean, from Turkey to the Azores.

The occurrence of earthquakes is quite adequately explained by the fact that the relative position of oceans and continents is not static: the continents are in motion. Thus, America is slowly moving away from Africa, and the Arab peninsula, whilst separating from Africa by the opening up of the Red Sea and the Gulf of Aden, is approaching Iran. The complementarity of the coastlines on both sides of the Atlantic had been discovered as far back as the 17th century. But one had to wait till 1912 for Alfred Wegener to elaborate and explain a coherent theory on the movement of continents, the continental drift. And it was only about 40 years later, around 1960, that this drift was actually proven by geophysicists, and developed in its current theory of plate tectonics.

The outermost crust of the earth, called the lithosphere, is about 100km thick; it is divided into large plates which are susceptible to horizontal movements, propelled by convection currents affecting the mantle below. Two neighbouring plates will converge or separate, taking the continents that they carry with them. The exploration of the ocean floor has led to the discovery of submarine mountain ranges, several thousands of kilometres long and a few hundred kilometres wide – the mid-ocean ridges. In their centre, they have an area of subsidence, or rift valley, where the separation of two neighbouring tectonic plates takes place. The speed of movement of a single plate may reach 6 to 8cm per year and the separation of two plates can therefore attain more or less double this distance, or 16cm per year. This causes the seismicity observed in the middle of oceans and the formation of submarine volcanoes, whose lava streams fill the voids caused by the separation of plates. Since the plates move apart along these ridges, they converge in other places. The confrontation of two converging plates causes a thickening of their rims and sometimes a submersion of one plate under the neighbouring one. Their zone of contact is the focus of very active seismicity and the heaviest earthquakes occur there. There are also zones where two neighbouring lithospheric plates move laterally along fault lines called transform faults.

Seismic shocks are relatively short; in most cases, their duration does not exceed one minute. But they can be terribly destructive, and they are almost always followed by other shocks, the aftershocks.

The scale of an earthquake is determined by its magnitude and by its macroseismic intensity.

The magnitude has been defined by Charles Richter as the decimal logarithm of the amplitude (in micrometres) of the largest wave registered on a special type of seismograph situated at 100km of the epicentre of the earthquake. The magnitude represents the scale of the phenomenon in its focus, where it cannot be measured directly; it is therefore calculated on the basis of the amplitude and of the period of soil movement registered on the seismograph. The most recent definition of magnitude (Kanamori) is most significant for the largest earthquakes. Although the definition of magnitude is limitless, a maximum limit is imposed in practice due to the solidity of the rocks of which the lithosphere is composed. And in fact, an earthquake with a magnitude above 9.5 has not been registered since the start of the measurements.

The macroseismic intensity is a measure of the felt effects of an earthquake on structures or natural sites and of the reactions of living beings. For a given location, it determines the effects of an earthquake against an intensity scale with twelve levels, the Mercalli or MSK scales. The scale currently used is the 1984 MSK scale, which has replaced the Modified Mercalli scale.

1. Tremors

They cause objects to fall, have a dynamic impact on structures, introduce components of horizontal acceleration and create shear forces. In most cases, ordinary structures are unable to withstand these effects undamaged. In earthquake zones, it is therefore necessary to design and construct buildings that take this into account. In addition, tremors affect the ground on which structures are built and can considerably reduce its bearing capacity.

2. Secondary effects

Structures can be damaged, destroyed or even buried by materials or water set in motion by an earthquake. Short circuits, breaks in reservoirs or pipelines carrying flammable substances, and overturned boilers or stoves can cause fires. The destruction of water reserves and distribution networks and of access roads hinders firefighting and emergency services. Quite often, therefore, secondary effects prove to be as destructive or even more so than the tremors themselves.

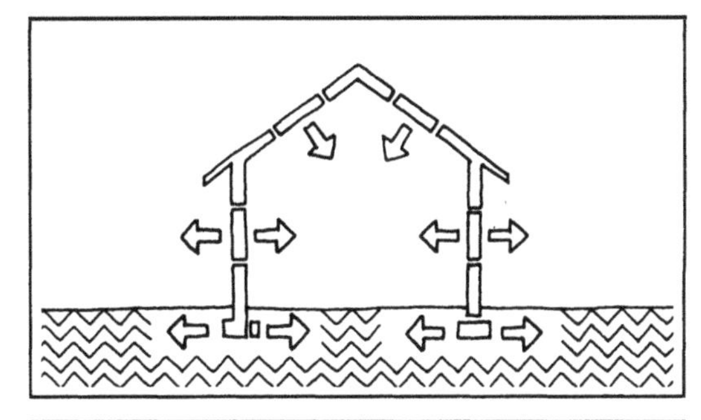

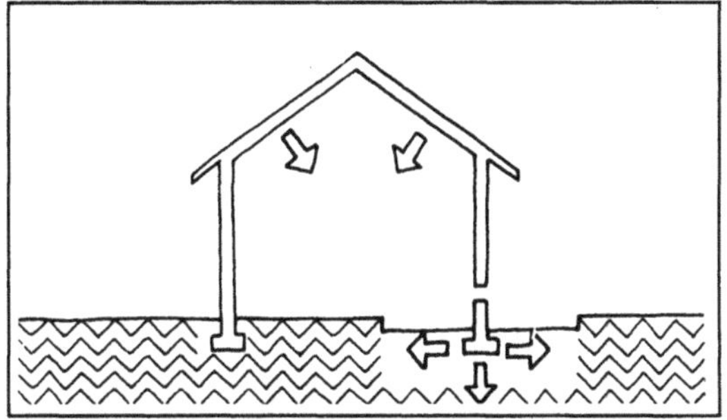

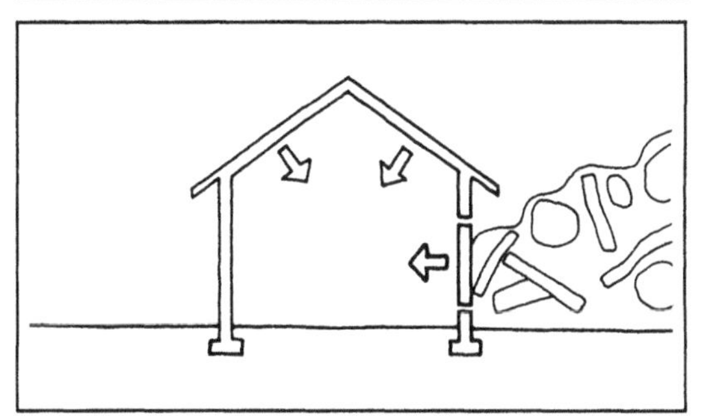

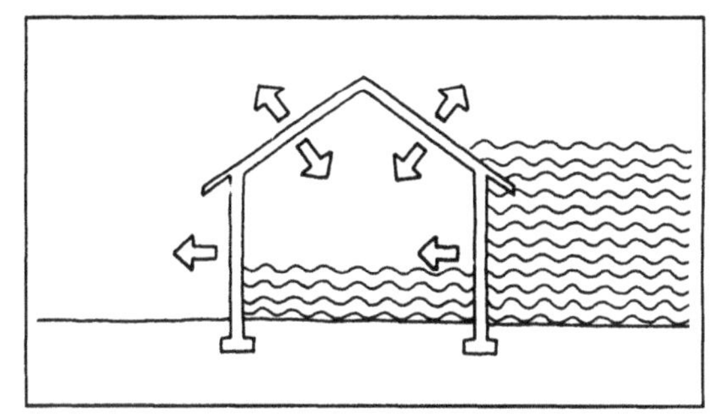

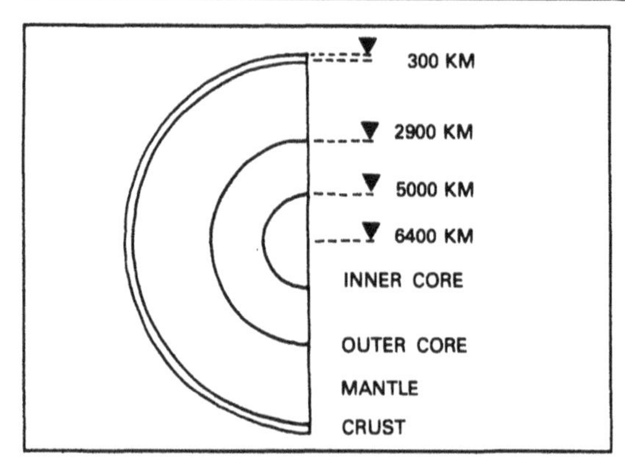

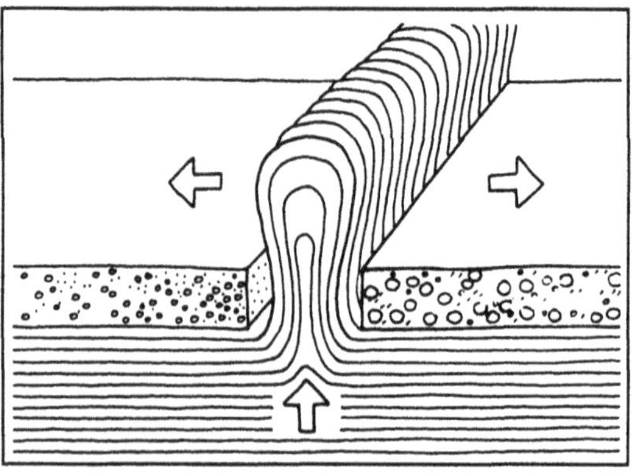

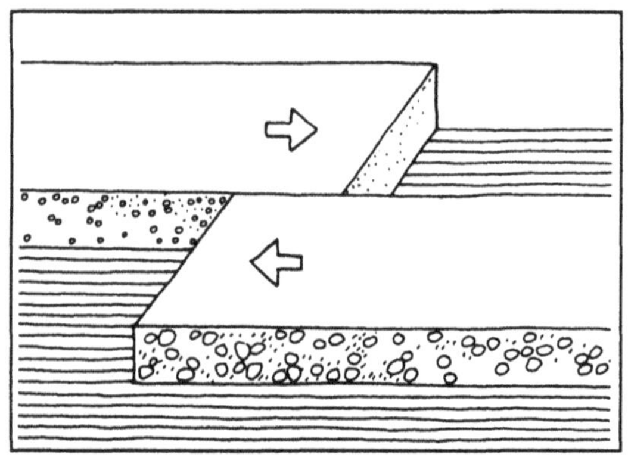

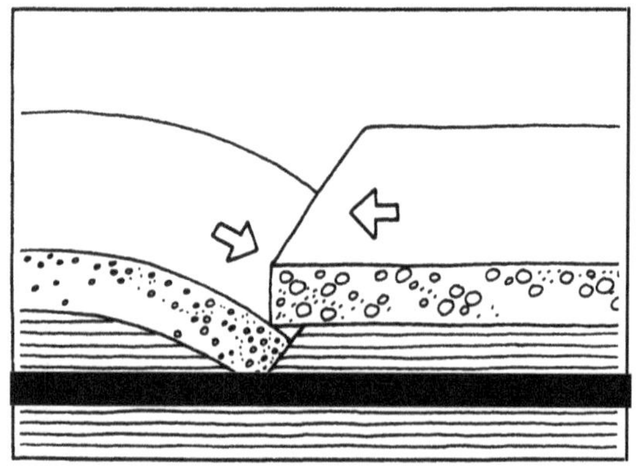

Every year, about one hundred large earthquakes with a magnitude of 6 or more are registèred around the world; about twenty of those attain a magnitude of 7 or more. If these earthquakes occur in very populated regions, they can be particularly fatal. Thus, earthquakes in 1976 in Guatemala, Friuli and Tangshan, caused hundreds of thousands of casualties and enormous material damage. The most destructive earthquakes are natural tectonic phenomena, brought about by the relative movement of two lithospheric plates. Volcanic activity is another natural cause of earthquakes.

Finally, human activity, such as mining or the filling of great dams can also cause earthquakes. Two types of seismicity exist therefore: a natural type, and one that is artificial or induced and of human origin.

Earthquakes of tectonic origin

These are the most common and also the most destructive kind of earthquakes. The formation and expansion of tectonic fractures release the energy accumulated during the deformation process. Five types of earthquakes can be distinguished:

- earthquakes caused by the divergence of plates, for instance those that occur along mid-ocean ridges, in Iceland and the Azores, and in the Afar region, although the latter are not submarine.
- earthquakes caused by the convergence of plates, leading to subduction. One of the plates submerges under the neighbouring one, which causes the most destructive earthquakes. This is the case in the Andean zone, where the Nazca plate penetrates under the South American plate. It also happens at the island arcs of the Pacific Ocean or the Caribbean.
- earthquakes caused by the sliding of plates in opposite directions; the major component of movement is parallel to the contact zone. This is the case of the San Andreas fault in California.
- earthquakes caused by normal faulting. These are mainly vertical movements, along fractures in the earth's crust. They occur for instance in subsidence zones such as the African rift in the region of the great lakes.
- in certain zones of complex structure, combinations of hard-to-analyse phenomena also generate earthquakes. This is the case of numerous earthquakes in continental zones.

An earthquake is experienced as a series of more or less intense vibrations of the ground and the structures. These vibrations originate in a point deep inside the earth, the focus or hypocentre, from where they are propagated. The point at the surface of the globe, vertically above the focus, is called the epicentre of the earthquake. The effects of the earthquake on a building are determined by the distance between building and focus, the nature and the layout of the terrain between building and focus, the morphology of the site, the nature and the intensity of the earthquake at the focus, and of course the nature of the building itself.

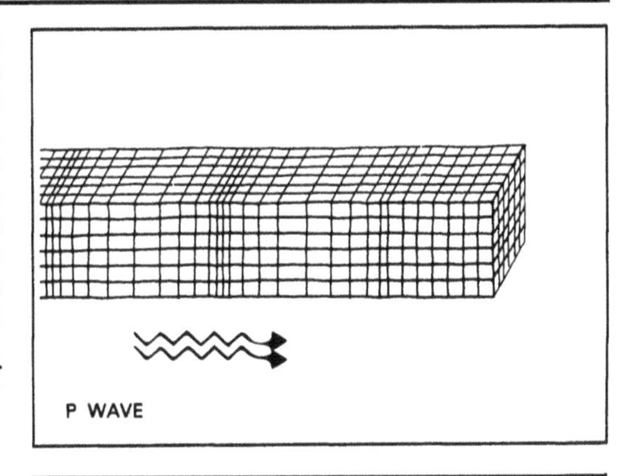
P WAVE

Waves

A propagation of, mostly elastic, waves takes place in the terrain situated between the focus and the place where the earthquake is felt.

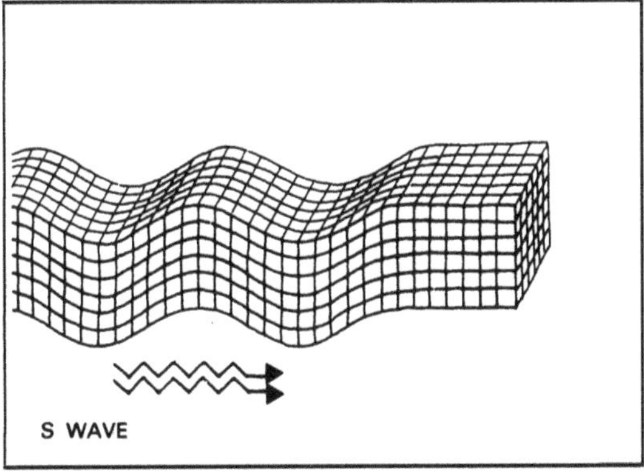
S WAVE

1. Body waves: primary and secondary waves

– 'P' or primary waves are longitudinal waves, and the ones most rapidly propagated, in solids as well as fluids. They successively compress and decompress the rock through which they pass. Their speed can range from a few hundred metres per second at the surface to more than 13 kilometres per second on the inside of the earth's mantle.
– 'S' or secondary waves are latitudinal waves; they are not as fast as 'P' waves. They are characterized by vibrations perpendicular to their direction and induce shear forces; they cannot be propagated in fluids. 'S' waves are the most destructive type.

The 'P' and 'S' body waves are reflected and refracted by the cleavage planes of rocks and by the ground surface. This can lead to interferences of waves, which can amplify the tremors or sometimes trap the waves in a certain area, where they cause resonance.

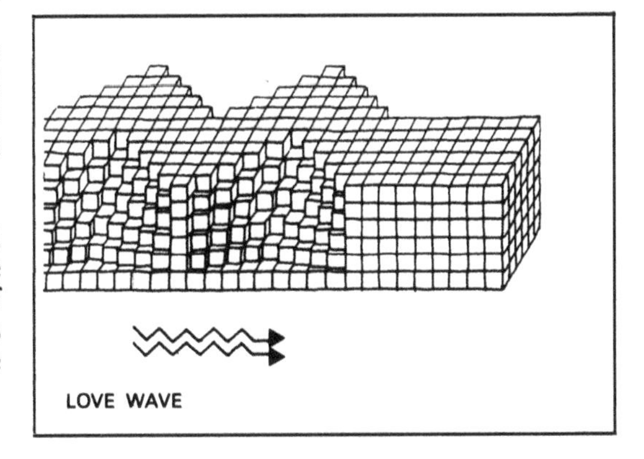
LOVE WAVE

2. Surface waves: Rayleigh and Love waves

These waves are slightly slower than the 'S' waves and are propagated close to the surface of the ground.

Rayleigh waves displace soil particles both vertically and horizontally, in a vertical plane parallel to the direction of propagation. Love waves are propagated without vertical displacement of solid particles.

The earth tremors felt in a given location therefore are the result of a combination of the various waves described above.

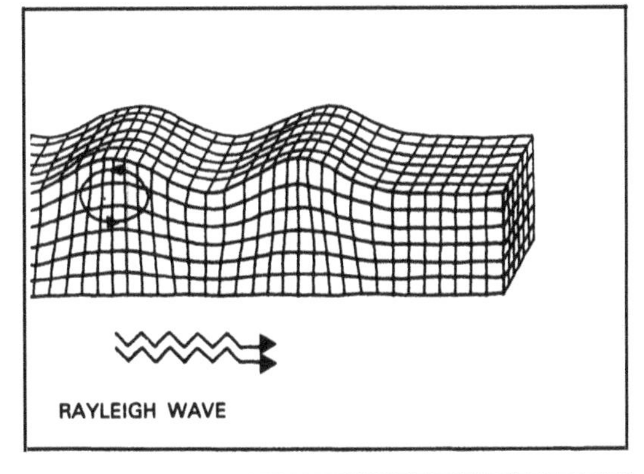
RAYLEIGH WAVE

When disturbances of seismic origin arrive in a given location, the soil particles there will undergo an acceleration, which is transferred to buildings. When the soil moves horizontally, the foundations of a building move with it, but the upper levels react with a certain delay. This causes a bending of walls and columns for which they are not designed. There is general agreement that horizontal accelerations

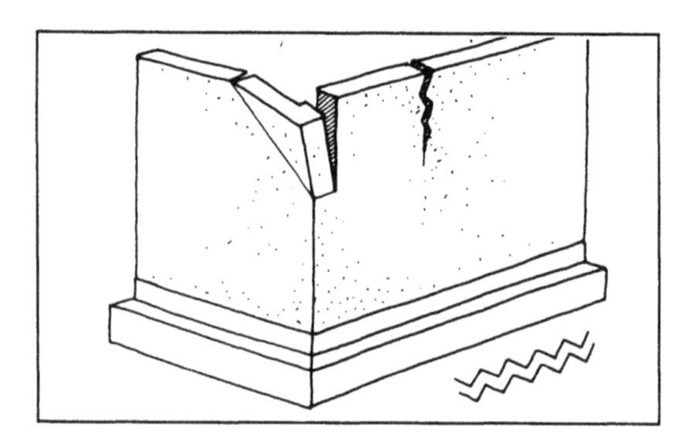

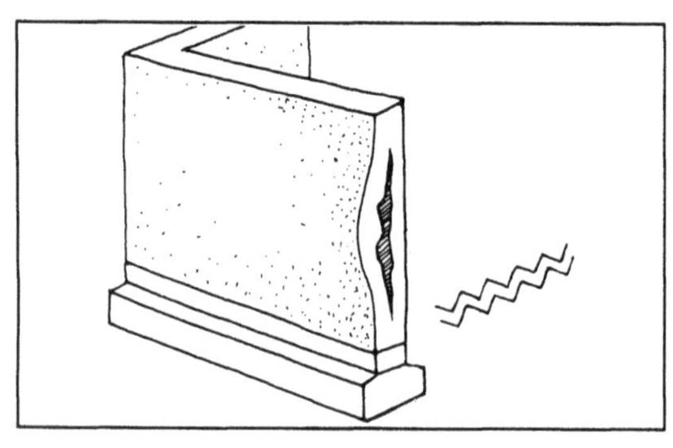

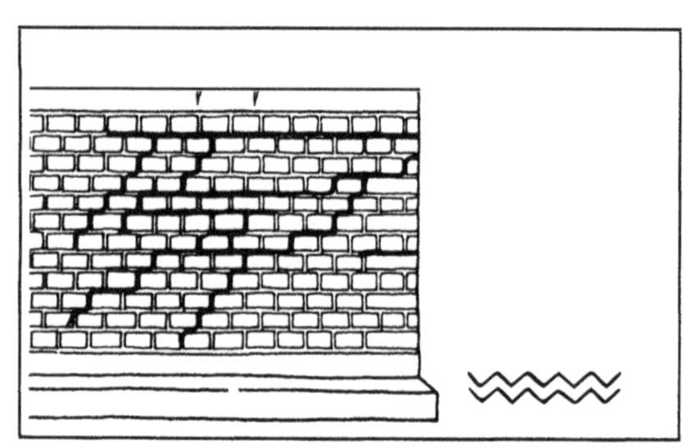

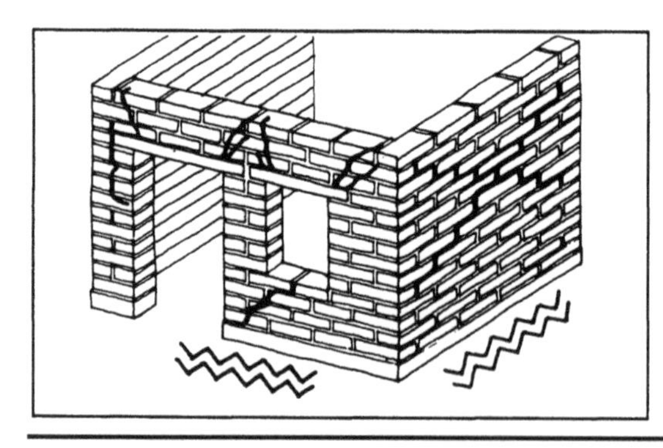

1. Impact on soils

Seismic vibrations cause shrinkage or expansion (e.g. P waves) and shear (e.g. S waves) of soils, with enormous effects. A large shrinkage of an area can cause damage as spectacular as the buckling of railway lines or underground pipelines. In neighbouring areas, the expansion will open cracks. Elsewhere, seismic vibrations tend to compact loose soils, which leads to settlement. And in some soils saturated with water, they can even produce liquefaction causing the loss of their bearing capacity, and, with the right kind of slope, they can lead to catastrophic landslides.

2. Impact on structures

The earthquake ground motion is the sum total of:

- a global acceleration, with three components: one in the vertical direction and two in perpendicular horizontal directions;
- a distortion (shrinkage-expansion and shear), also with three components.

This forces structures to vibrate, and then to sway freely; as a result their particles can be subject to movements with six components, of which three are longitudinal and three rotative.

Effects of the horizontal component

The longitudinal horizontal component may have different effects, according to the nature and the direction of the affected structural element.

- On a wall or isolated vertical panel parallel to the direction of this component (the in-plane walls), it causes expansion along one diagonal, and subsequent cracking along the other diagonal. Since the impact is vibrating and alternating, it leads to a characteristic diagonal crack pattern.
- On an isolated vertical panel perpendicular to the horizontal component (the out-of-plane walls), the oscillations cause bending.

Real buildings, composed of various exterior and interior walls and gables, will undergo a combination of the above effects.

are the most destructive ones and a tendency exists therefore to neglect the vertical ones. This takes into consideration that the usual static structural calculations contain a wide enough safety margin to deal with vertical forces of seismic origin. It happens quite often, though, that vertical accelerations do cause additional damage to structures and do increase destruction.

Effects of the vertical component

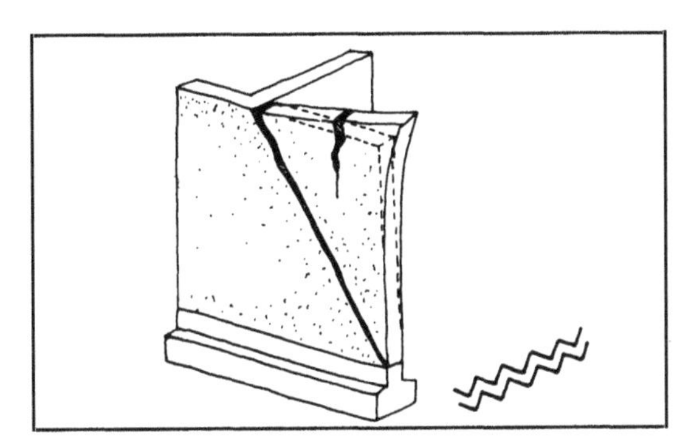

The vertical component has the impact of vertical expansion and contraction, against which the usual static structural calculations provide a safety margin, which in most cases may seem sufficient. But this component can have a dangerous direct effect on elements in flection and also on arches, where it develops inadmissable thrust. It is also very harmful to the stability of overhangs.

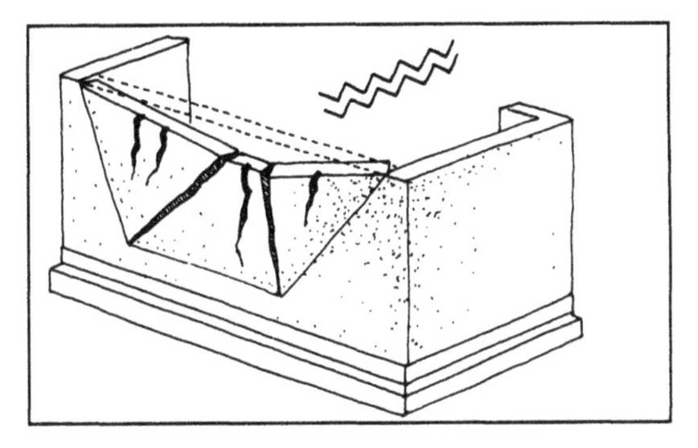

If one considers that the acceleration of the vertical component is often much stronger than that of the horizontal component, one should therefore not neglect its direct effects. The vertical component also aggravates the effects of the horizontal component, particularly on corners or borders.

Thus, earthquakes affect structures not only by an increase of the regular compressive and tensile stresses, but also by additional stresses of diagonal tension, in-plane flection or shear by twisting.

3. Behaviour of masonry

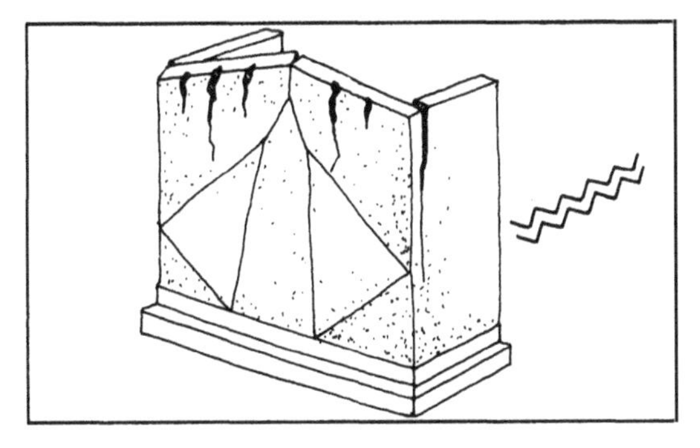

Earthquakes strongly affect masonry structures, particularly those in stone or earth. They have a very low tensile strength, as well as resistance to shear, and the stresses developed easily lead to their collapse. Since these are, in most cases, small or medium sized buildings, they have often not been built according to the regulations. Thus, when an earthquake strikes, these buildings are already weakened, the more so because their maintenance is frequently insufficient.

The walls collapse, and the ceilings and roofs will come down, which is extremely dangerous, particularly in the case of vaults and domes.

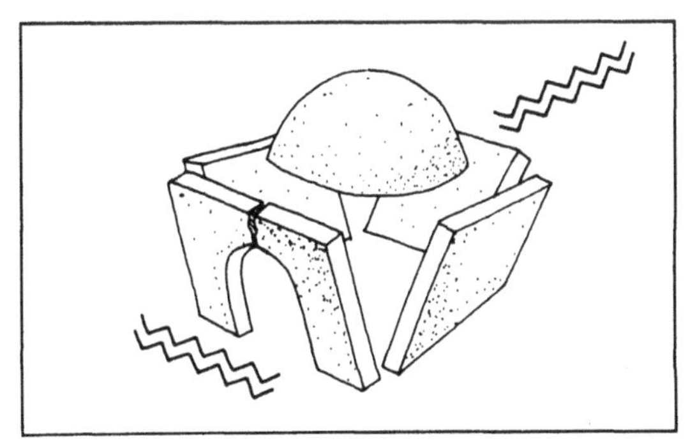

The pathology of earth buildings affected by earthquakes show that the material damage and the fatal consequences are often the result of careless construction, with many defects, and of defective maintenance. The weakness of earth as a building material is increased by ignorance and lack of supervision at all stages of the production of the material, of the design and of the construction of buildings. Earth structures, like other masonry structures, do not have much resistance to stresses of seismic origin. The poor bond between its various elements and the weakness of the bearing walls are severely punished. Even a low intensity earthquake is enough to produce cracks and to weaken the structure, thus reducing its resistance to the stresses mentioned before, and ultimately leading to its complete destruction.

The following section lists different damage factors related to buildings and their location. They are largely the same as those that reduce resistance to static impact; added to which are the elements of architectural design and of site conditions, whose effects are not well known. Current knowledge does not permit us to establish the relative importance of these various factors with much precision.

1. Site conditions

The level of damage is strongly influenced by the geological and topographical nature of a building site.

Free advice, above all of the wrong type, often remains strongest in people's minds; it is, for instance, often thought to be dangerous to locate a building close to an 'active fault'. In fact, when buildings are located above or very near to a fault, effectively known to be active, damage, and particularly that related to the vertical component, does increase. But, although one has not been able to make many observations in this respect, these seem to indicate that the zone of disturbance on both sides of the fault is less extensive than one would have thought beforehand. Finally, one should be aware that although the mechanism in the epicentre is a fault movement, the hypocentre is in most cases situated at great depth, at a considerable distance from the stricken locations. Not many faults that are visible at the surface and are indicated on geological maps can be guaranteed to become active again.

The topography of a site can cause wave amplifications; at certain frequency levels, these can be considerable. Thus, peaks and spurs, the edges of cliffs or embankments, or steep gradients are systematically the focus of heavier damage than flat or slightly sloping areas, with a regular topography.

As to the nature of the subsoil, it is found that loose and unstable soils coincide with heavy damage, whereas buildings founded on rock are generally less affected. Furthermore, the harmful character of soils whose granular structure stimulates liquefaction cannot be overlooked.

2. Materials

The poor quality of materials used in construction does increase damage.

When using earth, incorrect proportions, poor manufacture, or the irregular shape of the bricks reduce both the static and the dynamic mechanical qualities.

3. Shape: mass distribution and rigidity

Seismic accelerations transferred from the terrain to the foundations cause torque oscillations in structures. This is strongly increased when, at different levels, the centres of gravity do not coincide with the centres of torsion. This is why asymmetric buildings are more heavily damaged.

Asymmetric plans

Buildings with complex plans, and particularly those with several wings (L-, T-, H-, or cross-shaped plans) are subject to heavy damage: the individual responses of each part, which are all different, cause considerable torsion in the junctions.

Similarly, external additions to a building, such as added rooms, staircases or external terraces, are irregularities in the distribution of mass and rigidity; these horizontal overhangs are very susceptible to the impact of the horizontal components. Finally, asymmetric wind-bracing can be another source of damage.

Asymmetric elevations

The superposition of levels with different rigidities is extremely dangerous. Thus, stiff lower levels surmounted by a considerably more flexible upper part cause the 'whiplash' phenomenon. This can happen, for instance, in the case of retreating upper floors. The opposite layout, of a flexible level surmounted by a rigid block, is as dangerous or often even more so.

The corresponding behaviour, of an 'inverse pendulum', is shown to be disastrous. Overhangs, finally, are subject to maximum damage.

4. Foundations

In most cases, damage of seismic origin is not a case of what one would normally call foundation collapse. Design errors and defects of the foundations, however, diminish seismic resistance. Besides, it has been noticed that heavier damage occurs when the same structure is located upon two formations of a different geotechnical nature, when there are different foundation types or when the foundation slope is not uniform. Moreover, plastic soils are proven to be more dangerous than rigid soils or rock.

5. Masonry and structure

The most affected buildings are those with poorly executed masonry: poor bond, being out of plumb, and bad quality mortar are aggravating factors.

Similarly, very slim elements (in their height/thickness ratio), very long and unbraced elements, and walls only partially filled in with panels, all behave very badly.

Heavy damage systematically occurs when there is insufficient bond between foundations and walls, between adjoining walls or between walls and roof, and in the absence of tie-beams.

The development of cracks is stimulated by the presence of openings, particularly when these are large or close to corners.

6. Roofs and floors

The spectacular and dangerous collapse of floors and roofs are of course linked to the rupture of their bearing elements, walls and columns. But it appears that the type of construction can also contribute to the disaster. Arches and vaults behave quite badly, whereas domes, which are more symmetric and require more skill, seem to be more resistant.

With respect to flat roofs, weak anchoring of beams in the walls is very dangerous. And it appears that collapse is more frequent for beams of a large section than for those of a small section. This may seem paradoxical: if found to be true, it would provide an example of the peculiar dynamic calculations of paraseismic dimensions, compared to the usual static calculations. One notices very often cracks in the walls where they support beams or trusses. Excessive concentrations of loads stimulate damage in the case of an earthquake.

7. Humidity

The humidity of earth structures is very detrimental to their solidity in aseismic conditions; this is also the case in seismic conditions.

8. Age and lack of maintenance

The most heavily damaged houses are often those whose structures have been weakened at the user's initiative: excessive load of equipment, the addition of adjacent structures or of more openings. They also include the ones that have been badly maintained: unrepaired gutters, bad conservation of the wallbase and of corners, or degradation of surface rendering. And we have to add those houses which, already damaged once by a previous earthquake, were repaired badly or not at all.

Principles of earthquake-resistant engineering

The safety of the inhabitants of structures should come first in regions with a high seismic risk. It is absolutely impossible to protect buildings perfectly and completely, because of the intensity of the physical phenomena, the unpredictable nature of earthquakes, and the complexity of their effects. We should therefore aim to limit the scale of the damage and to prevent the collapse of buildings. Bearing this in mind, three important measures can be identified.

1. To guarantee a good location

One should take into account in particular: the nature of foundation soils, the topography, and the overall site conditions.

2. To build strong structures

'It is generally advisable to make the various parts of a structure behave as a strong entity, able to resist the seismic action homogeneously in all directions. If the structure itself does not directly provide this, it will be necessary to introduce ties or reinforcements, by means of elements that specifically aim to achieve that (in the case of masonry, for instance in the form of a geometrically and mechanically continuous frame of ties in three directions)'.

These few sentences by Jean Despeyroux, in 'Le projet de construction paraséismique' in the *Traité de Génie Paraséismique* edited under the direction of Victor Davidovici, completely define this essential measure.

In the chapter 'Enseignements tirés des séismes' in the same work, Victor Davidovici writes: 'Most individual houses are built of masonry. They are just the buildings that suffer most from earthquakes. Their rigidity and massive nature make them a focal point of strong forces, to which they are particularly vulnerable due to their poor mechanical characteristics and especially their poor tensile strength. It is also known that masonry is the most susceptible to accidental errors, but that these are most difficult to detect, once it is rendered. These types of buildings are therefore subject to rapid degradation'. In our eyes, this text presents very clearly the difficulties encountered in the search for appropriate structures.

3. To guarantee proper conservation of buildings

This is necessary because, as we have seen, the ageing of structures and the modifications made by the users are aggravating factors.

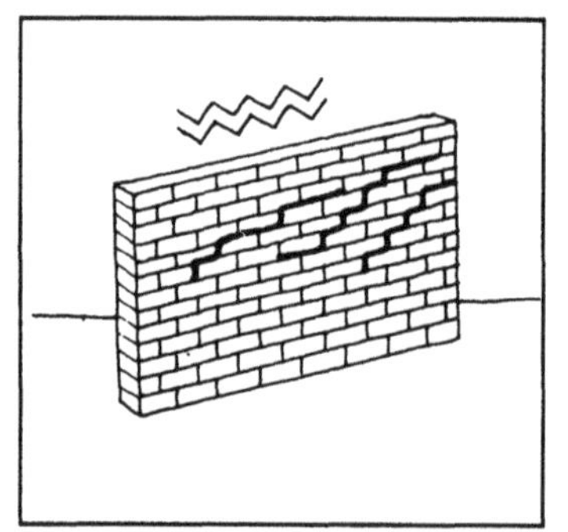

Action strategy

Earth structures are amongst the most vulnerable to earthquakes, but the study of their behaviour has attracted very little attention. The properties of the material and its structural response are not well known.

Some people classify structures made of clay, rammed earth or adobe amongst the least earthquake resistant. And for such structures, the only really reasonable solution is to prohibit their construction in zones with high earthquake risks. Although, strictly speaking, this point of view is entirely defendable, the problem is not as simple in reality. This is because earth housing is there in vast numbers, above all in regions with economic shortages, where prohibition is totally unfeasible for a long time to come. It is therefore necessary to find ways to improve the resistance of earth buildings.

To achieve this, it is advisable to define clearly the aims pursued. First of all, the reduction of losses in human life and physical assets whilst limiting the cost increase demanded by a suitable design. The greatest priority should go to the elimination of the risk of collapse of structures. For earthquakes of rather high intensity, the minimum level of protection should be to eliminate roof collapse and major structural damage.

Suitable principles and techniques must be widely disseminated and applied and be capable of being taken on by the greatest number of people. If, therefore, their implementation requires a sound and high level of technology, the resulting solutions should include a great deal of practical realism, to suit the local socio-economic context. The greatest lessons should be drawn from field observations of earthquakes occurring.

There have been numerous suggestions regarding the development of preventive or protective programmes for actual or potential disaster regions but some obstacles seem nearly unsurmountable. International colloquia (Albuquerque 1982, Lima 1983) have however allowed the comparison of ideas, technical solutions have been forwarded, and practical recommendations elaborated. This represents an interesting start, but in this field, one is up against technological, economical, socio-psychological and even cultural obstacles. Thus, certain shapes like cylinders, cones or domes are not well accepted in many regions. And it may therefore be a better idea to opt for structural improvements by suitable reinforcement in those cases.

Having defined the objectives and the principles of actions to be taken, the way towards their implementation still remains full of obstacles.

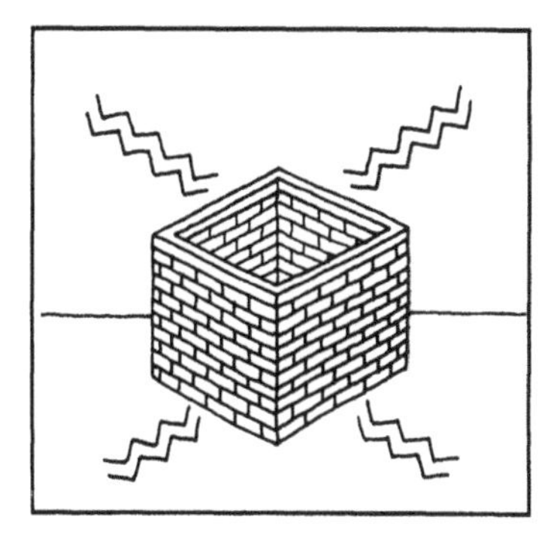

Amongst all earth constructions, existing recommendations for earthquake resistance refer primarily to adobe brick which is the most widely used material in the majority of regions with high earthquake risks. For earth masonry works, including monolithic structures in rammed earth, cob or direct shaping, one mostly refers to the recommendations established for adobe. The recommendations for timber frames with earth infill are those established for formworks. Hardly any information is currently available for buried structures. It has been demonstrated that well reinforced earth structures can stand up to fairly severe earthquakes. The recommendations are mostly of a qualitative nature and they are not supposed to replace the calculations required for a good design where these exist and are known to be reliable and realistic.

It is finally worth bearing in mind that 'it is better to give a single piece of advice to 100 people than 100 recommendations to a single person'.

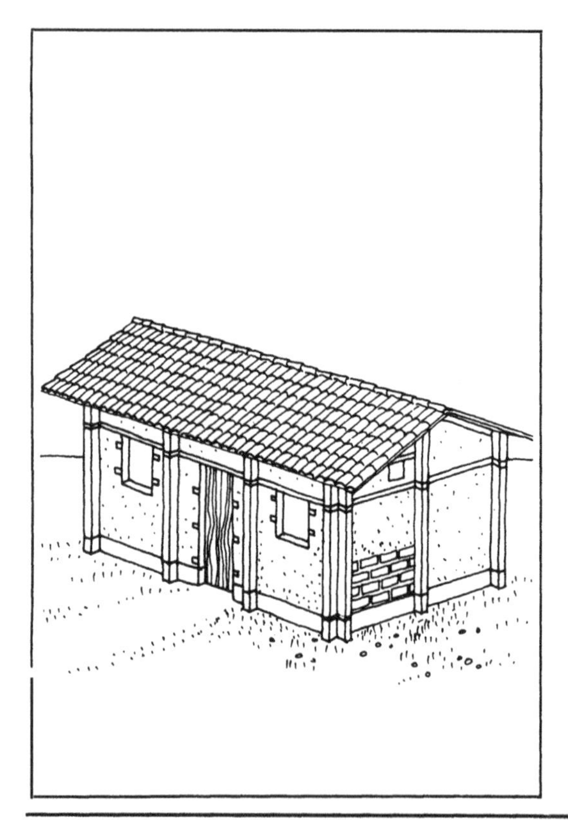

1. Site

Site conditions directly influence the intensity of earthquakes and they therefore have to be recognized and very thoroughly analysed. The location of a building in a region with high seismic risks requires the effective seismic risks to be established. This includes an inventory of known seismic activity, and a subsequent geological study essentially focusing on regional tectonics, on the determination of potentially active faults, on their types and indices of recent movement, and on the localization of sites with loose soils or that are sensitive to liquefaction. A geotechnical study will define the zones with a risk of soil movements, landslides, subsidence or caving in. It will then be possible to locate buildings on the best sites, in the least sloping zones, furthest away from active faults, peaks, spurs, edges of cliffs or embankments. It is sometimes recommended to avoid building on sites with a bearing capacity below 0.1MPa.

Such a recommendation can only be indicative, since the bearing capacity of a soil on its own, in relation to the foundation type, would not provide a criterion for seismic behaviour. It is the nature of the foundation site, on the other hand, which, by its filtration of certain frequencies, defines the frequency content of the shocks affecting the building.

And it has been observed that collapse is caused by the coincidence of the dynamic response of a structure with the frequency content of an earthquake.

2. Materials

In earthquake zones, materials have to be of a better quality. Some people advise using preferably thin bricks, with a length not exceeding twice their width plus the thickness of a vertical joint. These bricks should be well dried and cured.

The mortar used for laying the bricks should have the same composition as the adobe, and should have good consistency and adherence.

3. Shapes of structures

The use of dissymmetric shapes, in plan as well as elevation, and the dissymmetric distribution of mass or stiffness, should be prohibited. Plans should therefore be compact and of a square or circular shape.

4. Foundations

Foundations need to be impeccably designed and built. As much as possible, they should form a uniform body. One has to be careful not to place one building above two different geotechnical formations. The same structure should have the same type of foundation all over, and its foundation courses should have a uniform inclination. Foundations should be as stiff as possible (in a grid) and linked to the vertical tie-beams.

5. Masonry and structure

Masonry should be of excellent quality, carefully bonded, with strictly level courses of bricks all around the house, and perfectly plumb. Vertical joints should be strictly staggered and have a width of one to two centimetres. The composition of the masonry mortar should be such as to be free from shrinkage cracks. Before placing them, bricks should be wetted as required. The wall thickness should be at least 40cm, and wall height should not exceed six times their thickness. There should be no more than 3m between adjacent vertical tie-beams. No wall section should be longer than eight times its thickness between two vertical tie-beams or buttresses. One should avoid partial wall panels, and the link between adjacent sections should be very carefully executed. Buttresses should have generous dimensions, be placed symmetrically, and have adequate foundations. Horizontal as well as vertical reinforcement should be inserted in the mortar, every 50cm or so.

The total area of openings should be as small as possible, certainly no more than 15 to 20% of the wall surface. Their width should be limited as well, to no more than 35% of the length of the wall section. The width of wall piers between openings should be at least 90cm.

Lintels should have generous dimensions and be well embedded in the walls on each side. It would be preferable for the external horizontal tie-beam to act as lintel where there are openings.

Walls are to be topped by a continuous horizontal tie-beam, with high tensile strength and durability.

To increase a structure's torsional strength, one may incorporate diagonal tie-rods. And one should also include skilfully executed corner reinforcements.

6. Roofs and floors

Roofs have to suit the local climate and should be as light as possible. Extremely heavy earth roofs are to be avoided in areas with a high risk of earthquakes. A hipped roof distributes loads best. The thrust on the walls should be as small as possible. Roof trusses should be well braced against winds and supported on a horizontal bondbeam anchored in the walls. Beams are to be properly embedded. Domes and vaults are to be tied at their bases. The length of a vault should not exceed twice its width; vaults should not bear on end walls, but ideally end in half domes. One should not forget that vaults have a very poor earthquake resistance.

7. Protection against humidity

Earth structures should be perfectly protected against humidity with a proper wall base, a dampproof course and surface rendering. A ferrocement rendering which is strongly connected to the structure helps to resist earthquakes.

8. Maintenance and repairs

Buildings have to be scrupulously maintained. All damage due to earthquakes or otherwise should be repaired as quickly as possible, that is, if repairs are cheaper than total reconstruction. One should take advantage of the repairs to improve the earthquake resistant characteristics of a building.

The improvement of the earthquake resistance of structures implies the use of suitable construction techniques. These techniques are mostly drawn from masonry, but they have to take into account the specific qualities of the material earth as well. Its specific gravity, low mechanical strength and brittleness make earth a material very susceptible to seismic action.

If we add to the above the specific recommendations for masonry work, we can understand why most designers hesitate to use earth. Various research bodies have elaborated recommendations meant to improve the quality of brickwork, its bond, homogeneity and resistance to seismic forces. Recommendations have also been made with respect to earth masonry, but only a few of those have been systematically implemented.

Although it is probable, even certain, that the application of these measures will reduce the scale of the damage, one does not know yet by how much. A lot remains to be done. Without providing an opinion, we give here some examples of the contents of these recommendations. They show some aspects of the approach to the problem.

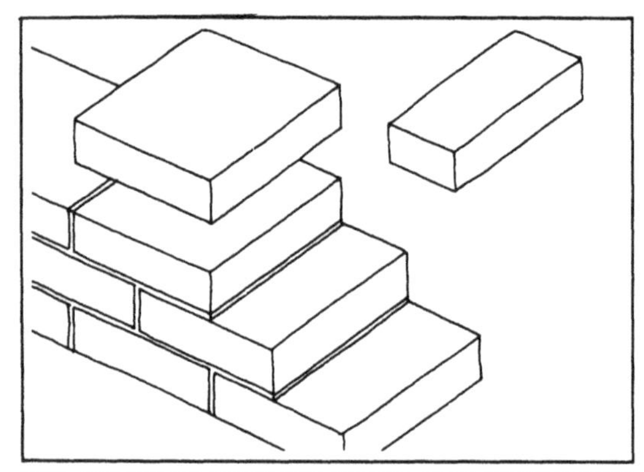

1. Shape and composition of bricks

A suggestion is to use square bricks, 40 × 38 × 10cm. These dimensions allow for many good bonds but require perfectly draughted plans. Besides, it is also possible to improve the behaviour of masonry made with this type of brick by adding some straw. To achieve this, it is sufficient for the improvement due to the reduction in mass not to be totally cancelled out by the reduction in resistance which occurs simultaneously.

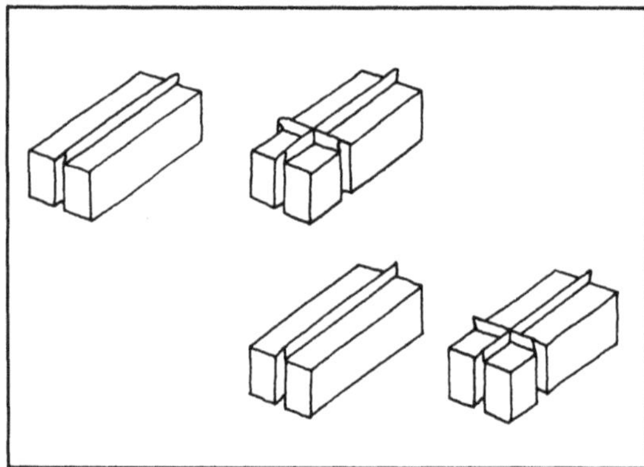

2. Interlocking bricks

In Mexico, tests have been made using interlocking bricks, bonded without mortar. Several buildings have been erected following this principle, which still can be improved upon.

Stabilized bricks of this type can be manufactured with very simple presses.

These bricks, however, have to be perfectly finished, and do still pose problems in storage and handling.

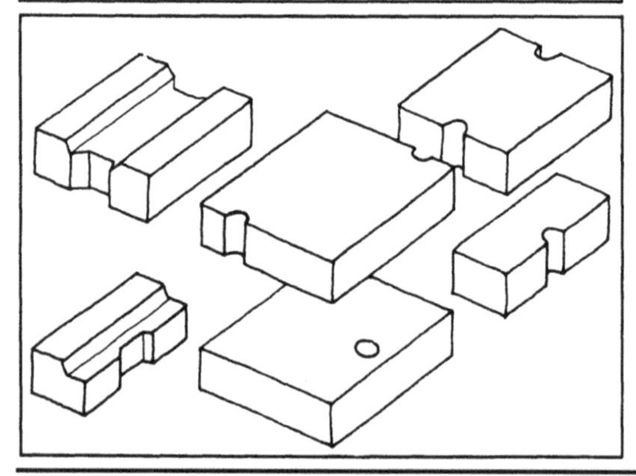

3. Bricks for reinforced masonry

Ordinary bricks can be used for reinforced masonry, but not without construction problems. It is preferable to use bricks with voids for the placement of horizontal and vertical reinforcement.

4. Mortar and bond

Good-quality mortar and bond improve the earthquake resistance. This can be also achieved by the use of a stabilized earth mortar instead of normal earth mortar. The stabilization of mortars improves the friction at the interfaces and makes them adhere better to the bricks. The use of excessively wet mortar should be avoided, as it causes shrinkage and microfissuration, and does not adhere well. The absence or poor execution of vertical joints substantially reduces the wall's compressive strength, and its bending and shear strength even more. The bond has to be designed in such a way that the position of the joints least resembles the characteristic diagonal crack pattern of damage caused by earthquakes.

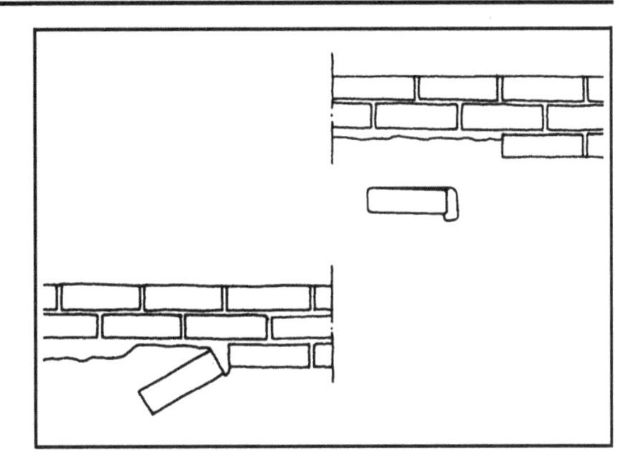

5. Reinforced masonry

Horizontal and vertical reinforcement are made of bamboo, eucalyptus, reinforcing bars and barbed wire. The reinforcement strongly increases the tensile and bending strength. It is possible to reinforce walls made with ordinary bricks, but it is preferable to use special bricks.

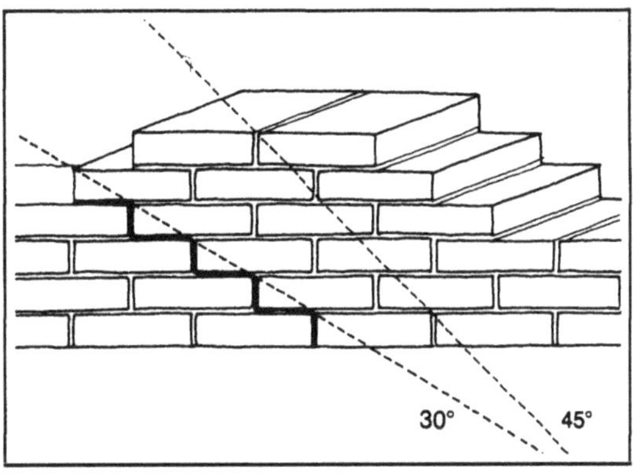

6. Ring-beams

These figure amongst the structural elements most resistant to earthquakes. They guarantee a proper transmission of forces, and thus achieve that the structure becomes a strong entity. Ring-beams at low and high levels have to be linked by vertical elements at corners and intersections. The absence of a ring-beam practically nullifies all other earthquake-resistant measures, particularly in the case of narrow or slender walls.

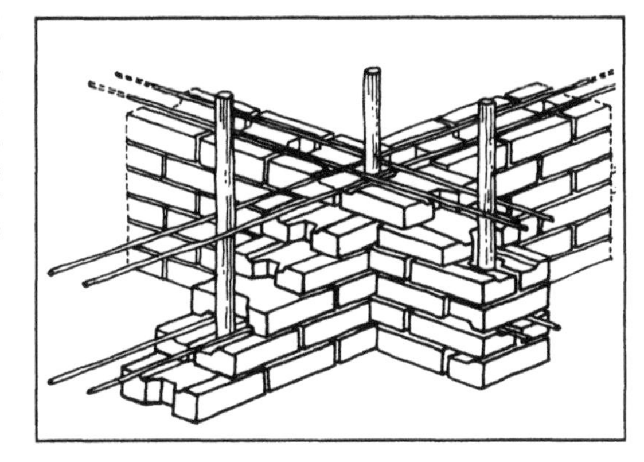

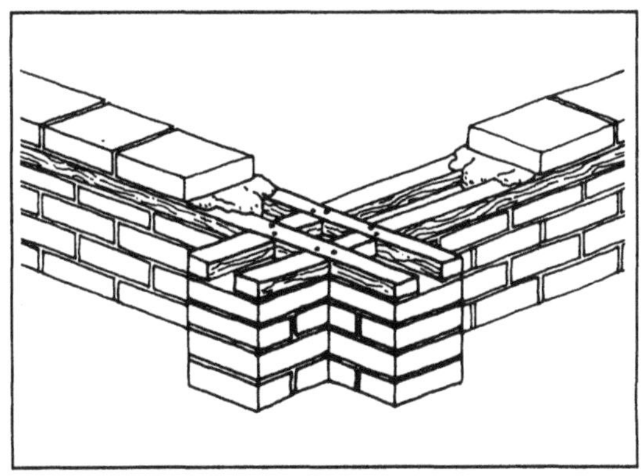

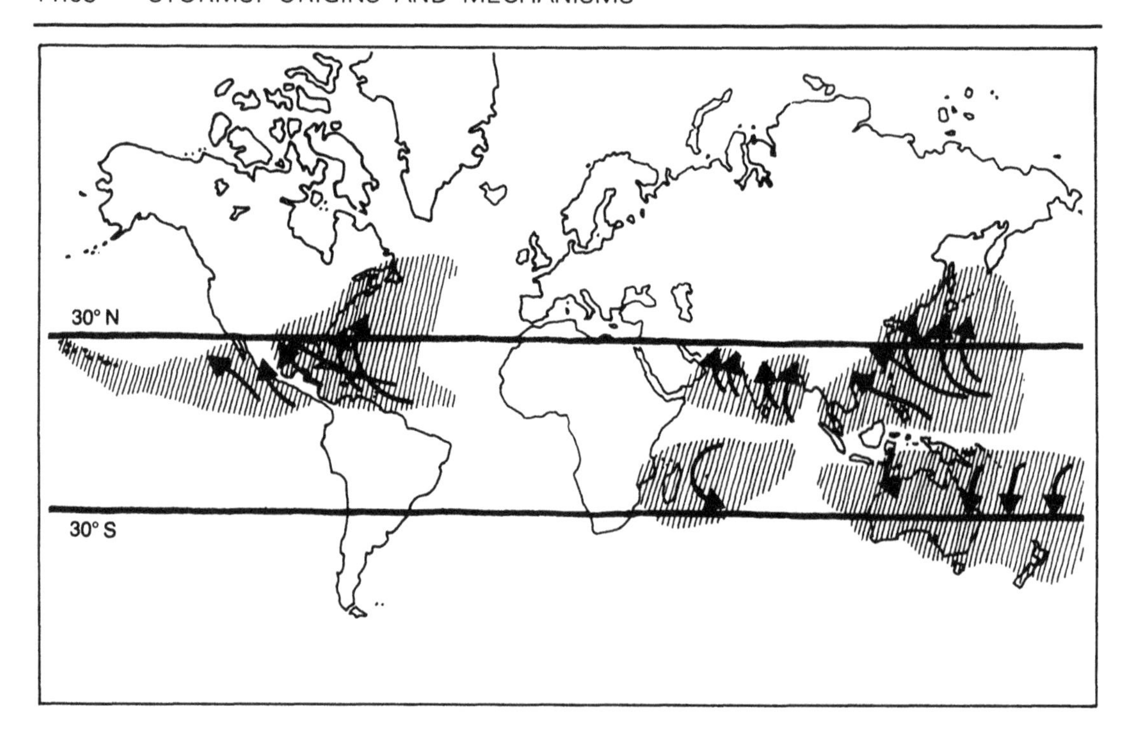

Hundreds of millions of people live in regions of the world exposed to cyclones or similar types of storms. Nevertheless, three-quarters of these people have not experienced a cyclone at first hand, as in each region exposed to them the likelihood of being at the centre of such a storm is low. It remains important, though, to be aware of the disasters they cause, particularly with respect to the stability of structures. Every year, cyclones devastate the homes of entire populations. On average, 23,000 people are killed and 2.6 million people stricken by this disaster every year. These figures are staggering and they refer primarily to the tropical regions of the globe. Cyclones are, in fact, storms of tropical origin; their name changes from region to region: in the Caribbean they are known as hurricanes, in China and Japan as typhoons, elsewhere as cyclones or tornadoes. Cyclones are the result of a combination of oceanic and atmospheric factors. When a mass of stable air lies over a warm sea the air rises, carrying a vast quantity of water vapour with it. This creates a low pressure zone surrounded by a high pressure zone, characterized by converging winds (blowing from the outside towards the centre) and ascending ones (in the centre), giving rise to a rotating motion. Once the water vapour has risen to a certain height, it condenses into rain. This change in state is accompanied by a release of heat energy, which encourages the air to rise further.

This sequence repeats itself, causing the cyclone to grow considerably in size and raising its speed of rotation to at least 120km/h. However, speeds have been recorded of around 320km/h (cyclone Camilla, 1969). A cyclone may build up for several days over warm sea air. Once in contact with land, it is very destructive before blowing itself out, because it is no longer fed with warm air and is broken up by the relief. Cyclones are, however, not the only destructive winds. There are also violent continental winds, such as the sandstorms in the deserts, which easily can attain a speed of 100km/h.

Mechanisms of destruction

1. Erosion of land

Violent winds can excavate the land around buildings. Foundations are exposed and their base undermined, leading to the collapse of the structure. This mechanism has been observed in Niamey, in Niger.

2. Wind pressure

Walls exposed to winds are subject to extremely powerful lateral thrusts which can overturn them. This pressure is at its highest in the centre of the exposed wall and decreases towards its corners; it is lower on the unexposed walls. These pressure differences create eddies at ground level, which may erode the building.

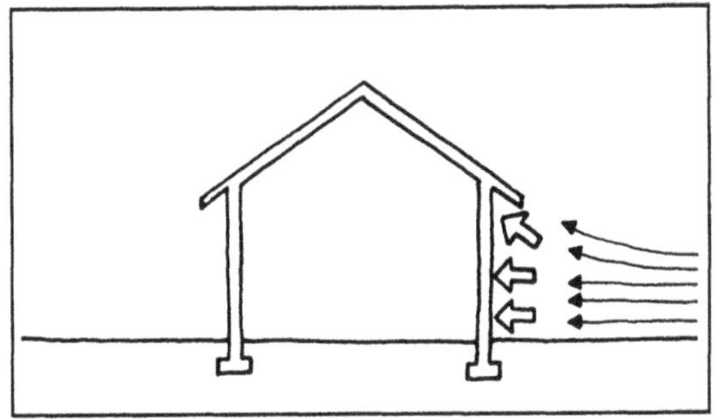

3. Erosion of foundations

Violent eddies which escape at high speed along the sides of buildings are often loaded with abrasive material (sand) and have a tendency to attack wall base and foundation structures, which can lead to the destruction of the building.

4. Suction on walls

Where walls are not directly exposed to wind (the sides and lee of a building), turbulence and low pressure can cause suction. Materials and even parts of the building may then be torn away.

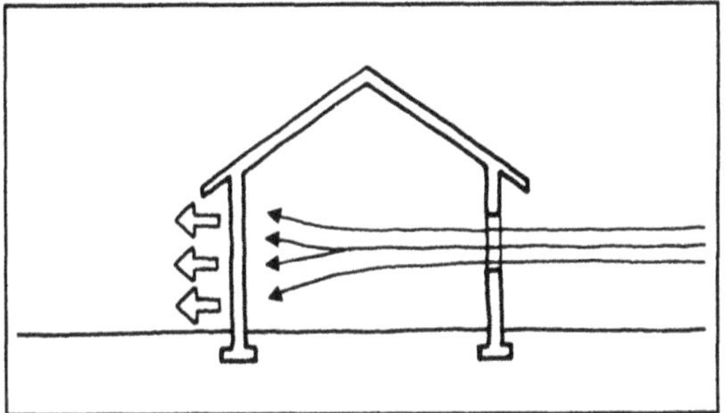

5. Suction on roofs

Depending on the closeness of the windflow to the surface of the roof, suction may occur on a directly-exposed beam. This phenomenon is related to the pitch of the roof, and is greater on horizontal surfaces (flat roofs) and on pitches less than 30°. Roof overhangs are notorious for eddies (vortex), causing local suction which can tear off all or part of the roof. This effect can be amplified by overpressure on the underside of the roof.

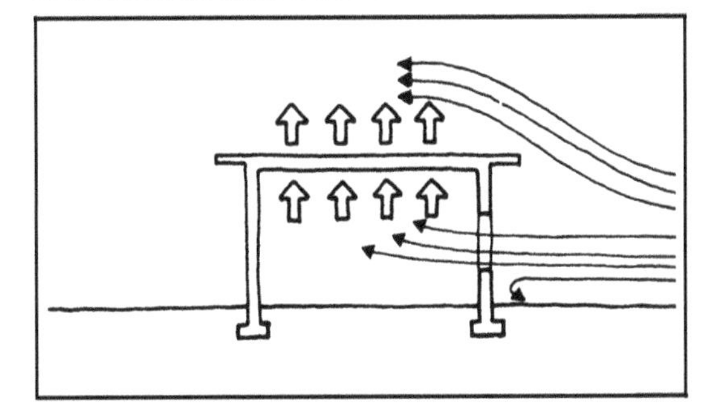

Complications

The mechanisms described above are applicable to isolated buildings of a simple design, with a rectangular groundplan and gable roof. When the architecture is more elaborate, with plans in U or L shape, with bays, porches, balconies and chimneys, or when buildings are grouped, the mechanisms are even more complicated.

The structural damage caused by cyclones is vast. Buildings suffer in general from careless design which is totally unsuited to withstand an event of such destructive power. Designers usually pay little attention to the loads exerted on buildings by winds. These loads are considered as secondary compared to those normally taken into consideration: the dead load of the building, live loads and overloads on floors, impact forces, loads due to soil pressure, etc. Stresses caused by wind, however, can far exceed those caused by normal loads.

These loads cannot usefully be compared to static loads, which is nevertheless frequently done. It is true that the stresses exerted by a cyclone are hard to assess, because of the wide range of possible violence and the many forms of turbulence. Furthermore the large majority of structures are not designed by professionals. Many are designed and built by their inhabitants or by small firms drawing on local knowledge or traditional building skills which ignore the wind resistance problem or resort to makeshift solutions. For instance, ties between walls and roofs are often very limited and cannot withstand tear. Scientific calculation and design methods do exist and are applied for large projects, but are neglected for small structures. The parameters used are very numerous and vary widely depending on the site and the layout of buildings. It would be impossible to examine each case and these methods are not suitable for low-income housing, since they often entail an increase in the study and design costs. Such luxury is useless if it is agreed that, statistically speaking, the probability of recurrence of extremely powerful cyclones is once every fifty years. Even so, this probable risk should be considered.

Some general design recommendations can be applied. These will not prevent all disasters, but will certainly limit their extent.

Recommendations

The pathological analysis of typical structural damage caused by cyclones shows that there is no standard 'cyclone-resistant' structure.

It is insufficient to simply adopt suitable building methods and regulations. In reality, there are other important parameters which must be added to the stresses caused by the wind, such as:

- the nature of the subsoil;
- the environment and the site (relief, vegetation, town planning);
- the architectural model (shape, volume, detail).

In general, the main weaknesses are the following:

- collapse of walls;
- failure of weak structural ties;
- non-existent or very minimal foundations;
- structural columns poorly anchored to the ground;
- roofs ripped off.

1. General recommendations

It is advisable to implement codes of good practice and to ensure the maintenance of structures.

The stability of the building as a whole will be considered as well as the wind resistance of each element of the building (foundations, walls, roofs, etc.). No lateral force should be able to move or turn the building over. Maintenance should attend to any deterioration of the ties between the various elements: foundations/walls and walls/roofs. And the undermining of the foundations by water, wind erosion or termites should be prevented.

In the case of cyclones, the action of violent winds is often associated with flooding. Attention should therefore be paid also to the recommendations concerning flooding.

2. Site

- Profit from natural protection. Locate the structure in the shelter of relief (hills) or vegetation (thickets, hedges) which face the prevailing winds.
- Avoid extremely hilly ground and steep grades which can accelerate wind speeds by about 50%.
- Avoid gaps or locations which can channel the wind.

3. Town planning

The proximity of buildings affects the speed of the wind. The tops of gable walls are susceptible to suction effects. Turbulence between two nearby parallel buildings can cause stress on elements such as boarding. Make sure it is fastened well. As a general rule, avoid building structures all in line in order to avoid channelling winds, thus speeding them up and increasing their destructive power.

4. Plot

Provide windbreaks in masonry or vegetation on the plot. Choose species with deep roots. Build open walls of the breeze block type, which divide the force of the wind. Provide good foundations in order to stabilize these solidly built enclosures. All adjoining structures such as barns, garages, garden sheds, outhouses should be solid so that they will not be blown over and hit the house. Avoid walls which are too thin when building such annexes. Remove all objects and debris which could be picked up by the wind from around the house.

5. Shape

Rounded shapes (domes, vaults) and cubes are to be prefered to rectangles. Use the following ratio: length/width = 1.5 and height/width = 1. Reduce volumes and limit the exposure of walls to the prevailing winds. Choose a location with a corner facing the prevailing winds.

6. Foundations

Foundations should allow good anchoring of the structure of the house. They should be deep so that they are not exposed by winds and built of strong and durable materials (stone, stabilized soil).

7. Walls

Raise the walls straight on top of the foundations. Make sure that there is a proper bond between foundations and walls. Whatever the material used, provide vertical reinforcement (iron bars, bamboo or other) anchored in the foundations. Provide horizontal reinforcement as well, particularly at the corners, which must be reinforced at all levels of the structure.

The walls should, if possible, be heavy and solid, in rammed earth, adobe or compressed blocks. Stabilize the earth, if the budget allows. The masonry bond should be carefully implemented (worked out on paper) to avoid cracking. Light walls in wattle and daub or in earth with straw should be well braced against wind and the framework anchored in the foundations. Make sure the structure is rigid by using dividing walls and ring-beams. Earth mortar can be used in construction.

8. Posts and pillars

Frames should be treated against rot, parasitic mosses, insects and termites. If the structure is on piling, ensure that corner supports are well anchored.

Pillars built into walls should be linked to the foundations and tied. Corner posts should be reinforced horizontally and vertically.

9. Openings

Use heavy doors and windows, well sealed into the thickness of the walls and provided with shutters to protect the glazing (shattering of glass, blows from objects). The shutters should be easy to handle and be airtight and waterproof. Avoid locating openings close to corners or the roof in order not to weaken the structure. Avoid piercing the exposed façade as this could raise pressures within the house. Openings in opposing façades should be in line to avoid pressure differences.

10. Roofs

Roofs should be heavy and as streamlined as possible: domes, vaults, cones and hipped roofs.

Choose a pitch of about 30° to reduce the high stresses to which flat roofs or those with a pitch of between 5 and 10° are prone. Parapets reduce the suction effect on flat roofs. Avoid roof overhangs of more than 50cm long which may be torn off. Ridge ventilators reduce internal pressure. Make sure that light roofs (thatch, tiles) are well anchored (netting or mortar).

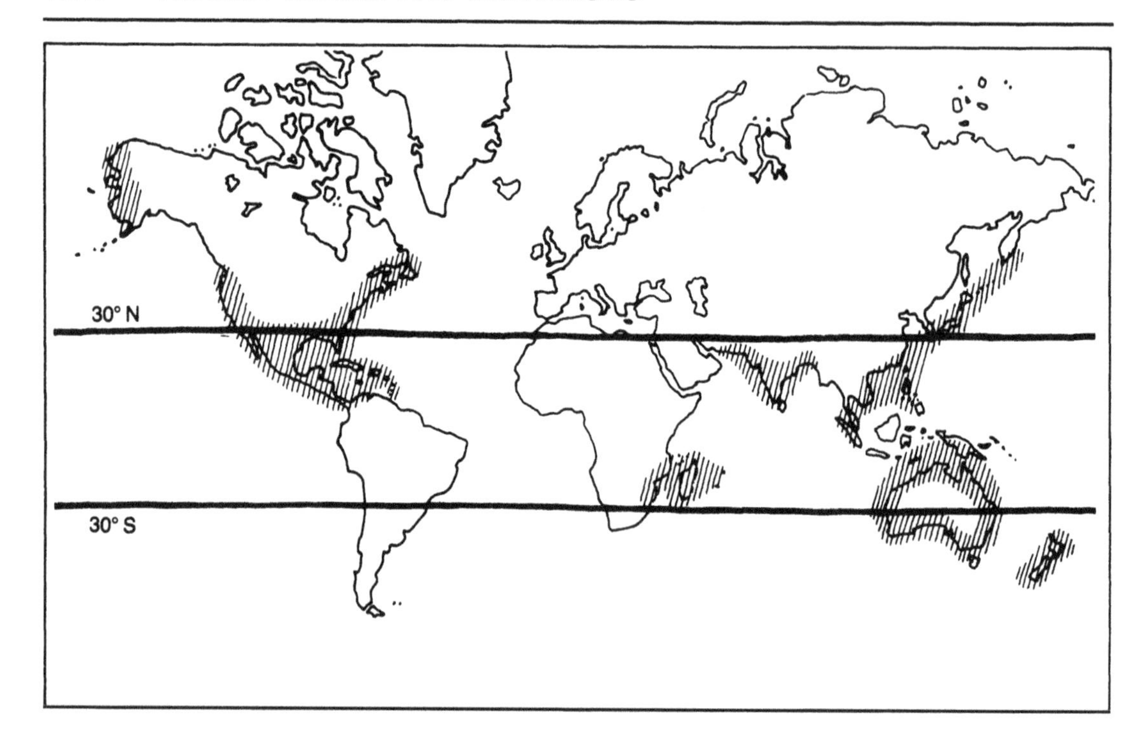

Floods have many causes; torrential rains, rising water courses, breaches of dams or dykes, earthquakes, volcanic eruptions and cyclones. The most deadly floods, however, are caused by tidal waves. These are the result of geological or atmospheric disturbances. The breaking of submarine faults or submarine earth slips, themselves often triggered by an earthquake, set the water in motion. Enormous surface waves spread out from the epicentre of the earthquake crossing the ocean until they reach a shore. The effects are devastating. This phenomenon is known by a variety of names such as tidal wave and tsunami (Japan). The Japanese term is the most widely used since Japan has particularly suffered from such disasters. During the Honshu tsunami on 15 June 1896, 26,000 people were killed by a surging wave rising 25 to 35 metres above the high tide mark. On average, the world experiences one tsunami per year, and it is often more destructive than a violent earthquake. Every ocean and most seas have been affected and are still exposed to them.

During the last two centuries, 300 tsunami have devastated coastal areas. But there are also 'seiches' or local waves in bays, natural and artificial lakes caused by landslips. The wave height of tsunami and seiches depends on the topography of the seabed and the exposed coastline. Sheltered coasts may be spared. The waves will not be particularly high in deep water but are suddenly amplified when they reach the coastline. They then turn into a real wall of water which pours itself devastatingly onto the shore, sweeping all structures before and sparing few human lives. There are also tsunami of atmospheric origin. Air pressure in the eye of cyclones is very low, creating a vertical suction which raises the water level. The winds increase the volume of water and the contact of the cyclone with the shore amplifies the height of the destructive surging wave. This sort of tsunami can attack several kilometres of coastline. Their force depends both on the air pressure in the eye of the cyclone and the wind speed, as well as its rate of advance and the level of the tide at the moment of impact. The height of the tidal wave which hit Andhra Pradesh in India in November 1970 was 20 metres.

Mechanisms of destruction

1. The force of the wave

A wave front several metres high will hurl itself with great force against the walls of buildings. The extremely violent shock may knock the structure over. This depends on the mass of the water and the speed with which the wave is advancing.

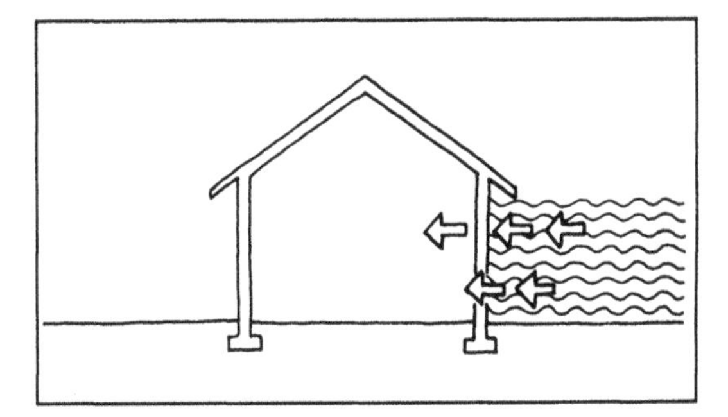

2. Impact of objects

Waves may carry objects or debris torn loose during their passage, such as tree trunks, rocks, and bits of buildings. The impact of these objects against a wall has the effect of a battering ram.

3. Hydraulic erosion

When water flows against weak foundations and wall bases, its speed and turbulence erodes the base of the building. It can also cause erosion of the ground under the building, which may cause it to be swept away.

4. Erosion by abrasion

The water may carry a heavy load of stones, gravel, sand and silt in temporary suspension. These materials may have an effect similar to sandblasting on the foundations and even wear them totally away. This may also affect the ground under the building. Structures which are no longer supported will collapse.

5. Pressure differences

The entry of water into a house and its rise into the living quarters may cause pressure differences on the walls. The creation of a pressurized air pocket under the ceiling may cause the roof to blow out. The collapse of parts of the building, walls, foundations, wall bases or roof hastens its eventual destruction.

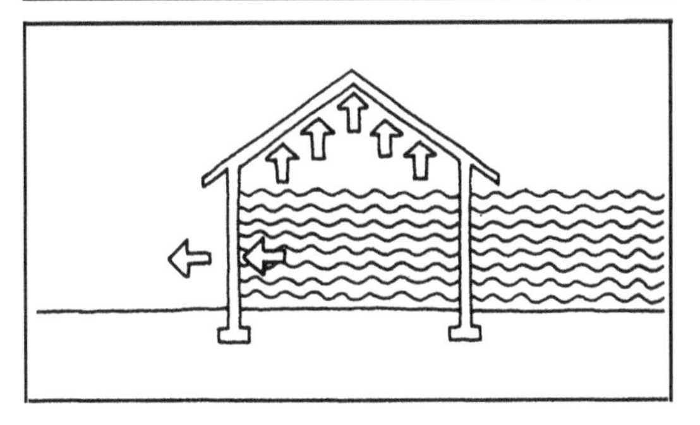

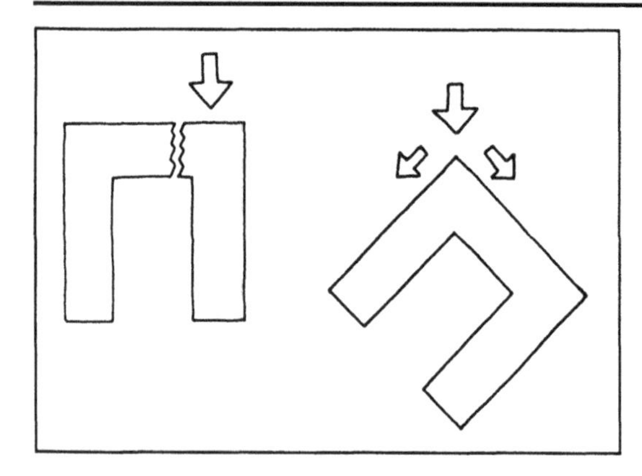

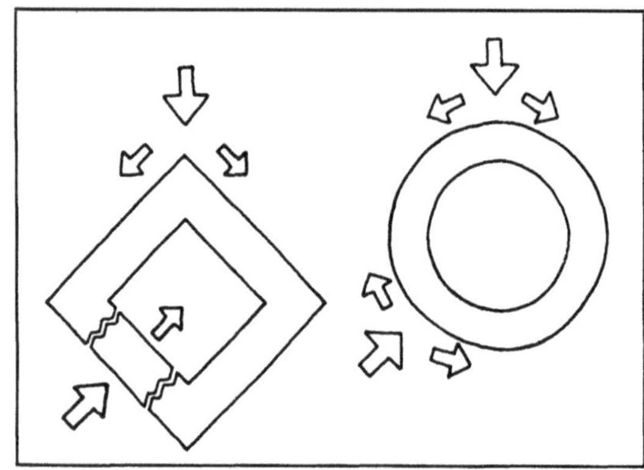

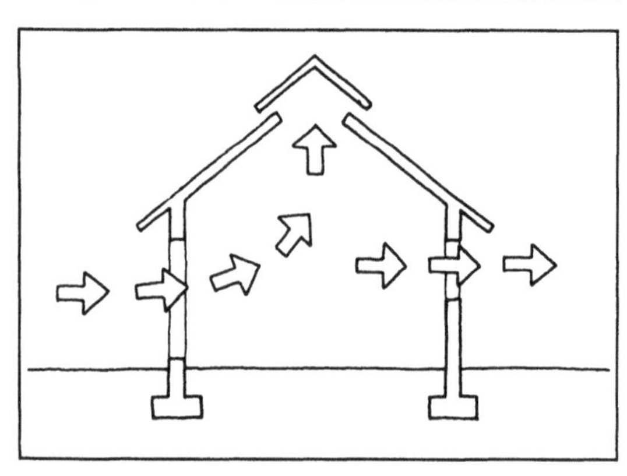

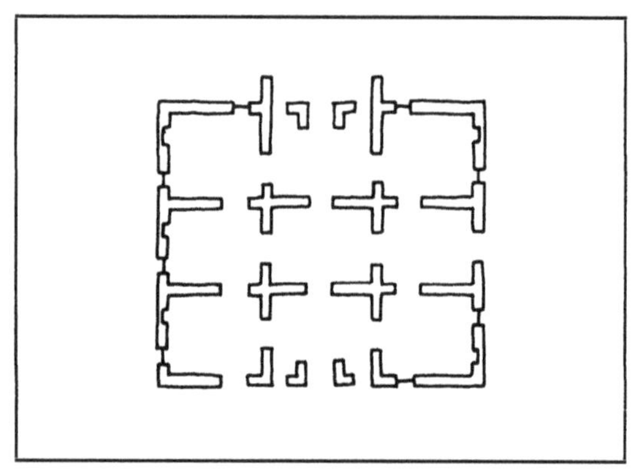

General recommendations

- Do not divide up land on the banks of water courses, lakes or bays as there is danger of landslips and local waves. Build at higher levels, on well-drained sites protected by vegetation.
- Surround the house with plantations and strong enclosures, to break the force of a wave.
- Do not build houses in one line, to avoid channelling and increasing the flow of the wave. Reduce the exposure of walls in the wave direction.
- Remove all heavy objects or debris, which could act as a battering ram, from around the house.
- Increase the level of walkways and pavements.
- Give preference to rounded shapes. For rectangular shapes, use a length/width ratio of 1.5 and a height/width ratio of 1.
- Respect a code of good practice: good anchoring, protection against termites and good maintenance.
- Match the weight of the house to the nature of the bearing soil. Use light structures on bad soils. Use heavy solid walls on good soil.
- Deep and strong foundations, well anchored; raised wall bases. Use good masonry and durable materials for those two structural elements.
- Reinforced and buttressed inside walls. Tie the foundations to the walls and the walls to the roofs.
- Make a rigid structure by using internal walls.
- Avoid single load-bearing brick pillars.
- Only use light materials for infill, where it is easy to repair.
- Limit water penetration by using heavy doors and windows, well anchored to prevent them being torn away.
- Align the openings on opposing walls to avoid pressure differences.
- Reduce the weight of floors and roofs, to avoid the danger of falling debris in the event of collapse.
- Provide openings in floors and roofs (stair cavities and skylights) to help occupants escape, as well as allowing air and water out (reduction of pressure differences).
- Provide access to raised terraces (staircase) to assist the flight of the occupants.

Recommendations regarding the use of earth

- Vertical reinforcement of enclosures in rammed earth, cob or wattle and daub using well-anchored stakes. Horizontal reinforcement using branches, barbed wire, etc.
- Walkways in stabilized or reinforced earth. Compact the earth very well.
- In areas exposed to exceptional tides, raise structures to above the level of the waters on dykes, embankments, landfills, wooden piles or reinforced concrete columns, etc.
- Prevent the undermining of part of the structure by water infiltration, since this reduces stability.
- In the case of heavy structures, give preference to heavy monolithic systems in rammed earth, cob or adobe. In the case of light structures, use a framework filled in with straw-clay or wattle and daub.
- Avoid earth foundations and wall bases. If there is no other choice, stabilize the earth as much as possible. Avoid the use of unstabilized earth for masonry mortars.
- Reinforce masonry of compressed blocks, adobe, rammed earth and others. Use extra care when constructing the framework of houses in straw-clay or in wattle and daub, or infilled with unfired bricks. The infills can be eroded without causing the destruction of the building.
- Use ring-beams and reinforcement in masonry.
- Avoid load-bearing pillars in earth masonry. Build such pillars of solid stabilized earth, with a good foundation.
- Anchor doors and windows with care. Locate them in the centre of wall panels.
- Avoid supporting roofs directly on unfired earth walls, which may be eroded and thus no longer able to support them.

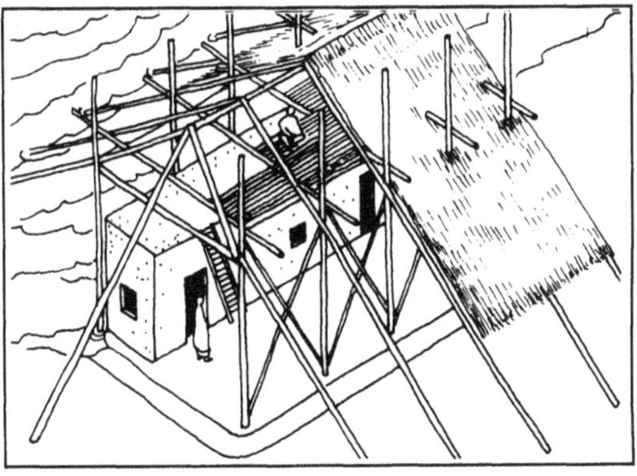

Afshar, F. *et al.* 'Mobilizing indigenous resources for earthquake construction'. In *International Journal IAHS*, New York, Pergamon Press, 1978.

An. 'Building in earthquake areas'. In *Overseas Building Notes*, Garston, BRE, 1972.

Barnes, R. 'Guidelines for anti-seismic construction'. In *Adobe Today*, Albuquerque, Adobe News, 1981.

Bolt, B.A. *Les tremblements de terre*. Paris, Pour la science, Belin 1982.

Carlson, G. *Earth blocks for Laos*. Washington, USAID/IVS, 1964.

CEMAT. *Fichas técnicas para la Vivienda popular de Zonas sismicas*. Guatemala, CEMAT, 1976.

CRATerre. 'Casas de tierra'. In *Minka*, Huankayo, Grupo Talpuy, 1982.

Cuny, F.C. *Improvement of adobe houses in Peru: a guide to technical considerations for agencies*. Dallas, Intertect, 1979.

Daldy, A.F. 'Small buildings in earthquake areas'. In *Overseas Building Notes*, Garston, BRE, 1972.

Davidovici, V. *Génie parasismique*. Paris, Presses de l'E.N.P.C., 1979.

Département des affaires économiques et sociales. *Comment Réparer les Bâtiments endommagés par un Séisme*. New York, Nations Unies, 1976.

Département des affaires économiques et sociales. *Construction d'Habitations à bon Marché à l'Epreuve des Séismes et des cyclones*. New York, Nations Unies, 1976.

Eaton, K.J. 'Buildings in tropical windstorms'. In *Overseas Building Notes*, Garston, BRE, 1981.

Eidgenössisches politisches Department. *Studie zu besserem baulichem Schutz indischer Dorfbewohner in zyklongefährdeten Gebieten*. Bern, EPD, 1978.

Escoffery Alemán, F.A. *Casa de Adobe resistente a Sismos y Vientos Como Solución al Problema de Vivienda Rural en Panamá*, Universidad Santa Maria la Antigua, 1981.

Hays, A. *De la terre pour bâtir*. Manuel Pratique. Grenoble, UPAG, 1979.

Homans, R. 'Seismic design for earthen structures'. In *Adobe Today*, Albuquerque, Adobe News, 1981.

Intertect; Carnegie-Mellon University. *Indigenous Building Techniques of Peru and their Potential for Improvement to Better withstand Earthquakes*. Washington, USAID, 1981.

Intertect; UNM. *Conference Report of the International Workshop Earthen Buildings in Seismic Areas*. Albuquerque, 1981.

Kliment, S.A.; Raufstate, N.J. *How Houses can Better Resist High Wind*.Washington, National Bureau of Standards, 1977.

Lainez-Lozada, Navarro, *et al.* *Comportamiento de las Construcciones de Adobe ante Movimientos Sismicos*. Lima, LLN, 1971.

Matuk, S. *Architecture Traditionnelle en Terre au Pérou*. Paris, UPA.6, 1978.

Ministerio de vivienda y construcción. *Mejores Viviendas con Adobe*. Lima, Ministerio de Vivienda y Construcción, 1975.

Morales Morales, R. *et al.* *Proyecto de Bloque Estabilizado. Estructuras*. Lima, Universidad Nacional de Ingenieria, 1976.

Normas de Diseño Sismo Resistente, Construcciones de Adobe y Bloque Éstabilizado, RNC. Lima, Resolucion Ministerial n° 1159–77/UC110, 1977.

Oficina de Investigación y Normalización. Adobe Estabilizado. Lima, Ministerio de Vivienda y Construcción, 1977.

Pichvaï, A. *Vers une Architecture Antisismique Appropriée. Constructions rurales en terre*. Bruxelles, ISAE La Cambre, 1983.

Pontificia Universidad Católica del Perú. *Memórias Seminário Latinoamericano de Construcciones de Tierra en Areas Sisimacas*. Lima, PUCP, 1983.

Reza Razani. 'Seismic design of unreinforced masonry and adobe low-cost buildings in developing countries'. In *IAHS International Conference of Housing Problems in Developing Countries*, Dharan, 1978.

Roubault, M. *Peut-on Prévoir les Catastrophes Naturelles?* Paris, P.U.F., 1970.

Santhakumar, A.R. 'Building practices for disaster mitigation'. In *International Journal IAHS*, New York, Pergamon Press, 1983.

UNAM. Arquitectura autogobierno. *Manual para la construcción de Viviendas con Adobe*. UNAM, 1979.

The Mosque of Bani in Burkina Faso, built entirely in adobe (Thierry Joffroy, CRATerre-EAG)

12. EARTH WALL FINISHES

Earth structures are not always built in the kindest of climates. In many countries, the raw earth must stand up to the rigours of cold and bad weather. In temperate and continental climates where rammed earth, adobe, cob and daub are very common, good architectural design and know-how can permit considerable economies by doing away with most of the rendering. Their use can be limited to the most exposed parts of the building. On the other hand the sight of the exposed earth is often unfavourably regarded, as people seem to associate it with poverty. Builders through the ages have sought to disguise it by covering it with a conventional decor. Decorated renderings, painted and moulded, enriched the earth with an attractive noble style. In other regions (the tropics) surface protection is indispensable, particularly when the architectural design neglects to provide elementary protection (base course and broad eaves), or when the technique is not sufficiently sophisticated.

Earth architecture often suffers from chronic defects caused by the effects of bad weather. Nowadays, when it is deemed desirable or necessary to protect earth walls a wide variety of technical solutions exist which can be easily adapted to numerous local contexts. These solutions are, however, often poorly applied, giving unsatisfactory results.

The chosen solution must above all be adapted to the local economy as the cost of protecting walls is often prohibitive. The conflict between technique and economy is all too often inevitable when dealing with this problem. The solutions used must be properly evaluated and implemented, if they are to be effective.

Earth and the construction techniques employing earth are burdened with a poor image. The main objection to them is that they do not perform well in severe weather conditions. This is admittedly true for simple earth structures which have not been built with the necessary care.

However, it is less true for structures built with care in good quality materials and having regard for the specific requirements of soil construction, which in particular afford protection to the material.

Certainly the very large majority of the world's earth dwellings, built for the most part in rural environments, all suffer from the same defects when exposed to bad weather: surface erosion, partial crumbling, unhealthy conditions because of constant humidity, walls hollowed at their base, etc. It is thus extremely desirable to propose effective solutions for the restoration and protection of these existing homes. These solutions can moreover be applied to homes which are still sound but which are liable to deteriorate if not effectively protected. Furthermore it would be best if these solutions were integrated into current or future construction in earth so that these typical defects could be banished forever.

Apart from this, a coherent and truly feasible revival of construction in earth can hardly be envisaged, if the material does not satisfy requirements as well as other modern materials. Soil must therefore in future produce buildings of obvious quality, without there being any doubts possible. This objective will be attained by improving the material itself and the construction techniques involved as well as by using a wide range of techniques which can bring about a marked reduction in the sensitivity to water of earth surfaces.

Like other surfaces in modern materials, earth surfaces must be capable of being given protective coatings which meet current specifications for wall facings. Earth will only enjoy a good image when it is regarded as a truly modern material. The necessity of protecting the earth used in the structure, long before being approached at the level of surface protection, remains subordinate to the quality of the material, design, and the construction procedures used for the structure. Between the stabilization of the material and the systematic employment of non-eroding rendering – the two most common approaches adopted – the range of solutions for the protection of surfaces is sufficiently wide to ensure durability, without having to resort to 'miracle cures'. However, of the numerous structures examined, stabilization and rendering only rarely appear to be satisfactory, and do not always provide a lasting solution.

The need for protection

Every wall built in earth should be able to stand up to damp and the direct action of water. This ability of an earth wall to withstand water is above all dependent on the quality of the soil itself, its grain-size distribution, structure, and porosity. It can be improved by adding a stabilizer under controlled conditions, or protected with protective coatings compatible with the material, or by taking protective measures in the design and construction stages, such as broad eaves, porches, and so on. The protection offered by renderings can take many forms, and varies considerably from region to region, as regional conditions impose specific protection requirements on the material.

Generally speaking, in temperate regions, when the soil is of a satisfactory quality, earth walls stand up to weather erosion insofar as they are built on good foundations or footings and are protected at the top. If the soil is of moderate quality, stabilization can bring some improvement without, of course, neglecting the basic protective measures at the top and the bottom of the wall. In dry climates, earth walls protected at the top by an edge of the roof, or some other 'hat', stand up well to water, if they are not liable to flood damage.

In any region where rainfall is low, or average, protection against rain can be afforded simply by taking architectural precautions. In contrast, in regions with high rainfall and driving rains with a virtually level trajectory (e.g. in the tropics) protective coatings are virtually indispensable.

This precaution is essential in regions where climatic vicissitudes go hand in hand with an architectural tradition which provides no protection either at the top or the bottom of the wall (e.g. mud architecture in the Sahel). Similarly coatings may be more or less necessary depending on the construction technique.

Thus, for monolithic soil walls (e.g. cob, rammed earth, clay-straw) which are suitably protected and uncracked, the need of protection is less than for walls in earth masonry (e.g. adobe, compressed blocks) where the water may penetrate at the brick mortar interface, i.e. the joints.

In principle, protective coating is not necessary for well-constructed structures in stabilized soil. Stabilized earth walls stand up well to bad weather for many years. Good base courses are, however, not pointless. It can be expected that a soil wall which is unprotected at the top will start to undergo local deterioration. The protective covering may thus be useful when the need becomes apparent, several years after construction. It should, however, be remembered that the systematic use of a coating may be burdensome and unnecessary when the protection proves to be pointless. Even so, this decision to protect the surface may be influenced by specific conditions relating to the use of the rooms and subsequent maintenance.

When it is decided to use a coating, an effort must be made to obtain the smoothest and finest outer surface possible. For example with rammed earth, stones will normally be placed in the centre of the wall and the outer surfaces rammed slightly more.

When it comes to providing a surface dressing, the stones will be concentrated on the outside edge of the wall and the wall may be less rammed in order to obtain a slightly open structure. After a first season of exposure to the elements, the stones will be exposed and will facilitate the adhesion of the rendering. In this case, the middle and the inside surface of the wall will be strongly rammed in order to provide the necessary strength.

Prior to deciding on the use of surface protection, three alternative solutions should be considered:

1. building a well-designed structure, provided with good foundations, and an anti-capillary barrier, overhanging roofs, gutters and downpipes, with good drainage and protected against the wind, and constructed according to the rules of the art;
2. stabilization, avoiding, if possible, stabilizing all the material (excessive cost) and limiting the stabilization to the outer surface of the walls; or by stabilizing the waterproof coating (wash, or distemper), rather than in the thickness of the wall;
3. application of a coating several years after the wall has become less smooth, giving good adherence. The deterioration of the surface of a wall in earth is in fact very rapid in the first two years but stabilizes rapidly.

It is above all advisable to protect structures against wind as this, associated with rain, can be particularly corrosive, even with very brief and short squalls. Finally a coating which is not suited to the earth, or which is ineffective because it has been poorly executed, can be more harmful than if it had not applied at all.

Functions and requirements

The main functions of a protective coating are the protection of the wall against bad weather and impact, the extension of the service life of the walls, the improvement of their appearance by hiding the imperfections of the rough work and giving them an attractive colour – without, however, masking defects – and improving thermal comfort. Finally, these functions should not allow the cost of the coating to get out of hand.

The functions may, moreover, result in contradictory requirements. For example a good coating should be impermeable to rain outside the structure but remain permeable to moisture from the inside. Renderings are very susceptible to climatic stresses (variations in temperature, sun, rain, and frost) and may deteriorate. However, they must not lead to the irreversible deterioration of the support (e.g. loss of wall material stuck to pieces of rendering).

A good protective coating should adhere well to the support without provoking the loss of wall material, be flexible in order to allow for the deformation of the support without cracking, be impermeable to rain, be permeable to water vapour in the wall itself, be frostproof and, finally, have a colour and texture compatible with the local environment.

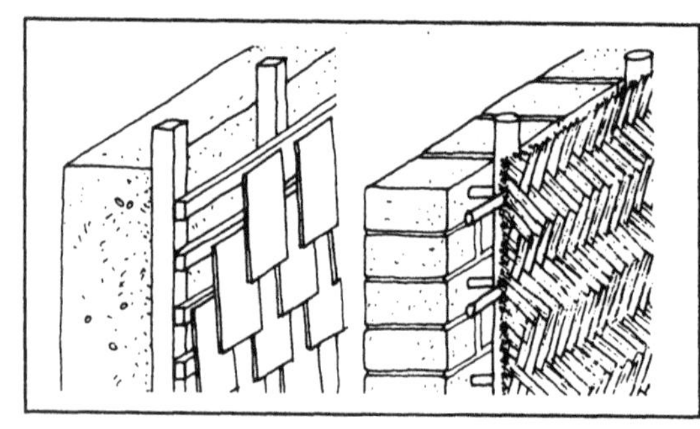

1. Weather boards

These are coverings attached to the wall and fixed to a secondary structure in wood or metal. Weather boards may be in any of a variety of materials, such as slabs of wood, planks, boards, tiles, cement-fibre elements, corrugated iron, external insulating systems, and so on.

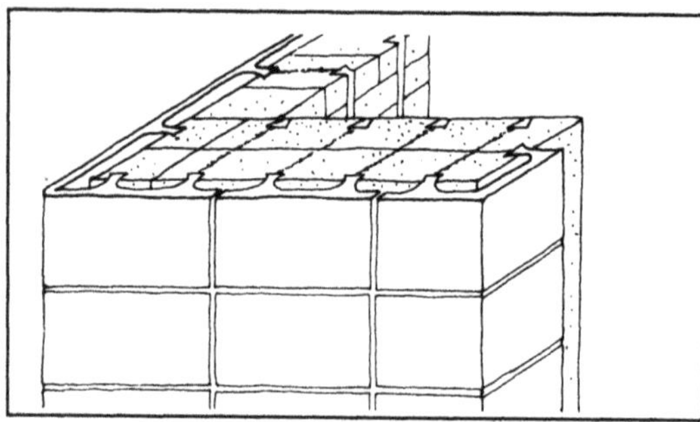

2. Cladding

Walls made of earth blocks and a cladding of prefabricated concrete elements were used in Germany in the 1920s. Both the earth and concrete products must be extremely well produced as design tolerances are very limited.

3. Facing

This protective procedure was known to the Mesopotamians, who applied glazed ceramic cones to the facing of the wall when still wet. The development of this system has been towards pebble and burnt brick facings, common in the Middle and Far East. The facing is applied while the rammed earth wall is still in the shuttering, or applied subsequently. The system may result in a mixed wall, the strength of which is not always uniform, and which may result in unequal subsidence of the wall and facing. The system is not suitable for use in earthquake areas.

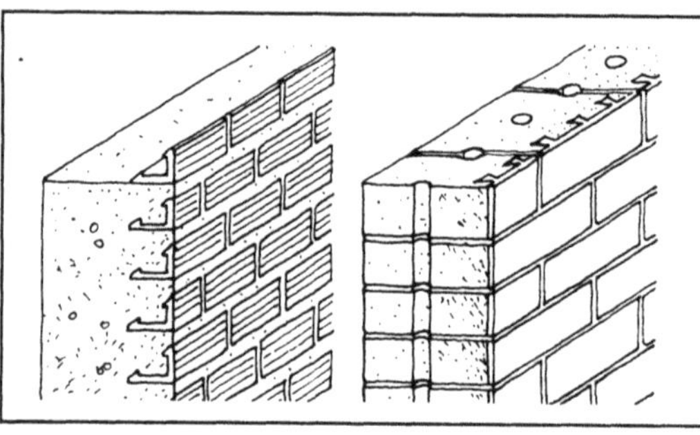

4. Integrated facing

Burnt clay elements which are either flat or L shaped are fitted as facing on the rammed earth wall during construction (German procedure) or are included in the earth block during moulding (procedure used by the EFPL at Lausanne). In the latter case, the adhesion of the facing to the earth block is ensured by dovetail fittings.

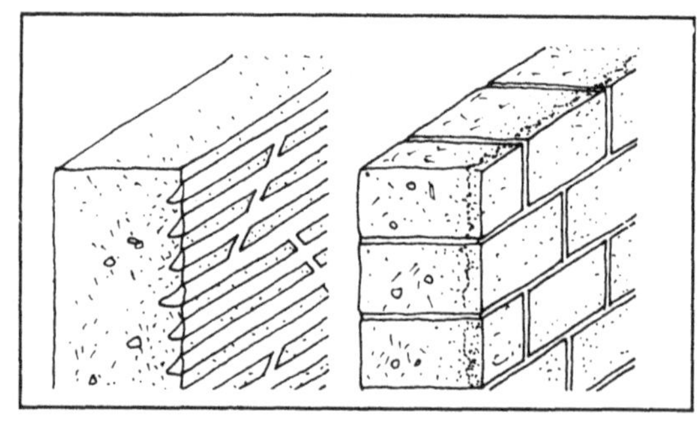

5. Twin layers

This is a surface stabilization system. With rammed earth this can take the form of the stabilization of the entire outer surface, while still in the formwork, or partial stabilization with layers of mortar or lime. The twin layer system has also been developed for soil blocks (Burundi, 1952 and EIER at Ouagadougou). It gives excellent results but is slow in production. The effect of the surface stabilization is limited to a thickness of 2 to 3cm.

6. Inlay

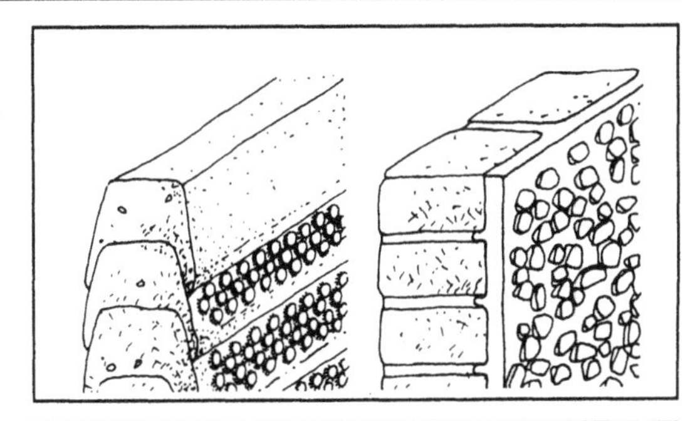

Here an outer wearing layer is made from inlaid elements which can range from pebbles or flakes of stones, potsherds or brick flakes, shells, bottle tops (seen in Mexico), bottoms of bottles, box tops (seen in Khartoum). The work is demanding and regards a large supply of the elements used. Only the most exposed walls are inlaid.

7. Surface treatment

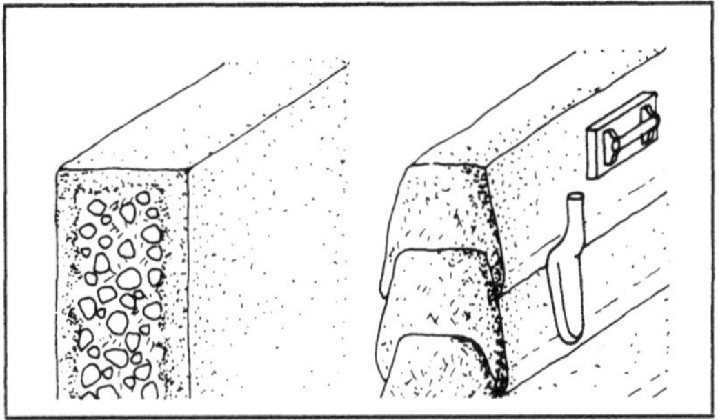

The exposed surface is carefully treated. French builders in rammed earth (pisé de terre) finish the wall with a 'fleur de pisé'. This involves the careful ramming of the outer surface with an extremely fine soil. The surface treatment may also be the finishing of the wall with a wooden paddle, as practised in Morocco. Such external tamping is also carried out in Yemen on cob structures. The surface of the wall can also be rubbed down with a stone for example. Such treatments reduce the porosity of the soil and are effective but should not be carried out when it is intended to apply a rendering.

8. Renderings

These may be in earth, stabilized earth, or a sand-based mortar to which a hydraulic binder has been added: cement or lime, or some other additive such as bitumen, resin, etc. Renderings can be applied in a single thick or thin layer, or in several layers. Multi-layer renderings perform extremely well but take longer to apply.

9. Paint

The coatings mentioned under this heading do in fact include conventional paints but also cover the distempers and washes. The latter are cement or lime slurries applied with a brush on walls properly prepared and hydrated in advance. It may also be bitumen applied in the form of a liquid cut-back. A spray-gun may also be used to apply them.

10. Impregnation

The soil is impregnated with a natural (e.g. linseed soil) or chemical (e.g. silicon) product which confers certain properties on the wall: impermeability, fixing of fine grains and powders, hardening of the exposed wall surface, colouring, and so forth. The impregnating product is applied either with a brush or with a spray.

The conventional renderings based on hydraulic binders such as cement and/or lime, plaster, and with or without additives, are well-known nowadays. When employed for the protection of earth buildings, this type of dressing often give adequate results, but it must be pointed out that numerous precautions must be taken. While these materials are often used with success, disappointments are also frequent. These failures appear to be mainly the result of the use of incorrect amounts of material and in particular a lack of know-how. Apart from the proportions used in mixing, which must be correct if the rendering is to be flexible and if it is to allow the passage of water vapour, problems often arise from the poor preparation of the support and careless application. The rendering and the soil must be compatible and the greatest care should be taken in choosing the ingredients, the proportions in which they are mixed, and the techniques used to prepare the wall and apply the rendering.

1. Non-hydraulic lime

The best results are obtained with hydrated slaked limes, in the form of extremely fine-ground powder or a paste prepared beforehand. The use of slaked lime as a surface rendering for soil structures is old and well-established in many countries. It must be remembered that the hardening of a rendering based on slaked lime is the result of slow carbonation by carbon dioxide in the air, and as a result these dressing should not be too sheltered. The long hardening process makes these renderings sensitive to atmospheric conditions, particularly frost and great heat. In many regions, lime dressings are modified during preparation, with various additives which can improve their quality. For example fresh bull's blood, leaving aside its importance in magical practice, improves the waterproofing qualities of the rendering. Other practices include the addition of natural soap which improves the workability of the mortar, facilitating mixing and application. Similarly in Morocco, lime renderings of the Taddelakt type were traditionally lubricated with yolk of egg, although nowadays soft soap is used, which improves waterproofing and facilitates polishing. The addition of a little molasses helps to harden the rendering. When slaked lime renderings are exposed to considerable stressing, a small amount of hydraulic lime or cement can be added. Only a small proportion may, however, be added to avoid excessive hardening or reducing permeability. Experimentation has made it possible to specify proportions for lime or sand based multi-layer renderings and mixed renderings based on lime, cement and sand.

	lime	cement	sand
1st layer	1		1–2
2nd layer	1		2.5–3
3rd layer	1		3.5–4
or			
1st layer	2	1	3–4
2nd layer	2	1	6
3rd layer	2	1	8

2. Hydraulic lime

A distinction is made between natural hydraulic limes and artificial hydraulic limes. The natural limes harden rapidly with water and slowly in air. This advantage reduces the sensitivity of the new rendering to damp and frost. The artificial hydraulic limes have properties similar to cement and their use should be avoided. In small proportions they can be useful, e.g. 1 part lime to 5 or 10 parts sand.

3. Cement

Cement mortars are too rigid and suffer from the defect of not adhering well to earth walls. Cracking, blow-up, and loosening in sheets are frequently observed symptoms. Their use is not advised and can at best be a make-shift solution, with proportions of the order of 1 part cement to 5 or 10 parts sand. It is better to add a little lime to them: 1 to 1 or 1 to 2 if at all possible. Cement renderings should be applied on a wire netting. This reduces cracking and breaking into slabs, but does not improve their adhesion.

4. Plaster

Plaster renderings are fairly compatible with soil walls but should by preference be used inside rather than outside. In dry climates they can be used outside as well. It is best to improve the adhesion of plaster to the earth by first applying a diluted wash of lime or cement. On outside walls slaked lime can be added to the plaster. This hardens the rendering and improves its water resistance. The rendering can be applied in two layers with 1 to 1.5 parts of slaked lime added to 10 parts of plaster and 7.5 to 10 parts of sand for the first layer. The same proportion of binder, but without sand, can be used for the 2nd layer. Waterproofing the surface with a fluoro-silicate solution after a period of a few days is desirable.

5. Pozzolana

Added to lime, pozzolana which contains enough silica produces a compound similar to Portland cement. Pozzolana-based renderings are, however, far more flexible than cement-based ones. They are often used for finishing flat earth-brickwork roofs and for vaulting.

6. Gum arabic

When added to soil, or even better to sand, gum arabic produces good protective coatings, which are hard, do not crack, and adhere well to earth walls. This product does not stand up well to water and it is therefore better to use it on the inside of the building. The colour obtained is a pastel red ochre. Gum arabic is used as a rendering chiefly in Sudan, but is becoming increasingly expensive.

7. Resin

As knowledge stands at present the use of resins, organic binders, and various mineral substances should preferably be limited to finishing the renderings described above.

8. Ready-to-use renderings

These renderings are prepared on the basis of an added dry mortar based on mineral binders. They are designed to be applied to other supports than earth walls in single layers but can be attractive if the support is properly prepared. Their use demands technical and strict, systematic experimentation. Combined systems such as mineral-organic products with a impregnation layer, finishing mortar based on mineral binders to which resins are added, and finishing layers using organic binders, are also worth considering, on condition though that the basic principles applicable to all renderings applied to earth walls are respected.

9. Plastic coating

The use of a plastic coating implies that the appearance of the support will not be preserved. The incorporation of reinforcement in waterproof plastic protection may be attractive, and depends on the configuration of the support. However, the danger of blistering and the impermeability to water vapour makes their use inadvisable.

Output

• Preparation of the support, scraping and removal of dust	½ day per m^2
• Preparation of the mortar + assistance to the mason	¼ day per m^2
• Application in three layers by the mason	¼ day per m^2
• Application of wash by one labourer	1/30 day per m^2
• Supervision by a foreman	1/20 day per m^2
TOTAL	1.1 day per m^2

These figures are based on observations on large-scale sites in tropical countries. The outputs given here are for qualified labour.

Cost

Cost is not particularly dependent on the type of rendering but rather on the organization of the work. The following figures are for work carried by direct employees or by a co-operative and were drawn up on the basis of various sites and for both interior and exterior rendering.

Type of structure	% of total cost without services
1. Very low-cost house	15
2. Small house with a minimum of equipment	20–25
3. Formwork dwelling	30
4. Covering for vaulting and domes	5

For 1, the 15% includes any incorporation of wire netting. The final price can be broken down as follows: materials 8%; equipment 8%; organization 8%; wages 76%. Labour costs can in fact reach 80 or even 90%. The degree of difficulty of the site may result in variations ranging from a factor × 1 to × 2.

For 2, a minimum of equipment means no electricity, a single tap, a minimum of rooms, no floor protection, no ceiling.

In case 3, the indicated value of 30 is not unreasonable even when the dwellings are no more than bare frames.

In 4 the covering is a damp-proofing coat.

Earth can undoubtedly be an excellent rendering or be one of the ingredients of a rendering. Even so earth is not a satisfactory basic component for first-class exterior renderings, particularly in rainy environments, unless it is improved by the use of a stabilizing additive. Earth renderings have been widely used and still are in many regions of the world.

The adhesion of earth renderings, whether used inside or outside, is virtually perfect. They are, however, above all a wearing layer which is the first affected by erosion and which can be cheaply replaced. The simple application of impregnation, a wash or grout, or paint can considerably improve these renderings, which are somewhat fragile. The earth rendering is often referred to by the term 'dagga', and the widespread employment of the term in the literature often leads to confusion.

1. Earth

When used as a rendering, the earth is first rid of all elements of a diam. > 2mm. Clayey and sandy soils will preferably be used (1 part of clay to 2 to 3 parts of sand). Preliminary tests are advisable in order to determine exactly what value should be given to the 2 to 3 parts of sand. Such tests examine cracking and adherence a few days after application. Clays which suffer from strong swell and shrinkage are not suitable. Clays of the kaolinite type are preferable. Lateritic clays often make good rendering in an attractive red or ochre. The main drawback of these earth renderings is that they are susceptible to cracking.

2. Water

No great problems with mixing water are encountered. The most crucial factor is the amount used, which is important in controlling the shrinkage and drying of the rendering. Observation has shown that it is best to use rainwater. This is because this water is depleted in cations and Na ions fixed in the clay-humus complex enter into solution and set off two reactions which result in the fixing of OH^- ions, and raising pH to 10. The abundant OH^- ions disperse the clay, which becomes more adhesive. Other improvements are also possible by adding deflocculants and dispersants to the water. By using less water and obtaining a dispersed and highly uniform mixture the rendering will be less subject to shrink and thus dry more quickly. The main deflocculants are sodium carbonate (Na_2CO_3) and sodium silicate ($Na_2O_xSiO_2$); between 0.1 and 0.4% should be added to the clay. Others products such as humic acid, tannic acid and horse urine can replace the water entirely.

3. Fibres

Fibres act as reinforcement. Fibres can be of many different origins: vegetable, such as straw from wheat, barley, winter barley, rice, millet; animal hair and fur; and even synthetic, such as polpropylene fibres. Common proportions are of the order of 20 to 30kg of fibres per m^3 of soil used. In the majority of earth renderings reinforced with fibres, they are cut to fairly short lengths. The finishing layer can also be reinforced with fibres which give an attractive texture to the rendering but which trap dust. Fibres can also be added as a light filler, such as wood shavings or sawdust. Wood waste fillers should, however, be first mineralized by soaking in milk of lime or a cement solution.

4. Stabilization

Virtually all products used for stabilizing soil in bulk are suitable for preparing renderings.

5. Cement stabilization

This is only really effective if the soil is very sandy. Proportions may vary from 2 to 15% of cement, depending on whether a mild improvement or true stabilization is desired. Cement-stabilized renderings should by preference be applied to stabilized surfaces. It is also possible to add between 2 and 4% of bitumen. This mixture tends to darken the dressing without spoiling the colour, but greatly improves water resistance.

6. Lime stabilization

Lime stabilization has its greatest effect on clayey soils when it used in large quantities, often over 10%. Similarly a lime-stabilized rendering is best applied to a stabilized surface. The addition of animal urine or dung can have truly astonishing effects on the rendering (less shrinkage, hardness and good permeability). The main drawback is the strong ammonia smell during mixing, which may upset some people.

7. Bitumen stabilization

Bitumen-stabilized soils should be neither too clayey nor too sandy and dusty. The quantity of bitumen ranges from 2 to 6%. They are usually cut-backs which should by preference be heated without, however, exceeding 100°C. Where bituminous emulsions are used the mixture must be made slowly in order to avoid any breakdown of the emulsion. The stabilizer can be prepared by adding four parts of bitumen to 1 part of kerosene, followed by heating and adding 1% paraffin wax. The kerosene can be replaced by coal creosote. The mixture described above can be replaced by 4.5 parts of cut-back or 3.5 parts of bitumen emulsion. Bitumen stabilization for renderings is particularly effective on soils which have already been reinforced with straw or even with dung. The bitumen is added only at the end, 2 to 3 hours before the rendering is applied. Mixtures of asphalt, gum arabic and caustic soda solution are also highly effective. The support should be properly prepared, brushed and moistened. Excellent results have been obtained with this type of rendering by the CBRI at Roorkee in India.

8. Natural stabilizers

These are highly diverse and often are the traditional stabilizer in numerous countries. Their effectiveness is extremely variable. Their effect is more the retarding of the decay in the material, without really ensuring the sustained lifetime of the rendering, but limiting the frequency with which it is remade. Traditional stabilizers include the juice of the agave and the opuntia cactuses, melted shea (karit) butter, often added to gum arabic; the juice from boiled banana stem; 15 litres of rye flour boiled in 220 litres of water, with the paste obtained being added to the soil; cowpats or horse droppings (1 part dung to 1 part clay and 5 to 15 parts of sand); gum arabic, which forms colloids with water; the sap of the fruit of the acacia scorpiodes (gonahier) boiled in water with several pieces of limonite (a type of laterite), which results in a rather effective water repellent; euphorbia latex precipitated with lime, the sap of the néré tree, obtained from a decoction of the powdered fruit, added to the soil and then applied as a wash to the soil rendering stabilized with néré; and peulh soap, a sort of casein, thinned and beaten like a paste. Other natural products have been tried at the Cacavelli Centre in Togo for the improvement of renderings. These include kapok oil obtained by roasting kapok seeds to obtain a powder form with a high lipid content. The powder is diluted with water and boiled for several hours. This is then mixed dry with the water, mixing water being added subsequently. The rendered wall is then distempered with two coats of kapok oil. Calcium palmitate, obtained by mixing fat lime and palmitic acid – a product obtained by reacting HCl with a native soap known as akoto – can also be used. The calcium palmitate is diluted in a small volume of water and the soil is mixed with the milk of lime obtained (10% by weight of the mixture). The Hausa of Africa use the natural potash which accumulates in dyeing trenches, or an infusion of carob-bean husks, or even of mimosa which the richest people import from Egypt. Without a doubt there are a great many other natural stabilizers.

9. Synthetic stabilizers

There are numerous chemical stabilizers and their effectiveness has as yet not been scientifically confirmed. These include the celluloses, polyvinyl acetate, vinyl chloride, the acrylics, sodium silicate, the quaternary amines, aniline, bentonite, soap stearates, casein glues, and paraffin. Others may be combinations of the above with, perhaps, the addition of natural products.

10. Application of earth renderings

When used indoors they give excellent results, although it is advisable to strengthen weak points of the building (internal and external corners, reveals, the bottom of walls) with a mortar of sand and lime. Outside, a single layer is not enough. At least two layers should be applied, and preferably three. First apply a rough-cast in a highly adhesive clayey soil which can be finished with a mortar consisting of one part lime and one part sand; followed by a 1.5cm thick layer of pargeting in clay soil and coarse sand, reinforced with fibres chopped to 3 to 5cm long; and finished with a top layer in clayey soil and sand, to which a light filler has been added (e.g. chaff or flax).

Opaque paints

The range of commercially available paints is extremely wide. First of all it can be said that when ordinary paints are used to protect earth walls they give apparently satisfactory results at first. Very quickly, however, deficiencies such as blisters and loss of adhesion become apparent. Paint can thus not be regarded as being a means to give a soil wall a lasting finish. They nevertheless afford temporary protection pending proper repair. Their use can be permitted indoors and for outside walls which are very well sheltered from the attack of the natural environment. Even so it is preferable to use them as a complement to the finishing layer of the rendering. The surface which is to be painted should be absolutely dry and all dust removed with a brush. Moreover, primers which penetrate deeply into the material give poor results. It is better to apply an impregnating layer, either in linseed oil, or in a very dilute lead-based paint, applied at a rate of 0.50 litres per m^2. When applying the finishing coat (in two coats), the best course is to consult the paint manufacturer for technical advice.

When walls are stabilized in bitumen, it is advisable to apply 2 to 3 layers of paint at least as bitumen exudations may appear. When all conditions are favourable, i.e. good quality paint, proper preparation of the support and good application, the paint may last between 3 and 5 years. American experiments have shown that better results are obtained on sandy earth walls than on clayey ones.

1. Industrial paints

Aluminium-based paints do not hold well when applied directly to earth. They may be applied to undercoats treated with bitumen or to bitumen-stabilized walls.

Casein-based paints give fairly satisfactory results on earth walls.

Primers can be used to impregnate the surface.

Lead-based paints can be used on a surface treated with linseed oil.

Oil-based paints perform only moderately.

Polyvinyl acetate emulsions can sometimes be satisfactory.

Water-bound distempers should not be used on crumbly walls.

Latex paints are quite efficient on stabilized soi walls.

Resin-based paints often give satisfactory results, but silicon paints are not very reliable.

Acrylic paints breathe, are elastic, water repellent, and stand up well to the alkalinity of earth walls.

On the other hand, watertight paints, alkyds, epoxies, and polyurethanes are to be avoided because they retain moisture.

Chlorinated rubber-based paints, which are elastic, and stand up well to heat, ultra-violet radiation and atmospheric conditions can be used to waterproof roofs but should not be used for soil walls.

2. Oils

Earth walls are an extremely porous support and absorb large amounts of non-oxidizing oils, such as sump oils. However, they do not perform satisfactorily as the impregnated layers should not go too deep. Linseed oil is oxidizing, reacts with air and becomes fixed. It is only slightly soluble in water and can be applied to moist soil. It is a very satisfactory primer for lead and oil paints, but is expensive. Castor oil has the same properties but is scarce and very expensive. It may be that fish oil is equally good. Palm and shea oil have both been studied in Ivory Coast but suffer from the fact that they are very viscous, making them difficult to apply, and giving rise to efflorescence.

3. Plant juices

The juice of the *Euphorbia lactea* is very well known in tropical countries. It can be used to protect earth walls but it is absolutely essential to add it to lime (precipitation). The juices of the agave and opuntia cactuses can be used but they are extremely poisonous and can harm the eyes. In Burkina Faso and Benin, and in Southern Ghana a red plant extract known as 'am' is employed. In Northern Nigeria 'laso' is used, which is an extract from a local vine called 'dafara' (*Vitis pallida*). 'Makuba' is also used, this being an extract from carob-bean husks. Banana juice can be used, but it must be boiled for a long time, consuming large amounts of fuel in the process; and there is no guarantee of it being effective.

4. Other natural products

Paints can also be made from glues based on cream cheese (6 parts) mixed with one part of quicklime and greatly diluted in water. Paints can also be made from whey, with 4 litres of whey added to 2kg of white cement. These formulations have been developed by the University of South Dakota.

5. Soil

Earth slurries can be used indoors and fixatives added to them. These slurries can lay dust and equalize surfaces. Outside they are not lasting, but can be improved by stabilizing with mineral binders (lime, cements) or organic binders (bitumens, plant juices, and various oils), or with acids. Even when improved these emulsions must, in general, be regularly refreshed.

Transparent paints

There is a strong trend at present towards the employment of products which aim at conserving the earth indefinitely, while preserving the appearance of the material. Unfortunately earth is a support very different from other industrial materials, and the results obtained with these 'magic' products are, to say the least, random, as real conditions are not the same as those in the laboratory. Numerous problems appear after a period of years.

The chemistry and the composition of the products are highly complex and it is advisable to carry out preliminary tests with a view to ensuring their effectiveness, at least in the medium term and for fairly harsh outside conditions. Many of these products, described as 'totally waterproof', can often resist only a low pore water pressure. Generally speaking, such transparent products help to reduce the deterioration of the wall at the wearing coat. Their quality depends on the depth of penetration (up to at least 2cm). These product may form a crust of treated soil, which has the effect of producing the disaggregation of the wall. This happens with sodium silicate and silicones in general.

With knowledge of these products – whether based on paraffin, wax, resin, or various minerals – as it now stands, their use is best limited to a treatment of the finishing course of thick renderings and on sheltered walls.

1. Surface water repellents

Silicons in a volatile solution require a suitably dry surface and their use is limited by the size of cracks, as these may not exceed 0.15mm, expecially on exposed walls. The only use for which they should be contemplated is for finishing renderings.

Silicons in an aqueous solution or emulsion can accommodate themselves to a certain moistness of the support, although the same reservations apply as above.

Metallic soaps, stearates, and polyoleofins must first be made the subject of special attention. Fluates, or more scientifically fluoro-silicates, form an artificial calcine by reacting with the calcium carbonate. They have absolutely no effect on soils stabilized with lime. They may be used with good effect on carbonated renderings using lime mortars.

2. Resin-based film-forming impregnation treatments

These may offer an attractive solution for sheltered outside walls if they can be strongly absorbed into the first few centimetres of the soil and if they do not form a thick surface crust. Checks should be made to ensure that permeability to water vapour is maintained and that the impregnation can be refreshed.

3. Waterproofing coatings

These products are based on resins in an organic solution or dispersed in water. Their efficiency is limited by cracks which may exist or appear in the support. The risk of blistering and deficiencies in permeability to water vapour makes their use highly unpredictable. They should in principle be avoided.

Limewashes made from non-hydraulic lime have been widely used in many regions since time immemorial. They represent a fairly cheap way of affording protection against the harmful effects of rain, particularly in the absence of sophisticated materials and where budgets are subject to severe restrictions. These limewashes are most suitable for affording protection indoors or on sheltered walls outside. They can, however, be easily improved so that they last several years.

1. Drawbacks

Limewashes are not particularly durable because they are easily washed off. The simplest must be refreshed periodically (once or twice a year), particularly in wet climates. Additives can make very significant improvements to them. These include vegetable drying-oils (linseed oil, nut oils, castor oil, croton oil, and hemp oil), glues, casein, salts which are more or less hydrated (zinc sulphate, potassium alum, sodium chloride), resins or oleo-resins and rubbers or water-soluble rubber resins. Such limewashes are also very sensitive to mechanical shock and only offer limited protection to abrasion.

2. Benefits

Limewashes made from non-hydraulic lime are cheap and stand up fairly well to alkalinity as well as bitumen exudations (on walls stabilized in bitumen). Light in colour, they reflect solar radiation. They can be easily tinted with oxides. They can be easily and quickly applied and do not require specialized labour to apply them, although they must be applied with care. Refreshing them is easy and their aging causes no major deficiencies in the support. The periodic refreshment rejuvenates the structure. They have the advantage of regulating the moisture balance between the support and its surroundings. Because of their constituents (quicklime or slaked lime, salts, formaldehyde), they have certain antiseptic properties. They bring light and hygiene to what would otherwise often be miserable and unhealthy slums.

3. Binder

The best results are obtained with non-hydraulic lime slaked in paste, using high yield, finely sieved quicklime. The wash is prepared a few days before application. Commercial slaked lime can also be used on condition that it is not too far gone in carbonation. The content of calcium and magnesium oxides should not be lower than 80%, while the carbon dioxide content should not be higher than 5%.

4. Preparation of the binder

The containers or trenches used for slaking quicklime should be considerably larger than the original quantity as the material increases greatly in volume (double). Attention should also be paid to the risk of burning as slaking quicklime generates high temperatures (120 to 130°C). The operation is best carried out at night (cool) with plenty of clean water available. All lumps are broken up and the lime mixed well until a uniform paste is obtained, which is brought to the desired consistency by adding suitable amounts of water. If slaked lime is used the quality of the sieving should be checked. The basic mixture is one volume of slaked lime to one volume of water. It may be necessary to add water to obtain the desired consistency.

5. Application

The limewash is applied to a clean, dust-free surface, which should be free of all crumbling, in at least 2 coats (3 or 4 coats being preferable). The first coat will be thin, but subsequent coats will be increasingly thick. A distemper brush can be used for the first coat, while for the second coat a broom can be used, and even a yard brush is suitable for the later coats. The distemper is applied when the wall is in the shadow and extremes of heat or cold should be avoided. Precautions should also be taken to avoid showers, which could wash the whitewash off the wall. The best method is 'al fresco' but this is very difficult to carry out on earth walls. Application to the dry surface will thus be the most common method used and care will be take to moisten the surface, preferably with clean milk of lime, but without soaking the surface. Excessively thick coats will flake off. Drying should be slow.

6. Simple test

A block is weighed in advance and then given two coats of limewash. The block is immersed in water for two days and then weighed again. If the difference in weight is of the order of a few tens of grams the limewash is good. If the difference is of the order of several hundred of grams the limewash should be rejected.

7. Fillers

Fillers are additives to the binders which give the limewash properties which the binder alone could not give it. The fillers discussed here are all compatible with lime.

Linseed oil increases the ability to stand up to variations in humidity and improves adhesion to the support. Should be added immediately before application.

Tallow is animal fat composed of glycerides, which gives greater plasticity when applying the limewash by increasing water resistance and adhesion. Proportions: add about 10% by weight of molten tallow to the lime. Tallow can be replaced by calcium stearate or linseed oil.

Skimmed milk or whey (10 days) increases the impermeability of the limewash. Add 1 part skimmed milk or whey to 10 parts of water used in the preparation of the whitewash, immediately before use.

Casein glue in powder, known as 'cold glue', acts as a fixative. The addition of formalin increases its strength. Dissolve this glue in boiling water until it becomes soft (2h) at the rate of 2.5kg of glue to 7 litres of water.

Animal glues improve the adhesion of limewash. They include glues made from skin and from bones.

Rye flour forms a vegetable glue soluble in warm water; requires the addition of zinc sulphate as a preservative when in paste form. Increases surface hardness and resistance to rubbing.

Alum is the double sulphate of potassium and hydrated aluminium. A small quantity in the form of a paste, made by grinding then boiling in water for a period of one hour, is added to the wash immediately before use. It increases workability, surface hardness and resistance to rubbing.

Sodium chloride (common salt) retains moisture in the limewash and facilitates the carbonation of the lime. It should be added slowly to the limewash prior to use (dissolves). Calcium salts and trisodium phosphate (Na_3PO_4) are also used.

Formaldehyde has the antiseptic and stabilizing properties of urea-formaldehyde. Dissolve in water and add slowly to a mixture of lime and casein glue or lime and trisodium phosphate. Does not keep well.

Molasses (syrupy residue remaining after the crystallization of sugar): 0.2% by weight added to lime speeds carbonation and increases strength.

Mineral fillers Inert fillers or soil (kaolin).

Colourings Exclusively minerals, in the form of moistened powders; add prior to use.

WHITEWASH FORMULATIONS

		QUICKLIME (CaO)								SLAKED LIME Ca(OH)								
		Q1	Q2	Q3	Q4	Q5	Q6	Q7	Q8	S1	S2	S3	S4	S5	S6	S7	S8	S9
QUICKLIME	(2)	20	20	20	20	20	20	20	20									
SLAKED LIME	(2)									25	25	25	25	25	25	25	25	25
PORTLAND CEMENT		2																<25
WATER	(1)	40	40	40	40	47	49	50	60	30	30	32	32	40	63	63	65	50
LINSEED OIL	(1)											1					1	
MELTED TALLOW						2		1.2										
SKIMMED MILK	(1)		6.5															
CASEIN											2.5							
ANIMAL GLUE				1.4														
RYE FLOUR							0.8											
ALUM					1								0.6		0.6	0.6	0.2	
SODIUM CHLORIDE		5	0.7				0.8								1.3	2.5		
ZINC SULPHATE			0.3															
TRISODIUM PHOSPHATE											1.5							
FORMALDEHYDE	(1)										1.9							
MOLASSES	(1)														7.8			
EVALUATION			D			D					D				D	D		D

1) All figures are in kilograms, except the liquids which are given in litres.
2) The selected reference quantities correspond basically to 20kg quicklime = 25kg of slaked lime.
3) D indicates the limewash is fairly durable.

Cement-washes and hydraulic limewashes

Simple washes of cement or hydraulic lime give good protection and improved durability to earth walls. They are generally feasible when little money is available and are not affected by the alkalinity of the wall. In good conditions of execution, they can help to reduce the quantity of stabilizer in the bulk of the wall. This type of wash requires a humid environment when hardening, and curing must be carefully supervised. They are thus more difficult to apply than the limewashes based on non-hydraulic limes. As cement and hydraulic lime are more finely ground than non-hydraulic lime, the water requirement should be reduced, with a cement/water ratio of the order of 1 to 1.5 compared to 0.78 for non-hydraulic lime. As a result workability and covering capacity are less. These washes are fairly resistant to the passage of water vapour and are therefore only suitable for use in regions where water vapour migration from the inside to the outside is not a problem (i.e. excellent in sub-tropical regions). They have a limited lifetime, this being between 5 and 10 years (on stabilized walls) and therefore require periodic refreshment. Paints based on white cements also exist and are available in various colours. These paints contain additives to improve their plasticity. These paints are only suitable for walls in stabilized earth and for very strong walls, but even here the results are rarely satisfactory. On weak walls, cement paints should not be used (flaking, blistering).

Generally speaking, when it is intended to give stabilized walls a cement-wash or hydraulic lime-wash finish they should be prepared with care (holes and cracks filled, dust removal) and thoroughly wetted. This moistening is indispensable when the mixture contains no calcium chloride (salt which retains water), the use of which is justified in hot and dry regions (not more than 5% of the mixture). Moistening facilitates application and prevents excessively rapid hydraulic shrinkage but may reduce the impermeability of the wash. The addition of hydraulic lime to the cement (max. 25%) does nothing for the wash but makes application easier (improved plasticity). Washes in cement or hydraulic lime should be applied in at least two coats, each 1 to 1.5mm thick. Three or four coats are even better. The first coat should be applied with a brush (not too hard) while subsequent coats can be applied with a broom or spray-gun. If the ground is smooth, a thinned coat should first be applied to serve as a size or primer for a subsequent thicker coat. The reverse applies if the surface is rough. The wash should be applied when the wall is in shadow. Once the last coat has dried it should be moistened in order to hydrate the cement and this should be repeated again just before nightfall. The second coat is best applied at the earliest 12 hours after the first, and if at all possible 24 hours later, after moistening the first coat again. These washes should be used in the 2 hours after their preparation and any leftovers should never be used on the following day. Colourings (3 to 7% of various oxides) or water repellents can be added, though this should be done in the last coat. The water repellent can be 2% of calcium stearate solution added to the cement, or 2% of copper sulphate solution in a concentration of 100g to 10 litres of water. It should never be forgotten that a cement-wash poorly applied to a badly prepared wall will have the tendency to peel, flake, blister, and lose all its protective power, not to mention the shabby appearance it will give to the building.

Formulations

1. One hundred parts of Portland cement to fifty parts of silica sand or any other hard fine sand. Calcium chloride equal to 4 per cent of the quantity of cement. As water repellent add a quantity of calcium stearate equal to 2% by volume of the cement. The sand is mixed in after first mixing the cement, the calcium chloride and the calcium stearate. The volume of water is more or less equal to the volume of cement, although this may vary according to site conditions.

2. Slurry made from lateritic soil, cement and water (Ivory Coast). Two 50-litre barrows of lateritic soil, one bag of cement, and 175 litres of water. The slurry covers at a rate of 2.5kg per m^2, or in other words 340g of cement per m^2, which is very economical.

Bitumen washes

Perfectly dry earth walls with well-prepared surfaces (harshness rubbed down, brushed, and dust removed) can be protected by giving a coating of bituminous products such as emulsion, cut-back, 'Flintkote', and so on. Local climatic conditions are very important as these products are more or less impermeable to water vapour. Moreover, the effectiveness of bituminous products for surface treatments is not very prolonged and care must be taken to ensure that periodic refreshment is carried out. Nevertheless these bituminous washes are among the least expensive and greatly improve the resistance to water erosion and surface abrasion of earth walls. A precondition to the success of these bituminous coatings is the dryness of the ground to which they are applied. If the ground is moist blisters and bubbles will soon appear leading quickly to the detachment of the coating and, even worse, the loss of material. Bituminous coatings are often resisted because of their sombre colour – black. This drawback can be cured by treating them with a finish. This finish may be coats of paint or cement or lime-based washes. Such finishes should be applied several months after the treatment of the bitumen wall so that any defects in the bitumen coating become apparent and in order to avoid bitumen exudations. Applications of a paint finish, particularly oil-based paints, to a bitumen-stabilized wall or a wall which has been given a bitumen coating, must be preceded by the application of an undercoat of bitumen-based aluminium paint. This paint is compatible with the emulsion of the wall: the flakes of aluminium in the paint spread and cover the wall, preventing any exudation of bitumen from the wall. Other treatments which can be immediately applied can also be considered. These include sanding with washed sand on the fresh bitumen coating. Study of these finishes for bitumen-stabilized walls has shown that generally speaking products applied with a brush, such as milk of lime, bitumen emulsions, polyvinyl acetate or styrene emulsions, remain highly permeable to water but hold back bitumen exudation. Distemper paints and alkyd emulsions are not advisable. Oil-based paints stand up well to water but not to exudation. Bitumen paints are satisfactory from both points of view.

Formulations

1. Paint based on coal-tar: one volume of Portland cement, with one volume of petrol, and four volumes of coal-tar. The tar does not have to be preheated. The cement and paraffin are mixed first and the tar is then added to them. The mixture is applied with a coarse brush to a fine primer coat made of a mixture of water and gas tar. The colour is black.

2. A liquid bitumen wash can be produced from 2 volumes of coarse benzol and 1 volume of bitumen dissolved in the benzoline, to which a small amount of resin and quicklime has been added. Apply with a brush or fine spray. This wash gives a brown colour.

3. A wash can also be prepared from 25kg of preheated asphalt and 50 litres of petroleum oil. The asphalt is added little by little to the oil and carefully mixed until totally dissolved. Once the mixture has cooled it is poured through a fine strainer into another container in order to eliminate any undissolved materials and foreign bodies. It can be applied with a pesticide spray at a rate of $100m^2$ per day per person. The colour is dark grey. A finishing coat in milk of lime to which an animal glue has been added gets rid of this undesirable colour.

The numerous symptoms of the deterioration of soil walls are based on mechanisms which are for the most part related to the influence of local climatic conditions, such as rain, frost, great heat, as well as the influence of the people concerned, such as a lack of maintenance, lack of knowledge about how to apply them, and mechanical shock. Observation of renderings indicates that:

Either the renderings are in good condition: they can be old and well-maintained, or new and little aged (less than five years old), recently applied. Or the renderings are in poor condition: they may be very old, 50 years old and their durability is praiseworthy, not to say enviable compared to modern renderings. These are usually renderings based on non-hydraulic lime. New renderings in a poor state are often in a very bad way after no more than 5 years, the shortest period possible in which a rendering can be evaluated. These renderings are often cement renderings. What happens, in fact, is that cement displaces the accumulated know-how acquired with older renderings. Cement may have remedied certain problems (e.g. speed of application, reliability) but does not really provide a lasting alternative. The loss of traditional know-how and the absence of modern know-how is thus often a major cause of deficiencies. The same applies to the maintenance of dwellings. Formerly renderings were considered more as wearing layers which have to be regularly maintained (e.g. refreshment of washes). It seems that home maintenance is slowly disappearing as a social custom in the majority of rich countries. The shift is due to the excessive confidence in certain products (e.g. cement, paints). In many of the developing countries maintenance of structures continues to be be a significant social link in the community – the festive regular renewal of the rendering of mosques in Mali, with the whole village taking part, is a good example.

1. Deficiencies and their causes

Conventional symptoms may range from a simple dirty streak (marring the appearance of the structure) to a change in composition. The main deficiencies are poor composition, a lack of flexibility or poor adhesion to the support, and poor waterproofing. The reasons for these defects include: the use of unsuitable materials, careless application, structural tension, a lack of maintenance or a defect in the supporting structure, such as subsidence, and cracking.

2. Symptoms

Defective renderings on soil walls are revealed by the fairly typical symptoms discussed below.

Crumbling The rendering can be easily scratched with a fingernail and disintegrates. It is mainly seen in accessible places such as the reveals of doors and windows.

Erosion Eroded renderings are thin and no longer protect the wall. The erosion may be even and the increasingly thin rendering tends to disappear. Erosion may also be localized and the rendering may be left as isolated patches.

Crazing The surface of the rendering cracks into an infinity of threadlike cracks, permitting the access of water.

Cracks These may be few in number but gaping or very numerous and more or less closed. Fine cracks or crazing may develop into larger cracks. There is a danger of penetration by water and frost.

Blistering Can be localized or overall swelling of the rendering, and is visible as a bump or series of bumps. The surface of the wall sounds hollow because the rendering is no longer attached to the wall. There is a danger that pieces of rendering fall off the wall.

Blowing The rendering is pitted with small craters with a diameter of no more than 20mm and of variable depth. Blowing occurs quite often on lime-based renderings. There is a danger of water penetration and frost damage.

Efflorescence The rendering is discoloured by small white or grey rings. These are crystalline or amorphous deposits of an alkaline or alkaline-earth character and include sulphates, carbonates, and nitrates. These accretions of salt may bring about the disintegration and loosening of the renderings.

Infiltration Water is trapped in the thickness of the rendering resulting in the appearance of efflorescence or triggering off cracking and loss of adhesion. Once this has started the rendering may disintegrate very rapidly.

Dark patches These may appear in the form of black or brown patches. They are the result of the decay of organic matter left after water has dried up or by a localized flow of water.

3. Mechanisms

Expansion Frost or alternating wetting and drying can cause the clay fraction to expand at the wall/rendering interface. If the rendering is too rigid, it will first crack and then crumble. Similarly on a heterogenous wall (e.g. large stones isolated in rammed earth) differences in the thermal expansion of the soil and the stone can cause localized failure.

Shrinkage When a rendering first dries it shrinks, putting the material under stress. There may be loss of adhesion and loosening. This happens when a rendering is too rigid and the support is too smooth. When the support is rough, the same rendering can cause cracking. The cracks may be clearly visible and few in number when the adhesion is low or numerous and fine when adhesion is high. The thicker and stronger the rendering the wider will be the cracks. They appear on the outer face of the rendering if the contraction is the result of exposure to wind or sun: the rendering is dry (evaporation on a well-moistened ground and saturated, which does not bring any suction into play). Cracks may also develop at the interface and advance towards the outer surface on dry walls which contain little water, resulting in capillary suction operating from the rendering/support interface. The most sensitive spots are recessed corners and the angles around bays.

Vapour pressure Water vapour may accumulate in the form of condensation at the rendering/ support interface. Blistering may be observed, heralding detachment. This phenomenon is typical for maritime and temperate climates where internal vapour pressure is higher than external vapour pressure. This differential pressure causes the vapour to migrate through the wall and the rendering, and these must be permeable, which is why waterproof or excessively thick renderings should be avoided. The opposite problem – internal condensation – may arise in tropical climates or in air-conditioned rooms.

Others Other mechanisms originating in the conventional defects of renderings are the infiltration of rainwater or drips entering via cracks (accumulation of damp), the use of unsuitable materials (poorly slaked lime, old cement), efflorescence caused by moisture or excessive smoothing of the rendering (appearance of laitance on the surface), attack by micro-organisms (lichens, algaes, mosses), and plants (creepers, ivy), sloppy application (poor preparation of the mortar or support, frost or very hot conditions), rain erosion, wind erosion, and damage due to mechanical impact.

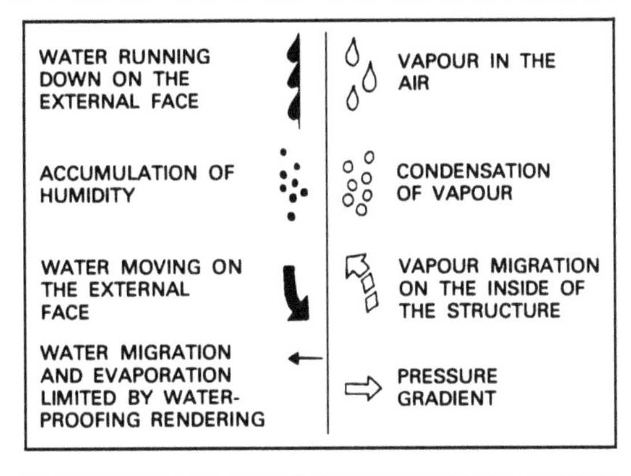

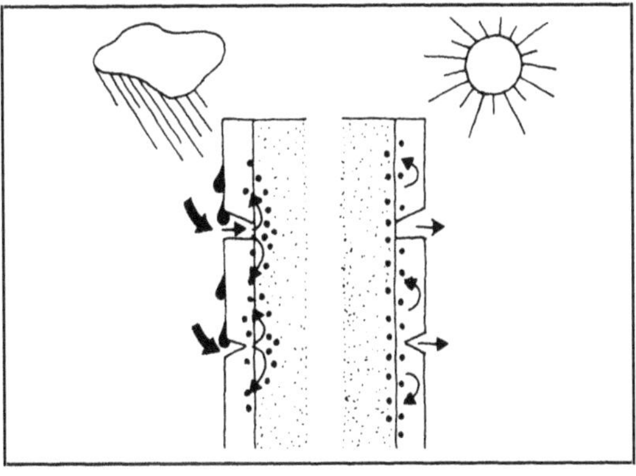

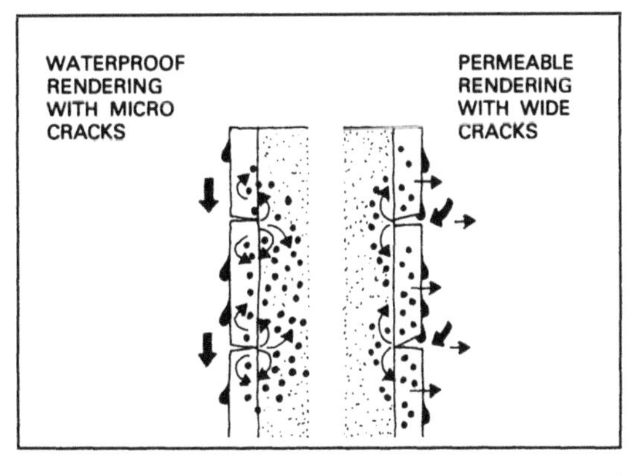

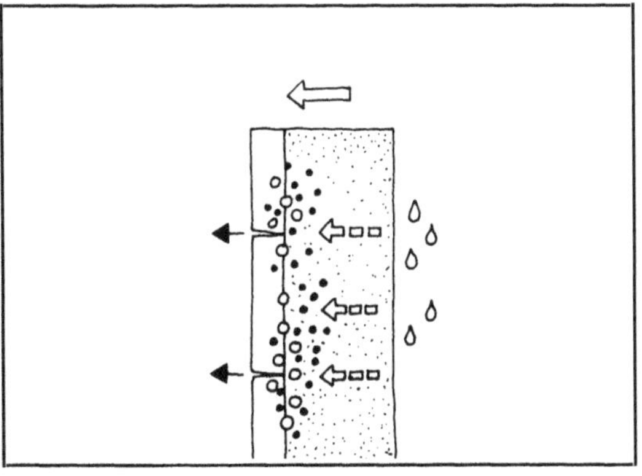

The main defects which can be observed in renderings are by and large due to:

- incorrect composition of renderings and defective constitution;
- hasty application, and a failure to respect the fundamental rules of the art;
- poor site conditions;
- incorrect or sloppy application;
- inadequate preparation of the support.

Taking heed of good practice, know-how, and the literature is absolutely essential. Moreover, the application of lasting renderings to earth walls demands special care and attention. The most essential elements are the correct preparation of the support, and the use of binders which are not too strong and which result in fat and plastic mortars.

1. Constitution of renderings

Single-coat renderings should be avoided, as they are too thick and too dense. Renderings based on mineral binders (lime, cement, and clay soils) should be applied in several coats, at least two – with an uncracked second coat – and preferably three, with a render, float, and set. This application in three coats is particularly important for conventional lime-based and cement-based renderings, the thickness of which diminishes with the float and the finishing coat.

Render (or in the US the anchor coat) This is the first coat applied to the support and provides the support for the float. It is made of a fairly fluid mortar, generously provided with binder, and very finely ground. It is energetically applied with a trowel, onto a well prepared support. Its thickness ranges from 2 to 4mm and the surface has a rough appearance.

Floating coat This is an intermediate coat which takes up any unevenness in the support. It gives the rendering its solidity and impermeability while staying permeable to migrations of water vapour. The floating coat is applied two to eight days after the render, in one or two layers, and has a final thickness of between 8 and 20mm. The floating coat will have no cracks and is level to the float. Application in several layers (if possible) makes it possible to plug fine cracks in the underlying coats. The amount of binder used is less and finer sand is used. The floating coat is grooved with the trowel or with a brush to improve the adhesion of the finishing coat. The curing of the floating coat must be perfect.

Finishing coat This coat finished the protective rendering and plugs any cracks in the floating coat; it is the decorative coat, with respect to both colour and texture. The finishing coat has the lowest binder content, as no cracks at all can be tolerated. Care should be taken that the finishing coat is not compressed too much with the float, because of the danger of causing the moisture to come to the surface and of crazing. This is the coat which may have to have to be redone from time to time.

2. When to apply

An earth wall should never be rendered before:

- shrinkage has stabilized; this may take several week or several months. Times may be 6 to 9 months for thick rammed earth, 3 to 4 months at least to a year for monolithic walls in cob or clay-straw; within a year of building for adobe or compressed block walls built when dry and not too late (2 to 3 months at least);
- any settlement in the wall has taken place; the formworks must therefore be complete, including any loads due to floorboards and the roof;
- migration of water and moisture vapour due to drying has stopped. The internal core of the wall should not contain more than 5% water by weight, and this can be a guide to when to start rendering works. Weather conditions when the site was in progress play an essential role.

3. Conditions of application

- Do not render in excessively cold weather or when it is very hot. Avoid working in driving rain as well as in sun and violent wind, and when it is very dry. The best climatic conditions are when it is moderately warm and slightly humid.
- Make horizontal and vertical joints with a view to completing panels of 10 to 20m^2 in size in one go. Walls should be finished on the same day they are started.
- Pay particular care to intersections, and reveals in bays. Where the support changes (e.g. earth and wood) the rendering should incorporate reinforcement at that point. Do not continue the rendering to ground level (capillary action); make a joint at the level of the footing.
- Avoid the excessively rapid drying of the rendering by spraying water on the surface (morning or evening) and by hanging up protective sheets against heat, and rain action which could wash off the rendering. Keep the environment humid.

- Make sure that the ingredients (binders and sands) are of good quality and that mixing is properly carried out.
- In warm climates, it is advisable to apply a wash to the rendering about three weeks after application, in order to plug any cracks.

4. Application techniques

- By hand, for earth-based renderings. Balls of mortar are thrown energetically against the wall and then smoothed with the palm of the hand, avoiding excessive finger pressure.
- Using conventional tools, trowels, floats, avoiding excessive compression.
- With a brush or broom: liquid render coat, or with an adjustable hand-operated plaster blowing machine, giving a Tyrolean finish.
- With a pneumatic blowing machine or rendering pump, making sure that the blowing pressure can be regulated as it must be neither too strong or too weak.

5. Preparation of the support

This preparatory phase in applying the rendering must be carried out with particular care.

Dust removal The support must be rid of all loose and crumbling material, and dust. It is carefully rubbed down and brushed with a metallic brush.

Moistening The support must not absorb the water contained in the rendering, as this may hinder setting and reduce the adhesion of the rendering. The support must therefore be moistened in order to avoid capillary suction, but not too much as a film of moisture could be created which reduces adhesion. The support should have a dull appearance, that is to say sprinkled until it flows off. This operation may require various applications of water. A distinction is made between supports stabilized through and through or only on their surface (moisten until rejection), and unstabilized supports, which should hardly be moistened at all (risk of causing the clay to swell). On non-stabilized supports, an impregnating wash can help create cracks to support the rendering.

Anchoring points On brickwork the joints must be roughened with a jointer. With rammed earth, brushing reveals the stones. Good anchoring points ensure good adhesion of the rendering.

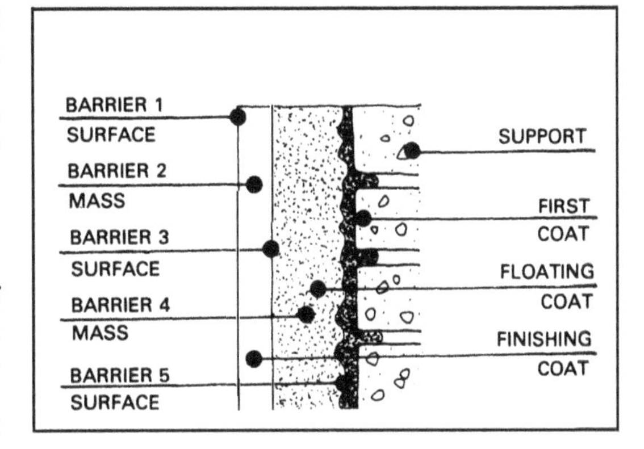

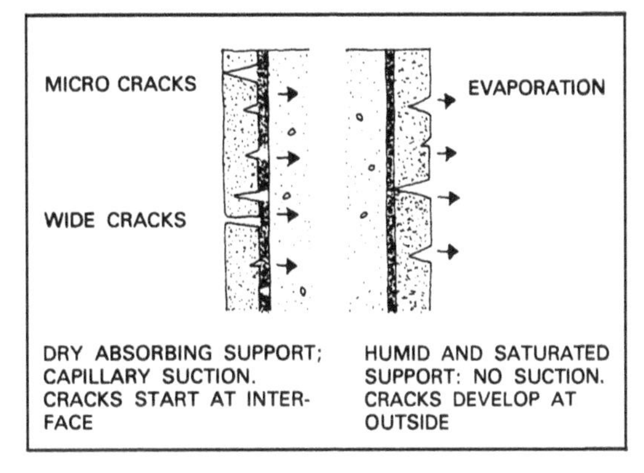

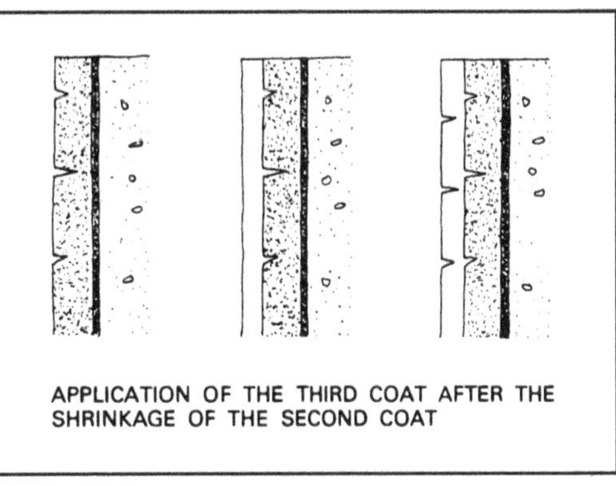

APPLICATION OF THE THIRD COAT AFTER THE SHRINKAGE OF THE SECOND COAT

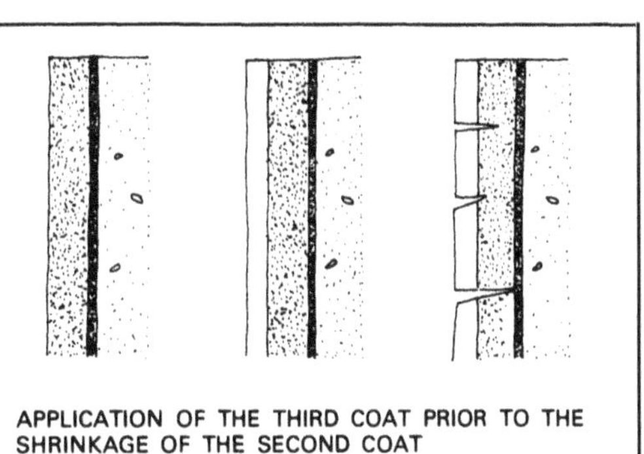

APPLICATION OF THE THIRD COAT PRIOR TO THE SHRINKAGE OF THE SECOND COAT

SUPPORTS					
MONOLITHIC			BRICKWORK		
SMOOTH LIKE RAMMED EARTH	POROUS LIKE COB	VERY POROUS LIKE STRAW-CLAY	SMOOTH LIKE BLOCKS	POROUS LIKE ADOBE	VERY POROUS LIKE SOD
1	2	3	4	5	6

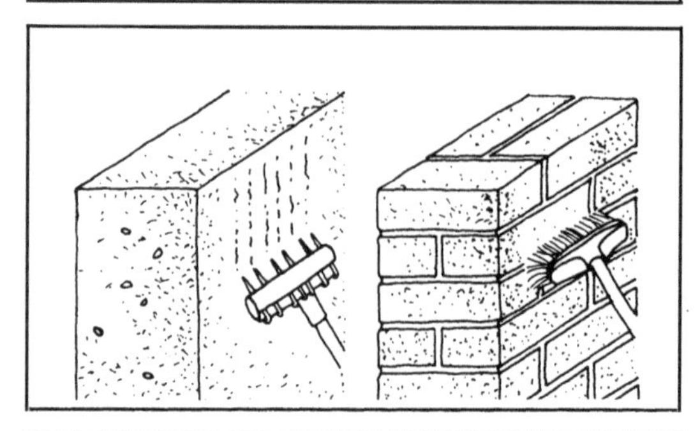

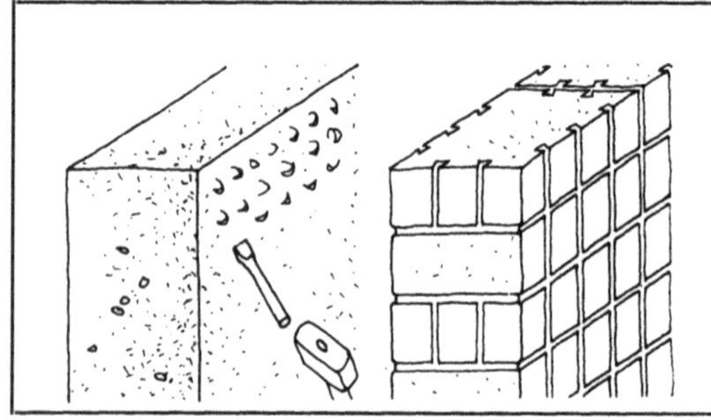

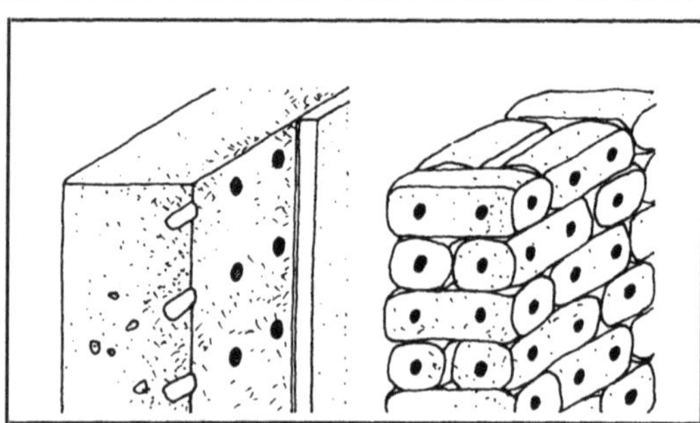

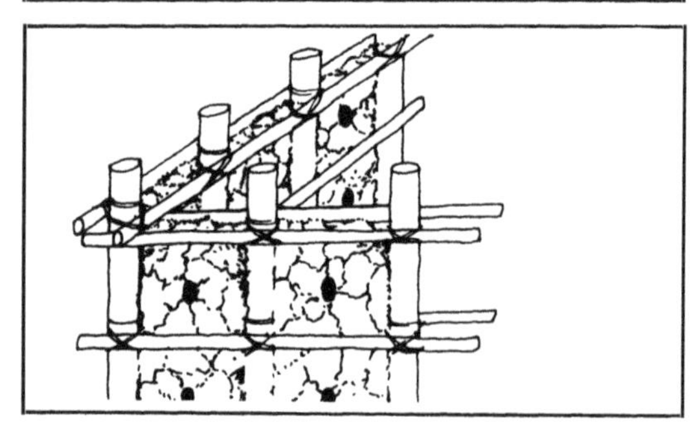

1. Glue

(Supports 1, 2, 4, 5). The use of white joinery glue (diluted polyvinyl acetate) has been tested by ITDG (Nigeria, Sri Lanka, Sudan, Egypt). This glue, diluted in water, is applied in two layers using a brush. Dust is fixed and the adhesion of the mortar of the rendering is facilitated. This glue surface treatment should by preference be used in conjunction with a fibre-reinforced rendering. Other dust fixatives can be used if they are compatible with the rendering.

2. Scraping and dust removal

(Supports 1, 2, 4, 5). Scraping the support is particularly important on supports which are at all crumbly, as this makes it possible to remove any materials lacking cohesion or which are not well fixed. On rammed earth, this exposes the sand and gravel skeleton holding the rendering. Dust removal is indispensable on the majority of earth supports. It can be done with a brush, when dry or when moist (without saturating the wall) or using compressed air blowers.

3. Grooving

(Supports 1, 2, 4, 5). On walls built in compressed blocks and adobes, the joints are scraped to a depth of 2 to 3cm and the rendering is anchored by the hollowed-out joints. The blocks can themselves can be grooved or chiselled. Grooving is a good way of ensuring anchoring on rammed earth and cob. The grooved surface can also be prefabricated with special moulds for blocks and formwork fitted with nailed, dovetailed battens for rammed earth.

4. Holes

(Supports 1, 2, 3, 4, 5, 6). This anchoring technique is particularly suitable for rammed earth, cob and straw-clay supports. It involves making slanting holes when the material is still moist or the formwork has just been removed. The holes should be at least 3cm and preferably 6cm deep. When building in balls or loaves of soil, the holes are made in the fresh material.

5. Piercing walls

(Supports 2, 3). This procedure is used in Gabon on houses built in cob between posts. The masses of clayey soil covering the netting are pierced with a dagger, from one side to the other. The rendering is applied on both the inside and the outside and the two layers are united by a sort of bridge of rendering.

6. Anchor points

(Supports 1, 2, 3, 4, 5, 6). The wall is encrusted with solid fragments, flakes of stone or broken pottery. This encrusting can be easily carried out on fresh cob, or even on daub. The fragments are set obliquely. On blockwork or adobe walls, the fragments are inserted into in the fresh mortar. Anchoring points of the same composition as the rendering can also be provided, e.g. strips of lime included in outer thickness of rammed earth.

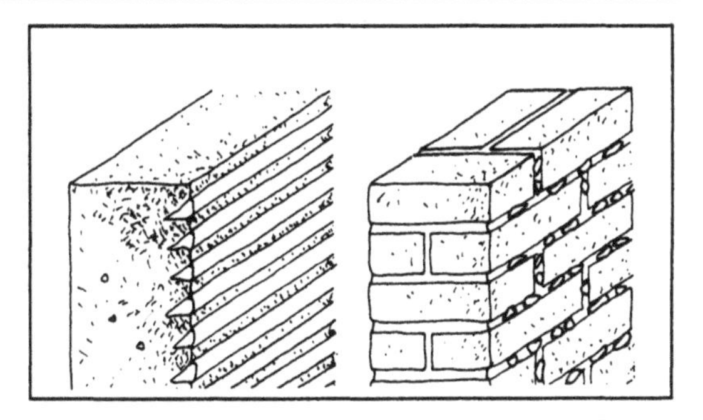

7. Nailing

(Supports 1, 2, 4, 5). The nails should preferably be galvanized and long (at least 8cm), with wide flat heads. They are inserted in the wall in a regular pattern, with about 10 to 15cm between each nail. As they can hinder the application of the rendering another method is to make holes in which the nails are placed so that they are level with the support, or to insert the nails after the application of the floating coat.

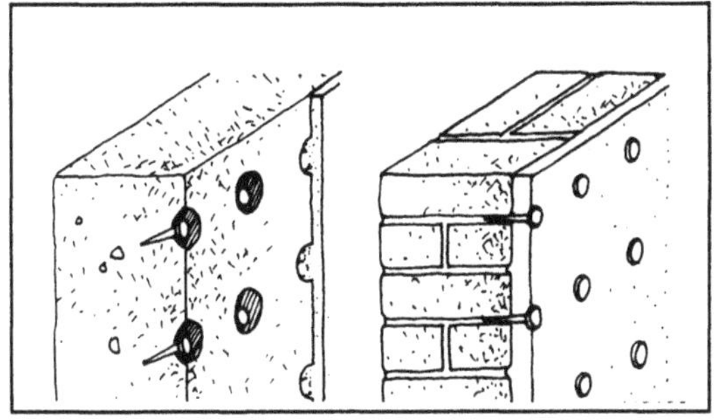

8. Lattice work

(Supports 1, 2, 3, 4, 5, 6). Conventional chicken wire can be used (hexagonal holes). It is best if the netting is galvanized (walls exposed to weather) although non-galvanized netting adheres better. The netting is fixed by nails twisted into the mesh and nailed in a regular triangular pattern. Steel wire can be woven onto nails driven into the wall.

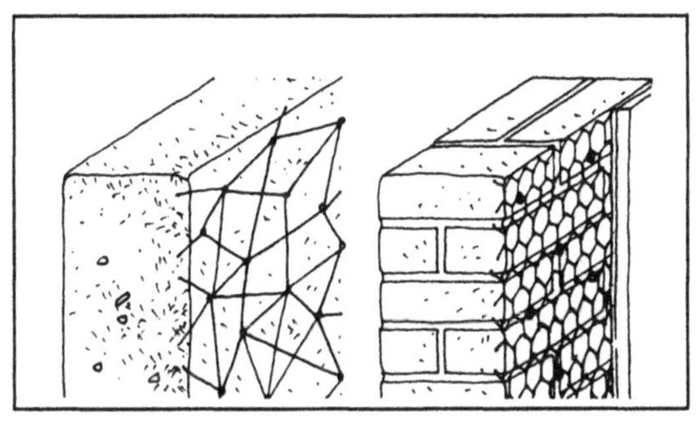

9. Wattle

(Supports 1, 2, 3). Some techniques leave the wattle exposed. This is the case for daub or cob on posts, or even rammed earth between reed formwork. Sometimes this is also done with heavy clay-straw, which is covered with plaited canes or woven reeds to anchor the rendering.

10. Fibres

(Supports 2, 4, 5). The University of Nairobi has tested a wall protection which combines cement and sisal fibres. The mixture is applied as a first coat and the short sisal fibres remain visible, facilitating the adhesion of the subsequent coats. Instead of sisal, other natural fibres can also be used (coir, hemp, etc.), synthetic fibres (polypropylene), animal hair, or woven materials (jute sacking).

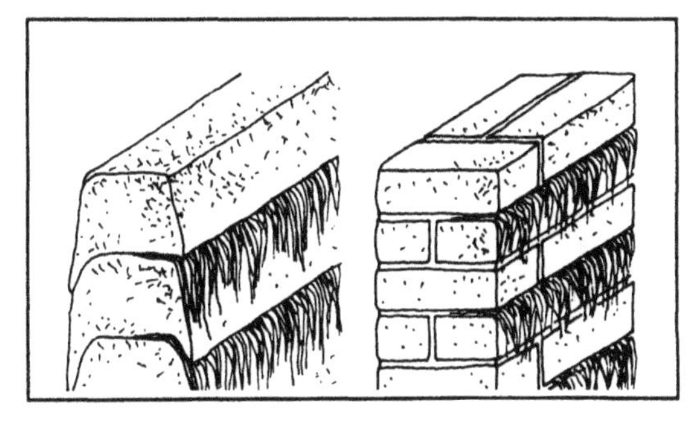

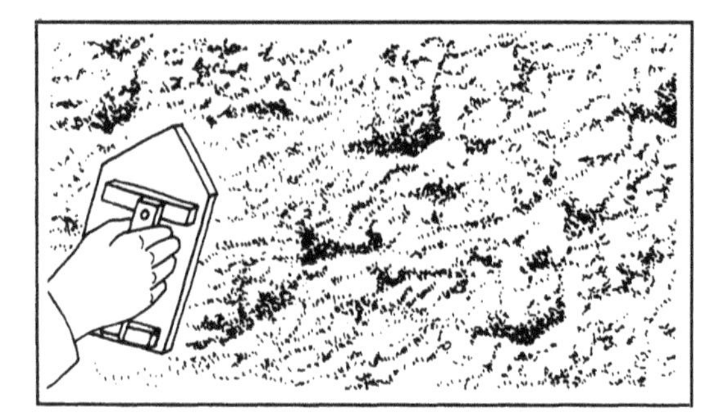

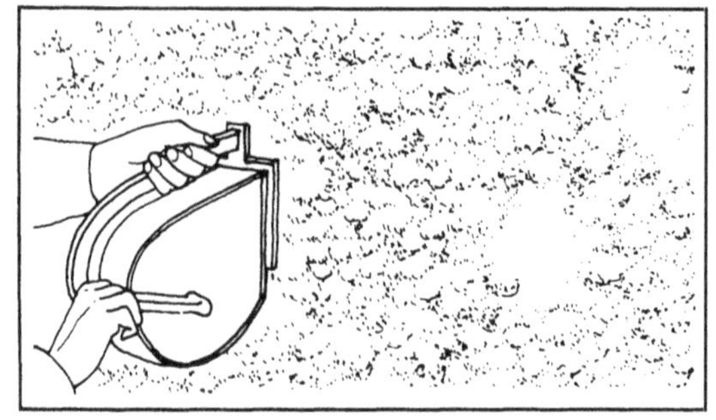

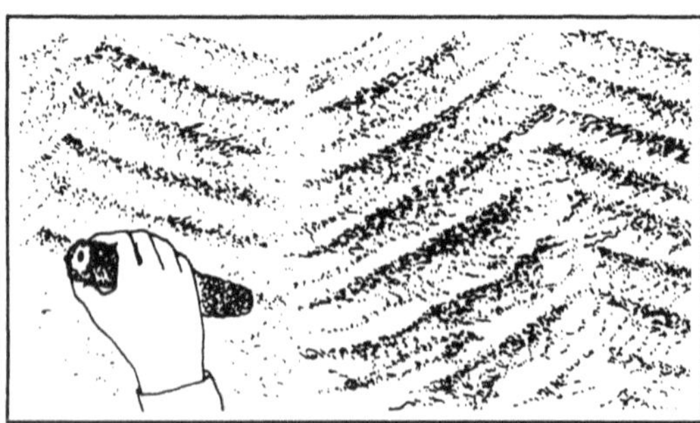

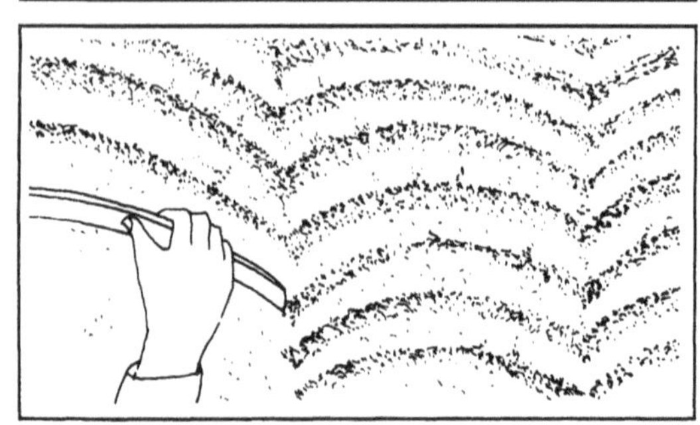

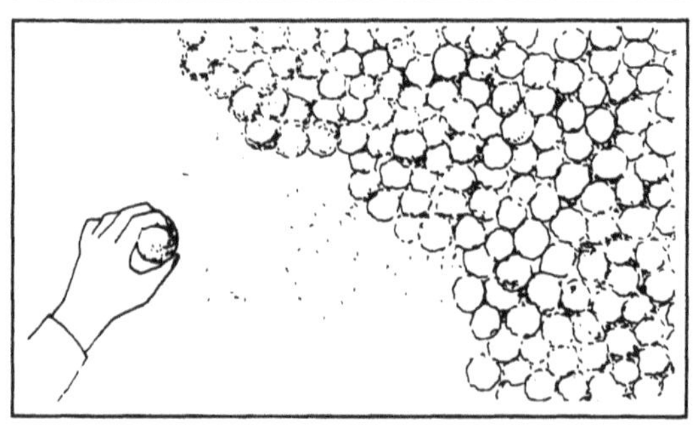

Finishes

Apart from their function as wall protection, both indoors and outdoors, surface coatings play a decorative architectural role. This aspect of the rendering in finishing and decorating building is apparent in many countries and has been exploited for as long as humans can remember. It includes customary techniques and motions just as much as the texture or grain of the finishing coat, reliefs worked in the bulk of the wall, colour, and flashing with various other materials. Finishes of the visible rendering are applied manually, either with traditional masonry tools (trowel, hawk, drag, etc.) or mechanically with blowing machinery.

The various treatments of the surface of the wall are carried out to obtain the final appearance, either before hardening (float-finished or rustic renderings), or after hardening (e.g. scraped or glazed renderings).

In general, interior finishes are often smooth to diminish the adhesion of dust produced by activities inside, whereas exterior finishes are more often rough, which has the advantage of being less susceptible to cracking or faults in the base.

Smooth finish Usually used indoors, smoothed with the laying-on trowel and the float.

Dragged finish The floated surface is scraped after hardening with a metal blade, or with a devil float (i.e. a float with steel nails protruding from its bottom). Once the rendering is dry, the dust can be removed with compressed air or by hosing down.

Rustic finish The rendering is applied by the trowelful, each covering the previous one. When slapped on with the trowel the rendering has a scattered irregular grain.

Grainy finish The rendering has a rough appearance resulting from being thrown or blown on (Tyrolean, fine sand, or plaster blowing equipment). Reduces cracking.

Crushed rustic The wall is first given a grainy finish and then crushed with a float or flattening tool.

Whipped finish The freshly applied rendering is whipped with a broom or with flexible fibres (e.g. palm twigs).

Aggregate finish Coarse sand, small stones, stone chippings, or shells are thrown against the fresh rendering. This pebble dash finish is well-known for its ability to reduce cracking.

Ornamentation

Decoration is the vehicle of a culture's value system and constitutes the identity of the community, passing on the symbols necessary to the moral and the ethical system of a people. The architecture of Africa is incomparably rich in this respect. The decoration is both aesthetic, magical or religious and apotropaic.(i.e. offering protection against demonic influences), or has a functional purpose.

The decoration calls form, relief of the wall, and colour (variety of natural or (nowadays) synthetic pigments), shadow and light into play. To mention just a few, there are the zoomorphic reliefs of Chan Chan (Gran Chim, Peru) and the geometrical floral and plant reliefs of urban housing in Niger and of numerous ethnic groups (Hausa, Dogon, Swahili, Ashante, Suku, Lobi, and so on). The variations on geometric decoration, painted, modelled, sculpted or moulded in the thickness of the wall are so numerous that they defy description. On the other hand some decoration is simpler, such as balls of earth (Sahara) placed on the surface, and which discourage cracking (repetition of hemispherical flattened balls) offering the thermal benefit of permanent shade and breaking water run-off. Ornamentation is also the ancient tradition of painted fresco work, the African murals of the Kru, Toma, Kisi, Ashanti, Ubangi playing primarily on contrasts of black and white, ochre and red; or the tradition of ceramic plating known to the Babylonians (the Ishtar gate) and still practised today in the zeligs of Morocco.

Numerous standardized tests exist which attempt to test the quality of renderings, and especially the behaviour of the inseparable support-rendering complex. The object of the rendering tests is thus basically to find a rendering the behaviour of which will be acceptable, in time, for a specific support and in relation to a number of performance criteria selected by the user (e.g. frequency of maintenance, resistance to climatic and mechanical agents). There are numerous laboratory tests, which are periodically revised by various research centres. They are, however, not universally applicable and their interpretation is hindered by a lack of correspondence between the ideal conditions of the laboratory and real conditions of use through the years. Ultimately, however, time is indeed the only real test. For rendering formulation tested by natural exposure, there will be no point in carrying out tests of the rendering on its support, to the extent, of course, that practical knowledge is confirmed. Among the usual tests for less well-known or new renderings, or where know-how could be improved, there are tests which examine the particular characteristics of the rendering. These tests can be carried out on the fresh mortar, on site, and attempt to check the constituents, setting time, workability, mechanical strength, and the capillary absorption of the mortar, and are conventional and easy to carry out; or alternatively they can be carried out on the hardened mortar. Other tests are aimed at testing the behaviour of the rendering-support complex.

1. Tests on hardened renderings

- Apparent bulk-density test.
- Tests on variations in mechanical characteristics when moist.
- Tests on compression strength, tensile strength, and shear strength.
- Dynamic and transverse moduli of elasticity.
- Tests on dimensional stability, changes in weight.
- Surface hardness tests, tests on layer thickness, depth of carbonation.
- Tests on water content, capillary water absorption, gravity water absorption.
- Tests on ability to withstand the diffusion of water vapour, and permeability to water vapour.
- Tests on water erosion, and run-off.
- Wearing resistance tests.
- Tests on ease of nailing, and maintenance.

All these tests are described in detail in the specialist literature.

2. Small-scale tests

These tests can be useful but are of limited application because they do not consider the rendered wall as a whole and concentrate on fragments. Although scientific, these tests come nowhere near to approximating real conditions.

Open porosity First of all the sample is dried (dry air or stove) until it attains a constant weight. It is then completely submerged (various procedures) then superficially dried (moist cloth) and weighed. Porosity is expressed as a percentage according to the relationship:

$$\frac{W' - W}{W} \times 100$$

(Where W = dry weight; W′ = absorbed weight of water).

Moisture content Measured either on the basis of the resistivity of the material with an apparatus equipped with 2 electrodes which are placed in the rendering, or with a sort of flat capacitor, which is applied to the rendering and gives a direct reading on a gauge.

Absorption capacity Water is forced into the surface of the rendering under pressure. The quantity of water crossing a predetermined area during a given time is recorded by means of a flat chamber held in place by means of a waterproof putty and connected to a recipient of fixed level (in order to ensure constant pressure). A rendered block can also be completely immersed in water and the difference in weight before and after immersion determined.

Erosion By spraying with a jet of water or by dripping water onto the material.

Adhesion tests Adhesion is measured by means of a dynamometer with gaiters which is used to tear loose a 50mm pellet from the surface of the rendering after first cutting it free with a corer to a depth slightly more than that of the rendering. The pellet is glued to a metal disc with a suitable glue. Adhesion is good, if breaking occurs in the rendering; not good if it takes place at the rendering-support interface.

- Another procedure is described by the Belgian norm B14–210. A round (Ø 8cm) or square (10×10 or 15×15cm) plate is glued to the hardened plaster with an epoxy resin. The plaster layer is cut along the glued plate as far as its base. Plate and plaster are subsequently pulled off by a manual or hydraulic apparatus.

– Yet another test for the adhesion of rendering to rammed earth has been tried in Morocco. A porous cement block is stuck to a sample of rendered rammed earth, and tensile force exerted on a ring located in the axis of the block. The test is for an adhesion between rendering and rammed earth of 1kg per cm^2.

3. Large-scale tests

Accelerated ageing This test should reflect local climatic conditions as far as is possible. The aging cycle by exposure to heat, rain and frost must be defined by the proper interpretation of the tests as it is a matter of gauging the responses to stresses rather than determining a state after ageing.

Natural ageing The behaviour of renderings with time is observed on small walls exposed to natural weather conditions. It is advisable to ensure that these test walls are properly oriented with respect to the prevailing rain and wind. This test has been tried in various countries, but has been carried out on a truly large scale in Senegal and in the USA and has been the object of study for several decades. Nevertheless the results are more accurate for perimeter walls than for those of dwellings, as no examination can be made of the migrating water vapour which affects most renderings.

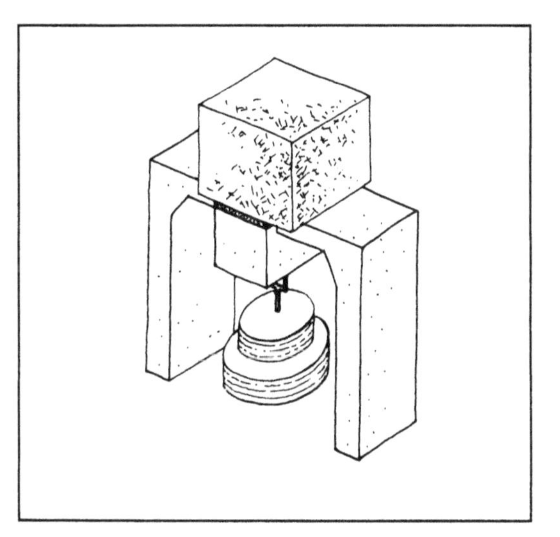

The test walls have a minimum exposed area of $1m^2$. They are subjected to the greatest climatic stresses in the region. They are covered with a waterproof cap projecting 10cm on each side, and fitted with a drip. The test walls are separated from the ground by a footing of at least 25cm high and provided with anti-capillary barrier. The rendering comes to within 2cm of the cap and reaches down to the footing, but does not touch it. At least one year and, more usually, two to three years are required to arrive at the first conclusions which do not consider the disorders suffered at the edges of the test walls.

4. Buildings or building panels

Natural exposure tests on buildings or building panels have been carrried out in the USA, Great Britain, and several other countries. None of these buildings have ever been normally occupied and the differing orientations of exposure make comparison impossible.

In fact the best place for experimentation and observation is the stock of existing buildings.

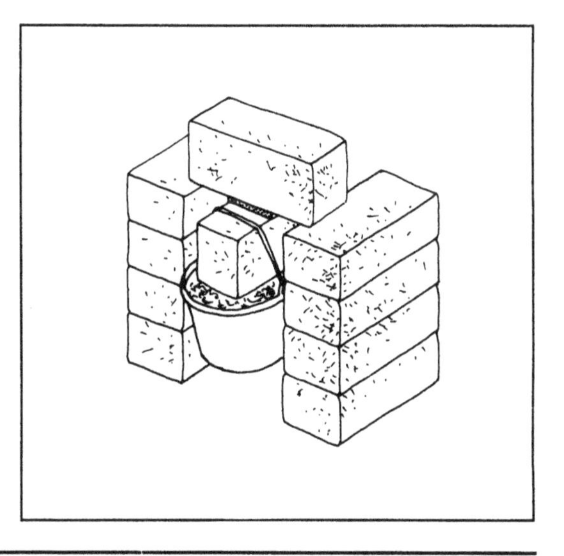

Afshar, F. *et al.* 'Mobilizing indigenous resources for earthquake construction'. In *International Journal of IAHS*, New York, Pergamon Press, 1978.
AGRA. *Recommandations pour la Conception des Bâtiments du Village Terre*. Grenoble, AGRA, 1982.
Ahmed Hassan Hamid. *Asphalt Based Coating*. Roorkee, CBRI, 1972.
Alcock, A. Swishcrete; *Notes on Stabilised Cement-earth Building in the Gold Coast. Kumasi, BRS, 1953.*
An. Maisons en Terre. Paris, CRET, 1956.
Aslam, M.; Satya, R.C. *Technical Note on Surface Waterproofing of Mudwalls*. Roorkee, CBRI, 1973.
BCEOM. *La Construction en Béton de Terre*. Paris, Service de l'habitat, 1952.
Bona, T. *Manuel des Constructions Rurales*. Librairie agricole de la maison rurale, 1950.
Brigaux, G. *La Maçonnerie*. Paris, Eyrolles, 1976.
Bureau de l'habitat rural. *Surfaçage des Parpaings de Terre et Badigeonnage*. Dakar, Direction de l'habitat et de l'urbanisme des TP et transports, 1963.
Chatterji, A.K. 'Les efflorescences dans les ouvrages en briques'. In *Bâtiment Built International*, Paris, CSTB, 1970.
CINVA. *Le Béton de Terre Stabilisé, son Emploi dans la Construction*. New York, UN, 1964.
CRATerre. 'Casas de tierra'. In *Minka*, Huankayo, Grupo Talpuy, 1982.
Cytryn, S. *Soil Construction*. Jerusalem, the Weizman Science Press of Israel, 1957.
Dayre, M. *Commentaires de la Fiche: "Laboratoire tiers monde" UPA 6, concernant la recherche "Protection du matériau terre"*. Grenoble, AGRA, 1982.
Dayre, M. *Conseils pour la Réalisation d'Enduits de Façade*. Privas, DDE Ardèche, 1982.
Delarue, J. 'Etude du pisé de ciment au Maroc.' in *Bulletin RILEM*, Paris, 1954.
Delaval, B. *La Construction en Béton de Terre*. Alger, LNTBP, 1971.
Denyer, S. *African Traditional Architecture*. New York, Africana, 1978.
Des Lauriers, T. *Projet Addis-Abeba*. Addis-Abeba, REXCOOP/MUDH, 1983.
Dethier, J. *Des architectures de Terre*. Paris, CCI, 1981.
Doat, P. *et al. Construire en Terre*. Paris, Editions Alternatives et Parallèles, 1979.
Dreyfus, J. *Manuel de la Construction en Terre Stabilisée en AOF*. Dakar, Haut commissariat en AOF, 1954.
Dreyfus, J. 'Peintures et moyens de protection divers pour construction en terre ou en terre stabilisée'. In *Peintures, Pigments, Vernis*.
Duriez, M.; Arrambide, J. *Etude sur Enduits et Rejointoiements*. Paris, Dunod, 1962.
Ephoevi-Ga, F. 'La protection des murs en banco'. in *Bulletin d'Information*, Cacavelli, CCL, 1978.
Fitzmaurice. *Manuel de Constructions en Béton de Terre Stabilisé*. New York, UN, 1958.
Gardi, R. *Maisons Africaines*. Paris–Bruxelles, Elsevier Séquoia, 1974.
Gratwick, R.T. *Dampness in Buildings*. London, Crosby Lockwood and Son, 1962.
Grésillon, J.M.; Dourthe, V. 'Un matériau pour les constructions rurales, la brique bi-couche'. In *Bulletin Technique*, Ouagadougou, EIER, 1981.
Groben, E.W. *Adobe Architecture: its design and construction*. New York, US Department of Agriculture Forest Service, 1941.
Guidoni, E. *Primitive Architecture*. New York, Harry N. Abrams, 1975.
Guillaud, H. *Histoire et Actualité de la Construction Terre*. Marseille, UPA Marseille Luminy, 1980.
Hammond, A.A. 'Prolongation de la durée de vie des constructions en terre sous les tropiques'. In *Bâtiment Build International*, Paris, CSTB, 1973.
Housing and home finance agency. 'A cheap coating for unstabilized earth walls'. In *Ideas and Methods Exchange*, Washington, Office of international affairs, 1961.
International Institute of Housing Technology. *The Manufacture of Asphalt Emulsion Stabilized Soil Bricks and Brick Maker's Manual*. Fresno, IIHT, 1972.
Kahane, J. *Local Materials, a self-builders manual*. London, Publications Distribution, 1978.
Kern, K. *The Owner Built Home*. New York, Charles Scribner's Sons, 1975.
Kienlin. 'Le Béton de Terre'. In *Revue Génie Militaire*, Paris, 1947.
L'Hermite, R. *Au Pied du Mur*. Paris, Eyrolles, 1969.
Laboratoire fédéral d'essai des matériaux et institut de recherches. *Directives pour l'Exécution de Crépissages*. Dübendorf, LFEMIR, 1968.
Letertre; Renaud. *Technologie du Bâtiment. Grosœuvre. Travaux de Maçonneries et Finitions*. Paris, Foucher, 1978.
Maggiolo, R. *Construcción con Tierra*. Lima, Comissión ejecutiva inter-ministerial de cooperación popular, 1964.
Manson, J.L.; Weller, H.O. *Building in Cob and Pisé de Terre*. BRB, 1922.
McCalmont, J.R. *Experimental Results with Rammed Earth Construction*. American Society of Agricultural Engineers. St Joseph, 1943.
McHenry, P.G. *Adobe Build it Yourself*. Tuscon, The University of Arizona Press, 1974.
Middleton, G.F. *Earth Wall Construction*. Sydney, Commonwealth Experimental Building Station, 1952.
Miller, L.A. & D.J. *Manual for Building a Rammed Earth Wall*. Greeley, REII, 1980.
Miller, T. *et al. Lehmbaufibel*. Weimar, Forschungsgemeinschaften Hochschule, 1947.

Ministère des affaires culturelles. *Vocabulaire de l'Architecture*. Paris, Ministère des affaires culturelles, 1972.
Morse, R. 'Plastic-C coating. Plastic-B/C/D'. Priv. com. New York, 1977.
Museum of New Mexico. 'Adobe past and present'. In *El Palacio*, Sante Fe, 1974.
Neubauer, L.W. *Adobe Construction Methods*. Berkeley, University of California, 1964.
Palafitte jeunesse. *Minimôme Découvre la Terre*. Grenoble, Palafitte Jeunesse, 1975.
Patty, R.L. 'Paints and plasters for rammed earth walls'. in *Agricultural Experiment Station Bulletin*. South Dakota State College, 1940.
PGC–CSTC. Priv. com. Brussels, 1984.
Plancherel, J.M. *Briques en Terre Séchée Revêtue de Planelles en Terre Cuite*. Lausanne, Ecole Polytechnique de Lausanne, 1983.
Pollack, E.; Richter, E. *Technik des Lehmbaues*. Berlin, Verlag Technik, 1952.
Simonnet, J. *Recommandations pour la Conception et l'Exécution de Bâtiments en Géobéton*. Abidjan, LBTP, 1979.
Soltner, D. *Les Bases de la Production Végétale*. Angers, Collection Sciences et Techniques Agricoles. 1982.
UNCHS. 'Construction with sisal cement'. In *Technical notes*. Nairobi, UNCHS (Habitat), 1981.
Van Den Branden, F; Hartsell, T. *Plastering Skill and Practice*. Chicago, American Technical Society, 1971.
Williams-Ellis, C.; Eastwick-Field, J. & E. *Building in Earth, Pisé and Stabilized Earth*. London, Country Life, 1947.
Wolfskill, L.A. *et al. Bâtir en Terre*. Paris, CRET.

Prototype of a rural school in Somalia, made of compressed earth blocks for a joint programme of UNESCO and the Ministry of Education (architects: Thierry Joffroy, Serge Maini) (Thierry Joffroy, CRATerre-EAG)

BIBLIOGRAPHY

A modern select bibliography of earth construction would comprise more than 10,000 titles. Most of these are difficult, if not impossible, to obtain. The bibliography given here covers only the most important works, which are obtainable through normal commercial channels.

German

Adam J.A. *Wohn und Siedlungsformen im Süden Marokkos.* Callwey Georg D.M., München, Germany, 1981. ISBN: 3–7667–0566–0.

Adam J.A., Farassat D., Wienands R., Wichmann H., Wright G.R.H., Hrouda B., Fiedermutz-Laun A., Wildung D. *Architektur der Vergänglichkeit. Lehmbauten der dritten Welt.* Birkäuser Basel, Stuttgart, Germany, 1983. 254 pages. ISBN: 3–7643–1283–1.

Frobenius-Institut *Aus Erde geformt. Lehmbauten in West- und Nordafrika.* Verlag Philipp von Zabern, Mainz, Germany, 1990. 200 pages. ISBN: 3 8053 1107 9.

Gardi R. – *Auch im Lehmhaus lässt sich leben.* Akademischer Druck, Verlagsanstalt, Berne, Switzerland, 1974. 248 pages.

Güntzel J.G. *Zur Geschichte des Lehmbaus in Deutschland. Band 1. Bibliographie Band* 2. Gesamthochschule Kassel, Universität des Landes Hessen, Kassel, Germany, 1986. 553 pages. ISBN: Band 1, Band 2: 3–922964–99–0.

Keppler M., Lemcke T. *Mit Lehm gebaut. Ein Lehmhaus im Selbstbau.* Müller C.F., Karlsruhe, Germany, 1986. 125 pages. ISBN: 3–924466–02–5.

Lander H., Niermann M. *Lehmarchitektur in Spanien und Afrika.* Karl Robert Langewiesche Nachfolger Hans Koster KG, Königstein, Germany, 1980. 132 pages. ISBN: 3–7845–7240–5.

Leszner T., Stein I. *Lehm-Fachwerk. Alte Technik, neu entdeckt.* Rudolf Müller, Köln, Germany, 1987. 156 pages + annexes. ISBN: 3–481–25491–1.

Minke G. *Lehmbau – Handbuch. Der Baustoff Lehm und seine Anwendung.* Ökobuch, Staufen, Germany, 1994. 321 pages. ISBN: 3–922964–56–7.

Niemeyer R. *Der Lehmbau und seine praktische Anwendung.* Ökobuch Verlag Gmbh, Grebenstein, Germany, 1982. 157 pages. ISBN: 3–922964–10–9.

Schillberg K., Knieriemen H. *Naturbaustoff Lehm. Moderne Lehmbautechnik in der Praxis – bauen und sanieren mit Naturmaterialen.* AT Verlag, Aarau, Switzerland, 1993. 160 pages. ISBN: 3 85502 466 9.

Schneider *J. Am Anfang Die Erde. Sanfter Baustoff Lehm.* Fricke im Rudolf Müller Verlag, Frankfurt am Main, Germany, 1985. 84 pages. ISBN: 3–481–50241–9.

Volhard F. *Leichtlehmbau: alter Baustoff – neue Technik.* Müller C.F., Karlsruhe, Germany, 1986. 159 pages. ISBN: 3–7880–7321–7.

Spanish

'Adobe en América y alrededor del mundo, historia, conservación y uso contemporáneo' In: *Exposición itinerante,* Proyecto Regional de Patrimonio Cultural y Desarrollo, PNUD, UNESCO, Paris, France, 1984. 74 pages.

Agarwal A. *Barro, Barro! Las posibilidades que ofrecen los materiales a base de tierra para la vivienda tercermundista.* Earthscan, London, United Kingdom, 1981. 100 pages. ISBN: 0–905347–20–X.
Bardou, P., Arzoumanian V. *Arquitecturas de adobe.* Editorial Gustavo Gili, Barcelone, Spain, 1979. 165 pages. ISBN: 84–252–0924–2.
Bauluz del Rio G., Bárcena Barrios P. *Bases para el diseño y construcción con tapaial.* Ministerio de Obras Públicas y Transportes, Madrid, Spain, 1992. 79 pages. ISBN: 84–7433–839–5.
CRATerre: Doat P., Hays A., Houben H., Matuk, S., Vitoux F. *Construir con tierra. Tomo 1 & 2.* ENDA America Latina, Fedevivienda, Dimensión Educativa, Bogota, Colombia, 1990. Tome 1: 220 pages, tome 2: 259 pages.
Font F., Hidalgo P. *El tapial. Una tècnica constructiva mil.lenària.* Fermín Font, Pere Hidalgo, Barcelone, Spain, 1991. 172 pages (Catalan).
Hernández Ruiz L.E., Márquez Luna J.A. *Cartilla de pruebas de campo para la selección de tierras en la fabricación de adobes.* CONESCAL, Mexico 1983. 72 pages. ISBN: 968–29–0055–7.
Medellin Anaya A., Renero J.L., Ipiña F., Castro de la Rosa S. *La casa de tierra.* Instituto Tamaulipeco de vivienda y Urbanización, Mexico, 1990. 93 pages.
Rodriguez E., Martinez R. *Suelo cemento. Fundamentos para la aplicación en Cuba.* Instituto nacional de la Vivienda, Cuidad Habana, Cuba, 1991. 209 pages.
Vildoso A., Monzón F.M.; CRATerre: Hays A., Matuk S., Vitoux F. *Seguir construyendo con tierra.* CRATerre, Lima, Peru, 1984. 236 pages.

French

Agarwal A. *Bâtir en terre. Le potentiel des matériaux à base de terre pour l'habitat du Tiers Monde.* Earthscan, London, United Kingdom, 1981. 115 pages. ISBN: 0–905347–19–6.
Andersson L.A., Johansson B., Åstrand J. *Torba stabilisée au ciment. Etude expérimentale d'un sol d'origine locale et développement de techniques pour sa mise en œuvre comme matériau de construction.* Université de Lund, Lund, Sweden, 1982. 72 pages + annexes. ISBN: 91–970225–0–0.
Bardou P., Arzoumanian V. *Archi de terre.* Editions Parenthèses, Marseille, France, 1978. 103 pages. ISBN: 2–86364–001–1.
Courtney-Clarke M. *Tableux d'Afrique. L'art mural des femmes de l'Ouest.* Arthaud, Paris, France, 1990. 204 pages. ISBN: 2–7003–0851–4.
CRATerre: *Le bloc de terre comprimée. Eléments de base.* GATE, Eschborn, Germany, 1991. 28 pages.
CRATerre: Doat P., Hays A., Houben H., Matuk S., Vitoux F. *Construire en terre.* Editions Alternatives, Paris, France, 1985. 287 pages. ISBN: 2 86 227 009–1.
CRATerre: Guillaud H. *Modernité de l'architecture de terre en Afrique. Réalisations des années 80.* CRATerre, Grenoble, France, 1989. 190 pages. ISBN: 2–906901–04–0.
CRATerre: Guillaud H. *La terre crue, des matériaux, des techniques et des savoir-faire au service de nouvelles applications architecturales. Encyclopédie du bâtiment. No 46.* Editions Techniques, éditions Eyrolles, Paris, France, 1990. 66 pages. ISBN: 141 472.
CRATerre: Houben H., Guillaud H. 'Traité de construction en terre'. In: *L'encylcopédie de la construction en terre.* Vol 1, Editions Parenthèses, Marseille, France, 1989. 355 pages. ISBN: 2–86364–041–0.
CRATerre-EAG: Houben H, ICCROM: Alva A. *5e réunion internationale d'experts sur la conservation de l'architecture de terre. Actes de colloques. Rome, Italie, 22–23 octobre 1987.* CRATerre, Villefontaine, France, 1988. Conference proceedings, 110 pages. ISBN: 92–9077·087–2.
CRATerre-EAG: *Blocs de terre comprimée. Manuel de conception et de construction.* GATE, Eschborn, Germany, 1995. 148 pages.
CRATerre-EAG: *Blocs de terre comprimée. Manuel de production.* GATE, Eschborn, Germany, 1995. 104 pages.
CRATerre-EAG: *Blocs de terre comprimée: choix du matériel de production.* CDI, Bruxelles, Belgium, 1994. 70 pages.
CRATerre-EAG: *Eléments de base sur la construction en arcs, voûtes et coupoles.* SKAT, Saint gallen, Switzerland, 1994. 27 pages.
CRATerre-EAG: 'Maisons en terre hier et auiourd'hui' In: *B.T. no 1002,* Publications de l'Ecole Moderne Française, Cannes la Bocca, France, 1988. 48 pages. ISSN: 0005–335X.
Development Workshop *Les toitures sans bois, guide pratique.* Development Workshop, Lauzerte, France, 1990. 77 pages. ISBN:2 906208 00 0.
Domain, S. *Architecture soudanaise. Vitalité d'une tradition urbaine et monumentale.* L'Harmattan, Paris, France, 1989. 191 pages. ISBN: 2–7384–0234–8.
Fathy H. *Construire avec le peuple. Histoire d'un village d'Egypte: Gourna.* Editions Jérome Martineau, Paris, France, 1970. 305 pages + annexes.
Gardi R. *Maisons Africaines. L'art traditionnel de bâtir en Afrique occidentale.* Elsevier Séquoia, Paris/Bruxelles, France/Belgium, 1974. 248 pages. ISBN: 2–8003–0046–9.

Jeannet J., Pignal B., Pollet G., Scarato P. *Le pisé. Patrimoine, restauration, technique d'avenir. Les cahiers de construction traditionnelle*. Editions CREER, Nonette, France, 1993. 122 pages. ISBN: 2–902894–91–0.
Lafond P., Amran El Maleh E. *Citadelles du désert*. Nathan Image, Paris, France, 1991. 137 pages. ISBN: 2–09–290096-X.
Loubes J.P. *Maisons creusées du fleuve jaune. L'architecture troglodytique en Chine*. Créaphis, Paris, France, 1988. 127 pages + annexes. ISBN: 2–907150–04–9.
Maas P. *Djenne chef d'œuvre architectural*. Institut des Sciences Humaines, Université de Technologie, Bamako/ Eindhoven, Mali/Belgium, 1992. 224 pages. ISBN: 90 6832 228 1.
Mester de Parajd C., Mester de Parajd L. 'Regards sur l'habitat traditionel au Niger' In: *Les cahiers de construction traditionnelle. Vol. 11* Editions CREER, Nonette, France, 1988. 101 pages. ISBN: 2–902894–57–0.
Mouline S., Hensens J. *Habitats des Qsour et Qasbas des vallées présahariennes*. Ministère de l'Habitat, Rabat, Morocco, 1991. 118 pages.
Olivier M. *La matériau terre, compactage, comportement, application aux structures en blocs de terre*. INSA, Lyon, France, 1994. 452 pages + annexes.
Seignobos C. *Nord-Cameroun, montagnes et hautes terres*. Editions parenthèses, Marseille, France, 1982. 192 pages. ISBN: 2–86364–015–1.

English

'*6th International Conference on the Conservation of Earthen Architecture. Adobe 90 preprints*". The Getty Conservation Institute, Marina Del Rey, USA, 1990. Conference proceedings. 469 pages. ISBN: 0–89236–181–6.
Bourdier J.P., Minh-Ha T.T. *African spaces. Designs for living in Upper Volta*. Africana Publishing Company, New York, USA, 1985. 232 pages + annexes. ISBN: 0–8419–0890–7.
Bourgeois J.L., Pelos C. *Spectacular vernacular. The adobe tradition*. Aperture, New York, USA. 1989. 191 pages. ISBN: 0–89381–391–5.
Changuion P. *The African mural*. New Holland, London, United Kingdom, 1989. 166 pages. ISBN: 1 85368 062 1.
Courtney-Clarke M. *African Canvas. The Art of West African Women*. Rizzoli International Publications, New York, USA, 1990. 204 pages. ISBN: 2–7003–0851–4.
Courtney-Clarke M. *Ndebele: the art of an African tribe*. Rizzoli International, New York, USA, 1986. Book: 290 × 285 mm, 200 pages, ill. ISBN: 0 8478 0685 5.
CRATerre: *Earth building materials and techniques. Select bibliography*. GATE, Eschborn, Germany, 1991. 121 pages.
CRATerre: *The basics of compressed earth blocks*. Gate, Eschborn, Germany, 1991. 28 pages.
CRATerre: Doat P., Hays A., Houben H., Matuk S., Vitoux F. *Building with earth*. The Mud Village Society, New Delhi, India, 1990. 284 pages.
CRATerre: EAG: Houben H., ICCROM: Alva A. *5th international meeting of experts on the conservation of earthen architecture. Proceedings of the international conference. Rome, Italy, 22–23 October 1987*. CRATerre, Villefontaine, France, 1988. Conference proceedings. 110 pages. ISBN: 92–9077–087–2.
CRATerre-EAG *Compressed earth blocks. Manual of design and construction*. GATE, Eschborn, Germany, 1995. 148 pages.
CRATerre-EAG: *Compressed earth blocks. Manual of production*. GATE, Eschborn, Germany, 1995. 104 pages.
CRATerre-EAG: *Compressed earth blocks: selection of production equipment* CDI, Bruxelles, Belgium, 1994. 70 pages.
CRATerre-EAG: Joffroy Th., Guillaud H. *The basics of building with arches, vaults and cupolas*. SKAT, Saint Gallen, Switzerland, 1994. 27 pages.
CRATerre-EAG: Odul P. *Bibliography on the preservation, restoration and rehabilitation of earthen architecture*. CRATerre-EAG; ICCROM, Villefontaine/Rome, France/Italy, 1993. 136 pages. ISBN: 92–9077–112–7.
Denyer S. *African traditional architecture: an historical and geographical perspective* Africana Publishing Company, New York, USA, 1978: Book 185 × 255 mm, 210 pages, ill., bibl. ISBN: 0–8419–0287–9.
Dmochowski Z. R. *An introduction to Nigerian traditional architecture. Northern Nigeria. Volume one. South-West and Central Nigeria. Volume two. The Igbo–speaking areas. Volume three*. Ethnograpicha, London, United Kingdom, 1990. Vol 1. 272 pages. ISBN: 0 905 788 26 5. Vol. 2: 298 pages, ISBN: 0 905 788 27 3; vol. 3: 245 pages, ill., graph. ISBN: 0 905 788 28 1.
Easton D. *Dwelling on earth. A manual for the professional application of earthbuilding techniques*. Napa, USA, 1991. 115 pages.
Edwards R. *Basic rammed earth. An alternative method to mud brick building*. The Rams Skull Press, Kuranda, Australia, 1988. 40 pages. ISBN: 0 909901 80 5.

Edwards R. *Mud brick techniques* The Rams Skull Press, Kuranda, Australia, 1990. 48 pages. ISBN: 0 909901 98 8.
Edwards R., Lin Wei-Hao. *Mud brick and earth building. The Chinese way*. The Rams Skull Press, Kuranda, Australia, 1984. 156 pages. ISBN: 0909901–34 1.
Fathy H. *Architecture for the poor. An experiment in rural Egypt*. The University of Chicago Press, Chicago, USA, 1973. 194 pages + annexes. ISBN: 0–226–23915–2.
Howard T. *Mud and man. A history of earth buildings in Australasia*. Earthbuild Publications, Melbourne, Australia, 1992. ISBN: 0 646 06962 4.
Iterbeke M., Jacobus P. *Soil-cement technology for low-cost housing in rural Thailand. An evaluation study*. PG-CHS-KULeuven, Leuven, Belgium, 1988. 154 pages + annexes. ISBN: 97–43200–53–1.
McHenry P. G. *Adobe and rammed earth buildings. Design and Construction*. Wiley-Interscience, New York, USA, 1984. 205 pages + annexes. ISBN: 0–8165–1124–1.
Middleton G. F., Schneider L. M. 'Earth-wall construction' in: *Bulletin no. 5. Fourth edition*, National Building Technology Centre, Chatswood, Australia, 1987. 67 pages. ISBN: 0–642–12289X.
Moughtin J. C. *Hausa architecture*. Ethnographica, London, United Kingdom, 1985. 175 pages. ISBN: 0–905788–40–0.
Mukerji K., Bahlmann H. 'Laterite for building' in: *Report 5*, IFT, Starnberg, Germany, 1978. 79 pages.
Mukerji K., CRATerre. *Soil block presses. Product information*. GATE, Eschborn, Germany, 1988. 32 pages.
Mukerji K., Wörner H., CRATerre. *Soil preparation equipment. Product information*. GATE, Eschborn, Germany, 1991. 19 pages.
Mukerji K., CRATerre. *Stabilizers and mortars (for compressed earth blocks). Product information*. GATE, Eschborn, Germany, 1994. 20 pages.
Norton J. *Building with earth. A handbook*. IT Publications, London, United Kingdom, 1986. 68 pages. ISBN: 0 946688 33 8.
Pearson G. T. *Conservation of clay and chalk buildings*. Donhead, London, United Kingdom, 1992. 203 pages. ISBN: 1 873 394 00 4.
Pudeck J., Stillefors B. *Adobe school furniture handbook. A manual on construction of furniture from mud*. SIDA, Stockholm, Sweden, 1993. 77 pages. ISBN: 91–586–6043–7.
Smith R.G., Webb D.J.T. 'Small-scale manufacture of stabilised soil blocks' in: *Technology series. Technical memorandum no. 12*, ILO, Genève, Switzerland, 1987. 147 pages + annexes. ISBN: 92–2–105838–7.
Stedman M. and W. *Adobe architecture*. The Sunstone Press, Santa Fe, USA, 1987. 45 pages. ISBN: 0–86534–111–7.
Stulz R., Mukerji K. *Appropriate building materials*. SKAT Publications, Saint Gallen, Switzerland, 1993. 434 pages. ISBN: 3–908001–44–7.
Tibbets J.M. *The earthbuilders' encyclopedia* Southwest Solaradobe School, Albuquerque, USA, 1988. 196 pages. ISBN: 0–9621885–0–6.
Verschure H., Mabardi J.F. *Project: Earth construction technologies appropriate to developing countries. Case studies: Kenya/Zambia; China/Thailand; New Mexico; Tunisia/Ivory Coast; Ecuador; Algeria; Chad; Morocco; Tanzania; Sudan; Jordan; Mayotte; Iran; Niger; Egypt*. KUL, PGCHS, CRA, HRDU, Leuven/Nairobi, Belgium/Kenya, 1983/1984. 19 reports and a conference proceedings.
Yesmeen Lari. *Traditional architecture of Thatta*. The Heritage Foundation, Karachi, Pakistan, 1989. 166 pages.

Italian

Ago F. *Moschee in adòbe. Storia e tipologia nell'Africa Occidentale*. Kappa, Rome, Italia, 1982. 146 pages.
Antongini G., Spini T. *Ill cammino degli antenati. I lobi dell'Alto Volta*. Laterza, Rome, Italia, 1981. 239 pages.
Bertagnin M. *Il pisé e la regola. Manualistica settecentesca per l'architettura in terra. Riedizione critica del manuale di Giuseppe Del Rosso dell'economica costruzione delle case di terra (1793)*. Edilstampa, Rome, Italia, 1992. 107 pages.
Galdieri E. *Le meraviglie dell'architettura in terra cruda*. Laterza, Rome, Italia. 1982. 305 pages.

Portuguese

7a conferência internacional sobre o estudo e conservação da arquitectura de terra. Comunicções Silves, Portugal, 24 a 29 de Outubro 1993. Direccão Geral dos edificios e Monumentos nacionais, Lisboa, Portugal, 1993. Conference proceedings. 659 pages.
Taipa em painéis modulados. MEC/SG/CEDATE, Brasilia, Brazil, 1988. 59 pages + annexes.
Braizinha J. J., Gonçalves, A.C., Lurdes Duarte M., El Basri J., Alegria J.A. *Batir en terre en Méditerranée. Construir em terra no Mediterrâneo*. Câmara municipal de Silves, Portugal, 1993. 120 pages.
Dethier J. *Arquitecturas de terra ou o futuro de uma tradição milenar, Europa – Terceiro Mundo – Estados Unidos*. Fundação Calouste Gulbenkian, Lisboa Portugal, 1993. 224 pages. ISBN: 2–85850–326–5.